STRUCTURAL FOUNDATION
DESIGNERS' MANUAL

STRUCTURAL FOUNDATION DESIGNERS' MANUAL

Curtins
CONSULTING ENGINEERS

W. G. CURTIN, MEng, PhD, FEng, FICE, FIStructE, MConsE

G. SHAW, CEng, FICE, FIStructE, MConsE

G. I. PARKINSON, CEng, FICE, FIStructE, MConsE

J. M. GOLDING, BSc, MS, CEng, MICE, FIStructE

Blackwell
Science

© Estate of W. G. Curtin, together with
G. Shaw, G. I. Parkinson, J. M. Golding 1994

Blackwell Science Ltd
Editorial Offices:
Osney Mead, Oxford OX2 0EL
25 John Street, London WC1N 2BL
23 Ainslie Place, Edinburgh EH3 6AJ
238 Main Street, Cambridge
  Massachusetts 02142, USA
54 University Street, Carlton
  Victoria 3053, Australia

Other Editorial Offices:
Arnette Blackwell SA
  224, Boulevard Saint Germain
  75007 Paris, France

Blackwell Wissenschafts-Verlag GmbH
  Kurfürstendamm 57
  10707 Berlin, Germany

  Zehetnergasse 6
  A-1140 Wien
  Austria

First published 1994
Reprinted 1994
Reissued in paperback 1997

Set by Setrite Typesetters Ltd, Hong Kong
Printed and bound in Great Britain
at the University Press, Cambridge

The Blackwell Science logo is a
trade mark of Blackwell Science Ltd,
registered at the United Kingdom
Trade Marks Registry

DISTRIBUTORS

  Marston Book Services Ltd
  PO Box 269
  Abingdon
  Oxon OX14 4YN
  (*Orders*: Tel: 01235 465500
            Fax: 01235 465555)

USA
  Blackwell Science, Inc.
  238 Main Street
  Cambridge, MA 02142
  (*Orders*: Tel: 800 215-1000
            617 876-7000
            Fax: 617 492-5263)

Canada
  Copp Clark, Ltd
  2775 Matheson Blvd East
  Mississauga, Ontario
  Canada, L4W 4P7
  (*Orders*: Tel: 800 263-4374
            905 238-6074)

Australia
  Blackwell Science Pty Ltd
  54 University Street
  Carlton, Victoria 3053
  (*Orders*: Tel: 03 9347-0300
            Fax: 03 9347-5001)

A catalogue record for this title
is available from the British Library

ISBN 0-632-04215-X

Library of Congress
Cataloging-in-Publication Data

Structural foundation designers' manual/Curtins Consulting
  Engineers plc. W. G. Curtin . . . [et al.].
      p.   cm.
  Includes bibliographical references and index.
  ISBN 0-632-04215-X
  1. Foundations. 2. Structural Design. I. Curtin,
  W. G. (William George) II. Curtins Consulting
  Engineers plc.
TA775.S75   1993
624.1′5–dc20   93-12682
                               CIP

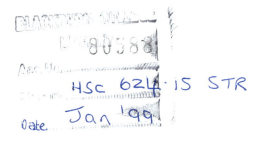

Dedication

This book is dedicated to Bill Curtin who died suddenly in November 1991 following a short illness.

Bill's contribution to the book at that time was all but complete and certainly well ahead of his co-authors. It is a source of sadness that Bill did not have the pleasure and satisfaction of seeing the completed publication but his input and enthusiasm gave his co-authors the will to complete their input and progress the book to completion.

CONTENTS

PREFACE

'Why yet another book on foundations when so many good ones are already available?' – a good question which deserves an answer.

This book has grown out of our consultancy's extensive experience in often difficult and always cost-competitive conditions of designing structural foundations. Many of the existing good books are written with a civil engineering bias and devote long sections to the design of aspects such as bridge caissons and marine structures. Furthermore, a lot of books give good explanations of soil mechanics and research – but mainly for *green field* sites. We expect designers to know soil mechanics and where to turn for reference when necessary. However there are few books which cover the new advances in geotechnical processes necessary now that we have to build on derelict, abandoned inner-city sites, polluted or toxic sites and similar problem sites. And no book, yet, deals with the developments we and other engineers have made, for example, in raft foundations. Some books are highly specialized, dealing only (and thoroughly) with topics such as piling or underpinning.

Foundation engineering is a wide subject and designers need, primarily, one reference for guidance. Much has been written on foundation construction work and methods – and that deserves a treatise in its own right. Design and construction should be interactive, but in order to limit the size of the book, we decided, with regret to restrict discussion to design and omit discussion of techniques such as dewatering, bentonite diaphragm wall construction, timbering, etc.

Foundation construction can be the biggest bottleneck in a building programme so attention to speed of construction is vital in the design and detailing process. Repairs to failed or deteriorating foundations are frequently the most costly of all building remedial measures so care in safe design is crucial, but extravagant design is wasteful. Too much foundation design is unnecessarily costly and the advances in

civil engineering construction have not always resulted in a spin-off for building foundations. Traditional building foundations, while they may have sometimes been over-costly were quick to construct and safe – on good ground. But most of the good ground is now used up and we have to build on sites which would have been rejected on the basis of cost and difficulty as recently as a decade ago. Advances in techniques and developments can now make such sites a cost-and-construction viable option. All these aspects have been addressed in this book.

Though the book is the work of four senior members of the consultancy, it represents the collective experience of all directors, associates and senior staff, and we are grateful for their support and encouragement. As in all engineering design there is no unique 'right' answer to a problem – designers differ on approach, priorities, evaluation of criteria, etc. We discussed, debated and disagreed – the result is a reasonable consensus of opinion but not a compromise. Engineering is an art as well as a science, but the art content is even greater in foundation design. No two painters would paint a daffodil in the same way (unless they were painting by numbers!). So no two designers would design a foundation in exactly the same manner (unless they chose the same computer program and fed it with identical data).

So we do not expect experienced senior designers to agree totally with us and long may individual preference be important. All engineering design, while based on the same studies and knowledge, is an exercise in judgement backed by experience and expertise. Some designers can be daring and others over-cautious; some are innovative and others prefer to use stock solutions. But all foundation design must be safe, cost-effective, durable and buildable, and these have been our main priorities. We hope that all designers find this book useful.

THE BOOK'S STRUCTURE AND WHAT IT IS ABOUT

The book is arranged so that it is possible for individual designers to use the manual in different ways, depending upon their experience and the particular aspects of foundation design under consideration.

The book, which is divided into three parts, deals with the whole of foundation design from a practical engineering viewpoint. Chapters 1–3, i.e. Part 1, deal with soil mechanics and the behaviour of soils, and the commission and interpretation of site investigations are covered in detail.

In Part 2 (Chapters 4–8), the authors continue to share their experience – going back over 30 years – of dealing with filled and contaminated sites and sites in mining areas; these 'problem' sites are increasingly becoming 'normal' sites for today's engineers.

In Part 3 (Chapters 9–15), discussion and practical selection of foundation types are covered extensively, followed by detailed design guidance and examples for the various foundation types. The design approach ties together the safe working load design of soils with the limit-state design of structural foundation members.

The emphasis on practical design is a constant theme running through this book, together with the application of engineering judgement and experience to achieve appropriate and economic foundation solutions for difficult sites. This is especially true of raft design, where a range of raft types, often used in conjunction with filled sites, provides an economic alternative to piled foundations.

It is intended that the experienced engineer would find Part 1 useful to recapitulate the basics of design, and refresh his/her memory on the soils, geology and site investigation aspects. The younger engineer should find Part 1 of more use in gaining an overall appreciation of the starting point of the design process and the interrelationship of design, soils, geology, testing and ground investigation.

Part 2 covers further and special considerations which may affect a site. Experienced and young engineers should find useful information within this section when dealing with sites affected by contamination, mining, fills or when considering the treatment of sub-soils to improve bearing or settlement performance. The chapters in Part 2 give information which will help when planning site investigations and assist in the foundation selection and design process.

Part 3 covers the different foundation types, the selection of an appropriate foundation solution and the factors affecting the choice between one foundation type and another. Also covered is the actual design approach, calculation method and presentation for the various foundation types. Experienced and young engineers should find this section useful for the selection and design of pads, strips, rafts and piled foundations.

The experienced designer can refer to Parts 1, 2 and 3 in any sequence. Following an initial perusal of the manual, the young engineer could also refer to the various parts out of sequence to assist with the different stages and aspects of foundation design.

For those practising engineers who become familiar with the book and its information, the tables, graphs and charts grouped together in the Appendices should become a quick and easy form of reference for useful, practical and economic foundations in the majority of natural and man-made ground conditions.

Occasional re-reading of the text, by the more experienced designer, may refresh his/her appreciation of the basic important aspects of economical foundation design, which can often be forgotten when judging the merits of often over-emphasized and over-reactive responses to relatively rare foundation problems. Such problems should not be allowed to dictate the 'norm' when, for the majority of similar cases, a much simpler and more practical solution (many of which are described within these pages) is likely still to be quite appropriate.

ACKNOWLEDGEMENTS

We are grateful for the trust and confidence of many clients in the public and private sectors who readily gave us freedom to develop innovative design. We appreciate the help given by many friends in the construction industry, design professions and organizations and we learnt much from discussions on site and debate in design team meetings. We are happy to acknowledge (in alphabetical order) permission to quote from:

- British Standards Institution
- British Steel
- Building Research Establishment
- Cement and Concrete Association
- Construction Industry Research and Information Association

- Department of the Environment/Inter-Department Committee on the Redevelopment of Contaminated Land
- Institution of Civil Engineers
- John Wiley & Sons.

We are thankful for the detailed vetting and constructive criticism from many of our directors and staff who made valuable contributions, particularly to John Beck, Dave Knowles and Jeff Peters, and to Mark Day for diligently drafting all of the figures.

We are indebted to Sandra Taylor and Susan Wisdom who were responsible for typing the bulk of the manuscript, with patience, care and interest.

AUTHORS' BIOGRAPHIES

W. G. CURTIN MEng, PhD, FEng, FICE, FIStructE, MConsE

Bill Curtin's interest and involvement in foundation engineering dated back to his lecturing days at Brixton and Liverpool in the 1950–60s. In 1960 he founded the Curtins practice in Liverpool and quickly gained a reputation for economic foundation solutions on difficult sites in the North-west of England and Wales. He was an active member of both the Civil and Structural Engineering Institutions serving on and chairing numerous committees and working with BSI and CIRIA. He produced numerous technical design guides and text books including *Structural Masonry Designer's Manual.*

G. SHAW CEng, FICE, FIStructE, MConsE

Gerry Shaw is a director of Curtins Consulting Engineers plc with around 40 years' experience in the building industry, including more than 30 years as a consulting engineer. He has been responsible for numerous important foundation structures on both virgin and man-made soil conditions and has been continuously involved in foundation engineering, innovative developments and monitoring advances in foundation solutions. He is co-author of a number of technical books and design notes and is external examiner for Kingston University. He acts as expert witness in legal cases involving building failures, and was a member of the BRE/CIRIA Committee which investigated and analysed building failures in 1980. He is co-author of both *Structural Masonry Designers' Manual* and *Structural Masonry Detailing Manual.* He is a Royal Academy of Engineering Visiting Professor of Civil Engineering Design to the University of Plymouth.

G. I. PARKINSON CEng, FICE, FIStructE, MConsE

Gary Parkinson is a director of Curtins Consulting Engineers plc responsible for the Liverpool office. He has over 30 years' experience in the building industry, including 25 years as a consulting engineer. He has considerable foundation engineering experience, and has been involved in numerous land reclamation and development projects dealing with derelict and contaminated industrial land and dockyards. He is co-author of *Structural Masonry Detailing Manual.*

JOHN GOLDING BSc, MS, CEng, MICE, FIStructE

John Golding currently works for CEDAC, an infrastructure services group within British Rail. He has spent 15 years with consulting engineers Pell Frischmann, Ove Arup, Harris & Sutherland, and Curtins. He is experienced in the design of commercial, residential, industrial and civil engineering structures, and their associated foundations. Many of these projects have been in difficult inner city sites, requiring a range of ground improvement treatments and other foundation solutions. He has been involved in, and has written papers on, research and innovative development for foundations, masonry, concrete and steelwork.

NOTATION

APPLIED LOADS AND CORRESPONDING PRESSURES STRESSES

Loads

$F = F_B + F_S$	foundation loads
F_B	buried foundation/backfill load
F_S	new surcharge load
G	superstructure dead load
H	horizontal load
H_f	horizontal load capacity at failure
M	bending moment
$N = T - S$	net load
P	superstructure vertical load
Q	superstructure imposed load
$S = S_B + S_S$	existing load
S_B	'buried' surcharge load (i.e. $\approx F_B$)
S_S	existing surcharge load
$T = P + F$	total vertical load
V	shear force
W	superstructure wind load

Pressures and stresses

$f = F/A$	pressure component resulting from F
$f_B = F_B/A$	pressure component resulting from F_B
$f_S = F_S/A$	pressure component resulting from F_S
g	pressure component resulting from G
$n = t - s$	pressure component resulting from N
$n' = n - \gamma_w z_w$	net effective stress
n_f	net ultimate bearing capacity at failure
$p = t - f$	pressure component resulting from P
$p_u = t_u - f_u$	resultant ultimate design pressure
p_z	pressure component at depth z resulting from P
q	pressure component resulting from Q
$s = S/A$	pressure component resulting from S
$s_B = S_B/A$	pressure component resulting from S_B
$s_S = S_S/A$	pressure component resulting from S_S
$s' = s - \gamma_w z_w$	existing effective stress
t	pressure resulting from T
$t' = t - \gamma_w z_w$	total effective stress
t_f	total ultimate bearing capacity at failure
v	shear stress due to V
w	pressure component resulting from W

General subscripts for loads and pressures

a	allowable (load or bearing pressure)
f	failure (load or bearing pressure)
u	ultimate (limit-state)
G	dead
Q	imposed
W	wind
F	foundation
P	superstructure
T	total

Partial safety factors for loads and pressures

γ_G	partial safety factor for dead loads
γ_Q	partial safety factor for imposed loads
γ_W	partial safety factor for wind loads
γ_F	combined partial safety factor for foundation loads
γ_P	combined partial safety factor for superstructure loads
γ_T	combined partial safety factor for total loads

Notation principles for loads and pressures

(1) *Loads* are in capitals, e.g.

P = load from superstructure (kN)

F = load from foundation (kN)

(2) *Loads per unit length* are also in capitals, e.g.

P = load from superstructure (kN/m)

F = load from foundation (kN/m)

(3) Differentiating between *loads* and *loads per unit length*.

This is usually made clear by the context, i.e. pad foundation calculations will normally be in terms of *loads* (in kN), and strip foundations will normally be in terms of *loads per unit length* (kN/m). Where there is a need to differentiate, this is done, as follows:

ΣP = load from superstructure (kN)

P = load from superstructure per unit length (kN/m)

(4) *Distributed loads* (loads per unit area) are lower case, e.g.

f = uniformly distributed foundation load (kN/m^2)

(5) *Ground pressures* are also in lower case, e.g.

p = pressure distribution due to superstructure loads (kN/m^2)

f = pressure distribution due to foundation loads (kN/m^2)

(6) *Characteristic* versus *ultimate* (u subscript).

Loads and pressures are either *characteristic* values or *ultimate* values. This distinction is important, since *characteristic* values (working loads/pressures) are used for bearing pressure checks, while *ultimate* values (factored loads/pressures) are used for structural member design. All ultimate values have u subscripts. Thus

p = characteristic pressure due to superstructure loads

p_u = ultimate pressure due to superstructure loads

GENERAL NOTATION

Dimensions

a	distance of edge of footing from face of wall/beam
A	area of base
A_b	effective area of base (over which compressive bearing pressures act)
A_s	area of reinforcement
	OR surface area of pile shaft
b	width for reinforcement design
B	width of base
B_b	width of beam thickening in raft
B_{conc}	assumed width of concrete base
B_{fill}	assumed spread of load at underside of compacted fill material
d	effective depth of reinforcement
D	depth of underside of foundation below ground level
	OR diameter of pile
D_w	depth of water-table below ground level
e	eccentricity
h	thickness of base
h_b	thickness of beam thickening in raft
h_{fill}	thickness of compacted fill material
h_{conc}	thickness of concrete
H	length of pile
	OR height of retaining wall
H_1, H_2	thickness of soil strata '1', '2', etc.
L	length of base
	OR length of depression
L_b	effective length of base (over which compressive bearing pressures act)
t_w	thickness of wall
u	length of punching shear perimeter
x	projection of external footing beyond line of action of load
z	depth below ground level

z_w	depth below water-table
ρ_1, ρ_2	settlement of strata '1', '2', etc.

Miscellaneous

c	cohesion
c_b	undisturbed shear strength at base of pile
c_s	average undrained shear strength for pile shaft
e	void ratio
f_{bs}	characteristic local bond stress
f_c	ultimate concrete stress (in pile)
f_{cu}	characteristic concrete cube strength
I	moment of inertia
k	permeability
K	earth pressure coefficient
K_m	bending moment factor (raft design)
m_v	coefficient of volume compressibility
N	SPT value
N_c	Terzaghi bearing capacity factor
N_q	Terzaghi bearing capacity factor
N_γ	Terzaghi bearing capacity factor
v_c	ultimate concrete shear strength
V	total volume
V_s	volume of solids
V_v	volume of voids
Z	section modulus
α	creep compression rate parameter
	OR adhesion factor
γ	unit weight of soil
γ_{dry}	dry unit weight of soil
γ_{sat}	saturated unit weight of soil
γ_w	unit weight of water
δ	angle of wall friction
ε	strain
μ	coefficient of friction
σ	(soil) stress normal to the shear plane
σ'	(soil) effective normal stress
τ	(soil) shear stress
ϕ	angle of internal friction

Occasionally it has been necessary to vary the notation system from that indicated here. Where this does happen, the changes to the notation are specifically defined in the accompanying text or illustrations.

PART 1

APPROACH AND FIRST CONSIDERATIONS

CHAPTER 1

PRINCIPLES OF FOUNDATION DESIGN

1.1 INTRODUCTION

Foundation design could be thought of as analogous to a beam design. The designer of the beam will need to know the load to be carried, the load-carrying capacity of the beam, how much it will deflect and whether there are any long-term effects such as creep, moisture movement, etc. If the calculated beam section is, for some reason, not strong enough to support the load or is likely to deflect unduly, then the beam section is changed. Alternatively, the beam can either be substituted for another type of structural element, or a stronger material be chosen for the beam.

Similarly the soil supporting the structure must have adequate load-carrying capacity (bearing capacity) and not deflect (settle) unduly. The long-term effect of the soil's bearing capacity and settlement must be considered. If the ground is not strong enough to bear the proposed initial design load then the structural contact load (bearing pressure) can be reduced by spreading the load over a greater area − by increasing the foundation size or other means − or by transferring the load to a lower strata. For example, rafts should replace isolated pad bases − or the load can be transferred to stronger soil at a lower depth beneath the surface by means of piles. Alternatively, the ground can be strengthened by compaction, stabilization, pre-consolidation or other means. The structural materials in the super-structure are subject to stress, strain, movement, etc., and it can be helpful to consider the soil supporting the superstructure as a structural material, also subject to stress, strain and movement.

Structural design has been described as: using materials not fully understood, to make frames which cannot be accurately analysed, to resist forces which can only be estimated. Foundation design is, at best, no better. 'Accuracy' is a chimera and the designer must exercise judgement.

Sections 1.2−1.6 outline the general principles before dealing with individual topics in the following sections and chapters.

1.2 FOUNDATION SAFETY CRITERIA

It is a statement of the obvious that the function of a foundation is to transfer the load from the structure to the ground (i.e. soil) supporting it − and it must do this safely,

for if it does not then the foundation will fail in bearing and/ or settlement, and seriously affect the structure which may also fail. The history of foundation failure is as old as the history of building itself, and our language abounds in such idioms as 'the god with the feet of clay', 'build not thy house on sand', 'build on a firm foundation', 'the bedrock of our policy'. The foundation must also be economical in construction costs, materials and time.

There are a number of reasons for foundation failure, the two major causes being:

(1) *Bearing capacity*. When the shear stress within the soil, due to the structure's loading, exceeds the shear strength of the soil, catastrophic collapse of the supporting soil can occur. Before ultimate collapse of the soil occurs there can be large deformations within it which may lead to unacceptable differential movement or settlement of, and damage to, the structure. (In some situations however, collapse can occur with little or no advance warning!)

(2) *Settlement*. Practically all materials contract under compressive loading and distort under shear loading − soils are no exception. Provided that the settlement is either acceptable (i.e. will not cause structural damage or undue cracking, will not damage services, and will be visually acceptable and free from practical problems of door sticking, etc.) or can be catered for in the structural design (e.g. by using three-pinned arches which can accommodate settlement, in lieu of fixed portal frames), there is not necessarily a foundation design problem. Problems will occur when the settlement is significantly excessive or differential.

Settlement is the combination of two phenomena:

(1) *Contraction of the soil* due to compressive and shear stresses resulting from the structure's loading. This contraction, partly elastic and partly plastic, is relatively rapid. Since soils exhibit non-linear stress/strain behaviour and the soil under stress is of complex geometry it is not possible to predict accurately the magnitude of settlement.

(2) *Consolidation of the soil* due to volume changes. Under applied load the moisture is 'squeezed' from the soil and the soil compacts to partly fill the voids left by the

retreating moisture. In soils of low permeability, such as clays, the consolidation process is slow and can even continue throughout the life of the structure (for example, the leaning tower of Pisa). Clays of relatively high moisture content will consolidate by greater amounts than clays with lower moisture contents. (Clays are susceptible to volume change with change in moisture content – they can shrink on drying out and *heave*, i.e. expand, with increase in moisture content.) Sands tend to have higher permeability and lower moisture content than clays. Therefore the consolidation of sand is faster but less than that of clay.

1.3 BEARING CAPACITY

1.3.1 Introduction

Some designers, when in a hurry, tend to want simple 'rules of thumb' (based on local experience) for values of bearing capacity. But like most rules of thumb, while safe for typical structures on normal soils, their use can produce uneconomic solutions, restrict the development of improved methods of foundation design, and lead to expensive mistakes when the structure is not *typical*.

For *typical* buildings:

(1) The dead and imposed loads are built up relatively slowly.
(2) *Actual* imposed loads (as distinct from those assumed for design purposes) are often only a third of the dead load.
(3) The building has a height/width ratio of between 1/3 and 3.
(4) The building has regularly distributed columns or load-bearing walls, most of them fairly evenly loaded.

Typical buildings have changed dramatically since the Second World War. The use of higher design stresses, lower factors of safety, the removal of robust non-load-bearing partitioning, etc., has resulted in buildings of half their previous weight, more susceptible to the effects of settlement, and built for use by clients who are less tolerant in accepting relatively minor cracking of finishes, etc. Because of these changes, *practical* experience gained in the past is not always applicable to present construction.

For *non-typical* structures:

(1) The imposed load may be applied rapidly, as in tanks and silos, resulting in possible settlement problems.
(2) There may be a high ratio of imposed to dead load. Unbalanced imposed-loading cases – imposed load over part of the structure – can be critical, resulting in differential settlement or bearing capacity failures, if not allowed for in design.
(3) The requirement may be for a tall, slender building which may be susceptible to tilting or overturning and have more critical wind loads.
(4) The requirement may be for a non-regular column/wall layout, subjected to widely varying loadings, which may

require special consideration to prevent excessive differential settlement and bearing capacity failure.

There is also the danger of going to the other extreme by doing complicated calculations based on numbers from unrepresentative soil tests alone, and ignoring the important evidence of the soil profile and local experience. Structural design and materials are not, as previously stated, mathematically precise; foundation design and materials are even less precise. Determining the bearing capacity solely from a 100 mm thick small-diameter sample and applying it to predict the behaviour of a 10 m deep strata, is obviously not sensible – particularly when many structures could fail, in serviceability, by settlement at bearing pressures well below the soil's ultimate bearing capacity.

1.3.2 Bearing capacity

Probably the happy medium is to follow the sound advice given by experienced engineers in the British Standard Institution's *Code of practice for foundations*, BS 8004. There they define *ultimate bearing capacity* as that value of the net loading intensity, for a particular foundation, at which the ground fails in shear as determined by either field loading tests or laboratory tests on undistorted samples. (*Ultimate* in this instance does *not* refer to ultimate limit state.)

The *net loading intensity* (net bearing pressure) is the additional intensity of vertical loading at the base of a foundation due to the weight of the new structure and its loading, including any earthworks.

The ultimate bearing capacity divided by a suitable factor of safety – typically 3 – is referred to as the *safe bearing capacity*.

It has not been found possible, yet, to apply limit state design fully to foundations, since bearing capacity and settlement are so intertwined and influence both foundation and superstructure design (this is discussed further in section 1.5). Furthermore, the superstructure itself can be altered in design to accommodate, or reduce, the effects of settlement. A reasonable compromise has been devised by engineers in the past and is given below.

1.3.3 Presumed bearing value

The pressure within the soil will depend on the net loading intensity, which in turn depends on the structural loads and the foundation type. This pressure is then compared with the ultimate bearing capacity to determine a factor of safety. This appears reasonable and straightforward – but there is a catch-22 snag. It is not possible to determine the net loading intensity without first knowing the foundation type and size, but the foundation type and size cannot be designed without knowing the acceptable bearing pressure.

The deadlock has been broken by BS 8004, which gives *presumed bearing values* (estimated bearing pressures) for different types of ground. This enables a preliminary foundation design to be carried out which can be adjusted, up or down, on further analysis. The presumed bearing value is defined as: 'the net loading intensity considered appropriate to the particular type of ground for preliminary design purposes and the value is based on either local experience or on calculation from strength tests or field loading tests using a factor of safety against shear failure'.

Foundation design, like superstructure design, is a trial-and-error method − a preliminary design is made, then checked and, if necessary, amended. Amendments would be necessary, for example, to restrict settlement or over-loading; in consideration of economic and construction implications, or designing the superstructure to resist or accommodate settlements. The Code's presumed bearing values are given in Table 1.1 and experience shows that these are valuable and reasonable in preliminary design.

1.3.4 Allowable bearing pressure
Knowing the structural loads, the preliminary foundation design and the ultimate bearing capacity, a check can be made on the *allowable bearing pressure*. The net allowable bearing pressure is defined in the Code as 'the maximum allowable net loading intensity at the base of the foundation' taking into account:

(1) The ultimate bearing capacity.
(2) The amount and kind of settlement expected.
(3) The ability of the given structure to take up this settlement.

This practical definition shows that the allowable bearing pressure is a combination of three functions; the strength and settlement characteristics of the ground, the foundation type, and the settlement characteristics of the structure.

1.3.5 Inclined loading
When horizontal foundations are subject to inclined forces (portals frames, cantilever structures, etc.) the passive resistance of the ground must be checked for its capacity to resist the horizontal component of the inclined load. This

Table 1.1 Presumed bearing values (BS 8004, Table 1)

NOTE. These values are for preliminary design purposes only, and may need alteration upwards or downwards. No addition has been made for the depth of embedment of the foundation (see 2.1.2.3.2 and 2.1.2.3.3).

Category	Types of rocks and soils	Presumed allowable bearing value		Remarks
		kN/m^2*	kgf/cm^2* $tonf/ft^2$	
Rocks	Strong igneous and gneissic rocks in sound condition Strong limestones and strong sandstones Schists and slates Strong shales, strong mudstones and strong siltstones	10 000 4 000 3 000 2 000	100 40 30 20	These values are based on the assumption that the foundations are taken down to unweathered rock. For weak, weathered and broken rock, see 2.2.2.3.1.12
Non-cohesive soils	Dense gravel, or dense sand and gravel Medium dense gravel, or medium dense sand and gravel Loose gravel, or loose sand and gravel Compact sand Medium dense sand Loose sand	>600 <200 to 600 <200 >300 100 to 300 <100 Value depending on degree of looseness	>6 <2 to 6 <2 >3 1 to 3 <1	Width of foundation not less than 1 m. Groundwater level assumed to be a depth not less than below the base of the foundation. For effect of relative density and groundwater level, see 2.2.2.3.2
Cohesive soils	Very stiff boulder clays and hard clays Stiff clays Firm clays Soft clays and silts Very soft clays and silts	300 to 600 150 to 300 75 to 150 <75 Not applicable	3 to 6 1.5 to 3 0.75 to 1.5 <0.75	Group 3 is susceptible to long-term consolidation settlement (see 2.1.2.3.3). For consistencies of clays, see table 5
Peat and organic soils		Not applicable		See 2.2.2.3.4
Made ground or fill		Not applicable		See 2.2.2.3.5

* 107.25 kN/m² = 1.094 kgf/cm² = 1 tonf/ft²
All references within this table refer to the original document

could result in reducing the value of the allowable bearing pressure to carry the vertical component of the inclined load. BS 8004 (*Code of practice for foundations*) suggests a simple *rule* for design of foundations subject to inclined loads as follows:

$$\frac{T}{T_a} + \frac{H}{H_a} < 1$$

where T = vertical component of the inclined load,
  H = horizontal component of the inclined load,
  T_a = allowable vertical load – dependent on allowable bearing pressure,
  H_a = allowable horizontal load – dependent on allowable friction and/or adhesion on the horizontal base, plus passive resistance where this can be relied upon.

However, like all simple *rules* which are on the safe side, there are exceptions. A more conservative value can be necessary when the horizontal component is relatively high and is acting on shallow foundations (where their depth/breadth ratio is less than 1/4) founded on non-cohesive soils.

In the same way that allowable bearing pressure is reduced to prevent excessive settlement, so too may allowable passive resistance, to prevent unacceptable horizontal movement.

1.4 SETTLEMENT

If the building settles excessively, particularly differentially – e.g. adjacent columns settling by different amounts – the settlement may be serious enough to endanger the stability of the structure, and would be likely to cause serious serviceability problems.

Less serious settlement may still be sufficient to cause cracking which could affect the building's weathertightness, thermal and sound insulation, fire resistance, damage finishes and services, affect the operation of plant such as overhead cranes, and other *serviceability* factors. Furthermore, settlement, even relatively minor, which causes the building to tilt, can render it visually unacceptable. (Old Tudor buildings, for example, may look charming and quaint with their tilts and leaning, but clients and owners of modern buildings are unlikely to accept similar tilts.)

Differential settlement, sagging, hogging and relative rotation are shown in Fig. 1.1.

In general terms it should be remembered that foundations are no different from other structural members and deflection criteria similar to those for superstructure members would also apply to foundation members.

From experience it has been found that the magnitude of *relative* rotation – sometimes referred to as angular distortion – is critical in framed structures, and the magnitude of the *deflection ratio*, Δ/L, is critical for load-bearing walls. Empirical criteria have been established to minimize cracking, or other damage, by limiting the movement, as shown in Table 1.2.

The length-to-height ratio is important since according to some researchers the greater the length-to-height ratio the greater the limiting value of Δ/L. It will also be noted that cracking due to hogging occurs at half the deflection ratio of that for sagging. Sagging problems appear to occur more frequently than hogging in practice.

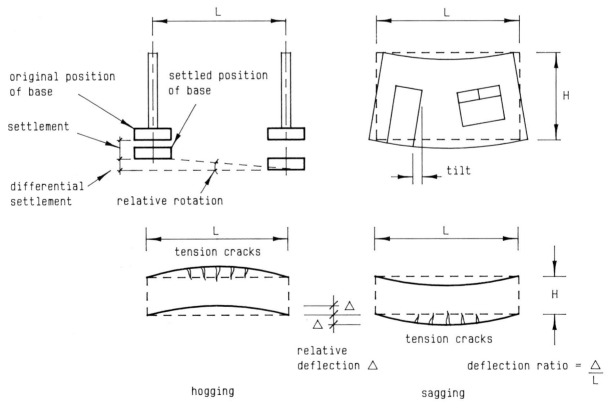

Fig. 1.1 Settlement definitions.

Table 1.2 Typical values of angular distortion to limit cracking (*Ground Subsidence*, Table 1, Institution of Civil Engineers, 1977)

Class of structure	Type of structure	Limiting angular distortion
1	Rigid	Not applicable: tilt is criterion
2	Statically determinate steel and timber structures	1/100 to 1/200
3	Statically indeterminate steel and reinforced concrete framed structures, load-bearing reinforced brickwork buildings, all founded on reinforced concrete continuous and slab foundations	1/200 to 1/300
4	As class 3, but not satisfying one of the stated conditions	1/300 to 1/500
5	Precast concrete large-panel structures	1/500 to 1/700

Since separate serviceability and ultimate limit state analyses are not at present carried out for the soil − see section 1.5 − it is current practice to adjust the factor of safety which is applied to the soil's ultimate bearing capacity, in order to obtain the allowable bearing pressure.

Similarly, the partial safety factor applied to the characteristic structural loads will be affected by the usual superstructure design factors and then adjusted depending on the structure (its sensitivity to movement, design life, damaging effects of movement), and the type of imposed loading. For example, full imposed load occurs infrequently in theatres and almost permanently in grain stores. Overlooking this permanence of loading in design has caused foundation failure in some grain stores. A number of failures due to such loading conditions have been investigated by the authors' practice. A typical example is an existing grain store whose foundations performed satisfactorily until a new grain store was built alongside. The ground pressure from the new store increased the pressure in the soil below the existing store − which settled and tilted. Similarly, any bending moments transferred to the ground (by, for example, fixing moments at the base of fixed portal frames) must be considered in the design, since they will affect the structure's contact pressure on the soil.

There is a rough correlation between bearing capacity and settlement. Soils of high bearing capacity tend to settle less than soils of low bearing capacity. It is therefore even more advisable to check the likely settlement of structures founded on weak soils. As a guide, care is required when the safe bearing capacity (i.e. ultimate bearing capacity divided by a factor of safety) falls below $125 \, kN/m^2$; each site, and each structure, must however be judged on its own merits.

1.5 LIMIT STATE PHILOSOPHY

1.5.1 Working stress design

A common design method (based on *working stress*) used in the past was to determine the ultimate bearing capacity of the soil, then divide it by a factor of safety, commonly 3, to determine the *safe bearing capacity*. The safe bearing capacity is the maximum allowable design loading intensity on the soil. The *ultimate bearing capacity* is the magnitude of the loading intensity at which the soil fails in shear. Typical ultimate bearing capacities are $150 \, kN/m^2$ for soft clays, $300-600 \, kN/m^2$ for firm clays and loose sand/gravel, and $1000-1500 \, kN/m^2$ for hard boulder clays and dense gravels.

Consider the following example for a column foundation. The ultimate bearing capacity for a stiff clay is $750 \, kN/m^2$. If the factor of safety equals 3, determine the area of a pad base to support a column load of $1000 \, kN$ (ignoring the weight of the base and any overburden).

$$\text{Safe bearing capacity} = \frac{\text{ultimate bearing capacity}}{\text{factor of safety}}$$

$$= \frac{750}{3} = 250 \, kN/m^2$$

$$\text{actual bearing pressure} = \frac{\text{column load}}{\text{base area}}$$

therefore,

$$\text{required base area} = \frac{\text{column load}}{\text{safe bearing capacity}}$$

$$= \frac{1000}{250} = 4 \, m^2$$

The method has the attraction of simplicity and was generally adequate for traditional buildings in the past.

However, it can be uneconomic and ignores other factors. A nuclear power station, complex chemical works housing expensive plant susceptible to foundation movement or similar buildings, can warrant a higher factor of safety than a supermarket's warehouse stacking tinned pet food. A crowded theatre may deserve a higher safety factor than an occasionally used cow-shed. The designer should exercise his or her judgement in choice of factor of safety.

In addition, while there must be precautions taken against foundation *collapse limit state* (i.e. total failure) there must be a check that the *serviceability limit state* (i.e. movement under load which causes structural or building use distress) is not exceeded. Where settlement criteria dominate, the bearing pressure is restricted to a suitable value below that of the safe bearing capacity, known as the *allowable bearing pressure*.

1.5.2 Limit state design

Attempts to apply limit state philosophy to foundation design have, so far, not been considered totally successful. So a compromise between *working stress* and *limit state* has developed, where the designer determines an estimated *allowable bearing pressure* and checks for settlements and building serviceability. The actual bearing pressure is then factored up into an *ultimate design pressure*, for structural design of the foundation members.

The partial safety factors applied for ultimate design loads (i.e. typically $1.4 \times$ dead, $1.6 \times$ imposed, $1.4 \times$ wind and 1.2 for dead + imposed + wind) are for superstructure design and should *not* be applied to foundation design for allowable bearing calculations.

For dead and imposed loads the actual working load, i.e. the unfactored characteristic load, should be used in most foundation designs. Where there are important isolated foundations and particularly when subject to significant eccentric loading (as in heavily loaded gantry columns, water towers, and the like), the engineer should exercise his discretion in applying a partial safety factor to the imposed load. Similarly when the imposed load is very high in relation to the dead load (as in large oil tanks of steel drums), the engineer should apply a partial safety factor to the imposed load.

In fact when the foundation load due to wind load on the superstructure is relatively small − i.e. less than 25% of (dead + imposed) − it may be ignored. Where the occasional foundation load due to wind exceeds 25% of (dead + imposed), then the foundation area should be proportioned so that the pressure due to wind + dead + imposed loads does not exceed $1.25 \times$ (allowable bearing pressure). When wind uplift on a foundation exceeds dead load, then this becomes a critical load case.

1.6 INTERACTION OF SUPERSTRUCTURE AND SOIL

The superstructure, its foundation, and the supporting soil should be considered as a structural entity, with the three elements interacting.

Adjustments to the superstructure design to resist the effects of bearing failure and settlements, at minor extra

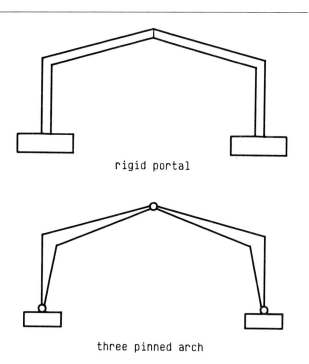

rigid portal

three pinned arch

Fig. 1.2 Rigid portal versus three pinned arch.

costs, are often more economic than the expensive area increase or stiffening of the foundations. Some examples from the authors' practice are given here to illustrate these adjustments. Adjustments to the soil to improve its properties are briefly discussed in section 1.8. The choice of foundation type is outlined in section 1.7. Adjustments and choices are made to produce the most economical solution.

1.6.1 Example 1: Three pinned arch

The superstructure costs for a rigid-steel portal-frame shed are generally cheaper than the three pinned arch solution (see Fig. 1.2).

Differential settlement of the column pad bases will seriously affect the bending moments (and thus the stresses) in the rigid portal, but have insignificant effect on the three pinned arch. Therefore the pad foundations for the rigid portal will have to be bigger and more expensive than those for the arch, and may far exceed the saving in superstructure steelwork costs for the portal. (In some cases it can be worthwhile to place the column eccentric to the foundation base to counteract the moment at the base of the foundation due to column fixity and/or horizontal thrust.)

1.6.2 Example 2: Vierendeel superstructure

The single-storey reinforced concrete (r.c.) frame structure shown in Fig. 1.3 was founded in soft ground liable to excessive sagging/differential settlement. Two main solutions were investigated:

(1) Normal r.c. superstructure founded on deep, stiff, heavily reinforced strip footings.
(2) Stiffer superstructure, to act as a Vierendeel truss and thus in effect becoming a stiff *beam*, with the foundation beam acting as the bottom boom of the truss.

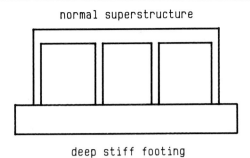

normal superstructure

deep stiff footing

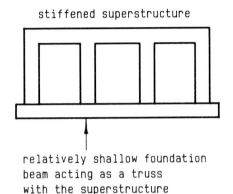

stiffened superstructure

relatively shallow foundation
beam acting as a truss
with the superstructure

Fig. 1.3 Stiff footing versus Vierendeel truss.

The *truss* solution (2) showed significant savings in construction costs and time.

1.6.3 Example 3: Prestressed brick diaphragm wall

A sports hall was to be built on a site with severe mining subsidence. At first sight the economic superstructure solution of a brickwork diaphragm wall was ruled out, since the settlement due to mining would result in unacceptable tensile stresses in the brickwork. The obvious solutions were to cast massive, expensive foundation beams to resist the settlement and support the walls, or to abandon the brickwork diaphragm wall solution in favour of a probably more expensive structural steelwork superstructure. The problem was economically solved by prestressing the wall to eliminate the tensile stresses resulting from differential settlement.

1.6.4 Example 4: Composite deep beams

Load-bearing masonry walls built on a soil of low bearing capacity containing *soft spots* are often founded on strip footings reinforced to act as beams, to enable the footings to span over local depressions. The possibility of composite action between the wall and strip footing, acting together as a deeper beam, is not usually considered. Composite action significantly reduces foundation costs with only minor increases in wall construction costs (i.e. engineering bricks

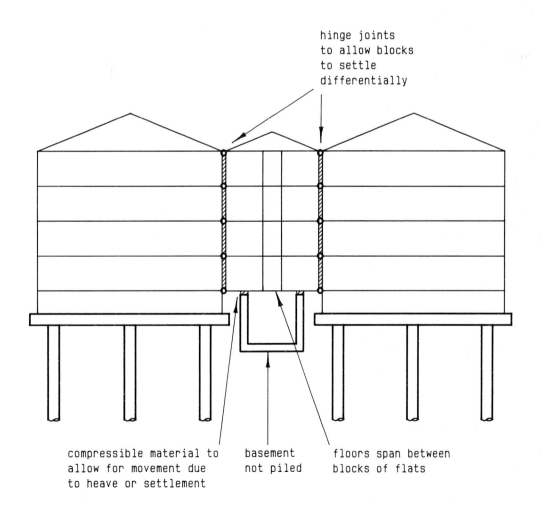

hinge joints
to allow blocks
to settle
differentially

compressible material to
allow for movement due
to heave or settlement

basement
not piled

floors span between
blocks of flats

Fig. 1.4 Buoyancy raft.

are used as a d.p.c. in lieu of normal d.p.c.s, which would otherwise act as a slip plane of low shear resistance).

1.6.5 Example 5: Buoyancy raft

A four-storey block of flats was to be built on a site where part of the site was liable to ground heave due to removal of trees. The sub-soil was of low bearing capacity overlying dense gravel. The building plan was amended to incorporate two sections of flats interconnected by staircase and lift shafts, see Fig. 1.4. A basement was required beneath the staircase section and the removal of overburden enabled the soil to sustain structural loading. To have piled this area would have added unnecessary expense. The final design was piling for the two, four-storey sections of the flats, and a buoyancy raft (see section 13.9) for the basement.

It is hoped that these five simple examples illustrate the importance of considering the soil/structure interaction and encourage young designers not to consider the foundation design in isolation.

Bearing capacity, pressure, settlement, etc., are dealt with more fully in Chapter 2 and in section B of Chapter 10.

1.7 FOUNDATION TYPES

Foundation types are discussed in detail in Chapter 9, and a brief outline only is given here to facilitate appreciation of the philosophy.

Basically there are four major foundation types: pads, strips, rafts, and piles. There are a number of grades within each type and there are combinations of types. Full details of the choice, application and design is dealt with in detail in later chapters. The choice is determined by the structural loads, ground conditions, economics of design, economics of scale of the contract and construction costs, buildability, durability — as is all structural design choice. Only a brief description is given in this section to help understand the soil behaviour.

1.7.1 Pad foundations

Pad foundations tend to be the simplest and cheapest foundation type and are used when the soil is relatively strong or when the column loads are relatively light. They are usually square or rectangular on plan, of uniform thickness and generally of reinforced concrete. They can be stepped or haunched, if material costs outweigh labour costs. The reinforcement can vary from nothing at one extreme through to a heavy steel grillage at the other, with lightly reinforced sections being the most common. Typical types are shown in Fig. 1.5.

1.7.2 Strip footings

Strip footings are commonly used for the foundations to load-bearing walls. They are also used when the pad foundations for a number of columns in line are so closely spaced that the distance between the pads is approximately equal to the length of the side of the pads. (It is usually more economic and faster to excavate and cast concrete in one long strip, than as a series of closely spaced isolated pads.)

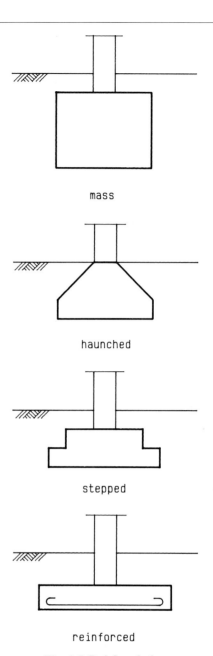

Fig. 1.5 Pad foundations.

They are also used on weak ground to increase the foundation bearing area, and thus reduce the bearing pressure — and the weaker the ground then the wider the strip. When it is necessary to stiffen the strip to resist differential settlement, then *tee* or *inverted tee* strip footings can be adopted. Typical examples are shown in Fig. 1.6.

1.7.3 Raft foundations

When strips become so wide (because of heavy column loads or weak ground) that the clear distance between them is about the same as the width of the strips (or when the depth to suitable bearing capacity strata for strip footing loading becomes too deep), it is worth considering raft foundations. They are useful in restricting the differential settlement on variable ground, and to distribute variations of superstructure loading from area to area. Rafts can be stiffened (as strips can) by the inclusion of tee beams.

Rafts can also be made *buoyant* by the excavation (dis-

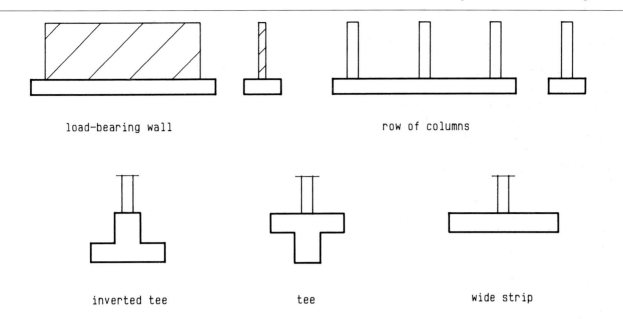

load-bearing wall

row of columns

inverted tee

tee

wide strip

Fig. 1.6 Strip footings.

placement) of a depth of soil, similar to the way that sea-going rafts are made to float by displacing an equal weight of water. A cubic metre of soil can weigh as much as three floor loads per square metre, so a deep basement excavation can *displace* the same weight of soil as the weight of the proposed structure. However where there is a high water-table then *flotation* of the raft can occur, if the water pressures exceed the self-weight!. Typical examples of rafts are shown in Fig. 1.7.

1.7.4 Piled foundations

Piles are used when they are more economical than the alternatives, or when the ground at foundation level is too weak to support any of the previous foundation types. Piles are also used on sites where soils are particularly affected by seasonal changes (and/or the action of tree roots), to transfer the structural loads below the level of such influence. Piles can transfer the structure load to stronger soil, or to bedrock and dense gravel. The structural load is supported by the pile, acting as a column, when it is end-bearing on rock (or driven into dense gravel), or alternatively by skin friction between the peripheral area of the pile and the surrounding soil (similar to a nail driven into wood).

The rapid advances in piling technology over the past two decades has made piling on many sites a viable alternative economic proposition and not necessarily a last resort. The reduction in piling costs has also made possible the use of land which previously was considered unsuitable for building. Recently the authors' practice, for example, economically founded a small housing estate on a thick bed of peat by the use of 20 m long piles to support the low-rise domestic housing. Typical examples of piling are shown in Fig. 1.8.

1.8 GROUND TREATMENT (GEOTECHNICAL PROCESSES)

Soil properties can change under the action of superstructure loading. It compacts, consolidates and drains, so it becomes denser, stronger and less prone to settlement. These improvements can be induced by a variety of geotechnical processes *before* construction. The ground can be temporarily loaded before construction (pre-consolidated), hammered by heavy weights to compact it (dynamic consolidation), vibrated to *shake down* and reduce the voids ratio (vibro stabilization), the soil moisture drained off (dewatering, sand wicks), the voids filled with cementitious material (grouting, chemical injection), and similar techniques.

Imported material (usually sandy gravel) can be laid over weak ground and compacted so that the pressure from column pad foundations can be spread over a greater area. Imported material can also be used to *seal* contaminated sites. Imported soils can also be laid and compacted in thin (say 150 mm) layers with polymer nets placed between each layer. The composite material, known as *reinforced earth*, has been widely used in retaining walls and embankments.

These techniques are discussed in detail in Chapter 8. The development of these techniques has made it possible to build economically on sites which, until recently, were too difficult and expensive to be considered as building land.

Temporary geotechnical processes can be used to ease excavation. Typical cases are:

(1) Temporary dewatering to allow the excavation to be carried out in the dry,
(2) Chemical injection, freezing, grouting and the like to maintain sides of excavations, etc.

Permanent processes are employed to improve the ground properties by:

(1) Compaction (making the soil denser and thus stronger), and
(2) Consolidation and drainage processes to reduce the magnitude of settlement. (Such measures are discussed in detail in Chapter 8.)

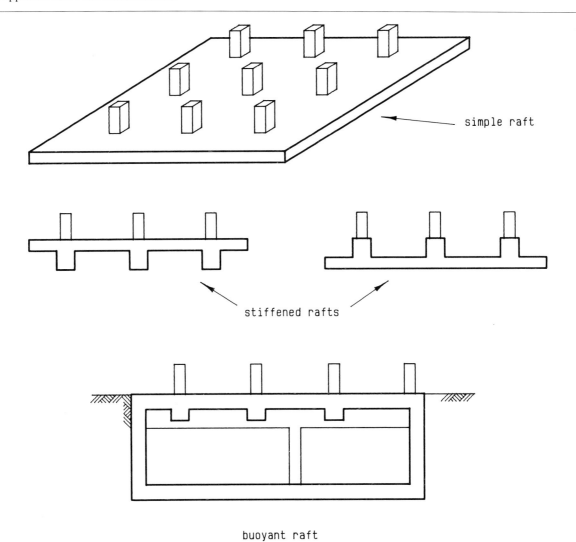

simple raft

stiffened rafts

buoyant raft

Fig. 1.7 Raft foundations.

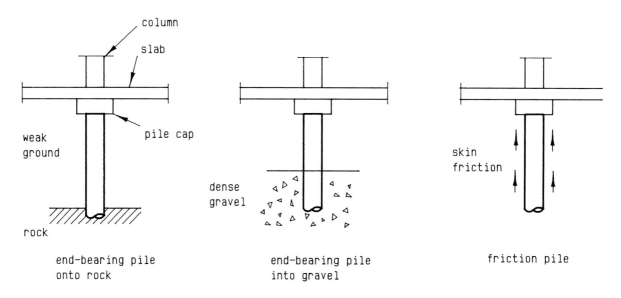

column

slab

weak
ground

pile cap

rock

end-bearing pile
onto rock

dense
gravel

end-bearing pile
into gravel

skin
friction

friction pile

Fig. 1.8 Piled foundations.

1.9 CHANGES OF SOIL PROPERTIES DURING EXCAVATION

The soil at level 1, below ground level — see Fig. 1.9 — is subject to pressure, and thus consolidation, due to the weight of the soil above, and is in equilibrium. If the overlying soil is removed to form a basement then the pressure, and consolidation effects, at level 1 are also removed. The unloaded soil, in this condition, is known as *over-consolidated*, and is likely to recover from the consolidation and rise in level (similar to the elastic recovery of contraction on a column when its load is removed).

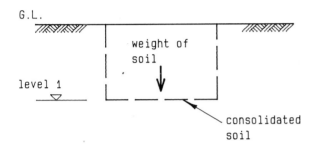

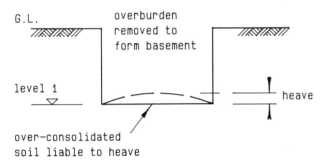

Fig. 1.9 Heave following removal of overburden.

1.10 POST-CONSTRUCTION FOUNDATION FAILURE

A foundation that has been designed well and has performed perfectly satisfactorily, may suffer distress due to nearby disturbance. Typical examples of such disturbance are piling for a new adjacent building; rerouting of heavy traffic; new heavy hammering plant installed in adjoining factories; and other activities which may vibrate or send impact shocks through the soil under the existing foundation, thus causing compaction and further settlement, which may be unacceptable.

Similarly, changes in the moisture content (by increasing it due to leaking mains and drains or by the removal of trees, or decreasing it by introducing drainage paths due to neighbouring excavation or by further growth of trees) can disturb the state of equilibrium of the soil/foundation interaction. An interesting case, investigated by the authors' practice, was the deforestation of land uphill of a factory.

The increased rain water run-off seriously affected the basement of the factory.

The construction and loading of new foundations may disturb existing buildings. The rising level of the water-table in cities due to the cessation of artesian well pumping is also causing problems (see Chapter 4 on topography, and CIRIA Special Publication 69, *The engineering implications of rising ground water levels in the deep aquifer beneath London*).

1.11 PRACTICAL CONSIDERATIONS

There are, in foundation design, a number of practical construction problems and costs to be considered. The chief ones are:

(1) The foundations should be kept as shallow as possible, commensurate with climatic effects on, and strength of, the surface soil; particularly in waterlogged ground. Excavation in seriously waterlogged ground can be expensive and slow.

(2) Expensive and complex shuttering details should be avoided, particularly in stiffened rafts. Attention should be paid to buildability.

(3) Reduction in the costs of piling, improvements in ground treatment, advances in soil mechanics, etc. have considerably altered the economics of design, and many *standard* solutions are now out-of-date. There is a need to constantly review construction costs and techniques.

(4) Designers need to be more aware of the assumptions made in design, the variability of ground conditions, the occasional inapplicability of refined soil analyses and the practicality of construction.

(5) The reliability of the soil investigation, by critical assessment.

(6) Effect of construction on ground properties, i.e. vibration from piling, deterioration of ground exposed by excavation in adverse weather conditions, removal of overburden, seasonal variation in the water-table, compaction of the ground by construction plant.

(7) Effect of varying shape, length and rigidity of the foundation, and the need for movement and settlement joints.

(8) After-effects on completed foundations of sulphate attack on concrete, ground movements due to frost heave, shrinkable clays, and the effects of trees; also changes in local environment, e.g. new construction, re-routing of heavy traffic, installation of plant in adjoining factories causing impact and vibration.

(9) Fast but expensive construction may be more economic than low-cost but slow construction to clients needing quick return on capital investment.

(10) Effect of new foundation loading on existing adjoining structures.

These practical considerations are illustrated by the following examples.

1.11.1 Example 6: Excavation in waterlogged ground

A simple example of excavation in waterlogged ground has probably occurred to many other practices as well as to the authors' practice. At the commencement of a 1–2 m deep underpinning contract in mass concrete, groundwater was found to be rising much higher and faster than previous trial pits had indicated. The circumstances were such that a mini-piling contractor was quickly brought onto site, and speedily installed what was, at face value, a more costly solution, but proved far less expensive overall than slowly struggling to construct with mass concrete while pumping. As will be well-known to many of our readers, few small site pumps are capable of running for longer than two hours without malfunctioning!

1.11.2 Example 7: Variability of ground conditions

On one site a varying clay fill had been placed to a depth of roughly 2 m over clay of a similar soft to firm consistency. Since a large industrial estate was to be developed on the site in numerous phases by different developers, a thorough site investigation had been undertaken. Nevertheless, on more than one occasion, the project engineer found himself down a hole of depth 2 m or greater, trying to decide if a mass concrete base was about to be founded in fill or virgin ground, and in either case whether it would achieve 100 kN/m^2 allowable bearing pressure or not. This emphasizes the old adage that the engineer should always physically see the ground – i.e. by examining the trial pits – with which he is going to be dealing.

1.11.3 Example 8: Reliability of the soils investigation

On one site a contractor quoted a small diameter steel tube pile length of 5 m (to achieve a suitable set), based upon a site investigation report. In the event his piles achieved the set at an average of 22 m (!), so obviously cost complications ensued. In addition to this, one of the main difficulties was convincing the contractor to guarantee his piles at that depth, as he was understandably concerned about their slenderness.

1.11.4 Example 9: Deterioration of ground exposed by excavation

An investigation by the authors' practice of one particular failure springs to mind as an example. Part of a factory had been demolished exposing what had been a party wall, but a 20 m length of this wall was undermined by an excavation for a new service duct and a *classic* failure ensued. The exposed excavation was then left open over a wet weekend, resulting in softening of the face and a collapse occurred early on the Monday.

So often the most catastrophic of failures are as a result of these types of classic textbook examples, which could be prevented by the most basic precautions.

1.11.5 Example 10: Effect of new foundation on existing structure

A new storage silo was to be constructed within an existing mill, and the proposal was to found it on a filled basement, in the same way that the adjacent silo had been 20 years before. The authors' practice was called in for their opinion fairly late in the day, with the steel silo already under fabrication.

After investigation of the fill, the client was advised to carry the new silo on small diameter piles through the fill down to bedrock. This would thereby avoid placing additional loading into the fill, and thus causing settlement of the existing silo.

1.12 DESIGN PROCEDURES

Good design must not only be safe but must aim to save construction costs, time and materials, and the following procedures should help to achieve this.

(1) On the building plan, the position of columns and load-bearing walls should be marked, and any other induced loadings and bending moments. The loads should be classified into dead, imposed and wind loadings, giving the appropriate partial safety factors for these loads.

(2) From a study of the site ground investigation (if available), the strength of the soil at various depths or strata below foundation level should be studied, to determine the safe bearing capacity at various levels. These values – or presumed bearing values from BS 8004 in the absence of a site investigation – are used to estimate the allowable bearing pressure.

(3) The invert level (underside) of the foundation is determined by either the minimum depth below ground level unaffected by temperature, moisture content variation or erosion – this can be as low as 450 mm in granular soils but, depending on the site and ground conditions, can exceed 1 m – or by the depth of basement, boiler house, service ducts or similar.

(4) The foundation area required is determined from the characteristic (working) loads and estimated allowable pressure. This determines the preliminary design of the types or combination of types of foundation. The selection is usually based on economics, speed and buildability of construction.

(5) The variation with depth of the vertical stress is determined, to check for possible over-stressing of any underlying weak strata.

(6) Settlement calculations should be carried out to check that the total and differential settlements are acceptable. If these are unacceptable then a revised allowable bearing pressure should be determined, and the foundation design amended to increase its area, or the foundations taken down to a deeper and stronger stratum.

(7) Before finalizing the choice of foundation type, the preliminary costing of alternative superstructure designs should be made, to determine the economics of increasing superstructure costs in order to reduce foundation costs.

(8) Alternative safe designs should be checked for economy, speed and simplicity of construction. Speed and economy can conflict in foundation construction – an initial low-cost solution may increase the construction period.

Time is often of the essence for a client needing early return on capital investment. A fast-track programme for superstructure construction can be negated by slow foundation construction.

(9) The design office should be prepared to amend the design, if excavation shows variation in ground conditions from those predicted from the site soil survey and investigation.

CHAPTER 2

SOIL MECHANICS, LAB TESTING AND GEOLOGY

SECTION A: SOIL MECHANICS

2.1 INTRODUCTION TO SOIL MECHANICS

Since most foundation designers have an understanding of soil mechanics testing it is not proposed, in this chapter, to go into great detail on the topic. There are, in any case, numerous textbooks, proceedings of international conferences and learned papers on the subject.

It is aimed therefore to give a recapitulation (and greater confidence) for the experienced designer, and perhaps a sense of proportion to those young engineers who appear to think it is a branch of applied mathematics. The subject is of vital importance to the designer and contractor. The designer *must* know the strength, stability and behaviour of the soil under load and the contractor must equally know what he will have to contend with in construction. Soil mechanics is a serious and valuable scientific attempt to determine the soil's type and properties.

The subject grew out of separate inquiries into a variety of early foundation failures, together with the new need to found heavier loads on poorer soils. The early pioneers of the subject, such as Terzaghi, collected and collated this dispersed information to establish a scientific, organized discipline. After the Second World War, the desperate need for reconstruction focused more widespread interest in the subject, and by the mid- to late 1950s many universities had started courses and research. Today it is accepted as normal that it forms part of an engineer's training. The earlier hostility to this relatively new science by older engineers, and the uncritical acceptance of it as 'gospel' by young engineers, has since developed into healthy appreciation of its value, and the need for experience and judgement in its application by many designers.

When practical designers criticize passive acceptance of inapplicable theory they can be accused (admittedly by second-rate academics and researchers) of being reactionary, and anti-scholarship and research − this is not the case. Terzaghi himself stated, after criticizing some teaching, that:

'as a consequence, engineers imagined that the future of science of foundations would consist in carrying out the following programme − drill a hole in the ground. Send the soil samples to a laboratory, collect the figures, intro-

duce them into equations and compute the results. The last remnants of this period are still found in attempts to prescribe simple formulas for computing settlements − no such formulas can possibly be obtained except by ignoring a considerable number of factors.'

He has also said:

'Rigour is often equated with mathematics but there is at least as much rigour in observing and recording physical phenomena, developing logical argument and setting these out on paper.'

Casgrande criticized those teachers:

'who had not the faculty to train their students to critical, independent thinking. Such ideas are then dragging through his life [the student] like invisible chains, hampering his professional progress.'

Professor Burland, in his Mash lecture, said:

'the greatest problem lies in the fact that all too often the boundaries between reality, empiricism and theory become thoroughly confused. As a result the student can quickly lose confidence, believing that there is no secure basic frame of reference from which to work − the whole subject becomes a kind of 'black art' ... an attitude widely prevalent today amongst general practitioners.'

He also said:

'soil mechanics is a craft as much as a science. A distinctive feature of a craftsman ... is that he 'knows' his material. He may not be able to quote its Young's modulus, yield strength, etc., but he knows from handling it and working it far more about its likely behaviour than would be revealed by measuring a dozen difficult properties.'

It is reassuring to designers that such eminent experts express these views.

Soil mechanics tests determine the soil's classification, its bearing capacity, settlement characteristics, its stability and pressures within it, and finally its ease or difficulty of excavation and treatment.

2.2 PRESSURE DISTRIBUTION THROUGH GROUND

The pressure distribution of concentrated loads on, say, concrete padstones or masonry walls is often assumed to disperse through 45° planes as shown in Fig. 2.1 (a).

Since

$$\text{stress} = \frac{\text{load}}{\text{area}} \quad \text{and} \quad \text{area} = z^2$$

then

$$\text{stress} = \frac{\text{load}}{z^2}$$

and a pressure distribution/depth results in the graph shown in Fig. 2.1 (b). In most soils, a dispersion angle of 60° from the horizontal plane is a more commonly accepted value. The use of a dispersion angle is an oversimplified approach which can produce incorrect results, but helps to understand the principles. A redefined and more accurate method developed by Boussinesq is more generally adopted.

The vertical stress, p_z, at any point beneath the concentrated load, P, at a depth, z, and a radius r is given by the equation:

$$p_z = \frac{3P}{2\pi z^2} \times \frac{1}{\left[1 + \left(\dfrac{r}{z}\right)^2\right]^{1/2}}$$

This results in the pressure distribution graph shown in Fig. 2.2.

The solving of the equation for a number of different depths and plan positions is obviously laborious without the aid of a computer, and designers tend to use pressure contour charts as shown in Fig. 2.3.

While the 60° dispersal is an assumption, it should be appreciated that the Boussinesq equation is also based on assumptions. The assumptions are that the soil is elastic, homogeneous and isotropic – which, of course, it is not, and it also assumes that the contact pressure is uniform which it is often not. Nevertheless the assumptions produce reasonable results for practical design and more closely correlates with pressure distribution in the soil, than the 60° dispersal assumption.

The three exceptions to the Boussinesq equation occur:

(1) When a soft layer underlies a stiff layer leading to a wider spread of lateral pressure,
(2) When a very stiff foundation does not transfer uniform pressure to the soil, and
(3) For those occasional soils with high vertical shear modulus, which tend to have a narrower spread of lateral load.

The variation of vertical stress across a *horizontal* plane within the soil subject to uniform vertical contact pressure is not uniform. Figure 2.4 shows the variation of pressure

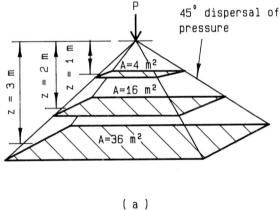

(a)

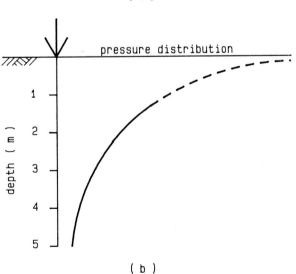

(b)

Fig. 2.1 Variation of vertical stress with depth (45° dispersal assumption).

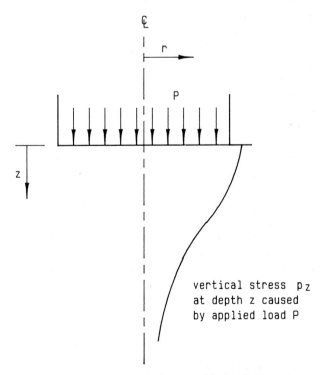

Fig. 2.2 Variation of vertical stress with depth (Boussinesq assumption).

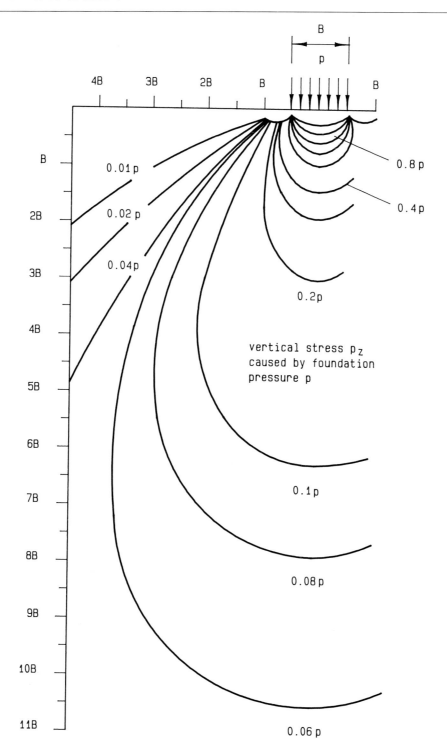

Fig. 2.3 Vertical stress contours beneath an infinite strip (Weltman & Head, *Site Investigation Manual*, CIRIA SP25 (1983), Fig. 72).

along a horizontal plane due to a uniform contact pressure under a raft or strip, assuming again a 45° dispersal of stress for simplicity.

The simplification shows the maximum pressure under the centre of the raft, or strip, and diminishing pressure towards the edge. This may help to clarify the cause of strip footings sagging when supporting a uniformly distributed load, and a uniformly loaded raft deflecting like a saucer. Figure 2.4 also shows that the soil is subject to vertical stress (and thus settlement) beyond the edge of the foundations.

An existing building, close to a new raft foundation, may suffer settlement due to the new loaded foundation. Figure 2.3 shows the stress variation across a horizontal plane based on the Boussinesq equation.

2.3 BEARING CAPACITY

2.3.1 Introduction to bearing capacity

A simplistic explanation, to ensure the understanding of the basic principles of bearing capacity, is given below.

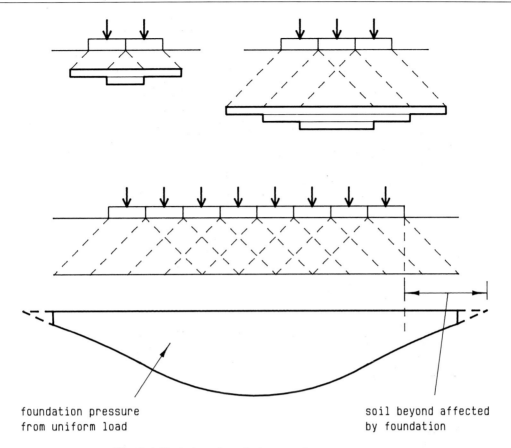

foundation pressure
from uniform load

soil beyond affected
by foundation

Fig. 2.4 Variation of vertical stress along a horizontal plane.

The loaded foundation in Fig. 2.5 (a) pushes down a triangular wedge of soil, the downward, P, is resisted by the upward reactions, $P/2$ on each triangle. The reactions can be resolved parallel and perpendicular to the boundary planes, AC and BC, (Fig. 2.5 (b)) into compressive and shearing forces P_σ and P_τ. These forces are resisted by the soil's shear strength, τ, and its compressive strength, σ (see Fig. 2.5 (c)). The soil will tend to fail in shear long before it fails in compression.

The shearing resistance of the soil, τ, is a factor of its cohesion, c, and its internal friction (dependent on the angle of internal friction, ϕ).

Coulomb's equation states that:

$$\tau = c + \sigma \tan \phi$$

where σ = normal pressure across the shear plane.

In a friction-less clay:

$$\tau = c$$

In a non-cohesive sand:

$$\tau = \sigma \tan \phi$$

Many soils are rarely solely cohesive or frictional but are a mixture of both, such as silty sands, sandy clays, etc.

As an example, determine the shear resistance of a soil with $c = 100\,\text{kN/m}^2$, and $\phi = 20°$, subject to a normal pressure of $200\,\text{kN/m}^2$.

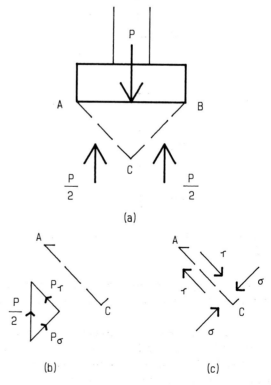

Fig. 2.5 Normal and shear stresses for a triangular wedge of soil.

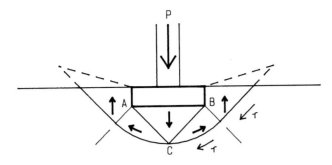

Fig. 2.6 Triangular wedge action in cohesive soils.

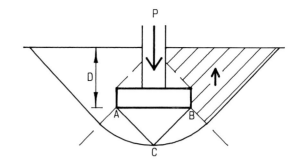

Fig. 2.8 Effect of depth of base on bearing capacity.

$$\tau = c + \sigma \tan \phi$$
$$\tau = 100 + (200 \times \tan 20°)$$
$$\tau = 173 \, kN/m^2$$

The simple triangular wedge action shown in Fig. 2.5 is mainly confined to frictional non-cohesive soils. In mainly cohesive soils the triangular wedge in pushing down tends to disturb and displace soil on both sides of the wedge (see Fig. 2.6) and further soil shear resistance will be mobilized along the planes of disturbance.

2.3.2 Main variables affecting bearing capacity

(1) The surface area of the wedge resisting the foundation load depends on the size of the foundation and its shape, as is shown in Figs 2.7 (a), (b) and (c).

Figures 2.7 (a) and (b) show diagrammatically that the larger a square base then the greater the surface area of the wedge, and that a strip footing has less surface area per unit area of foundation (see Fig. 2.7(c)).

(2) The bearing capacity of a foundation is affected by its depth, D, and the density of the soil (see Fig. 2.8).

Comparing Fig. 2.8 with Fig. 2.6, it will be noted that there is a greater volume of soil to push up, and the shear planes are longer. Furthermore, the greater the density (the weight) of the soil then the greater the force necessary to push it up.

(3) In any horizontal plane at or below foundation level there is an existing pressure due to the weight of soil above the plane. This existing *overburden* pressure will vary with the density and weight of the soil and the percentage of water within the soil.

(a) Total overburden pressure, s, equals pressure due to weight of soil and water (and any other existing surcharge loads) before construction.

(b) Effective overburden pressure, s', equals the total overburden pressure, s, minus the porewater pressure (usually equal to the head of water above the plane).

At a depth z_w below the water-table, $s' = s - \gamma_w z_w$

where γ_w is the unit weight of water.

As an example, determine the effective over-burden pressures at the levels of water-table, proposed foundation base, and 1 m below proposed foundation, shown in Fig. 2.9. The sand has a dry unit weight of $17.5 \, kN/m^3$ and a saturated unit weight of $20 \, kN/m^3$.

At level 1 ($z_w = 0$):

$$s' = s = \gamma_{dry} D_w = 17.5 \times 0.5 = 8.75 \, kN/m^2$$

At level 2 ($z_w = D - D_w$):

$$s' = \gamma_{sat}(D - D_w) + \gamma_{dry}(D_w) - \gamma_w(D - D_w)$$
$$= 20(1.0 - 0.5) + 17.5(0.5) - 10(1.0 - 0.5)$$
$$= 10 + 8.75 - 5 = 13.75 \, kN/m^2$$

At level 3 ($z_w = 2.0 - D_w$):

$$s' = \gamma_{sat}(2.0 - D_w) + \gamma_{dry}(D_w) - \gamma_w(2.0 - D_w)$$
$$= 20(2.0 - 0.5) + 17.5(0.5) - 10(2.0 - 0.5)$$
$$= 30 + 8.75 - 15$$
$$= 23.75 \, kN/m^2$$

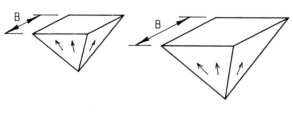

square bases

(a)   (b)

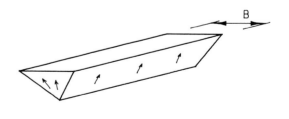

strip footing

(c)

Fig. 2.7 Effect of base size and shape on soil wedge.

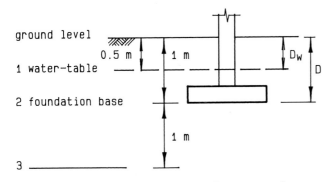

Fig. 2.9 Variation of effective overburden pressure for a pad foundation.

2.3.3 Bearing capacity and bearing pressure

In the previous section both bearing pressure and capacity were discussed. It is important to differentiate between the two.

The *bearing capacity* is the pressure the soil is capable of resisting.

The *bearing pressure* is the pressure exerted on the soil by the foundation.

Both terms have sub-divisions as follows:

(1) The *total bearing pressure*, t, is the total pressure on the ground due to the weight of the foundations, the structure and any backfill.

(2) The *net bearing pressure*, n, is the net increase in pressure due to the weight of the structure and its foundation, i.e. $n = t - s$.

(3) The *total ultimate bearing capacity*, t_f, is the total loading intensity at which the ground fails in shear (*Note* 'ultimate' does *not* refer to ultimate limit state in this context.)

(4) The *net ultimate bearing capacity*, n_f, is the net loading intensity at which the ground fails in shear, i.e., $n_f = t_f - s$.

(5) The *net allowable bearing pressure*, $n_a = n_f/($factor of safety$)$. The factor of safety is determined by the designer's experience and judgement, the magnitude and rate of settlement and the structure's resistance, or susceptibility, to settlement. It is common in practice to adopt a factor of safety of 3 for normal structures.

2.3.4 Determination of ultimate bearing capacity

As discussed above the bearing capacity depends on such factors as the soil's shear strength and the foundation's size and shape. Terzaghi, some 50 years ago, developed mathematical solutions to cover all these variations. The solutions were modified by experiments, and further modified by Brinch Hansen. For shallow foundations, using dimensionless coefficients, N_c, N_q and N_γ (given in Fig. 2.10), the net and total ultimate bearing capacities are, respectively,

(1) Strip footings

$$n_f = cN_c + s'(N_q - 1) + 0.5\gamma BN_\gamma$$
$$t_f = cN_c + s'(N_q - 1) + 0.5\gamma BN_\gamma + s$$

(2) Square or circular bases

$$n_f = 1.3cN_c + s'(N_q - 1) + 0.4\gamma BN_\gamma$$
$$t_f = 1.3cN_c + s'(N_q - 1) + 0.4\gamma BN_\gamma + s$$

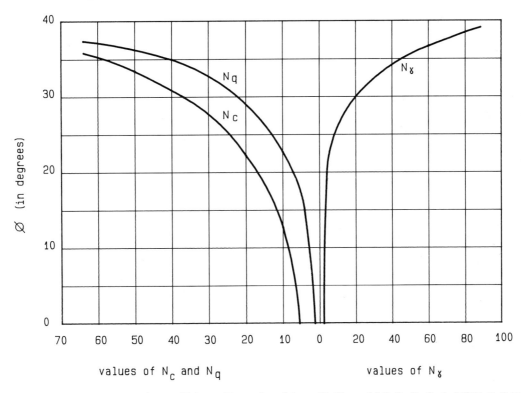

Fig. 2.10 Terzaghi's bearing capacity coefficients (Reproduced from K. Terzaghi & R. B. Peck (1967) *Soil Mechanics in Engineering Practice*, **by permission of John Wiley and Sons, Inc., © authors)**

For sands and gravels, when non-cohesive, the term cN_c in the above equations is equal to zero.

The net ultimate bearing capacity, n_f, for such soils is:

$$n_f = s'(N_q - 1) + 0.5\gamma BN_\gamma \quad \text{for strips, and}$$
$$n_f = s'(N_q - 1) + 0.4\gamma BN_\gamma \quad \text{for square bases.}$$

For pure cohesive soils, where $\phi = 0°$, $n_f = cN_c$ for both strips and square bases. For $\phi = 0°$, N_c is generally taken as 5.14.

EXAMPLE 1

A strip footing of width $B = 1.5$ m is founded at a depth $D = 2.0$ m in a soil of unit weight $\gamma = 19$ kN/m^3. The soil has a cohesion $c = 10$ kN/m^2 and an angle of internal friction of $\phi = 25°$. No groundwater was encountered during the site investigation.

For a strip footing the total ultimate bearing capacity is given by:

$$t_f = cN_c + s'(N_q - 1) + 0.5\gamma BN_\gamma + s$$

Since there is no groundwater, the effective overburden pressure equals the total overburden pressure.

$$s' = s = \gamma D = 19 \times 2.0 = 38 \text{ kN/m}^3$$

From Fig. 2.10, $N_c = 25$, $N_q = 13$, $N_\gamma = 10$. Thus:

$$\begin{aligned}
t_f &= cN_c + s'(N_q - 1) + 0.5\gamma BN_\gamma + s \\
&= cN_c + s'N_q + 0.5\gamma BN_\gamma \\
&= 10(25) + 38(13) + 0.5(19 \times 1.5 \times 10) \\
&= 250 + 494 + 142.5 \\
&= 886.5 \text{ kN/m}^2
\end{aligned}$$

Applying a factor of safety of 3, this gives a total allowable bearing pressure

$$t_a = \frac{t_f}{3} = \frac{886.5}{3} = 295 \text{ kN/m}^2$$

EXAMPLE 2

A strip footing of width $B = 1.0$ m is to be founded at a depth $D = 1.5$ m below the surface of a cohesionless sand with dry and saturated unit weights $\gamma_{dry} = 16$ kN/m^3 and $\gamma_{sat} = 18$ kN/m^3, and an angle of internal friction of $\phi = 30°$.

The net ultimate bearing capacity is

$$n_f = s'(N_q - 1) + 0.5\gamma BN_\gamma$$

From Fig. 2.10, $N_q = 22$ and $N_\gamma = 20$.

The net ultimate bearing capacity at depth D is to be checked, assuming the groundwater is

(1) below 3 m depth,
(2) at 1.5 m depth,
(3) at 0.5 m depth.

(1) *Groundwater below 3 m depth*

Effective overburden, $s' = \gamma_{sat}D = 16 \times 1.5 = 24$ kN/m^2

Unit weight, $\gamma = \gamma_{dry} = 16$ kN/m^3

$$n_f = s'(N_q - 1) + 0.5\gamma BN_\gamma$$

$$\begin{aligned}
&= 24(22 - 1) + 0.5(16 \times 1.0 \times 20) \\
&= 664 \text{ kN/m}^2
\end{aligned}$$

(2) *Groundwater at 1.5 m depth*

$$s' = 24 \text{ kN/m}^2 \text{ as in (1).}$$

When groundwater is present at or above foundation level, the unit weight γ in the second half of the bearing capacity equation should be the submerged unit weight.

$$\begin{aligned}
\gamma &= \gamma_{sat} - \gamma_w = 18 - 10 = 8 \text{ kN/m}^3 \\
n_f &= s'(N_q - 1) + 0.5\gamma BN_\gamma \\
&= 24(22 - 1) + 0.5 (8 \times 1.0 \times 20) \\
&= 584 \text{ kN/m}^2
\end{aligned}$$

(3) *Groundwater at 0.5 m depth*

$$\begin{aligned}
s' &= \gamma_{dry}D_w + \gamma_{sat}(D - D_w) - \gamma_w(D - D_w) \\
&= 16(0.5) + 18(1.5 - 0.5) - 10(1.5 - 0.5) \\
&= 17 \text{ kN/m}^2 \\
\gamma &= 8 \text{ kN/m}^3 \quad \text{as in (2)} \\
n_f &= s'(N_q - 1) + 0.5\gamma BN_\gamma \\
&= 16(22 - 1) + 0.5(8 \times 1.0 \times 20) \\
&= 416 \text{ kN/m}^2
\end{aligned}$$

There are underlying and well-known approximating assumptions in all the equations both in this section and the previous sections. Typically these are:

(1) ϕ and c are well-known from tests and are constant for a given soil,
(2) That the loads imposed on the ground are known with exactitude, and
(3) The effect of settlement on the structure is not considered.

As in all structural design, the engineer will therefore apply the results of calculations with judgement and experience.

It has not yet proved possible to apply limit-state philosophy to bearing capacity. Simply applying a partial safety factor to ultimate bearing capacity and checking for serviceability, i.e., prevention of undue settlement, does not go all the way to producing good design. This is considered further in the following sub-sections.

In general, however, when the bearing capacity is low the settlements tend to be high, and, conversely, when the bearing capacity is high the settlement is more likely to be low.

2.3.5 Safe bearing capacity – cohesionless soils

It is extremely difficult to obtain truly undisturbed samples of cohesionless soils (sands and gravels), and furthermore, shear tests, which fully simulate in situ conditions, are not without difficulties. The angle of internal friction, ϕ, is more often determined by the various penetration tests, and these too can give varying results. From Fig. 2.10, it will be seen that for small increases in ϕ there are large increases in both N_q and N_γ, leading to a large increase in net ultimate bearing capacity, n_f.

For example,

when $\phi = 30°$, $N_q = 22$ and $N_\gamma = 20$
when $\phi = 33°$, $N_q = 30$ and $N_\gamma = 30$

Thus, for a 3 m square base founded in sand of unit weight $\gamma = 20 \, kN/m^3$ with an effective overburden pressure $s' = 20 \, kN/m^2$, then:

For $\phi = 30°$, $n_f = s' (N_q - 1) + 0.4\gamma B N_\gamma$
$$= 20(22 - 1) + 0.4(20 \times 3 \times 20)$$
$$= 420 + 480 = 900 \, kN/m^2$$

For $\phi = 33°$, $n_f = 20(30 - 1) + 0.4(20 \times 3 \times 30)$
$$= 580 + 720 = 1300 \, kN/m^2$$

So a 10% increase in ϕ results in approximately a 40% increase in n_f. However, foundation design pressure on non-cohesive soil is usually governed by acceptable settlement, and this restriction on bearing pressure is usually much lower than the ultimate bearing capacity divided by the factor of safety of 3. Generally only in the case of narrow strip foundations on loose submerged sands is it vital to determine the ultimate bearing capacity, since this may be more critical than settlement.

In practice settlements are limited to 25 mm by use of charts relating allowable bearing pressure to standard penetration test results, as shown in Terzaghi & Peck's chart in Fig. 2.11 and reproduced with an example in Appendix N.

2.3.6 Safe bearing capacity – cohesive soils
It is easier to sample and test clay soils. The test results can be more reliable – provided that the moisture content of the test sample is the same as the clay strata in situ. As water is *squeezed* (or drained) from the soil then the value of c increases. But since the drainage of water from the clay is slow then so too is the increase in c, so that generally the increase in bearing capacity is ignored in foundation design. The value of c from undrained shear strength tests is therefore adopted in most designs.

Unlike non-cohesive soils, the bearing capacity, and not settlement, is found to be the main design factor in the foundation design of light structures founded on firm clay. Applying a factor of safety of $2.5-3.0$ to the ultimate bearing capacity usually restricts settlement to acceptable levels. Where there is no experience of the behaviour of the soil under load, the clay is less than firm, or the structures are heavy, then settlement estimates should be made.

2.3.7 Safe bearing capacity – combined soils
Soils such as silts, sandy clays, silty sands and the like possess both c and ϕ properties. Reasonable soil samples can be taken for testing, usually by triaxial compression tests. The ultimate bearing capacity results obtained from such tests are divided by a factor of safety based on experience and judgement and the design for settlement (as is shown later).

2.4 SETTLEMENT

2.4.1 Introduction to settlement
Soils, like other engineering materials, contract under load. This contraction, known in foundation engineering as *settlement*, must be determined and checked, so that either its magnitude will not affect the superstructure, or the superstructure design should *build-in* flexibility to accom-

modate the settlement. In the same way as the magnitude of a beam's deflection depends on the strength/stiffness of the beam and the load on it, so too does settlement depend on the strength/stiffness of the soil and the load (bearing pressure) on it. Limiting beam deflections to acceptable levels is done by either reducing the load or strengthening/stiffening the beam, and so too settlement is limited in design, by either restricting the load (bearing pressure), or strengthening/stiffening the material (by geotechnical processes).

Just as steel and concrete beams deflect by different amounts, so too does the *magnitude* of settlement differ between cohesive and non-cohesive soils. The *rate* of deflection of a prestressed concrete beam differs from that of a steel beam, the prestressed beam is affected by long-term creep. Similarly the rate of settlement differs between cohesive and non-cohesive soils.

If the whole structure settled evenly there would be little problem, but, as shown in Figs 2.3 and 2.4, even uniform pressure at foundation level results in non-uniform pressure within the soil, leading to differential settlement and sagging (or hogging) as shown in Fig. 1.1. The situation is worse when the foundation loading is not uniform.

The settlement of soils under load is somewhat analogous to squeezing a saturated sponge. If the sponge shown in Fig. 2.12 is contained in a sealed and flexible plastic envelope it will deform by spreading. The water in the sponge will be under pressure. But in the strata it is difficult for the soil to spread, and if the sponge is restrained the water pressure will be greater. If the plastic is punctured the water will at first spurt out, reduce gradually to a trickle, and when there is equilibrium of pressure between the sponge and the loaded pressure on it, then the drainage of moisture will cease (see Fig. 2.13).

If the load is increased then again water will drain from the sponge, settlement increases, and finally reaches equilibrium again.

Apart from drainage of moisture from the sponge other actions take place. The sponge particles are compressed and

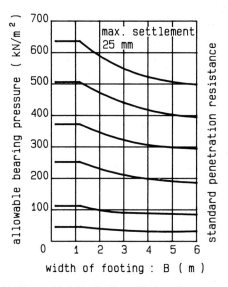

Fig. 2.11 Terzaghi & Peck allowable bearing pressure/SPT chart. (Reproduced from K. Terzaghi & R. B. Peck (1967) *Soil Mechanics in Engineering Practice*, by permission of John Wiley and Sons, Inc., © authors)

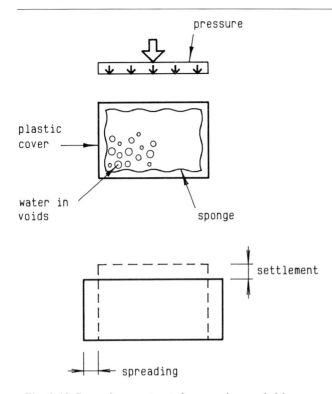

Fig. 2.12 **Squeezing a saturated sponge in a sealed bag.**

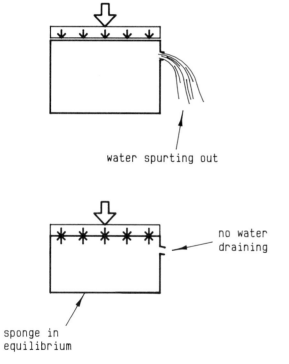

Fig. 2.13 **Squeezing a saturated sponge in a punctured bag.**

pushed into closer contact — similar to elastic contraction. The spreading of the soil, shown in Fig. 2.12, is indicative of a Poisson's ratio action. With the reduction in volume the sponge is becoming more compact and therefore stronger, more able to resist the load pressure, and settles relatively less with increased pressure. A settlement/time graph under increasing pressure, say σ_1, σ_2, etc., would then be as shown in Fig. 2.14.

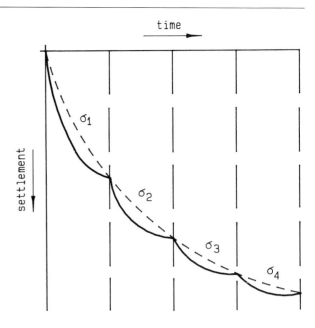

Fig. 2.14 **Settlement with time under increasing pressure.**

This is what happens in practice. The mechanics are outlined in the following sub-sections.

2.4.2 Void ratio
Soils are not totally solid, but comprise a mixture of soil particles and water below the water-table, or soil, air and water above the water-table.

Figure 2.15 (a) shows the actual soil, and Fig. 2.15 (b) shows a convenient idealized form. The ratio of the voids to the solids, i.e.

$$\text{void ratio, } e = \frac{\text{volume of voids}}{\text{volume of solids}} = \frac{V_v}{V_s}$$

All readers will have experienced the effect of differing void ratios in practice. Where a road repair has been undertaken, backfilled and resurfaced, on a route used regularly, they will have noticed after a few weeks that the repair has consolidated under vehicle loading, and has become an irritating rut in the road. The poorly compacted backfill started off with a relatively high voids ratio; loading has led to compaction and settlement, until the voids ratio has reduced to a similar level to the rest of the road construction.

2.4.3 Consolidation test
This is basically a refined *squeezing the sponge* exercise, and is shown in Fig. 2.16.

A pressure, known as the consolidation pressure σ', is applied, and by reading the dial gauge the settlement is noted at time intervals until full consolidation is reached, normally after 24 hours. The water in the soil squeezes out through the porous discs, the sample contracts, and the new void ratio, e, can be determined. The test is repeated with increasing increments of σ' (i.e. σ'_1, σ'_2, σ'_3, etc.) and the change in e (i.e. e_1, e_2, e_3, etc.) noted.

A typical graph of void ratio to consolidation pressure generally results in a curve (see Fig. 2.17).

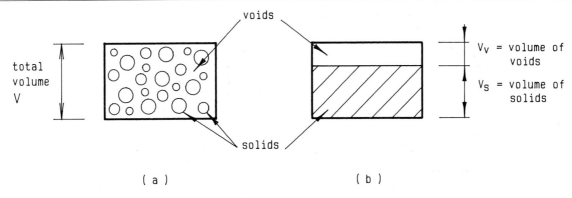

Fig. 2.15 Void ratio in soils.

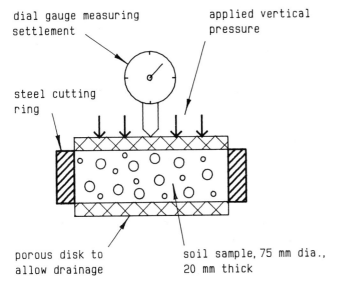

Fig. 2.16 Consolidation test.

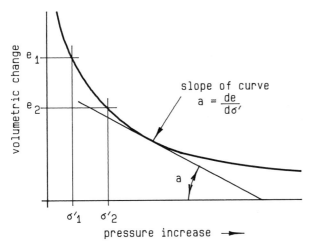

Fig. 2.17 Variation of void ratio with increasing pressure.

The slope of the $e-\sigma'$ curve, a, decreases with increase in pressure (since the soil is becoming more and more dense); consequently a is not constant. In calculations, however, the pressure range, from initial to final result, is such that a is often assumed constant, i.e. the $e-\sigma'$ curve between the two pressures is a straight line. Therefore:

$$a = \frac{e_1 - e_2}{\sigma'_1 - \sigma'_2}\ m^2/kN$$

2.4.4 Coefficient of volume compressibility

This coefficient is important in calculating settlement.

The compression of a soil, per unit thickness, due to a unit increase in effective pressure is represented by

$$m_v = \frac{\text{volumetric change}}{\text{unit of pressure increase}}$$

If H_1 = original thickness, and H_2 = final thickness, then, since the area is constant,

$$\text{volumetric change} = \frac{H_1 - H_2}{H_1}$$

But the change in height is due to the change in void ratio, i.e.

$$\text{volumetric change} = \frac{e_1 - e_2}{1 + e_1}$$

Now

$$a = \frac{e_1 - e_2}{d\sigma'}$$

therefore

$$\text{volumetric change} = \frac{a\,d\sigma'}{1 + e_1}$$

hence

$$m_v = \frac{\dfrac{a\,d\sigma'}{1 + e_1}}{d\sigma'}$$

$$= \frac{a\,d\sigma'}{1 + e_1} \times \frac{1}{d\sigma'}$$

$$= \frac{a}{1 + e_1}\ m^2/kN$$

Determining a from experiments, and knowing m_v, the pressure increase $d\sigma'$, and the thickness of the strata H_1, then:

$$\text{settlement, } \rho_1 = m_v\,d\sigma' H_1$$

If there are, say, four strata, then the total settlement = $\rho_1 + \rho_2 + \rho_3 + \rho_4$, where ρ_1, ρ_2, etc. represent the settlement of the individual stratas.

Typical values of m_v for clays are given in Table 2.1.

Table 2.1 Typical values of m_v

Soil	m_v (m²/kN × 10^{-3})	Compressibility
Soft clay	2.0–0.25	very high to high
Soft-to-stiff clay	0.25–0.125	medium
Stiff-to-hard boulder clay	0.125–0.006 25	low to very low

For example, for a clay strata, $m_v = 0.2 \times 10^{-3}$ m²/kN, the thickness of the strata $H_1 = 1.5$ m and the change in pressure in the strata d$\sigma' = 100$ kN/m².

To determine the settlement, ρ_1,

$$\begin{aligned}\rho_1 &= m_v d\sigma' H_1 \\ &= (0.2 \times 10^{-3})\,(100)\,(1.5 \times 10^3) \\ &= 30\,\text{mm}\end{aligned}$$

2.4.5 Magnitude and rate of settlement

Soils are of course not solid (as are steel, granite or similar materials) but, like a wet sponge, are a mixture of soil particles and water in the voids between the particles. The magnitude of the settlement depends not only on the bearing pressure but also on the amount of water in the soil (its void ratio). The rate of settlement depends on how fast the water can be squeezed from the soil (its permeability). Sands generally contain less water than clays, and the water can escape faster. So sands settle less and faster than clays (see Fig. 2.18).

In sands the bulk of the settlement occurs during construction, but clays continue to settle long after construction is complete.

The short-term settlement of sands is termed *immediate settlement*. The long-term settlement of clays is termed *consolidation*. Because of the similarity between the settlement/time graph for sand and clay and the stress/strain graphs for steel and concrete, it is tempting to postulate a modulus of elasticity for soils. Thus mathematical theories can be proposed which, while elegant, can bear little relation to facts.

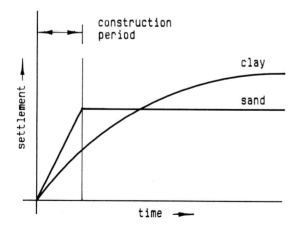

Fig. 2.18 Settlement of sands and clays with time.

2.4.6 Settlement calculations

The structural loads used in foundation design for settlement calculations (and bearing pressure) should be the actual loads, and *not* those factored up to give ultimate loads.

Estimation of magnitude and particularly the rate of settlement is one of the most difficult engineering design estimates – accurate forecasts are practically impossible, and engineering experience and judgement are essential. Trial hole inspection, to study the horizontal, vertical and inclined drainage paths, is essential in order to make adjustments to calculated results. Vegetation roots which have decayed leave drainage paths which can be undetected in sample tests. Examination of the settlement behaviour on comparable structures on similar soils is advisable, and the insertion of movement joints in the structure (to form controlled cracks) where damaging differential settlement is likely, is good design policy.

As the soil moisture is drained away under foundation loading, the soil becomes denser and stronger until equilibrium is reached and settlement ceases. If, later, the soil is further loaded by increasing the structural loads, or new structures are added on, or there are further soil moisture reductions, then further settlement will start. The pressure within the strata varies, see Fig. 2.1, and though the strata can be sub-divided into thin layers for purposes of calculation, this still gives only the settlement of that strata at one particular point, and not along the strata.

The sample tested in the consolidation test is allowed to drain on the vertical axis only, whereas in situ there are other drainage paths. The sample (supposedly undisturbed) is, relative to the strata, very thin, and may not be representative of the strata in situ.

The pressure in the strata is not always that due to the total load assumed in design. The design load at foundation level must cover the case of full imposed load, yet, for say an office block, this may occur for only a quarter of the time, so the pressure in the strata causing settlement is an estimate.

Certainly the use of finite-element analysis and computers can eliminate the need for laborious calculations but they do not necessarily produce the right *answer*.

It cannot be over-emphasized that it is the magnitude of *differential* settlement that mainly causes structural damage, rather than the magnitude of overall settlement. Particular settlement calculation checks should be made where the foundation loading is not uniform, where the strata varies in thickness, where the structure is particularly susceptible to differential settlement, and where there is no previous experience of the soil from which to work.

Because of so many variables, exact estimates are difficult, and it is usual to quote settlements to the nearest:

- 5 mm where the settlement is 25 mm or less.
- 25 mm where the settlement is up to 150 mm.
- 50 mm where the settlement is greater than 150 mm.

To reduce the effects of differential settlements to an acceptable level the designer can:

(1) Avoid the adoption of structures and foundations sensitive to settlement.
(2) Employ ground improvement techniques.
(3) Transfer, by piling, the load to strong strata.
(4) Build in jacking pockets to re-level the structure.
(5) Use deep basements of cellular construction.
(6) Use rigid rafts or strip beams.

In addition to the measures of structural/foundation interaction design, given in sections 10.5 and 10.6, the designer can *let* the structure settle differentially and control the cracking by inserting movement joints (which in effect are controlled cracks). This method is often the most economic solution. The placing of such joints is based on experience and some guidance is given below:

(1) Separate tall heavy blocks from low, light ones.
(2) Decrease the centres of joints positioned for structural differential movement due to thermal moisture and other movements.
(3) Place joints at stress concentrations (i.e. top of door to bottom of window-sill above).
(4) Place joints at changes of plan shape.

2.5 ALLOWABLE BEARING PRESSURE

In most structural designs a factor of safety is applied to ultimate strength to produce a design safe strength, and then checked for serviceability resulting in, sometimes, a further restriction to produce an *allowable* strength. In foundation design the soil's ultimate bearing capacity is determined and a factor of safety is applied to give a *safe* bearing capacity, so that the soil does not fail in shear. This safe bearing capacity is checked for the possibility of undue settlement, and to control this it may be necessary to reduce the safe bearing capacity to an *allowable* bearing pressure, to limit undue settlement to the structure.

Safety factors, as in all structural design, are necessary to allow for uncertainties, so judgement and experience are necessary in the choice of magnitude of the factor. Safety factors account for:

(1) Variations in the shear strength within and between the strata.
(2) Variation in the reliability of experimental and theoretical determination of ultimate bearing capacity.
(3) Variation of shear strength during and after foundation construction.
(4) Consideration of the *serviceability limit* of settlement.
(5) The life of the structure, i.e., a lower factor of safety may be adopted for temporary works.

Common values are 2.5–3.0 to cover these variations. This is reduced to 2.0 when the strata is uniform, reliable and differential settlement is not critical. This can be further reduced to 1.5 for temporary works when unaffected by significant settlement.

2.6 CONCLUSIONS

Soil mechanics is not an *exact* science (but neither is much of structural design). Some engineers dismiss the subject as *academic* and of no practical value — such an attitude can lead to over-design or even foundation failure. The good designer will know the subject, appreciate its limitations, and apply sound judgement and experience in design. The engineer, needing detailed information on specific matters, is referred to the numerous excellent textbooks on the subject.

SECTION B: LABORATORY TESTING

2.7 INTRODUCTION TO LABORATORY TESTING

Soil mechanics tests determine a soil's classification, its bearing capacity, its settlement characteristics, its stability and pressures within it, and finally, its ease, or difficulty, of excavation and treatment.

2.8 CLASSIFICATION – (DISTURBED SAMPLE TESTS)

2.8.1 Particle size and distribution
Soils vary enormously in formation, chemical composition, density and even colour. The main factor affecting their physical behaviour is the size of the soil's particles and this characteristic is used to determine the classification.

For example, clay particles are relatively minute (less than 0.002 mm) and the particles *stick* together — they are *cohesive* (as every site engineer knows, clay can stick like glue to gumboots). Sand particles are 30–1000 times bigger than clay particles and they interlock — they possess *internal friction* (dry sand does not stick to gumboots). Clay particles are practically impervious, while sands and gravels possess high permeability.

The cohesion of clay particles and the friction between sand particles have an important affect on the soil's strength, stability and behaviour, as does its permeability and variation in moisture content. Silt particles are intermediate between clays and sands; gravel particles are bigger than sands, and cobbles and boulders are greater than gravels. There is further sub-division into coarse and fine soils, and, within a soil type, into fine, medium and coarse.

The relationship between particle or grain size, and the main descriptive divisions for soils, together with their approximate permeability, are shown in Table 2.2.

Soils are frequently variable — the particles vary in size — and are often mixtures of differing soils. The variation in particle size is termed *grading*. When there is uniformity of particle size the soil is described as *uniformly graded*, and when it varies widely it is termed *well-graded* (see Table 2.2 and Fig. 2.19).

Figure 2.20 shows, diagrammatically, that well graded soils and compacted soils tend to be denser and therefore stronger than uniform uncompacted soils.

Table 2.2 Soil descriptions and particle sizes (Weltman & Head, *Site Investigation Manual*, CIRIA SP 25 (1983), Table 6)

Grain size (mm) (log scale)	200　　60　20　6　2　0.6　0.2　0.06　0.02　0.006　0.002				
Basic soil type	Boulders / Cobbles	Coarse / Medium / Fine — GRAVELS	Coarse / Medium / Fine — SANDS	Coarse / Medium / Fine — SILTS	CLAYS
	VERY COARSE SOILS	COARSE SOILS		FINE SOILS	
Drainage properties	High permeability generally $k > 10^{-5}$ m/s (fine sands) Maximum can approach 1 m/s			Low permeability poor drainage $10^{-6} > k > 10^{-8}$ m/s	Practically impervious $k < 10^{-8}$ m/s

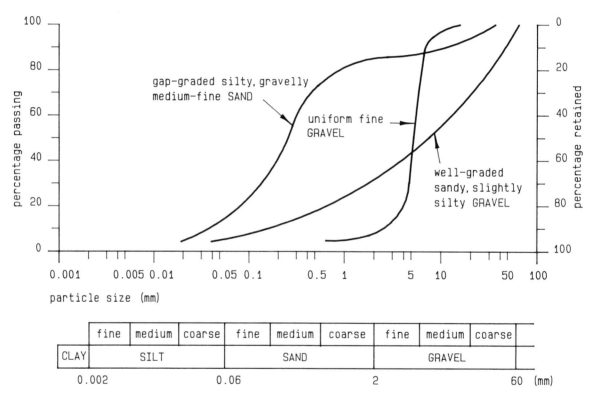

Fig. 2.19 Grading curves for coarse grained soils (Weltman & Head, *Site Investigation Manual*, CIRIA SP25 (1983), Fig. 61).

When gravel contains a proportion of sand it is described as *sandy gravel*, and clays with a silt content are described as *silty clays*. A typical classification of coarse soils is shown in Table 2.3.

A more detailed identification and description of soils from BS 5930 is given in Table 2.4, and the classification system is given in Table 2.5.

2.8.2 Density

The denser the soil then generally the stronger it is likely to be. There are in situ and laboratory tests to determine

Table 2.3 Classification of coarse soils (Weltman & Head, *Site Investigation Manual*, CIRIA SP 25 (1983), Table 5)

Material	Composition (by weight)
Slightly sandy GRAVEL	up to 5% sand
Sandy GRAVEL	5% to 20% sand
Very sandy GRAVEL	over 20% sand
GRAVEL/SAND	about equal proportions of gravel and sand
Very gravelly SAND	over 20% gravel
Gravelly SAND	5% to 20% gravel
Slightly gravelly SAND	up to 5% gravel

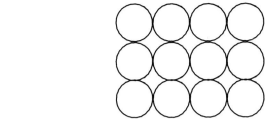

UNIFORM GRADED GRAVEL

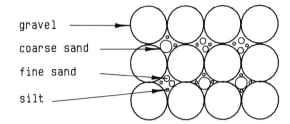

gravel

coarse sand

fine sand

silt

WELL GRADED, SANDY, SLIGHTLY
SILTY GRAVEL

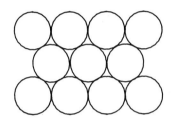

COMPACT GRAVEL

Fig. 2.20 Effect of grading on density.

density, and it is also important to evaluate the moisture content of the sample. This is performed by weighing the soil before and after drying.

2.8.3 Liquidity and plasticity

A clay, depending on its moisture content, can be in three physical states, i.e., solid, plastic or liquid. The divisions between the three states are known as the *plastic limit* and *liquid limit* which are the moisture contents at which, by defined tests, the soil changes physical state. Knowing these *consistency limits* allows the soil to be classified according to its position on the plasticity chart shown in Fig. 2.21. Silty soils are usually found to be plotted below the A-line, and clayey soils above it.

The plasticity can be a useful guide to the compressibility and liability to shrinking of clays and silts.

2.8.4 General

Soils within a strata having the same particle size distribution, moisture content, density, etc. will tend to have the same engineering properties and behaviour. The disturbed sample tests (on particle size, consistency limits, etc.) are relatively cheap and quick to carry out, and can give good guidance on the degree and magnitude of the test programme of the

more expensive and time-consuming undisturbed sample testing. Undisturbed samples are tested for shear strength, permeability, settlement, etc., and are relatively expensive.

2.9 UNDISTURBED SAMPLE TESTING

Undisturbed is a misnomer, for the soil sample is not only disturbed in obtaining it from a borehole or trial pit, but also there can be further disturbance in extruding it from the tube sampler (an *undisturbed* sample however, is less disturbed than a *disturbed* sample). Before testing, the sample should be examined for its soil fabric and possible disturbance. A lateral slice can be cut off to check further the fabric, note any organic matter, root holes (direction and distribution) and inclusions of other material which may affect the performance of drainage paths.

Testing apparatus is becoming increasingly more sophisticated and reliable. However, poor laboratory techniques, use of incorrect loading rates and drainage conditions, and other lack of care, will produce results which will be unreliable. As in structural design, final calculations are checked against preliminary estimates, so too should the results of soil tests be checked against expectations from the borehole logs, site tests and inspection of trial pits.

2.9.1 Moisture content

While moisture content can be determined from good disturbed samples, it is usually better practice to determine it from undisturbed samples, since disturbing the sample may alter its moisture content so that it is unrepresentative of the soil's in situ condition.

2.9.2 Shear strength

The shear strength of silts and clays is vitally important since it determines their bearing capacity. There are two main types of test:

(1) *Unconfined compression test*. This is the simpler test carried out on a 40 mm diameter cylindrical section cut from the sample and subjected to axial compression. The test cannot be carried out on sands and gravels, or on very weak silts and clays which are too soft to stand under their own weight in the apparatus.

(2) *Triaxial compression test*. The sample is subjected to axial and all-round lateral compression (i.e. on the three axes). A wider range of clay and silt soils can be tested under varying conditions, and can determine the cohesion and angle of shearing resistance of the soil. Under stress the soil's moisture will tend to be squeezed out of the sample and thus alter its density, strength, etc., and the contraction (consolidation) of the sample and the 'drainage' of the sample can be controlled to simulate expected site conditions.

The shear strength of granular soils such as sands and fine gravels is sometimes determined by the shear box test, but it is often more reliable to obtain data from the in situ SPT and vane tests (see section 3.6).

The shear strength of clay is related to its cohesion,

Table 2.4 Field identification and description of soils (BS 5930, Table 6)

	Basic soil type	Particle size (mm)	Visual identification	Particle nature and plasticity	Composite soil types (mixtures of basic soil types)	
Very coarse soils	BOULDERS		Only seen complete in pits or exposures.	Particle shape: Angular Subangular Subrounded Rounded Flat Elongate	Scale of secondary constituents with coarse soils	
	COBBLES	200	Often difficult to recover from boreholes.		Term	% of clay or silt
Coarse (non-cohesive) soils (over 65% sand and gravel sizes)	GRAVELS	60 coarse 20 medium 6 fine 2	Easily visible to naked eye; particle shape can be described; grading can be described. Well graded: wide range of grain sizes, well distributed. Poorly graded: not well graded. (May be uniform: size of most particles lies between narrow limits; or gap graded: an intermediate size of particle is markedly under-represented.)		slightly clayey } GRAVEL or slightly silty } SAND	under 5
					— clayey } GRAVEL or — silty } SAND	5 to 15
				Texture: Rough Smooth Polished	very clayey } GRAVEL or very silty } SAND	15 to 35
	SANDS	coarse 0.6 medium 0.2 fine 0.06	Visible to naked eye; very little or no cohesion when dry; grading can be described. Well graded: wide range of grain sizes, well distributed. Poorly graded: not well graded. (May be uniform: size of most particles lies between narrow limits; or gap graded: an intermediate size of particle is markedly under-represented.)		Sandy GRAVEL } Sand or gravel and important second Gravelly SAND } constituent of the coarse fraction (See 41.3.2.2) For composite types described as: clayey: fines are plastic, cohesive silty : fines non-plastic or of low plasticity	
Fine (cohesive) soils (over 35% silt and clay sizes)	SILTS	coarse 0.02 medium 0.006 fine 0.002	Only coarse silt barely visible to naked eye; exhibits little plasticity and marked dilatancy; slightly granular or silky to the touch. Disintegrates in water; lumps dry quickly; possess cohesion but can be powdered easily between fingers.	Non-plastic or low plasticity	Scale of secondary constituents with fine soils	
					Term	% of sand or gravel
					sandy } CLAY or gravelly } SILT	35 to 65
	CLAYS		Dry lumps can be broken but not powdered between the fingers; they also disintegrate under water but more slowly than silt; smooth to the touch; exhibits plasticity but no dilatancy; sticks to the fingers and dries slowly; shrinks appreciably on drying usually showing cracks. Intermediate and high plasticity clays show these properties to a moderate and high degree, respectively.	Intermediate plasticity (Lean clay) High plasticity (Fat clay)	— CLAY : SILT	under 35
					Examples of composite types (Indicating preferred order for description) Loose, brown, subangular very sandy, fine to coarse GRAVEL with small pockets of soft grey clay Medium dense, light brown, clayey, fine and medium SAND	
Organic soils	ORGANIC CLAY, SILT or SAND	Varies	Contains substantial amounts of organic vegetable matter.		Stiff, orange brown, fissured sandy CLAY Firm, brown, thinly laminated SILT and CLAY	
	PEATS	Varies	Predominantly plant remains usually dark brown or black in colour, often with distinctive smell; low bulk density.		Plastic, brown, amorphous PEAT	

References within this table refer to original document.

Compactness/strength		Structure			Colour
Term	Field test	Term	Field identification	Interval scales	
Loose	By inspection of voids and particle packing.	Homogeneous	Deposit consists essentially of one type.	Scale of bedding spacing	Red Pink Yellow Brown Olive Green Blue White Grey Black etc.
Dense		Interstratified	Alternating layers of varying types or with bands or lenses of other materials. Interval scale for bedding spacing may be used.	**Term** — **Mean spacing (mm)**	
				Very thickly bedded — over 2000	
Loose	Can be excavated with a spade; 50 mm wooden peg can be easily driven.	Heterogeneous	A mixture of types.	Thickly bedded — 2000 to 600	
		Weathered	Particles may be weakened and may show concentric layering.	Medium bedded — 600 to 200	
Dense	Requires pick for excavation; 50 mm wooden peg hard to drive.			Thinly bedded — 200 to 60	
				Very thinly bedded — 60 to 20	
Slightly cemented	Visual examination; pick removes soil in lumps which can be abraded.			Thickly laminated — 20 to 6	Supplemented as necessary with: Light Dark Mottled etc.
				Thinly laminated — under 6	
Soft or loose	Easily moulded or crushed in the fingers.	Fissured	Break into polyhedral fragments along fissures. Interval scale for spacing of discontinuities may be used.		and
Firm or dense	Can be moulded or crushed by strong pressure in the fingers.				Pinkish Reddish Yellowish Brownish etc.
Very soft	Exudes between fingers when squeezed in hand.	Intact	No fissures.		
Soft	Moulded by light finger pressure.	Homogeneous	Deposit consists essentially of one type.	Scale of spacing of other discontinuities	
Firm	Can be moulded by strong finger pressure.	Interstratified	Alternating layers of varying types. Interval scale for thickness of layers may be used.	**Term** — **Mean spacing (mm)**	
Stiff	Cannot be moulded by fingers. Can be indented by thumb.			Very widely spaced — over 2000	
Very stiff	Can be indented by thumb nail.	Weathered	Usually has crumb or columnar structure.	Widely spaced — 2000 to 600	
Firm	Fibres already compressed together.			Medium spaced — 600 to 200	
				Closely spaced — 200 to 60	
Spongy	Very compressible and open structure.	Fibrous	Plant remains recognizable and retains some strength.	Very closely spaced — 60 to 20	
Plastic	Can be moulded in hand, and smears fingers.	Amorphous	Recognizable plant remains absent.	Extremely closely spaced — under 20	

Table 2.5 British soil classification system (BS 5930, Table 8)

Soil groups (see Note 1)			Subgroups and laboratory identification				
GRAVEL and SAND may be qualified Sandy GRAVEL and Gravelly SAND, etc. where appropriate (See 41.3.2.2)			Group symbol (see Notes 2 & 3)	Subgroup symbol (see Note 2)	Fines (% less than 0.06 mm)	Liquid limit (%)	Name
COARSE SOILS less than 35% of the material is finer than 0.06 mm	GRAVELS More than 50% of coarse material is of gravel size (coarser than 2 mm)	Slightly silty or clayey GRAVEL	G — GW / GP	GW / GPu GPg	0 to 5		Well graded GRAVEL / Poorly graded/Uniform/Gap graded GRAVEL
		Silty GRAVEL / Clayey GRAVEL	G-F — G-M / G-C	GWM GPM / GWC GPC	5 to 15		Well graded/Poorly graded silty GRAVEL / Well graded/Poorly graded clayey GRAVEL
		Very silty GRAVEL / Very clayey GRAVEL	GF — GM / GC	GML, etc. / GCL GCI GCH GCV GCE	15 to 35		Very silty GRAVEL; subdivide as for GC / Very clayey GRAVEL (clay of low, intermediate, high, very high, extremely high plasticity)
	SANDS More than 50% of coarse material is of sand size (finer than 2 mm)	Slightly silty or clayey SAND	S — SW / SP	SW / SPu SPg	0 to 5		Well graded SAND / Poorly graded/Uniform/Gap graded SAND
		Silty SAND / Clayey SAND	S-F — S-M / S-C	SWM SPM / SWC SPC	5 to 15		Well graded/Poorly graded silty SAND / Well graded/Poorly graded clayey SAND
		Very silty SAND / Very clayey SAND	SF — SM / SC	SML, etc. / SCL SCI SCH SCV SCE	15 to 35		Very silty SAND; subdivided as for SC / Very clayey SAND (clay of low, intermediate, high, very high, extremely high plasticity)
FINE SOILS more than 35% of the material is finer than 0.06 mm	Gravelly or sandy SILTS and CLAYS 35% to 65% fines	Gravelly SILT / Gravelly CLAY (see Note 4)	FG — MG / CG	MLG, etc. / CLG CIG CHG CVG CEG		<35 / 35 to 50 / 50 to 70 / 70 to 90 / >90	Gravelly SILT; subdivide as for CG / Gravelly CLAY of low plasticity / of intermediate plasticity / of high plasticity / of very high plasticity / of extremely high plasticity
		Sandy SILT (see Note 4) / Sandy CLAY	FS — MS / CS	MLS, etc. / CLS, etc.			Sandy SILT; subdivide as for CG / Sandy CLAY; subdivide as for CG
	SILTS and CLAYS 65% to 100% fines	SILT (M-SOIL) / CLAY (see Notes 5 & 6)	F — M / C	ML, etc. / CL CI CH CV CE		<35 / 35 to 50 / 50 to 70 / 70 to 90 / >90	SILT; subdivide as for C / CLAY of low plasticity / of intermediate plasticity / of high plasticity / of very high plasticity / of extremely high plasticity
ORGANIC SOILS	Descriptive letter 'O' suffixed to any group or sub-group symbol.			Organic-matter suspected to be a significant constituent. Example MHO: organic SILT of high plasticity.			
PEAT	Pt Peat soils consist predominantly of plant remains which may be fibrous or amorphous.						

NOTE 1. The name of the soil group should always be given when describing soils, supplemented, if required, by the group symbol, although for some additional applications (e.g. longitudinal sections) it may be convenient to use the group symbol alone.

NOTE 2. The group symbol or sub-group symbol should be placed in brackets if laboratory methods have not been used for identification, e.g. (GC).

NOTE 3. The designation FINE SOIL or FINES, F, may be used in place of SILT, M, or CLAY, C, when it is not possible or not required to distinguish between them.

NOTE 4. GRAVELLY if more than 50% of coarse material is of gravel size. SANDY if more than 50% of coarse material is of sand size.

NOTE 5. SILT (M-SOIL), M, is material plotting below the A-line, and has a restricted plastic range in relation to its liquid limit, and relatively low cohesion. Fine soils of this type include clean silt-sized materials and rock flour, micaceous and diatomaceous soils, pumice, and volcanic soils, and soils containing halloysite. The alternative term 'M-soil' avoids confusion with materials of predominantly silt size, which form only a part of the group.

Organic soils also usually plot below the A-line on the plasticity chart, when they are designated ORGANIC SILT, MO.

NOTE 6. CLAY, C, is material plotting above the A-line, and is fully plastic in relation to its liquid limit.

References within this table refer to the original document.

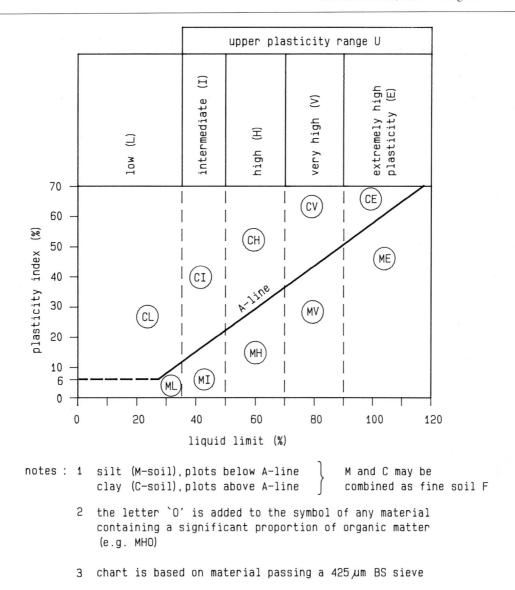

Fig. 2.21 Plasticity classification chart (Weltman & Head, *Site Investigation Manual*, CIRIA SP25 (1983), Fig. 62).

which is usually constant and mainly unaffected by the foundation pressure. The shear strength of sand is related to its internal friction and is affected by foundation pressure — for example, the greater the pressure on two sheets of sandpaper then the more difficult it is to slide them apart. The shear strength of soils is highly important in determining their bearing capacity.

Many soils are a mixture of sand, clay and silt (see Table 2.2), and will possess both frictional and cohesive properties.

2.9.3 Consolidation tests (oedometer apparatus)

A lateral slice of the soil sample is enclosed in a metal ring and loaded. The magnitude and rate of consolidation (contraction under load) is noted, and used to predict the settlement behaviour of the foundation. (The stiffer the clay then the less it will compress; typical values are given in Table 2.6.) In many cases the settlement behaviour of the soil has a more critical influence on foundation design than bearing capacity — the soil may not fail in bearing, but the structure may fail due to unforeseen differential settlement.

Table 2.6 Typical values of compressibility of cohesive materials (Weltman & Head, *Site Investigation Manual*, CIRIA SP 25 (1983), Table 10)

Clay type	Compressibility	Coefficient of volume compressibility m_v $(m^2/kN) \times 10^{-3}$
Very heavily overconsolidated clays, sun weathered rocks, some tills	Very low	<0.05
Heavily overconsolidated clays, some tills, hard London clay	Low	0.05 to 0.1
Overconsolidated clays such as upper London clays, some glacial clays	Medium	0.1 to 0.3
Normally consolidated clays (e.g. alluvial or estuarine)	High	0.3 to 1.5
Highly organic alluvial clays and peats	Very high	>1.5

2.9.4 Permeability tests

Permeability is the rate at which fluid passes through the material, and thus affects the drainage and rate of consolidation of the soil. Relatively permeable soils are tested in a *constant head permeameter*, where a constant head of water is maintained across the sample. For less permeable soils a *falling head permeameter* is used. In some soils the permeability differs in the vertical and horizontal planes, and the laboratory tests are susceptible to errors. In such cases some designers prefer to carry out site tests, particularly when full details of the soil are necessary, as in impounded reservoirs, earth dam construction and similar projects.

2.9.5 Chemical tests

It is often advisable to determine the sulphate and chloride content of the soil and ground water, and the pH value, in order that the concrete properties and mix proportions are adapted to ensure durability. In dealing with contaminated or filled sites, the reader should consult the further guidance given in Chapters 5 and 7.

2.10 SUMMARY OF TESTS

A brief summary of tests for simple foundations and excavations is given in Table 2.7 and a fuller schedule is given in Table 2.8.

2.11 ANALYSIS OF RESULTS

A senior soils engineer, of the soil investigation firm, should from his study of the test results, borehole logs and other data, be able to give firm recommendations, agreed with the design engineer, in a soil report on the following:

(1) Soil classification, density, compaction, moisture content, plastic and liquid limits, the soil's permeability, and the effect of any variation in level of groundwater.
(2) Soluble sulphate and chloride content, pH value, corrosive action from soil and/or wastes, methane gases.
(3) Presence of peat, possibility of running sand, presence of possible cavities, boulders or other obstructions.
(4) Strength, shear value and cohesion (drained and undrained); *bearing capacity*.
(5) *Settlement characteristics* — magnitude and rate.
(6) Need for any type of geotechnical processes to improve the soil or ease excavation.
(7) Possible difficulties in excavation.
(8) Whether, in clays, the soil is naturally consolidated or overconsolidated.

It is strongly advisable that the soils engineer should discuss his report with the structural design engineer where there is a possibility of conflict, dispute or difference of

Table 2.7 Summary of tests on soils (Weltman & Head, *Site Investigation Manual*, CIRIA SP 25 (1983), Table 9)

Geotechnical problem	Soil type	Classification tests	Other laboratory tests	Remarks
Bearing capacity	Soft to firm CLAYS	Moisture content, liquid and plastic limit. Bulk density.	Triaxial compression tests — generally unconsolidated undrained. Laboratory vane tests in soft clays.	—
	Firm to stiff CLAYS	Moisture content, liquid and plastic limit. Bulk density.	Triaxial compression tests — generally unconsolidated undrained or consolidated undrained for effective stress parameters.	Sample size and anisotropy effects can be important in stiff fissured clays.
	Gravelly CLAYS	Moisture content and liquid and plastic limit on material passing a 425-micron sieve. Bulk density.	Triaxial unconsolidated undrained compression tests on 100 mm diameter specimens. Multi-stage tests.	—
	SANDS	Maximum and minimum densities. Particle size distribution.	Possibly shear box for range of densities.	Presence of secondary constituents (e.g. organic or clay pocket) have a marked effect on bearing capacity. Bearing capacity usually determined from in situ tests (SPT or static cone).

Table 2.7 continues

Table 2.7 (*cont*)

Geotechnical problem	Soil type	Classification tests	Other laboratory tests	Remarks
Bearing capacity (*cont*)	GRAVELS	Maximum and minimum densities. Particle size distribution.	Possibly shear box for range of densities.	For most projects field tests are used rather than laboratory tests.
	WEAK ROCKS	Bulk density, specific gravity, moisture content, point load tests, disc test, petrological examination	Uniaxial compression tests may be appropriate in some situations.	Shear box test on appropriate discontinuities may sometimes be useful.
Settlement	CLAYS	Moisture content, liquid and plastic limit. Specific gravity.	Consolidation tests. Stress path triaxial.	Swelling parameters may also be of importance.
	SANDS	–	Stress path triaxial.	Analysis usually based on in situ tests (SPT or static cone).
	GRAVELS	–	–	No appropriate test.
	WEAK ROCK	Moisture content Specific gravity Petrological examination Uniaxial compression strength	–	Modulus of deformation tests are unlikely to be representative of the field condition. Long term creep tests on large specimens may sometimes be appropriate
Excavation	CLAYS	Moisture content, liquid and plastic limit. Bulk density.	Consolidated undrained and consolidated drained triaxial tests for effective stress parameters.	–
	SANDS and GRAVELS	Particle size distribution.	–	–
	ROCK	Moisture content, point load index, disc tests, uniaxial compressive strength	Uniaxial compression tests.	Laboratory tests mainly to establish ease or difficulty of excavation
Earth pressures and stability	CLAY	Moisture content, liquid and plastic limits. Bulk density.	Consolidated undrained and consolidated drained triaxial compression tests – for effective stress parameters.	Fully softened or residual shear strength parameters may be appropriate for stiff fissured clays if long term stability is required.
	SANDS and GRAVELS	Particle size distribution	–	Shear box may be considered if representative density can be approximated.
	WEAK ROCKS	Moisture content Bulk density.	Shear box tests on discontinuities.	Residual strength may be appropriate.

Table 2.8 Detailed schedule of soil tests (BS 5930, Table 4)

Category of test	Name of test	Where details can be found	Remarks
Soil classification tests	Moisture content	BS 1377	Frequently carried out as a part of other soil tests. Read in conjunction with liquid and plastic limits, gives an indication of the shear strength of cohesive soil.
	Liquid and plastic limits (Atterberg limits)	BS 1377	Used to classify cohesive soil and as an aid to classifying the fine fraction of mixed soil.
	Cone penetration limit	BS 1377, [108]	An alternative method of determining the liquid limit of cohesive soil.
	Linear shrinkage	BS 1377	
	Specific gravity	BS 1377	Used in conjunction with other tests, such as sedimentation and consolidation. Values commonly range between 2.55 and 2.75, and a more accurate value is required for air voids determination. Only occasional checks are needed for most British soils for which a value of 2.65 is assumed unless experience of similar soils shows otherwise. However, determination of specific gravity may be necessary where spoil heap material is concerned, and the test is normally required where soils have come from sites overseas.
	Particle size distribution: (a) sieving (b) sedimentation	BS 1377	Sieving methods give the grading of soil coarser than silt. The proportion passing the finest sieve represents the combined silt/clay fraction. The relative proportions of silt and clay can only be determined by means of sedimentation tests which should be carried out when there is a real need for this information.
Soil chemical tests	Organic matter	BS 1377 BS 1924	Detects the presence of organic matter able to interfere with the hydration of Portland cement in soil:cement pastes.
	Sulphate content of soil and groundwater	BS 1377	The tests assess the aggressiveness of soil or groundwater to buried concrete. (See remarks on test for pH value.)
	pH value	BS 1377	To measure the acidity or alkalinity of the soil or water. It is usually carried out in conjunction with sulphate content tests. This test and the one above should be performed as soon as possible after the samples have been taken.
	Carbonate content	[87]	Reference describes method using the Collins' calcimeter. Useful for estimating the chalk content of soils.
	Chloride content	BS 1881: Part 6	

Table 2.8 continues

Table 2.8 (*contd*)

Category of test	Name of test	Where details can be found	Remarks
Soil compaction tests	Dry density of soil on site	BS 1377	Measures the mass of solids per unit volume of soil. Most of these tests are used to establish the dry density of soil, either naturally occurring or compacted fill, to which direct access may be obtained. Some, however, can be applied to samples obtained from depth, and these tests are used when the density of the soil is required in conjunction with other tests.
	Dry density/moisture content	BS 1377	This test indicates the degree of compaction that can be achieved at different moisture contents.
	Relative density of cohesionless soil	[109]	Methods for determining the maximum dry density of the soil other than those given in the ASTM are currently in use in the UK. One such method is the use of the vibrating hammer method, specified in BS 1377. This and other alternatives are quite valid provided the method used is clearly stated.
Pavement design tests	California bearing ratio (CBR)	BS 1377, [110]	This is an empirical test used in conjunction with the design of flexible pavements. The test can be made either in situ, see **29.4**, or in the laboratory.
	TRRL frost heave test	[111]	A laboratory test used to determine the susceptibility to frost heave of a specimen of compacted soil.
Soil strength tests	Triaxial compression:		By far the most commonly used of these tests is the standard undrained test. There is a large amount of experience in its use and many partly empirical methods are available to utilize the parameters so obtained in the design of foundations and other sub-structures. The remaining tests also have their own uses which will be found fully described in the references quoted. The tests are normally carried out on nominal 100 mm or 40 mm diameter specimens, as appropriate.
	(a) undrained	BS 1377	
	(b) undrained with measurement of pore water pressure	[110, 112]	
	(c) consolidated undrained	[110, 112]	
	(d) consolidated undrained with measurement of pore water pressure	[110, 112]	
	(e) consolidated drained	[110, 112]	
	(f) multi-stage triaxial test	[112, 113, 114, 115]	Several techniques have been used for both drained and undrained tests, details of which will be found in the references. The test is useful where there is a shortage of specimens, and its main use is with 100 mm nominal diameter specimens, only one of which can be prepared from each sampling tube.
	(g) free end test	[116, 117]	The use of lubricated end plattens in the free end test leads to improved uniformity of stress and deformation in all types of triaxial test. It also enables the height-to-diameter ratio of all samples to be reduced to 1 : 1, thus enabling several 100 mm diameter specimens to be cut from each standard sampling tube, which is useful when a limited number of undisturbed samples are available.

Table 2.8 continues

Table 2.8 (*contd*)

Category of test	Name of test	Where details can be found	Remarks
Soil strength tests (*cont*)	Unconfined compressive strength	BS 1377	This simple test is a rapid substitute for the undrained triaxial test, although it is suitable only for saturated non-fissured cohesive soil.
	Laboratory vane shear	[118]	For soft clay, an alternative to undrained triaxial test where the preparation of the specimen sometimes has an adverse effect on the measured strength of the soil.
	Direct shear box: (a) immediate (b) consolidated immediate (c) drained	[105, 110]	In the measurement of the shearing resistance of soil, these tests are an alternative to triaxial tests, although the latter have now largely superseded them. One of their main disadvantages is that drainage conditions cannot so easily be controlled. Another is that the plane of shear is predetermined by the nature of the test. One of their advantages is that specimens of non-cohesive soil can be more readily prepared than in the triaxial test. The small 60 mm square shear box is suited only to soil containing particles not larger than will pass a 3.35 mm sieve complying with the requirements of BS 410. For coarser soils, the large 305 mm square box [105] should be used. This is suitable for soil all of whose particles will pass a 37.5 mm sieve complying with the requirements of BS 410.
	Residual shear strength: (a) multiple reversal shear box (b) triaxial test with pre-formed shear surface (c) shear-box test with pre-formed shear surface (d) ring shear test	[119, 120] [119, 121] [122, 123] [124]	The residual shear strength of clay soil is increasingly used in slope stability problems. The multiple reversal shear box test is the one which is most commonly used, although the ring shear test would appear to be the more logical. This latter test tends to give lower parameters than the former.
Soil deformation tests	Consolidation: (a) one-dimensional consolidation properties (oedometer test) (b) triaxial consolidation (c) Rowe consolidation cell	BS 1377 [110, 112] [125]	These tests yield soil parameters from which the amount and time scale of settlements can be calculated. The simple oedometer test is the one in general use and although reasonable assessment of settlement can be made from the results of the test, estimates of the time scale have been found to be extremely inaccurate with certain types of soil. This is particularly true of clay soil containing layers and partings of silt and sand, where the horizontal permeability is much greater than the vertical. In these cases, more reliable data may be obtained from tests in the Rowe cell which is available in sizes up to 250 mm diameter and where a larger and potentially more representative sample of soil can be tested. Another alternative is to obtain values of the coefficients of consolidation, C_v, from in situ permeability tests and combine them with coefficients of volume decrease, m_v, obtained from the simple oedometer test, see **21.4.7**.

Table 2.8 continues

Table 2.8 (*contd*)

Category of test	Name of test	Where details can be found	Remarks
Soil deformation tests (*cont*)	Elastic modulus	BS 1377	Values of the elastic modulus of soil can be obtained from the stress/strain curve from undrained triaxial compression with the cell pressure equal to the overburden pressure. Experience, however, shows that the results so obtained are often very much lower than the actual values. It is now generally considered that the plate bearing test or back analysis of existing structures yield more reliable results.
Soil permeability tests	Constant head permeability test Falling head permeability test	[110]	The constant head test is suited only to soils of permeability roughly within the range 10^{-4} m/s to 10^{-2} m/s. For soils of lower permeability, the falling head test is applicable. For various reasons, laboratory permeability tests often yield results of limited value and in situ tests are generally thought to yield more reliable data.
	Triaxial permeability test	[112]	
	Rowe consolidation cell	[125]	The Rowe consolidation cell allows the direct measurement of permeability by constant head with a back pressure and confining pressures more closely consistent with the field state, and by both vertical and radial flow.
Soil corrosivity tests	(a) Bacteriological tests (b) Redox potential	CP 1021 CP 1021	Undisturbed specimens required in both cases; air-sealed and in sterilized containers for the bacteriological tests.

All references within this table refer to the original document.

opinion on the recommendations. This is particularly important when other specialists (geologists; piling engineers; mining and brine extraction experts; ground treatment specialists in stabilization, dewatering, compaction, etc.) are called in, since specialization can lead to limited outlook, conflicting advice and a tendency to ignore alternatives. Over-reliance on impressive scientific specialist reports can fog engineering judgement. The engineer should exercise his judgement on the reliability, relevance and practicality of the information and make his own interpretation and recommendations.

2.12 FINAL OBSERVATIONS ON TESTING

It is hoped that this very brief description of soil mechanics and testing will show the importance of the subject – and also its limitations. Soil strata vary in composition and degree of consolidation, they are liable to change in properties with variation of moisture content, and may further change under foundation pressure. It is essential to use engineering judgement, based on experience and knowledge, in applying the results of small samples, of varying degrees of disturbance, taken from isolated boreholes and tested in laboratory apparatus designed to simulate the site conditions of the in situ and variable strata. To accept uncritically the results of too few and unreliable tests would be akin to accepting the computer print-out of an untested finite-element analysis program based on unverified, theoretical and over-simplified assumptions for a real structure.

However, rather than rejecting soil mechanics, it should be appreciated that, for example, design engineers can leap to false conclusions after a casual inspection of a trial pit. Soil mechanics tests can act as a safety net, and alert the engineer to re-examination of his possibly false assumptions.

The laboratory test data must be checked against the borehole and trial pit logs, site tests, site investigation, any specialists' reports and, wherever possible, previous experience of similar local soil. Where there is conflict between the engineer's estimate from his observations and the results of testing, the engineer must re-examine his predictions and have a check carried out on the tests and the test procedure.

Laboratory testing is costly and time-consuming – its justification is more economical design, better pre-planning and costing of construction, and a reduction in the possibility of foundation failure.

SECTION C: GEOLOGY

2.13 INTRODUCTION TO GEOLOGY

The subject of geology is very briefly treated here to refresh designers' memories, increase awareness of its relevance, assist in choice of sites, help in site investigations and to know when to call on specialist advice. There is an increasing awareness that at the very least some knowledge of engineering geology is essential for sound assessment and application of soil mechanics. Many foundation failures have been due to ignorance of geology, and not due to inadequacies in the study of soils. To restrict the size of this manual the authors have had to limit the discussion of this important subject; nor is there any discussion on rock mechanics, since the overwhelming bulk of building structures, as distinct from civil engineering works, are founded on soils. Where rock is encountered in building structure foundations the strength of even weak, fractured or decomposed rock is not usually a serious foundation design problem.

2.14 FORMATION OF ROCK TYPES

As the original molten mass of the earth cooled to form a hard, dense crust, *igneous* rock was formed. This contained all the mineral elements to form sand, clays, silts, chalk, etc., and under erosion and weathering (see section 2.16) formed sediments of these materials. These sediments under high pressure over a long length of time created *sedimentary* rocks, i.e., sandstone, chalk, limestone, etc. Under the action of extreme heat (from phenomena such as volcanic activity) and exceptionally high pressures these rocks could change, metamorphose, to form *metamorphic* rocks – limestone changed to marble, clay deposits metamorphosed to slate. As the earth continued to cool and shrink it 'crinkled' (like a drying orange) to form hills and mountains, and these sedimentary and metamorphic rocks were again attacked by weathering and erosion. Vegetation and forests grew on some of the sedimentary rocks, and as the forests decayed they formed layers of peat which were sometimes metamorphosed to coal and other deposits.

2.15 WEATHERING OF ROCKS

Young engineers 'know' that rocks weather – because they've been told so – but it can be difficult to believe that such strong dense material can be worn away by rain, wind and sun. They believe it when they examine old gravestones in a cemetery where inscriptions are difficult to read due to weathering of the stone, and in excavation they find that rock overlain by soil invariably has the top metre or so shattered, disintegrated, etc., due to weathering. Those interested in mountaineering can see ample evidence of weathering and erosion in even the old rocks, in a temperate climate, in Snowdonia and the Lake District. The evidence is even more striking in the relatively new mountains of the Alps and the Himalayas, where glaciation and extreme cold is wearing away the rock more swiftly.

2.16 AGENTS OF WEATHERING

2.16.1 Temperature
The mineral constituents of igneous rocks have varying thermal coefficients of expansion and contraction, so that under heat and cold they expand or contract differentially which sets up internal strains and stresses in the rock, causing it to shatter and fracture.

2.16.2 Water
Rain water enters the fractures in rocks, freezes to ice and expands and levers the rock apart and deepens the cracks. More rain can enter, penetrate deeper, freeze and expand and break off chunks of rock.

Heavy rain and floods can roll and wash the lumps of rocks to streams and down rivers and finally to the sea. In this transportation the rocks are rolled along against other rock particles, etc. and become more and more broken down into fragments – boulders, cobbles, sand particles and mud. The load-carrying capacity of a river is approximately proportional to the square of its velocity, so as the river reaches the plains or dries up in the summer, the boulders are deposited and will remain until the next flood. When the river reaches the sea and the velocity drops, the cobbles are deposited on the beach, the sand is deposited further out and the clay particles further still. (In a silt content test for concreting sands, when the sand is stirred up in water and allowed to settle, the coarser grains of sand settle first, followed by the finer grains and finally by the silt.) The calcium content dissolves, is absorbed by marine life to form their skeletons, and on their death they sink to the floor of the ocean to form beds of chalk.

2.16.3 Wind
Sand-blasting is an effective technique for scraping off the surface of dirty deteriorated stone masonry. Sandstorms are erosive as is evident from the scouring of the Egyptian Sphinxes and other stone artifacts. A measurement of building exposure is the *driving rain index* – the combination of rain and wind velocity.

Storms at sea erode the coastline where sea cliffs are subject to a barrage of beach cobbles, hurled by the wind. Sand, drifted by the wind, forms sand dunes.

2.16.4 Glaciation
In previous Ice Ages, deep rivers of ice (glaciers) spreading from both the north and south poles have eroded deep valleys and transported large quantities of stone and soil huge distances. At the ends of the glaciers the melt water has formed large outwash plains of *boulder clay* – i.e. fine particles of clay containing some boulders. When the glaciers terminated for any length of time, a jumble of boulders, clay, stones and sand have left an undulating mass, termed a *moraine*. These terminal moraines are highly variable in content, and are practically impossible to investigate with precision.

The Ice Ages, being relatively recent geologically, have spread their deposits over earlier sedimentary and other rock.

2.17 EARTH MOVEMENT

The earth is not static. Great land masses have split apart —
England was once connected to Europe. Land masses, in
splitting, move relative to one another, as suggested by the
plate theory. As the movement takes place, earthquakes
occur in such areas as the San Andreas fault in California,
USA.

2.17.1 Folds, fractures and faults

As the earth contracts, the strata are subject to lateral
pressure causing it to fold — like a tablecloth pushed from
both ends. The peaks of the folds are termed *anticlines* and
the inverts *synclines* (see Fig. 2.22).

The anticline, in tension, is seriously weakened and cracks.
(The London basin was once covered by a dome or anticline
of chalk which has been eroded back to the North and South
Downs (see Fig. 2.23).)

The syncline, under enormous compression, can crack
and shatter. This folding can lift strata up thousands of
metres from the sea bed (and the discovery of marine fossils
on mountain tops caused Victorian Christian fundamen-
talists problems with theology!). The folding and resulting
stresses create joints in the rock at right angles to the
bedding plane and can form planes of structural weakness,
and are more prone to attack by weathering and erosion.

The fracture and movement of rock is termed *faulting* and
the plane of fracture is termed a *fault* (see Fig. 2.24).

2.17.2 Dip and strike

The slope of the folded rock is known as the *dip*. Dip is the
angle of maximum slope, and *strike* is the direction at right

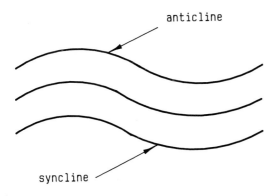

Fig. 2.22 Synclines and anticlines.

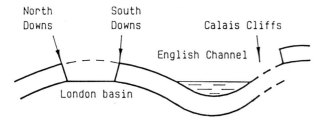

Fig. 2.23 Erosion of anticline to form London basin.

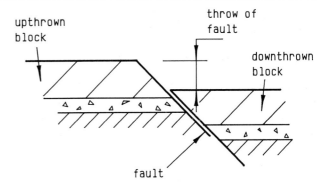

Fig. 2.24 Faulting.

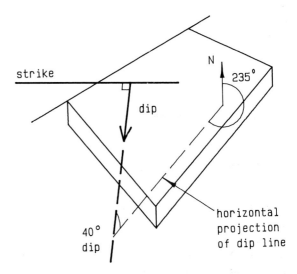

Fig. 2.25 Dip and strike.

angles to the dip (see Fig. 2.25). The dip angle is expressed
in degrees from the horizontal and its compass orientation
should also be stated. In Fig. 2.25 the strata dips 40° at 235°
to N.

2.17.3 Jointing

Joints are fractures in the rock where the rock, either side of
the fracture, has not moved differentially as occurs in
faulting. Joints are due to the contraction in cooling of
igneous and volcanic rocks, the shrinkage in drying out of
sedimentary rock (particularly chalk and limestone), and
the fracture of the rock in folding, particularly in domed
anticlines. The joint patterns are frequently a mesh of
cracks often at right angles to each other and perpendicular
to the bedjoints. Joint patterns can cause areas of weakness
in the strata, and provide easy access to the ingress of water
and accelerate the weathering process, see Fig. 2.26.

2.17.4 Drift

Drift is the term used for superficial (surface) deposits
overlying the solid rock. The drifts may be deposits from
glaciers, rivers (alluvium), old lagoons and beaches, etc.
The drift covers, or blankets, the underlying rock which
may be faulted, folded, eroded and otherwise weakened,
and examination of the drift alone could lead to false
conclusions about the ground behaviour. Drift can vary in
thickness from a few metres to 30 m or more.

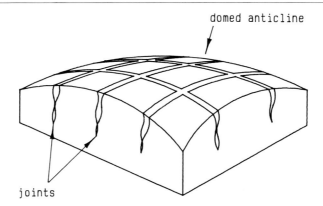

Fig. 2.26 Jointing.

Geological maps, for many areas, are of two types: one, showing the type and condition of the underlying rock, is known as a *solid* geological map, and the other, showing the type, depth etc., of the overlying deposits, is known as the *drift* geological map. It is advisable in site investigation, particularly for heavy structures on shallow drifts, to study both types of map.

2.18 ERRORS IN BOREHOLE INTERPRETATION

Some typical errors, due to ignorance of geology, are given below:

(1) Mistaken bedrock. Boulders in boulder clay are assumed to be bedrock, see Fig. 2.27.
(2) Mistaken strata formation. An unchecked fault, see Fig. 2.28.
(3) An unchecked dip. The retaining wall shown in Fig. 2.29 was not designed to take extra pressure from rock inclined to the wall.
(4) Folded strata mistaken for level strata, see Fig. 2.30.

It was decided to use piles for the structure, and because of false interpretation the piles had to be extended beyond their estimated length, resulting in a large claim for extras on the contract.

(5) Highly variable borehole information, see Fig. 2.31.

When boreholes show little correlation and high variability this frequently indicates morainic deposits (i.e. terminal moraines).

(6) Drift underlain by uninvestigated rock.

Figure 2.32 shows clay overlying coal seams and Fig. 2.33 shows clay overlying chalk. In both cases the clay was found to be firm and consistent. It was thoroughly tested and assumed to overlie firm strata.

In both cases the assumption was wrong – and caused extensive foundation problems, extra costs and site delays. The coal seam was later found to be extensively *bell-worked* (see Fig. 2.34) and this frequently occurs where coal seams are at relatively shallow depths from the surface.

In the case of the chalk the site was riddled with *swallow-holes* (sink holes) – see Fig. 2.35. Swallow-holes frequently occur at the intersections of joints in chalk and limestone, where groundwater can seep through easily to lower bedjoints. As the water seeps through it dissolves the chalk thus forming a shaft in it. The water having travelled down through the strata may then travel along the strata on a weak bedding joint. Again the chalk is dissolved and underground caverns and caves are formed (to the delight of potholers!).

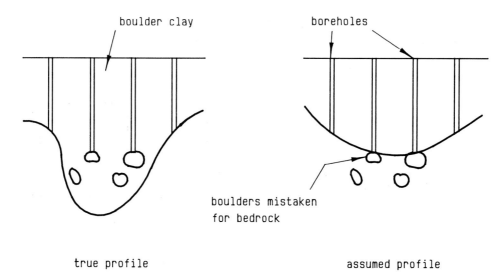

Fig. 2.27 Mistaken bedrock.

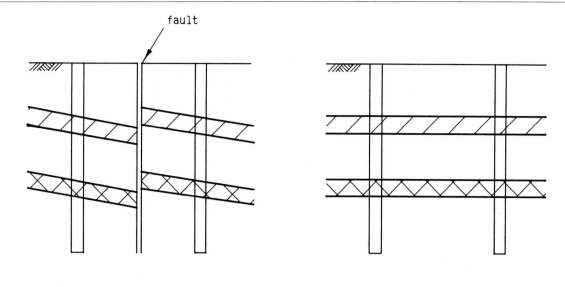

fault

true profile   assumed profile

Fig. 2.28 Unchecked fault.

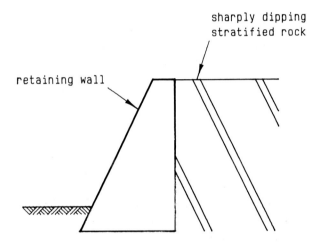

sharply dipping
stratified rock

retaining wall

Fig. 2.29 Unchecked dip causing overloading of retaining wall.

Often the shafts are filled with a mixture of stones and gravel transported by rain run-off, and may not be detectable from casual inspection of the ground level.

(7) Slope failure. Where clay overlies sloping slate or similar rock the ground may be stable before construction and the soil stiff, dense and strong, but construction work or foundation loading may disturb the equilibrium (see Fig. 2.36).

The removal of passive resistance due to excavating the trench for services may result in the clay strata sliding over the smooth, and possibly wet, surface of the slate. Such trenches should only be opened in short sections, provided with extra strong walings and strutting, and backfilled as quickly as possible with lean concrete.

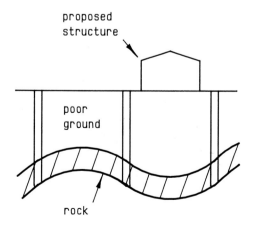

proposed
structure

poor
ground

rock

true profile   boreholes

assumed profile

Fig. 2.30 Folded strata mistaken for level strata.

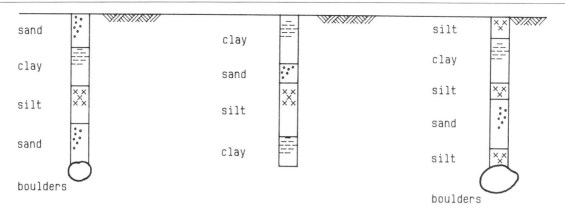

Fig. 2.31 Highly variable borehole information.

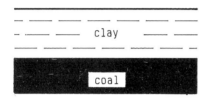

Fig. 2.32 Clay overlying *sound* coal seam.

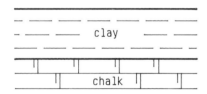

Fig. 2.33 Clay overlying *sound* chalk.

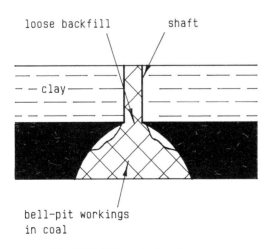

Fig. 2.34 Bell-worked coal seam.

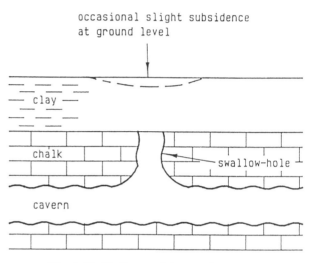

Fig. 2.35 Chalk containing swallow-holes.

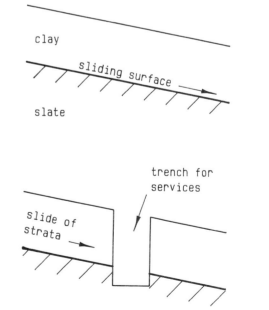

Fig. 2.36 Slope failure due to trench construction.

(8) Soil creep and landslides. Soil can *creep* (i.e. the upper layers move downhill) even on slopes as little as 1 in 10, particularly when the sub-soil is stiff, fissured clay. The moving layer can vary from 200 mm to several metres in depth. The soil can remain static for years, then, without apparent warning, start to creep again; this is often due to excessive increase in groundwater due to unusually heavy rainfall, interference with the natural drainage, or new construction works affecting stability.

Warning signs of a creeping slope area are tilted boundary walls, fences, trees and sometimes a crumpled appearance of the ground surface. Such sites should be avoided where possible. If there is no option but to build on such sites then attention must be given to the uphill drainage of the site, the use of raking piles to increase passive resistance, excavation being kept to a minimum, retaining structures designed for high surcharge, and similar precautions.

2.19 GEOPHYSICAL INVESTIGATION

Geophysical investigation, in addition to the normal boring and sampling, employs specialist techniques not commonly used by designers. Satellites and aerial photographic techniques can record the energy of the electromagnetic spectrum; infra-red photography aids the assessment of moisture contents and flooding danger; seismic reflection and refraction surveys determine depths of strata as do electromagnetic techniques. Many of these techniques were developed to aid the exploration for oil, natural gas and mineral deposits, and have since been applied to site investigation for major civil engineering works.

Some site investigation boreholes should, in discussion with the geologist, be left open to allow further penetration for taking rock cores. The top metre or so of the bedrock is frequently severely weathered, and it can be difficult to withdraw good undisturbed samples of weathered rock.

2.20 EXPERT KNOWLEDGE AND ADVICE

Most experienced engineers have sufficient knowledge of geology to interpret geological maps, records and local knowledge. Furthermore, geological causes of failures to building foundations (as distinct from some civil engineering foundations) are fortunately relatively rare. However, when the geological conditions are suspect or beyond the experience of the engineer, then he should seek advice from expert geologists. Even though this advice may be affirmation, reassurance or confirmation, it is still nevertheless advisable to obtain it. The ever-increasing breadth of knowledge required by senior designers increases the difficulty of acquiring specialized, deep expertize, and designers should not feel inadequate in seeking such assistance.

CHAPTER 3

GROUND INVESTIGATION

3.1 INTRODUCTION

This chapter is a summary of experience in dealing with a large variety of ground conditions on which to build a wide range of structures. It may help young engineers who tend to deal with soil properties, geotechnical engineering and superstructure design only – but sometimes give too little attention to ground investigation on which such engineering topics are dependent. This chapter may also be a helpful recapitulation for the experienced engineer.

The ground or sub-strata material needs to be considered as part of the structure, for, like the superstructure, it will be subject to stress, strain and deformation and also possibly to deterioration. If that part of the 'structure' is defective or fails then experience shows that it can be the most expensive structural failure to remedy. Furthermore no matter how well or expertly the superstructure is designed, if the foundation fails it is possible that the superstructure will also fail. Foundation failure is one of the largest causes of cost claims. For example, in 1976, a year of severe drought, in Britain, 90% of all claims under the National House Building Council (NHBC) guarantee for new houses arose from foundation problems. It has been stated that piling contractors do not have adequate site investigation details for over half the projects for which they are invited to tender. Before a foundation can be *designed* it is necessary to know what load the ground can support, how it will react under the load, both in the short-term and over the structure's life, and also the effect of this new loading on adjoining structures. Without this information, safe and economic design is difficult and may not be possible. Further, the design should be practical and buildable so the designer should be aware of the contractor's likely construction methods and possible problems. (Site construction progress can be slow until the foundations are complete, i.e. the building is 'out of the ground'.)

Before a foundation can be *constructed* the contractor needs to tender for the project, plan methods of excavation, temporary works and ground treatment, and be forewarned of possible problems, etc., to enable him to construct the foundations skilfully, safely and with rapid progress. Standard forms of agreement between the design engineer and his client usually state that the designer should exercise 'reasonable skill and ability' and meet the standard of a 'reasonably competent practitioner'. The *designer* is not expected to be an expert in construction or a specialist in ground treatment. The designer should be wary of non-standard forms with clauses which increase his duty of care to 'fitness for purpose'. The fact that a cause of failure could not have *reasonably* been foreseen is no defence with such a clause nor would professional indemnity insurers accept any obligation.

The ground information is obtained by means of a site investigation. Site investigation, like X-rays and other tests on a sick patient, is not an exact science. The investigation of the ground – as laid down by geological processes, sometimes modified by previous construction, mining, etc., and possibly subject to future change – requires detailed planning, careful collection of information, testing and analysis, to be as reliable as possible. Most importantly, it requires the application of engineering knowledge, judgement and experience.

The results need to be reported clearly, precisely and without ambiguity – but it should be appreciated that the result of the most thorough investigation is an *estimate* and not necessarily an accurate *forecast*. (It has been stated, cynically, that 'there is only one way to determine the exact soil conditions and that is to dig it all out, examine it and replace it – the designer would then be faced with the problem of building on a fill!')

Even the most thorough, detailed and careful survey and investigation can sometimes lead to false conclusions. Isolated pockets of peat, meandering channels of loose, saturated sands, fissures, filled-in shafts and wells, etc., can remain undetected. It is always advisable, therefore, to include in the project estimate a contingency item to cover the possible additional expense of dealing with unforeseen foundation construction difficulties. The engineer should remember that no two samples of soil will have identical properties and that most of the soil tested and reported on will now be in the testing laboratory and not left on site. The soil encountered on site is likely to differ (in varying degrees) to that previously tested and due allowance should be made for this at all stages of the design and construction process.

Since investigation, analysis and reporting (i.e. interpretation of the results of the investigation) should be based on readily available knowledge and established soil investigation procedures – it may be difficult to plead ignorance in a later dispute over a failure. The designer must obtain the available and relevant data from reliable sources and must interpret that data not necessarily with over-sophisticated mathematics but with sound judgement and skill.

BS 5930[1] recommends that 1% of the capital cost of the project should be spent on investigations – engineers can find it difficult to get their clients to agree to spending half this amount! Engineers need to educate their clients on the costly results of *cheap*, inadequate soil surveys.

Though a percentage cost is useful as a preliminary estimate it must be appreciated that it is not a true guide for every site and sufficient funding should be allocated to site investigation to ensure both economic foundation design and construction. Sites and the structures built on them are so varied that it is not possible to fix firm cost percentages without details of the site and proposed structure. For example, a low-rise housing estate to be built on well-known and tested London clay overlying Thames ballast is likely to incur less investigation cost than that for a multi-storey, heavily loaded structure on a suspect, highly variable glacial deposit. Contractors tendering for excavation without adequate (or with suspect) site information may gamble, 'load' the tender or claim high rates for 'extras' and variations.

Clients who object to the cost of a survey or foundation design should be informed of the risks of such cost-cutting. Clients rarely accept the responsibility of the risk or refuse the additional finance for survey and design.

Delays in construction due to inadequate investigation can easily cost more than any money 'saved' by cheap surveys. Extra over-costs in amending foundation design or construction methods to cope with undetected problems can substantially exceed the total cost of an inadequately funded investigation. Time spent in advising the client to provide adequate funding is therefore worthwhile and often essential. In addition, it is advisable that the client should be made aware that the survey cost begins as an estimate which may need revising as the investigation proceeds.

Many clients are in a hurry for early handover of the completed project (with the increasing need for early return on capital investment) and can find time spent on site investigation to be an irksome and unnecessary delay to construction start. The engineer should resist any temptation to skimp the survey and have regard for his client's long-term interests.

When analyzing tenders it is important to have a clear understanding of the kind of sampling and test regime which will actually be required. In the following example (Table 3.1 (a)), Firm 1 appears cheaper initially because their set-up on site and rates of boring and sampling are lower. But when an analysis is undertaken to include the anticipated laboratory testing requirements (Table 3.1 (b)), Firm 2 is the more cost-effective, due to their lower testing rates.

3.2 THE NEED FOR INVESTIGATION

Site investigations can determine the soil properties and behaviour which will affect the choice and design of the foundations, the method of construction, and can also affect the design of the superstructure as an economic and viable

Table 3.1

(a) Site investigation tenders as received

	Firm 1 (£)	Firm 2 (£)
Set-up, etc.	400	500
Boring to z metres	350	600
Obstructions (6 hours)	Rate only	Rate only
Tests on site		
10 standard penetration tests	100	150
20 disturbed	30	10
10 water	40	10
20 undisturbed	150	300
Laboratory tests		
6 No. sulphates	Rate only	Rate only
6 No. pHs	Rate only	Rate only
3 No. particle size distribution	Rate only	Rate only
6 No. PL/LL	Rate only	Rate only
6 No. triaxial tests	Rate only	Rate only
Engineer's site visit	100	75
Comprehensive report	250	150
Insurance	50	–
	1470	1795

(b) Analysis of site investigation tenders

	Firm 1 (£)	Firm 2 (£)
Set-up, etc.	400	500
Boring to z metres	350	600
Obstructions (6 hours)	180	200
Tests on site		
10 standard penetration tests	100	150
20 disturbed	30	10
10 water	40	10
20 undisturbed	150	300
Laboratory tests		
6 No. sulphates	150	75
6 No. pHs	50	15
3 No. particle size distribution	200	50
6 No. PL/LL	200	150
6 No. triaxial tests	250	50
Engineer's site visit	100	75
Comprehensive report	250	150
Insurance	50	–
	2500	2335

proposition. So the designer, the contractor and the client all have a 'need to know'.

Site investigations are also necessary prior to carrying out remedial measures to a failed existing foundation.

3.2.1 The designer's need

The following information does not cover all of the designer's needs but it may assist him in producing the most economical design:

(1) Is the site suitable for the proposed structure, i.e. can it be built economically on the soil or should an alternative location be investigated or has the right price been paid for the land in the first instance?

(2) The load-bearing capacity, settlement and behaviour characteristics of the soil.

(3) The effect of the new foundation loading on adjoining structures and sub-structures.

(4) The presence of aggressive chemicals in the soil, e.g. high sulphate content which could attack concrete.

(5) Possible changes in settlement behaviour, i.e. future and past mineral extraction, changes in permeability and moisture content, danger of running sand.

(6) Shrinkage and swelling characteristics, frost heave susceptibility and vibration sensitivity of the soils.

(7) Water-table fluctuations, tidal effects, sub-surface erosion, seasonal and possible long-term variations.

(8) Change in behaviour of the soils due to exposure during foundation construction.

(9) The advisability and economy of ground treatment.

3.2.2 The contractor's need

Similarly the following information assists the contractor in producing the most economical construction:

(1) The stability of the soil during excavation and foundation construction, i.e. soft mud and similar material will not support heavy piling frames without matting.

(2) The amount of timbering and shoring necessary to support the sides of excavations.

(3) The need for geotechnical processes such as dewatering, freezing and chemical injection.

(4) The presence of any fill material which must be treated (if polluted) or removed.

(5) The presence of useful excavated material such as broken rock for hardcore, sand for concreting or suitable backfill material.

(6) The suitability of the ground at excavation inverts as a base for poured concrete.

(7) The need for special plant such as rippers and drills for decomposed rock, or draglines and grabs where the ground is too weak to support scrapers.

(8) The ground levels relative to a known datum. (This is particularly important for piling operations where pile cut-off levels are specified.)

(9) The need for any special precautions due to ground conditions, e.g. dangerous shafts, running sand, etc.

(10) The position and other details of existing services, old foundations, etc.

3.2.3 The client's need

The client needs to know:

(1) If it is worth buying the site.

(2) If the foundations will be slow and expensive to construct.

(3) If the soil conditions are such that there are planning constraints on his proposed building.

(4) If the site contains contaminates for which he is legally responsible.

(5) If the soils on the site are combustible.

(6) If methane gas or other dangerous gases exist beneath the site.

(7) If the site is subject to flooding, subsidence or landslides.

(8) If the developable area is likely to be restricted by mineshafts or other sterilized zones.

3.2.4 Site investigation for failed, or failing, existing foundations

Failure of existing foundations are often due to changes in local environment such as re-routing of heavy traffic, leaking drains and water mains, new adjoining construction work (e.g. piling, inadequately shored excavations), new fast-growing tree planting, extra load on sub-soil from new buildings and similar. Before carrying out a soil investigation it is usually worthwhile examining such possible causes of failure in the same way that a desk study and a site walkabout should precede any other soil investigation.

3.3 PROCEDURE

The stages of a ground investigation are given in Table 3.2.

A ground investigation consists, basically, of four main operations:

(1) Study of existing information (known as *desk-top* study) and preliminary site reconnaissance (site *walkabout*).

(2) Soil investigation and testing.

(3) Analysis and appraisal of results.

(4) Writing and distribution of soil reports.

In the same way that structural design is a continuous decision-making process and interactive with detailing, other members of the design team, building control and services authorities and the client, so too is the site investigation.

Decisions must be made:

(1) At the start of the survey to determine objectives and methods to achieve the objective.

(2) On choice of site equipment, where and how best to use the equipment.

(3) On choice of samples to be tested, how to test and interpretation of the tests.

(4) On methods of analysis and recommendations to lead to efficient and economic design and construction.

There should be interaction between the designer and site investigator:

(1) The preliminary design should give the investigator an indication of the proposed positioning of the structure on the site, an estimate of foundation loading, any special requirements of basements, services, vibrating or stamping plant and similar information.

(2) The investigator should report periodically to the designer on his findings so that he can, if necessary, amend the scope of the site investigation, alter the position of the building, amend the foundation loading, or alter the preliminary foundation proposals.

(3) When the designer has the final site investigation information he can refine and finalize his design.

Table 3.2 Stages of a ground investigation

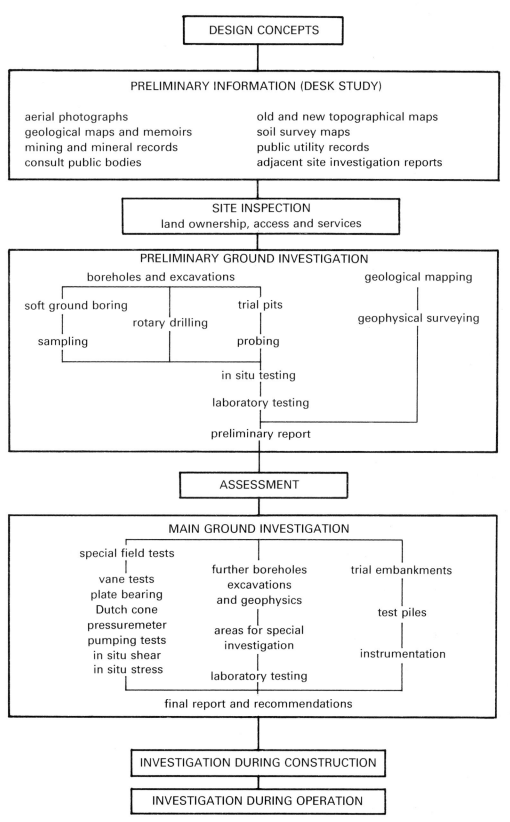

N.B. all stages demand consultation with
the design engineer

(4) The site investigator, given the final design, can refine his report.

(5) Either the designer alone or in collaboration with the site investigator can then write the final report.

(6) Both the designer and investigator should monitor the progress of foundation construction and post-construction structural behaviour. This will determine whether the ground conditions were as predicted; whether there were any unexpected excavation problems; whether the magnitude and rate of settlement was as calculated; whether movement joints performed satisfactorily and did the structure remain fully serviceable (i.e. no cracking, undue settlement, etc.). It is difficult for a busy designer to find time to go back and examine past projects, but from long experience it has been found beneficial for progress in foundation design to make time to go back and look critically at past projects.

3.3.1 Site survey plan

If a site survey has not been done, or provided by the client, then a topographical survey should be carried out. The survey should show the site location and access, give site boundaries, building lines, position of proposed structure, levels and contours, bench marks and survey stations or reference points. In addition, information should be shown on such conditions as previous workings, overhead lines, underground services, evidence of drainage or flooding, condition of adjacent structures and other easily detectable and useful evidence.

3.3.2 Study of existing information

There is often quite a surprising amount of information available for many sites and the surrounding area — even *green* sites in undeveloped areas — and a study of this information can be invaluable in planning an efficient and economical soil survey.

Valuable sources of information are listed below:

(1) Ordnance Survey maps (old maps are often useful in providing information on any previous use of the site which may not appear on revised up-to-date maps).

(2) Geological survey maps, both solid and drift; the Institute of Geological Science Records; The Soil Survey of England and Wales (Rothamstead Experimental Station); The Land Utilization Survey; British Coal can often supply information on proposed mining, present, past and abandoned workings, and finally the Institution of Mining and Metallurgy may have records of other extractions such as tin mining in Cornwall and brine extraction in Cheshire.

(3) Aerial survey photographs which may be of use can be obtained from the DoE's Central Register of Air Photographs.

(4) Local authorities' building control offices and inspectors often have detailed information on any previous use of the site, local conditions and records of previous investigations.

(5) Local contractors frequently know of behaviour and

construction difficulties of excavation, together with records of ground condition and type in the locality.

(6) Local people, such as miners, quarry workers and grave-diggers, can be helpful (sometimes they can be 'over-helpful' in telling the investigator what they think he wants to know with the temptation to embroider their information).

(7) Local Planning Authorities. It is essential to contact them to determine any planning conditions or restrictions for the proposed structure and such matters as rights of light, way and support. They can also advise on site access for plant and transport, noise and other nuisance restrictions. They may also have information on existing or proposed services below ground level, i.e. water mains, sewers, other service pipes, etc., and similar information on overhead power lines, and can put the investigator into contact with the public utility authorities.

(8) Public Services Authorities. Utility services such as British Telecom, British Gas, water authorities, Electricity Boards and similar usually keep up-to-date records.

(9) Local street and area names. Sometimes local place names may indicate previous use. Typical names are Brick Kiln Lane, Quarry Bank and Marsh Street.

Since many people in the above list of information sources are busy it has been found from experience that it can be quicker and more efficient to go and see them to discuss the site rather than engage in long drawn-out correspondence. It is not uncommon in such discussions to discover valuable information which may have been unexpected, not known to exist or not asked for.

For further information see Reference 2.

3.3.3 Preliminary site reconnaissance and site *walkabout*

With the above information, presented clearly in an easily digested report, the senior design engineer should visit the site and the immediate neighbourhood to develop a *feel* for the site. It is sometimes advisable for the senior engineer to visit the site *before* the 'study of existing information' so that he can advise his assistants on important points, such as where there is a particular need for detailed study and the like in carrying out the investigation.

The senior engineer would note the soil type and condition in any adjoining cuttings (road, rail and stream banks), adjacent buildings showing signs of foundation distress, uneven ridge lines, tilting or settled boundary walls, unstable or creeping slopes, depressions in the ground and their possible cause, type and changes in vegetation on *green* sites, previous use and ground behaviour of abandoned sites and similar points. Typical warning signs of possible foundation difficulties are:

(1) Unused sites in built-up pre-war housing estates which can indicate that local builders had encountered site problems.

(2) Flat, rubble-strewn derelict sites in inner-city housing

areas which may be riddled with backfilled basements, cellars and bomb craters (unexploded bombs remain a distinct possibility).

(3) Dry, firm ground in summer which is sprouting marsh grass may be a quagmire in winter.

(4) Undeveloped areas around the outskirts of towns and not encroaching on green-belt boundaries which can indicate problem sites.

(5) Backfilled quarries; domestic refuse and industrial waste tips.

(6) Bumpy, irregular ground surface which can be indicative of glacial terminal moraine deposits.

(7) Evidence of 'bell-working' where coal seams are near ground level.

(8) 'Blow holes' in chalk soils.

(9) Subsidence in areas of brine extraction.

(10) Evidence of erosion or deposition. Where structures are to be founded on coasts, estuaries or tidal rivers, then full hydrographic information on extremes of tides, velocity of currents, seasonal levels, flooding danger, etc., must be obtained.

(11) Warm soils in winter months or burnt shales indicating possible combustion.

Problems of confined access, overhead cables or steeply sloping sites should be noted since this can affect the soil investigation equipment and the contractor's excavation and piling plant.

Knowledge of the position and type of the proposed structure is important so that particular attention can be given to areas where deep excavations for basements, heavy loads and the like are to be located.

It is useful for the senior engineer when visiting the site to be assisted by a young engineer to make notes of his observations and to take photographs and soil samples. This saves the senior engineer time and gives the young engineer valuable experience. The senior engineer should write up his notes and report his findings while they are fresh in his mind. Where possible the findings from the study and reconnaissance should be shown diagrammatically on the site survey plan. This enables a clearer image of site conditions and aids the planning of the soil survey.

This section should be read in conjunction with Chapter 4 on site topography.

3.4 SOIL INVESTIGATION

A soil survey can range from a few trial pits inspected by the designer and the soil untested by laboratory analysis to an extensive borehole investigation with deep and numerous bores and extensive sampling and testing of the soil usually by specialist investigation contractors.

The factors affecting the investigation are the amount of existing information available, the known uniformity or likely variability of the sub-soil in the area, the foundation loading and the type of structure, the general topography and likely groundwater conditions of the site.

Subsidiary factors such as the amount of time and money

available, the site access and other matters should not inhibit the planning of a thorough (and as reliable as is reasonably possible) investigation.

No matter what kind of investigation is carried out, the authors, from experience, recommend the digging of trial pits as a first stage. Trial pits have over the past few decades fallen into almost contemptuous dismissal by some with the increased sophistication of boring techniques, increased cost of labour in digging pits and increased awareness of the limitations of pits (e.g. they do not detect underlying soft soils which can be affected by foundation loading). But during the same period there has been increased adaptability, mobility, etc., of relatively small excavators. Such machines can easily excavate and backfill a dozen pits, or trenches, in a day to a depth of 3−6 m for a hire charge of about the equivalent of a labourer's weekly wage.

3.4.1 Borehole layout

Three bores are the minimum necessary to determine the dip of a plane strata (where known with confidence to be plane) and as a rough guide this is the minimum for a proposed investigation (it is almost self-evident not to have too many!). The more bores drilled then the more is known about the soil and the risks of meeting difficulties and the greater surety and economy of the foundation design. But obviously once enough is known to design an economical foundation then any further bores are an added-on cost to the project. This assumes, of course, that the stratum are accurately recorded, described and positioned, etc. by a competent supervisor during the drilling operations. Inadequate or inexperienced supervision could lead to expensive errors.

On large sites, say for an industrial estate, when the structures' positions have not been defined it is advisable to establish a grid as shown in Fig. 3.1 (c). As stated above, the spacing of the grid depends upon the site study and reconnaissance. A common grid spacing is about 30 m but, if the site is well-known and of uniform strata, the spacing may be increased and, if the site is unknown, suspect and variable, the spacing should obviously be decreased. Where the findings are not uniform and difficulties are unknown, or are expected, then the grid centres should be closed up. Where the site has been mined, an irregular grid is advisable since the workings may be on a regular grid.

The boreholes enable soil profiles (cross-sections) to be drawn noting the strata classification, thickness and level, and samples taken from the borehole enable the properties of the soil in each strata to be examined. The bores can also enable observations to be made on groundwater levels and variations. The depth of the borehole depends on:

(1) The foundation load. Light, single-storey structures founded on known firm ground of thick strata need investigation to a depth of about 3 m − and this can be done effectively by trial pits. Tall, heavily loaded structures may need bores taken down to proven firm soil of adequate strata thickness.

(2) The width of the structure. At a depth of 1.5 times the

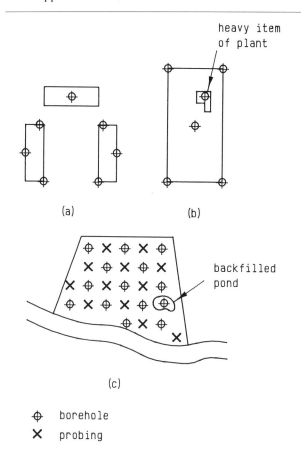

Fig. 3.1 Typical borehole layouts for (a) multi-storey flats, (b) factory building, (c) large development area where building layout is not decided.

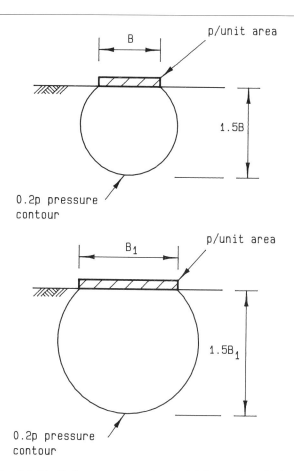

Fig. 3.2 Vertical pressure at a depth of 1.5 times foundation width.

width of the structure the vertical pressure on the soil can be about 20% of the foundation pressure. Closely spaced (i.e. at centres less than about 4 times their width) strip or pad foundations due to pressure distribution overlap would have the same pressure effect at such a depth as a raft foundation. The wider the structure the deeper the effect of vertical pressure (see Fig. 3.2) and it may be necessary to bore down to 1.5 times the width of the structure.

(3) Whether there is a possible need for piling. Then the bores should be taken down to 3 m below preliminary estimated pile base level.

(4) Whether there is a possible need for foundations to be taken down to bedrock. It is advisable to prove that it is in fact bedrock and not boulders (in glacial or flood deposits or quarry backfill) or relatively thin layers of cemented rock-hard soils (shales in mining areas). This can mean that drilling should continue for at least 3 m into the rock. There have been a number of spectacular failures in mistaking isolated boulders as bedrock.

3.4.2 Trial pit layout

Trial pits should be located near to the proposed or existing foundations but not so close as to adversely affect foundation excavation or to disturb existing underground services and drains. They should straddle the proposed site of the building to give cross-sections along the major axes. Generally five or more pits are necessary.

Trial pits yield such information as soil classification, how well the sides of the excavation stand up, the position of the water-table, whether seepage of groundwater will be a problem, the ease of *level, ram and trim*, the invert of the excavation, possible deterioration of the soil on exposure to the atmosphere, the presence and depth of fills, and the ease or difficulty of excavation. (Boreholes can discover sandstone, for example, and contractors will tend to price with high excavation rates yet the trial pit excavator may well be able to excavate the rock easily.) Percussion boring may compress thick layers of peat into thin slices and it is not uncommon to receive descriptions such as 'sand with *traces* of peat' when trial pits would disclose the layer of peat within the stratum of sand. For this reason it is good practice to excavate several trial pits in the vicinity of proposed boreholes so as to check the correlation of the findings of the two techniques. It is easier to take good undisturbed soil samples from a trial pit than a borehole; to carry out in situ tests (such as the standard penetration test and shear vane test) and to give the soil the apocryphal *kick with the heel* to estimate its strength.

Trial pits should be excavated down to at least the expected excavation level and on difficult sites (subject to thorough boring, sampling and testing) the information obtained can be used as a useful additional aid to foundation design and construction. They can also provide a visual check on the likely reliability of test information. If the sides of the pit are liable to collapse and access is required, then propping should be carried out to protect the investigator, or the sides should be battered or stepped by the excavator.

Where the site is open to access by children or animals, the pits should be backfilled or protected at the end of each day. Where it is necessary to check, over a period of time, seepage or deterioration, the pit should be planked over and covered with tarpaulins or otherwise adequately protected.

The position, ground level and invert level of the pits should be noted together with the findings of soil classification and properties and levels of the strata. Colour photographs of the sides of the pits can be useful and the photographs have increased value if a ranging rod is included to confirm the scale. Where the presence of services is suspected, trial pits can be used to detect them, preferably by careful hand-digging.

Where, from past experience, the ground is known to be firm clay or dense gravel of considerable depth, then trial pits may be all that is necessary to investigate the suitability of the site for a lightly loaded structure. They must be dug to an adequate depth, to prove the stratum and to detect soft lenses or layers likely to be affected by the foundation loading. Trial pit information is also invaluable in determining the borehole grid layout.

3.4.3 Hand augers

Hand augers are still sometimes used in preliminary reconnaissance since the equipment is light, cheap and immediately available, and so that overall, time can be saved in planning a full survey. They can, in soft to firm soils, bore a hole about 150 mm diameter to a depth of 3–4 m and provide disturbed samples of the soil. They can be used in restricted spaces, which is useful in investigating foundation failure below a confined basement. However the work can be physically hard, somewhat slow and very difficult, or impossible, in stony clays and gravels.

3.4.4 Boring

Most bores are carried out using light cable percussion plant backed up, when necessary, with rotary coring and other equipment and attachments. The cable percussion rig commonly uses an 8 m high tripod and employs a friction winch to raise and lower the boring tubes and tools. The rotary coring is used when hard shales, boulders or rock is encountered.

There is an increasing variety of plant, sampling methods and tools, with particular advantages in cost, quality of sampling, speed of operation, use in conditions of limited access or headroom, etc., and the choice of rig is affected by the likely soil conditions to be encountered. Further details are given in Reference 3.

3.4.5 Backfilling of trial pits and boreholes

If bores and particularly pits are positioned sufficiently close to the proposed structure so as to affect foundation excavation then they should be carefully backfilled. A strip footing founded on firm clay and passing over an inadequately compacted backfilled trial pit is effectively passing over a *soft-spot*. Also, a borehole can sometimes act as an artesian well or as a seepage point. Trial pits or trenches

should be backfilled in layers with controlled compaction. Boreholes should be backfilled, as the casing is withdrawn, with selected excavated material and punned with a weighted shell. Grouting boreholes is sometimes necessary with 4:1 cement:bentonite. The quality of backfilling of trial pits is however often unreliable and if the pits are close to the foundation they should be re-excavated along with the foundation excavation and backfilled again after completion of foundation construction.

3.4.6 Soil sampling

Samples of the soil are taken from boreholes and trial pits so that the soil can be described and tested. There are two types of samples:

- *Disturbed samples*. Samples taken from boring tubes or hand excavated from the sides and bottom of trial pits where the soil structure is *disturbed* i.e. broken up, cut, pressed, etc. These samples are placed in airtight jars (similar to screw lid jam-jars), labelled to identify the borehole or pit number, the position of the sample, the number given to it in the records, and the date taken. Failure to label samples in standard format will obviously lead to confusion back at the laboratory so the label must be secure and the information noted on it must be legible and written in waterproof ink.

  Disturbed samples are tested to determine, mainly, the type and description of the soil. The sampling and testing of disturbed samples is relatively inexpensive and the test results are used to determine the test programme of undisturbed samples.

  If the disturbed samples are to be used to determine the moisture content of the soil it is important that the sample jar should be completely filled by the sample to prevent it drying out. As a further precaution the air-tight cap should be wound round by a water-resistant tape.

- *Undisturbed samples*. The term *undisturbed* is somewhat of a misnomer for even with refined equipment it is difficult to obtain a true undisturbed sample. Certainly, undisturbed samples are generally superior to disturbed samples in representing more closely the actual in situ structure and moisture content of the soil. The soil structure and moisture content are important factors in soil strength and behaviour under load. The sample tubes are trimmed, at the ends, of disturbed soil. The ends are then covered by foil and waxed before screwing on the tube cap or lid. Labels, giving the same information as for disturbed samples, should be placed both inside the cap and outside the tube.

  Undisturbed samples are tested to determine mainly the strength and behaviour of the soil. Undisturbed samples are relatively expensive to obtain and test and it is generally not necessary to test all the samples. Nevertheless it is advisable to obtain at least one sample for each stratum at each borehole. The test programme is fully determined after study of borehole logs and soil profiles.

3.4.7 Storage of samples

Preferably samples should be sent to the testing laboratory immediately – and this, of course, is not always possible. If they are just left lying around the site they could be subject to drying out, impact, etc. so they should be carefully stacked and stored in a cool and somewhat moist site hut or container box.

3.4.8 Frequency of sampling

The soil investigation engineer, preferably with the design engineer's report on site study, reconnaissance and trial pit findings if available, can decide on an economic frequency of sampling. Generally undisturbed soil samples should be taken at 1.5 m centres and at change of strata level and disturbed samples taken at 1 m centres. This is not a rigid rule and should be varied to suit soil and foundation conditions. When trial pits have not been excavated these centres should be halved from ground level to 2–3 m below the anticipated depth of foundation excavation. It is at or near ground level that the soil is usually most variable due to exposure to weather, change in moisture conditions and variations in the water-table level.

The foreman driller should keep a log noting the type (classification) of soil, its depth, change of strata level, position of obstructions, changes of soil conditions within a strata, groundwater level, seepage and similar information. Experienced and reliable foremen drillers are becoming, unfortunately, rarer and it is essential that the soil survey investigator backs up the foreman's observations by

adequate inspection visits by site supervision engineers. The log should give a continuous description of the soil in the borehole from ground level to base of bore. It is important that the foreman is aware of the standard classification and description used in References 1 and 3 and does not solely employ (the often colourful) local terms such as *cow-belly*, *sludge*, *mucky clay*, *cobbly clay*. While these terms may be well-known to local engineers they can be unfamiliar and totally misleading to others. The local terms are often an invaluable guide to experienced local engineers in describing the soil and its properties and it would be a pity in some cases if these were to die out. Where there is a mixture of clay, silt and sand the MIT (Massachusetts Institute of Technology) classification should be used (see Fig. 3.3).

3.4.9 Appointment of specialist soil investigator

Most design offices do not have sufficient demand for soil investigations to warrant the capital costs of obtaining site and laboratory equipment, nor the current costs of employing site and laboratory personnel. It is therefore generally necessary to appoint specialist firms – and this may not always be as easy as it might appear.

The work should be carried out by competent soil survey specialists of good reputation, staffed by experienced engineers (and drillers) who will not only supervise the borings but also the testing and can be relied upon to report accurately and advise soundly on their findings. The specialist firm should carry adequate indemnity. In the past a number of excellent firms have been driven out of business by cut-

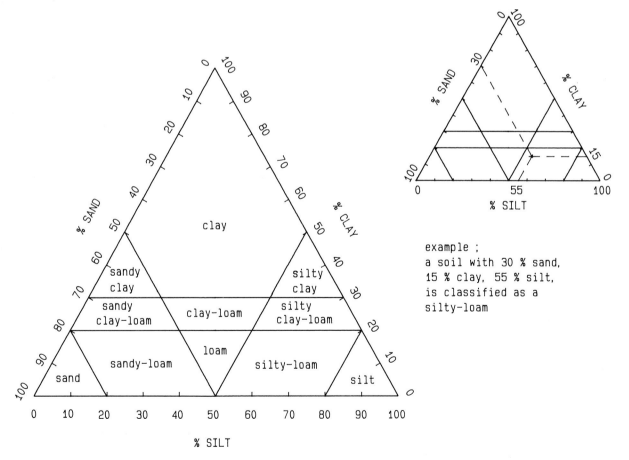

example ;
a soil with 30 % sand,
15 % clay, 55 % silt,
is classified as a
silty-loam

Fig. 3.3 The MIT classification for clay, silt and sand.

throat competition from 'cowboy' firms savagely under-cutting sensible rates. This is a deplorable situation which could cost the client, in the end, far more than he has *saved* by employing such firms. (On more than one occasion the authors' practice has been asked to investigate foundation failures and found that borehole logs are a complete fabrication – because they were not done!)

There should be detailed discussion between the design engineer and the soil specialist on the survey specification, cost and time. Soil specialists may not have wide experience on foundation design, behaviour of structures, economics of alternative designs, construction difficulties, etc., so the discussion is essential for good investigations.

It is also strongly advisable for the design engineer to inspect the boring during progress to see for himself the condition of the soil samples and sampling methods.

3.5 SITE EXAMINATION OF SOILS

Trial pits allow the soil to be examined in situ. Similarly the soil can be examined from borehole samples which may be of a disturbed nature. Examination methods to identify and describe the soil should be based on the guidance given in BS 5930[1] (see Table 2.4).

3.6 FIELD (SITE) TESTING OF SOILS

No matter how carefully soil samples are taken, stored, transported to a laboratory and tested, some disturbance is possible and even likely – and therefore many engineers prefer the alternative of testing the soil in situ. As with sampling techniques there have been advances in sophistication and variety of field testing techniques and the most common types are briefly described here.

Site testing has come a long way from kicking the clay at the bottom of a trial pit with the heel of the investigator's shoe – though this can still be a useful, if crude, assessment when carried out by an experienced engineer familiar with local conditions.

In foundation design less is known of soil as a structural material than is known of concrete and steel, it is not possible to analyse and forecast, with certainty, the stresses in the soil or the soil's reaction to those stresses, and the foundation loading can only be a reasonable assessment.

Foundation design is therefore based not solely on analysis but also needs the application of sound engineering judgement.

In a sensible and valuable search to understand the material it must be tested and some researchers have devoted their careers to this essential cause. In each of the following field and laboratory tests there has been extensive research, literally thousands of learned papers and many international conferences – some devoted to just one test, for example, see References 4 and 5. It is not possible therefore in a book on foundation design to discuss fully in depth any one test; discussion is limited to the broader considerations. Furthermore the site and laboratory testing of soils is the contractual responsibility of the soil survey specialist. Hence the following sections outline and summarize the tests and the main references are given for designers wishing for more detailed information. Experience is necessary to estimate what and how to test, the test results need engineering judgement in assessing their application and relevance and in forecasting estimated behaviour – for none of the tests give scientifically *accurate* results applicable to the actual strata under the real pressure. The theories, as in structural theory, are based on simplifying assumptions not fully related to the reality of practice. But to dismiss tests and theory and rely on outdated *rules of thumb* methods is inappropriate to modern structures and is as foolish as blind faith in *science*.

3.6.1 Standard Penetration Test (SPT)

The SPT is a useful method of indicating the relative density of sand and gravels. It is based on the fact that the denser the sand or gravel the harder it is to hammer a peg into it. A standard weight is dropped a defined distance on a tube, with either a split tube or a cone head (cone penetration test, CPT), placed in the borehole. The tube is driven 450 mm into the soil and the number of *hammer-blows* taken to drive the tube into the last 300 mm of soil is termed its N value. Care in interpreting the result is particularly necessary where boulders, very coarse gravel or bricks in backfill may be present, for the measurement may be of the resistance of the obstruction and not of the soil.

Approximate values of the relationship between sand properties and N values are given in Table 3.3 and a summary of the test is given in Table 3.4.

Table 3.3 Relationship between N values and sand properties (Reference 9)

	Very loose	Loose	Medium dense	Dense	Very dense
SPT N value (blows/0.3 m)[a]	<4	4–10	10–30	30–50	>50
CPT cone resistance (MN/mm²)[b]	<5	5–10	10–15	15–20	>20
Equiv. relative density (%)[c]	<15	15–35	35–65	65–85	85–10
Dry unit weight (kN/m³)	<14	14–16	16–18	18–20	>20
Friction angle (degrees)	<30	30–32	32–35	35–38	>38
Cyclic stress ratio causing liquefaction (τ/σ')	<0.04	0.04–0.10	0.10–0.35	>0.35	–

[a] At an effective vertical overburden pressure of 100 kN/m²

[b] There is no unique relationship between CPT and SPT values – it should be reassessed at each site

[c] Freshly deposited, normally consolidated sand

Table 3.4 Standard Penetration Test (Weltman & Head, *Site Investigation Manual*, CIRIA SP 25 (1983), Section 4.1.1)

Method	Application	Advantages	Disadvantages
Standard Penetration Test	Derivation of a standardized blow count from dynamic penetration in granular soils (silts, sands, gravels) and in certain cases, other materials such as weak rock or clays containing gravels which are not readily sampled by other means. Convenient both above and below the groundwater table. The blow count (*N* value) may be used directly in empirical formulae for bearing capacity and settlement estimates: relative density and estimation of φ. Approximate values of cohesion may be inferred using empirical relationships.	Simple, robust equipment. Procedure is straighforward and permits frequent tests. A highly disturbed sample obtained when the shoe is used, permitting identification of the soil. A number of empirical relationships exist to convert the *N* value to approximate various soil characteristics or indications of performance. Widespread use. Inexpensive.	Simplicity of the equipment belies sensitivity to operator techniques, equipment, malfunctions and poor boring practice. Equipment and technique are not standardized internationally. Tests below 6 m in water-bearing sands may not be fully representative and in other materials as the depth increases. If solid cone used instead of the shoe to prevent damage, the results may not be comparable. Test values may vary with diameter of borehole. Results require interpretation. Test insensitive in loose sands. Misleading results in fissured clays. *N* values are affected if a sample liner is used with a 38 mm diameter spoon.

3.6.2 Vane test

If a garden spade is driven into clay and then rotated it will effectively shear the clay and the higher the shear resistance of the clay then the greater the force (torque) required to rotate the spade. This is the principle of the vane test.

The vane is a cruciform of four blades fixed to the end of the boring tube's rod. It is pushed into the undisturbed soil at the base of the borehole or trial pit and the torque required to rotate the vane is measured. Table 3.5 gives a summary of the test.

When the height of the vane is twice its diameter, D(m), the relationship between shear strength of the soil, τ, and the maximum applied torque, M(kN m), is generally:

$$\tau = \frac{M}{3.66\,D^3}\ \text{kN/m}^2$$

3.6.3 Plate bearing test

A plate, of known area, can be placed at the bottom of a trial pit or borehole and loaded. The settlement of the soil under load can be measured and also the pressure required to cause shear failure of the soil. The test is summarized in Table 3.6.

3.6.4 Pressuremeters

A pressuremeter could be considered as basically a vertical plate test. If an expanding cell is placed in a borehole and pumped up to exert pressure against the sides of the bore then the stronger the soil the greater the pressure required to expand the cell. Summaries of different pressuremeters are given in Table 3.7.

3.6.5 Groundwater (piezometers and standpipes)

The presence of moisture in, and the magnitude of moisture content of, soils has a pronounced effect on soil properties and behaviour. Since the moisture content can vary so too can the soil. It is essential therefore to investigate the groundwater conditions and possible variation. Groundwater variations are likely on coastal, estuarine and tidal river sites; sites subject to artesian conditions and variable water-table levels; sites with permeable granular soil where bored piles or bentonite diaphragm walls are to be used, and particularly sites founded on fills.

The rate of seepage of groundwater into pits and bores together with level and variations in level should be recorded. Piezometers or standpipes should be employed when groundwater problems are anticipated.

A standpipe can at its simplest be the open borehole, and the outline of the test is summarized in Table 3.8.

Piezometers, of varying sophistication, are basically perforated tubes lined internally with porous tubing, and details are summarized in Table 3.9.

3.6.6 Other field tests

There are a number of developments, refinements and adjustments to the above tests as well as geophysical tests, aerial infra-red photography, video photography in boreholes, etc. These newer tests can sometimes be less expensive, less time-consuming and yield more information than the traditional tests. The interested reader should refer to specialist soil mechanics literature for details.

Table 3.5 Vane test (Weltman & Head, *Site Investigation Manual*, CIRIA SP 25 (1983), Section 4.12)

Method	Application	Advantages	Disadvantages
Vane test	Measurement of undrained shear strength of clays and measurement of remoulded strength. The results should be used in conjunction with laboratory derived values of cohesion and measurement of plasticity index in order that an assessment of the validity of the results may be made.	Permits in situ measurement of the undrained strength of sensitive clays with cohesions generally up to $100\,kN/m^2$. The remoulded shear strength may also be measured in situ. Causes little disturbance to the soil. Can be used direct from the base of a borehole. Results are direct and immediate. Tests can be rapid. Small hand-operated vane test instruments are available for use in side or base of excavations.	The results are affected by silty or sandy pockets or significant organic content in the clay. There is some dependence on the plasticity index (PI) of clay. Anisotropy effects can give rise to values of cohesion unrepresentative of the engineering problems being studied. Poor maintenance of equipment gives excessive friction between rods and guide tubes, or in bearings. To be used in conjunction with careful soil description and backed up with high-quality sampling and laboratory testing. Results are in terms of total stress only. Specialist technicians required.

Table 3.6 Plate bearing test (Weltman & Head, *Site Investigation Manual*, CIRIA SP 25 (1983), Section 4.1.3)

Method	Application	Advantages	Disadvantages
Plate bearing test	For determination of elastic modulus and bearing capacity of soils and weak rock, with minimum disturbance.	Gives close simulation of actual loading condition typically found in foundations. The loaded volume of soil or rock is large by comparison with other tests, and therefore more representative. There is close control of loading intensity, rate and duration. More representative results than laboratory testing. Can be carried out in pits or boreholes.	A number of tests are required to obtain coverage with depth for application to foundation designs. Upward seepage pressures at the test level reduce effective stress and have significant effects. Specialist technicians are necessary. An expensive and time-consuming test. Equipment not widely available. Scale effects should be considered. Possibility of ground disturbance during excavation. Excavation causes unavoidable change in ground stresses which may be irreversible. Large-diameter hole desirable for tests in boreholes. Results difficult to interpret in some soil types.

Table 3.7 Pressuremeter test (Weltman & Head, *Site Investigation Manual*, CIRIA SP 25 (1983), Section 4.1.3)

Method	Application	Advantages	Disadvantages
Pressuremeter test	Three similar main types of pressuremeter are available: (a) The Menard pressuremeter, installed into a borehole. (b) The Camkometer, self-boring type. (c) The Stressprobe, pressed into the soil from the base of the borehole. The Menard pressuremeter is particularly suitable in weak rock, for modulus creep pressure and limit pressure. The Stressprobe and Camkometer give similar information, the former being particularly suited to measuring the shear strength of stiff clays, the latter also containing a porewater pressure transducer to enable effective stress measurements to be carried out. It is suitable in clays, silts and sands. Lateral stress and K_0 (coefficient of earth pressure at rest) measurement are possible. Becoming more widely used and expected to be used more extensively in future.	In situ low-disturbance measurement of important soil and weak rock parameters. Less expensive than direct bearing tests and larger volume of rock stressed than laboratory testing methods. Depth limitations vary with subsoil, but could be carried out at any depth in appropriate circumstances. Direct bearing capacity measurements can be taken. Rapid test procedure.	In some soils and rocks the operation of the equipment can be uncertain, particularly granular soils. In some weak rocks, unstable walls can give rise to results which are difficult to interpret (Menard type). The loading direction is radial, in a horizontal plane, which may not correspond to the condition in the foundation considered. Where porewater pressures are not measured, drainage conditions have to be assumed. A large number of tests with depth are required if the results are to be used for typical foundation designs. Tests not suited to coarse granular materials. Drilling disturbance cannot be detected and may lead to unusually low results (Menard type). Specialist technicians necessary.

Table 3.8 Open borehole test (Weltman & Head, *Site Investigation Manual*, CIRIA SP 25 (1983), Section 3.3.1)

Method	Application	Advantages	Disadvantages
Open borehole test	For estimation of permeability in medium and course grained soils and fissured or fractured rock where appropriate. The approximate particle size of granular soils may be estimated from the results (e.g. using the Hazen[11] formula). Broken or fissured zones in rock may be identified. Seepage conditions likely during construction and under foundations may be estimated. The need for dewatering schemes may be assessed.	Relatively low cost method of obtaining permeability information and additional grain size information. Conventional equipment utilized. No specialized personnel necessary. Widely used. Yields more reliable data than laboratory tests in some cases.	Methods very approximate particularly falling or constant head tests where sedimentation or loosening can occur. Rising head tests are markedly affected by poor boring techniques which leave loosened soil, or by piping, should it occur during the test. Results may require close scrutiny, particularly in variable sub-soil. Accurate permanent groundwater levels necessary. Hydraulic fracturing can occur.

Table 3.9 Piezometer test (Weltman & Head, *Site Investigation Manual*, CIRIA SP 25 (1983), Section 3.3.2)

Method	Application	Advantages	Disadvantages
Constant head test from piezometers	For estimation of permeability and consolidation parameters in fine grained soils. When combined with laboratory determined values of m_v (coefficient of volume compressibility), better estimates of c_v (coefficient of consolidation) may be made.	Large-scale determination of permeability and consolidation parameters. Generally more reliable than laboratory values in alluvial soils.	The tests are most conveniently carried out with a positive head. Swelling conditions produced are not appropriate to the foundation problem. The tests are time-consuming and expensive. The radial drainage conditions require to be carefully assessed relative to the stratigraphy detail at the test location and the full-scale drainage conditions. Specialist technicians required. The groundwater must be at equilibrium in the borehole before starting the test.

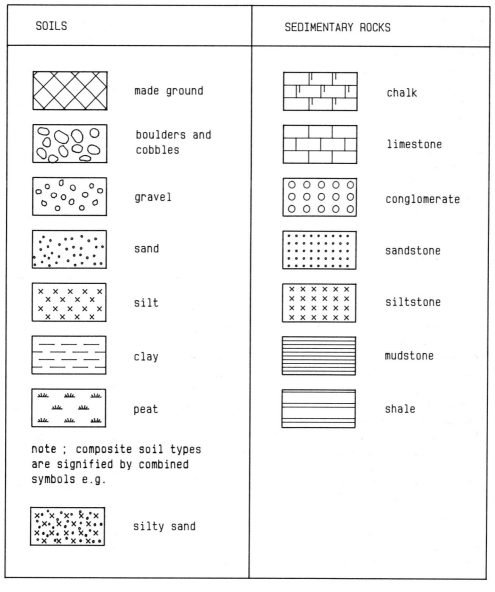

Fig. 3.4 Recommended symbols for soils and rocks (BS 5930, Table 11).

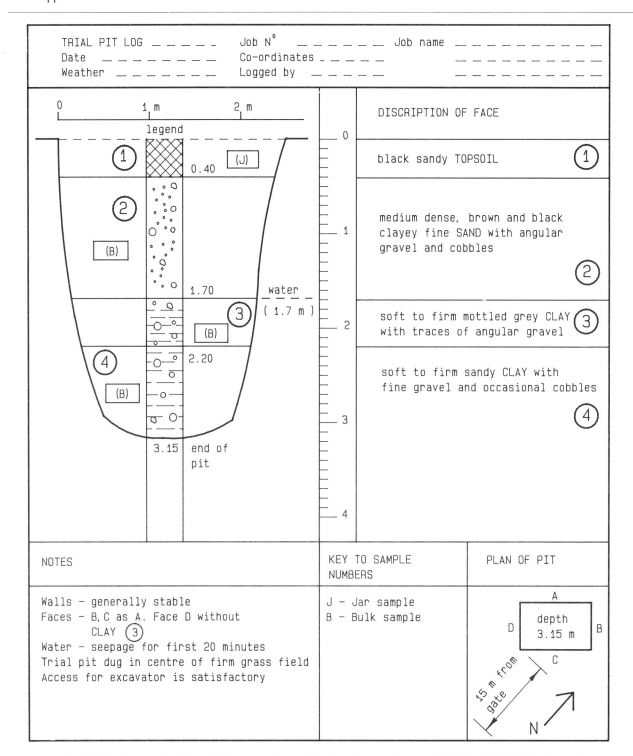

Fig. 3.5 Typical trial pit log (Weltman & Head, *Site Investigation Manual*, CIRIA SP25 (1983), Fig. 66).

3.7 RECORDING INFORMATION – TRIAL PIT AND BOREHOLE LOGS AND SOIL PROFILES

Before embarking on expensive laboratory testing of soil samples it is advisable to record (log) the information gained on site in order to plan the test programme. To facilitate the reading of logs and boreholes the soils and rocks should be indicated by standardized symbols. Widely accepted diagrammatic symbols are given in Fig. 3.4.

A typical trial pit log of the engineer's observations is given in Fig. 3.5.

A borehole log should give details of the foreman driller's log, the observations of the supervising engineer and the results of any site tests. A typical borehole log is shown in Fig. 3.6.

Trial pits, trenches and boreholes should be given reference numbers, located on plan, their ground level noted and the date of excavation recorded. It is advisable to record the following additional information:

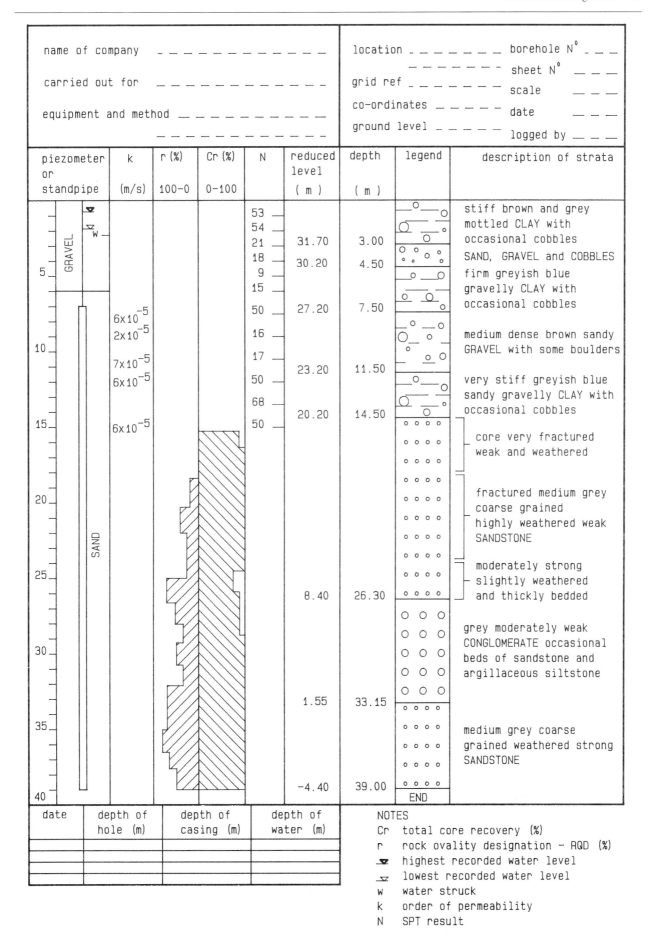

Fig. 3.6 Typical borehole log (BS 5930, Fig. 30).

(1) Type of rig, diameter and depth of bore or width of bucket.
(2) Diameter and depth of any casing used and why it was necessary.
(3) Depth of each change of strata and a full description of the strata. (Was the soil virgin ground or fill?)
(4) Depths at which samples taken, type of sample and sample reference number.
(5) In situ test depth and reference number.
(6) The levels at which groundwater was first noted; the rate of rise of the water; its level at start and end of each day. (When more information on permeability, porewater pressure, and the like is required, then it is vitally important that the use of piezometers should be considered.)
(7) Depth and description of obstructions (i.e. boulders), services (drains) or cavities encountered.
(8) Rate of boring or excavation (useful to contractors and piling sub-contractors as such information gives some guidance in ease of excavation or pile driving).
(9) Name of supervising engineer.
(10) Date and weather conditions during investigation.

3.8 SOIL SAMPLES AND SOIL PROFILES

It is a wise precaution to take more soil samples than necessary to determine the ground conditions (and increasing the frequency of samples does not proportionally increase the cost of the soil survey). It is not however necessary to test every single sample. If the surface soil is weak and underlain by good rock or dense gravel there may be little point in testing the weak surface soil if piling down to the good strata is proposed.

Soil profiles (section through boreholes) are extremely helpful in enabling the designer to visualize the ground conditions. This valuable aid is, in the authors' opinion, too often given inadequate attention in site investigations. Many foundation failures can be traced back to faulty visualization of the ground conditions due to inadequate soil profiles or misinterpretation of them. A typical soil profile is shown in Fig. 3.7.

Most experienced designers would tend to study the soil profile first before reading the site report, studying the test results and checking other data. This makes for efficiency, better assessment of site conditions, improved judgement of data, it warns of problems and can indicate the need for possible further investigation.

Some typical misinterpretations or inadequate data leading to false conclusions and similar errors are shown in Fig. 3.8 (see also Figs 2.27, 2.28 and 2.30).

3.9 PRELIMINARY ANALYSIS OF RESULTS

It is often necessary in practice to save time to issue a preliminary report before the results of laboratory tests are available or even planned and programmed. This must be done by an experienced engineer who would appreciate that boreholes give only information on the soil at the *boreholes* and not factual information on the soil between them. The

engineer interpolates what *might* be the soil profile between bores.

The formulation of an accurate (as possible) 'picture' of the ground and water conditions is necessary for a good analysis and all the data from the preliminary investigations, history of the site, local experience, borehole logs, field test results, etc., must be collected, sorted, appraised and assessed. If the soil investigation engineer cannot make firm reliable recommendations he must either ask for further information or qualify his preliminary recommendations to the designer (and state why he qualifies his recommendation).

The preliminary analysis must produce adequate data for preliminary foundation design, if necessary, and draft information for the contractor's initial costing. Though the contractor normally has a contractual responsibility to examine the site, it is sensible to give him the information acquired.

The analysis should also enable the contractor to assess the need for specialist operations such as dewatering. The final results, after laboratory testing, should be given to such specialists along with the invitation to tender for the project. (It is a wise precaution to state in the invitation to tender the specified results and *not* the method of operation. For example, unscrupulous dewatering sub-contractors may tender to *hire* dewatering equipment to the site and not necessarily quote for dewatering it.)

3.10 SITE INVESTIGATION REPORT

The report should contain the information gained in reconnaissance, survey, investigation, testing and soil survey recommendations and the design engineer's recommendations. Since the report is the property of the client his permission should be obtained for its distribution to invited main and appropriate specialist sub-contractors and any public authority collecting soil data.

The report will contain a mass of information which must be presented in an orderly, easily digested manner and written in clear, unambiguous, good English. Since most of the intended readers are mainly visually orientated, the use of photos, maps, soil profiles, borehole logs and other visual aids is to be recommended as is the tabulation of test results and other information. The report is not a thesis nor a scientific treatise, but a factual report with comments, opinions and recommendations based on the interpretation of the facts from experience. The facts and opinions must be clearly separated. Since the report is likely to be subject to hard and frequent usage it is advisable to bind it between stiff covers rather than merely stapling a mass of A4 sheets.

The script, drawings and layout should be checked and re-checked just as carefully as calculations and drawings from the design office.

A recommended procedure is as follows:

(1) Collect data, categorize it and rough out a preliminary draft.
(2) Edit the draft and seek methods of visual presentation and tabulation.
(3) Polish re-draft and check for improvements in presentation, check for typing errors and appearance.

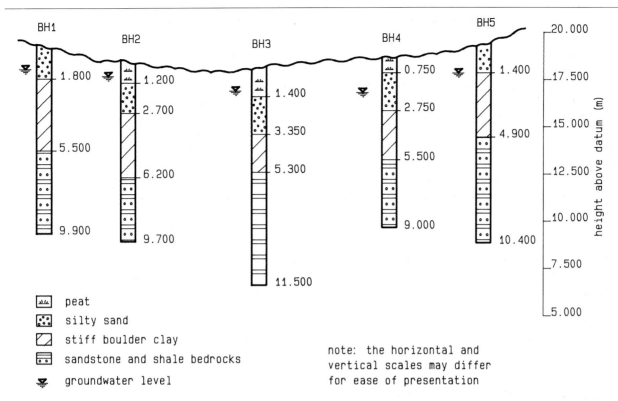

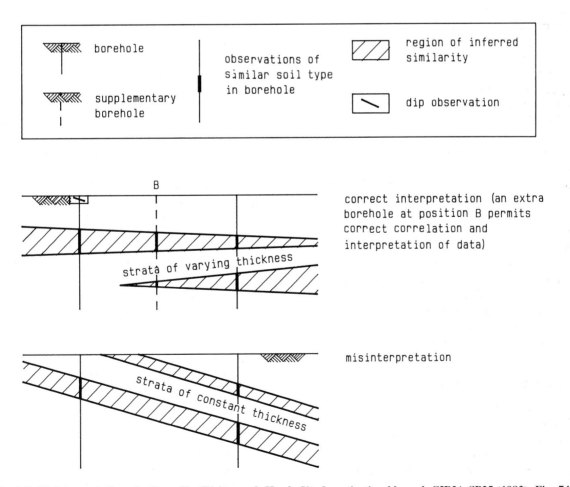

Fig. 3.7 **Soil profile for a typical site.**

Fig. 3.8 **Misinterpretation of soil profile (Weltman & Head, *Site Investigation Manual*, CIRIA SP25 (1983), Fig. 74).**

3.10.1 Factors affecting quality of report

The restraints of time and funding that need to be allowed for in the investigation have been discussed in earlier sections and there are other factors which can affect the quality of the investigation, recommendations and the engineering judgement. Among those which may affect some engineers are:

(1) Uncritical acceptance of well-presented opinion, results of sophisticated (but not necessarily relevant) tests and over- and unqualified respect for some specialists.
(2) Allowing site difficulties to dictate the investigation in an attempt to keep the investigation simple and cheap.
(3) Lack of recognition that piling and other foundation techniques can be used to economic advantage even on good sites.
(4) Lack of recognition that some fills, possibly upgraded by ground improvement techniques, can provide an adequate and economic bearing strata.
(5) Lack of appreciation that advances in structural design can accommodate relatively high settlements.
(6) Under-estimation of the importance of the designer, at least, visiting the site during the investigation or dismissal of trial pits as unscientific or out-dated.

3.10.2 Sequence of report

Foundation reports follow the normal sequence of items of engineering reports in having a title, contents list, synopsis, introduction, body of the report, conclusions and recommendations. Lengthy descriptions of tests and similar matters are best dealt with in appendices and the test results tabulated in the body of the report. The client tends to read the synopsis and recommendations; the main and subcontractors concentrate on the body of the report and the design office on its conclusions and recommendations.

If the brief imposed such limitations on cost and time allocation for the investigation that the engineer was not able to carry out an adequate survey he should tactfully point this out. It should also be made clear in such cases that the engineer is qualifying his conclusions and recommendations − this is unfortunately advisable in the present litigatious climate.

3.10.3 Site description

This, as far as possible, should be given on small-scale plans showing site location, access and surrounding area. The proposed position of the buildings and access roads should be shown. The site plan should also show the general layout and surface features, note presence of existing buildings, old foundations and previous usage, services, vegetation, any subsidence or unstable slopes, etc.

Written description of the site exposure (for wind speed regulations) should be given together with records of any flooding, erosion and other geographical and hydrographic information.

Geological maps and sections should, when they are necessary, be provided, noting mines, shafts, quarries, swallow-holes and other geological features affecting design and construction.

Photographs taken on the site, preferably colour ones, can be very helpful and should be supported by aerial photographs if considered necessary.

3.10.4 The ground investigation

(1) *Background study and location of holes.* This should give a full account of the desk-top study, examination of old records, information from local authorities, public utilities and the like, and the field survey. It should detail the position and depth of trial pits and boreholes, equipment used and in situ testing and information.
(2) *Boreholes, trial pits and soil profiles.* This section will be mainly a visual presentation of the logs and profiles together with colour photographs of the trial pits. Where possible written information should be given in note form on the soil profiles.
(3) *Soil tests.* This should list the site and laboratory tests drawing attention to any unusual, unexpected or special results. The results of the tests should be tabulated, for ease of reference, and diagrams of such information as particle size distribution, pressure−void ratio curves and Mohr's circles should be given. If such form of presentation is not fully adequate then test descriptions and results should be given in an appendix.

3.10.5 Results

This must give details of ground conditions, previous use of site, present conditions, groundwater and drainage pattern.

The tests must give adequate information to determine the soil's bearing capacity, settlement characteristics, behaviour during and after foundation construction and, where necessary, its chemical make-up and condition.

3.10.6 Recommendations

This is both comment on the facts and also opinions based on experience; the difference should be made clear. Since the discussion is usually a major part of the report it should be broken down into sections for ease of reference and readability.

The first section should briefly describe the proposed main and subsidiary structures and their loading, a description and assessment of the ground conditions and the types of appropriate foundations.

The second section should advise on foundation depths, pressures, settlements, discuss alternatives giving advantages, disadvantages and possible problems keeping in mind cost and buildability considerations.

Typical main recommendations are:

(1) Safe bearing capacities at various depths, estimates of total and differential settlement and time-span of settlement.
(2) Problems of excavation (fills, rock, water ingress, toxic and combustible material).
(3) Chemical attack on concrete and steel by sulphates and chlorides or acids within soil.
(4) Flotation effect on buoyant or submerged foundations.
(5) Where the proposed structure houses plant which could

vibrate or impact shock the soil, the effect on the soil must be assessed.

(6) Details of any necessary geotechnical processes to improve the soil's properties.

(7) Where piling is necessary, information must be given on founding level, possible negative skin friction, obstructions, appropriate type and installation of piles and the effects of piling on adjacent constructions and existing buildings.

(8) Where foundation is subject to lateral loading, the magnitude and position of the loading must be given together with the skin friction between the soil and the soil's passive resistance.

(9) Where retaining walls are required, information is needed on active pressure, passive resistance, surcharge, factor of safety against slip circle failure, possible landslides or slips.

(10) Where road construction is involved requiring CBR values, etc., though this is outside the scope of this book.

The final section should give firm recommendations on the foundation type or types to be adopted.

3.11 FILLS (MADE GROUND)

Filled ground can vary from carefully backfilled selected material placed in relatively thin layers which have been properly compacted to indiscriminate tipping of domestic or industrial waste. Fills vary so widely that it is impossible to standardize investigation procedure. Details are given in Chapters 5 and 7 where health and safety aspects relating to the carrying out of the investigation are highlighted and discussed and further reference should be made to these chapters before embarking upon any ground investigation.

Fills can contain highly toxic chemicals, dangerous asbestos, voids caused by rusting containers and old cars, biodegradable materials, obstructions such as old girders and existing foundations, waste from collieries and gas works with high sulphate content, material liable to spontaneous combustion when exposed to the atmosphere — and similar horrors. Boring through and sampling such material can be difficult, hazardous and, more, be unreliable in forming an assessment of behaviour and properties of such fills.

Such sites would not have been considered suitable a decade ago, but with the growing demand for building land, the drive for inner-city regeneration and increased resistance to encroaching on *green belts* around cities, such sites do now have to be considered. Encouragement to develop such sites has been given by central government in the form of Derelict Land Grants. Obviously for such sites the preliminary investigation is of even greater importance though it can be more difficult. For derelict industrial sites efforts should be made to contact the former owners and for abandoned inner-city sites the local authority may have old records.

Where the depth of fill is relatively shallow, about 5 m as in old filled-in cellars and basements, the probable best method of investigation is by trial pits dug by excavators.

When the depth of possibly contaminated fills exceeds 5 m, it is expensive to excavate trial pits and it is better to employ specialist soil survey firms who should attempt to identify the material, its toxicity, concentration and extent.

The taking of samples (unless the fill is uniform) is difficult and site and laboratory testing of the normal small samples does not generally enable a reasonable assessment of the fill's strength and behaviour to be made.

The main problem of foundation design is usually the estimate of likely large and/or differential/settlement and the presence of aggressive chemicals which could attack the foundation concrete and could cause health or environmental hazards.

The test for settlement is better carried out by site tests on a larger test area than a plate test. There is an increasing use of refuse skips of base area of about $2-5\,\text{m}^2$. These are placed on a levelled area of the fill covered with a levelled, 100 mm, layer of sand to ensure uniform pressure. The skip can be either filled with water or damp sand, of known density, and the settlement of the fill measured over a period of a month or so.

The presence of aggressive chemicals is determined by chemical analysis of samples from the pits or bores. It is even more important on contaminated sites to determine the possible changes in water movement since deep-lying toxics can be leached to the surface and attack the foundations.

Though the foundation costs on such sites are very likely to be higher than normal sites, this can be compensated for by lower land costs and the possible award of Derelict Land Grants. As is shown in later chapters, such sites can be successfully reclaimed and developed using ground treatments and foundation techniques such as vibro-stabilization, dynamic consolidation, preloading, buoyant rafts, piling, etc.

Such sites are a challenge to the designer's ingenuity and provide job-satisfaction in changing an eyesore into a social amenity. The authors' experience on such sites, for example, the Liverpool International Garden Festival site, Birkenhead Docks, the abandoned Tate and Lyle works at Liverpool and many others, testify to this.

3.12 LEGAL ISSUES

As stated earlier site investigation is not an exact science; it provides a reasonable estimate and predictions and not an accurate forecast. Therefore, on occasions, unexpected difficulties can occur causing increased costs and construction delay. Most major clients appreciate this possibility and are aware of the need for contingency funds. However some clients (and some contractors) are 'claim-happy' and may be liable to proceed with litigation. Provided that the engineers have been prudent, have given normal professional skill and thoroughness, have advised the client on limitations and of the need to amend the brief, then it is very unlikely that claims against the engineer would be substantiated. Should a claim be likely or threatened the engineer should immediately advise his indemnity insurers.

Since boring, sampling, testing and analysis has become,

and are continuing to be more highly specialized and sophisticated, specialist engineering skills beyond the experience and knowledge of many structural designers are called for. It is therefore advisable to employ specialist firms. Reputable, experienced soil mechanics firms, provided with adequate indemnity, should be invited to quote for the survey. While attention and study should be given to the rates for boring, extra-overs for drilling through boulders, costs for sampling, standing time, etc., care must be exercised in not attaching over-importance to individual rates – it is analogous to ask a designer to quote for A3, A2 details and A4 calculation sheets and to use such quotes in assessing the design fee. It is the thoroughness and reliability of the survey that is important and value-for-money takes precedence over individual rates.

3.13 TIME

The last part of the structure that can be designed and detailed are the foundations but they are the first working details the contractor needs. So time is usually of the essence. Generally if enough is known of the site to be assured that it is suitable to build on then planning and design of the structure can start while the site and soil is being examined. If the site is suspect or likely to require exorbitantly expensive foundations it is right to delay design until sufficient information is available to decide on the feasibility of the project.

Obtaining preliminary information from the sources mentioned in section 3.3.2 can be a slow process so the earlier it is started the better.

If the client is negotiating to buy the land he often urgently needs an approximate cost of the proposed foundation. Frequently cheap land is only cheap because it is thought that the foundation costs are likely to be high. It is not advisable for the engineer to commit himself to an unequivocal foundation cost but rather to provide an estimate based on the available knowledge at the time of estimate and to inform the client that adjustment to the estimate may be necessary when the results of the full investigation are available.

Foundation construction can be the major cause of delaying completion of the project and thus expense to the client (in delay on return on capital) and to the contractor (adding to his site overheads).

3.14 CONCLUSIONS

It should be apparent from the foregoing discussion that neither ground investigation nor soil mechanics is an exact science – but no engineering activity is *exact*, and soil mechanics tends to be somewhat less exact than structural design.

In order to simplify structural design, simplifying assumptions have to be made to develop theories – e.g., the material is assumed to be perfectly elastic, the loadings are known with exactitude, the end conditions are firmly postulated, etc. These simplifications do not reflect the practicalities of construction but are helpful to the designer in his assessments to which he applies factors of safety.

Exactly the same process occurs in soil mechanics with a range of simplifying assumptions being made in the sampling, testing and interpreting of results in order to obtain soil parameters for bearing pressure and settlement calculations.

Similarly, the testing should be subject to engineering assessment and not accepted passively and uncritically. The concrete cube test is a somewhat simple and crude assessment of concrete strength but its correlation with the strength of the *real* concrete in the actual structure is reasonably well established from long experience. Few engineers would order the demolition of recently built concrete simply because an occasional cube failed to reach a specified strength. The engineer would probably check first the test procedure, method, etc., then check the materials, mixing, etc. on site and finally examine the concrete in the structure.

Table 3.10 Field classification of clays (Weltman & Head, *Site Investigation Manual*, CIRIA SP 25 (1983), Table 4)

Consistency	Field test	Undrained shear strength range (kN/m^2)	Equivalent N^a value (very approximate)
Very soft	Exudes between fingers when squeezed in hand	under 20	under 2
Soft (soft to firm)	Moulded by light finger pressure	20 to 40 (40 to 50)	2 to 4
Firm (firm to stiff)	Can be moulded by strong finger pressure	50 to 75 (75 to 100)	4 to 8
Stiff	Cannot be moulded by fingers. Can be indented by thumb.	100 to 150	8 to 15
Very stiff or hard	Can be indented by thumb nail	over 150	over 15

[a] Such a relationship should only be used as a preliminary evaluation of clay consistency, and should be reassessed at individual sites

Similarly many engineers would check, by physical feel, the strength of clay on site in addition to relying on tests. A typical example is shown in Table 3.10.

As with the interpretation of concrete cube results it is not likely that an experienced engineer would condemn a soil on the result of one test alone without examining the other data as well as the test and sampling procedure method. Nor would an experienced engineer reject the results of a sound test merely because it contradicted his preconceived assessment of the soil strength and characteristics.

This discussion is not meant in any way to denigrate or dismiss the very valuable research and theoretical analyses carried out with devotion over the past 50 years. But for such work it is likely that with heavier structures built on poorer ground there would have been far more foundation failures.

The object of this discussion is to caution young engineers not to be 'blinded by science' but to use critical assessment in applying the results to design.

3.15 FURTHER INFORMATION

In addition to the information on field testing given in section 3.6 and in the references at the end of the chapter, a further CIRIA publication *The Standard Penetration Test (SPT): methods and use* is currently being prepared for publication.

The sampling, testing and appraisal of soils are constantly being improved and updated. At the time of writing BS 5930 *Site Investigations* is currently under revision; there is also a substantial cross-industry initiative on site investigation being led by Professor Stuart Littlejohn of Bradford University, involving the major construction institutions, learned societies, trade associations and client organisations.

3.16 REFERENCES

1. British Standards Institution (1981) *Code of practice for site investigations*. BS 5930, BSI, London.
2. Dumbleton, M.J. & West, G. (1976) *Preliminary sources of information for site investigation in Britain*. Transport and Road Research Laboratory Report LR403, revised edition, Department of the Environment.
3. Weltman, A.J. & Head, J.M. (1983) *Site Investigation Manual*. CIRIA (Construction Industry Research and Information Association) Special Publication 25.
4. Proceedings of the European Symposium on Penetration Testing (II), Amsterdam, May 1982.
5. Institution of Civil Engineers (1970) *Conference on In Situ Investigation in Soils and Rocks*.
6. British Standards Institution (1990) *Methods of test for soils for civil engineering purposes*. BS 1377, BSI, London.
7. British Standards Institution (1986) *Code of practice for foundations*. BS 8004, BSI, London.
8. Terzaghi, K. & Peck, R.B. (1967) *Soil Mechanics in Engineering Practice*, 2nd edn. John Wiley & Sons, New York.
9. Nixon, I.K. (1982) *The Standard Penetration Test: A State-of-the Art Report*, Proceedings of the European Symposium on Penetration Testing (II), May 1982, Vols 1, 3 to 24. Amsterdam and Meigh, A.C. (1987) *Cone Penetration Testing*, (CIRIA), Butterworths.
10. Muir, R.J. & Wood, D.M. (1987) *Pressuremeter Testing*. (CIRIA), Butterworths.
11. Hazen, A. (1982) *Some physical properties of sand and gravels with special reference to their use in filtration*, Massachusetts State Board of Health, 24th Annual Report.

PART 2

SPECIAL AND FURTHER CONSIDERATIONS

CHAPTER 4

TOPOGRAPHY AND ITS INFLUENCE ON SITE DEVELOPMENT

4.1 INTRODUCTION

Topographical and physical features on a site can be a product of the sub-soil and below-ground conditions. An engineering interpretation of topographical features can give an important insight into these conditions and highlight potential problems for development. For example, the types of vegetation which have established themselves on the site, the vigour of their growth and colour variation, all express information on quality and type of soils and water content below surface level. Natural ground slopes can provide information on likely soil strengths; ground contours, depressions and ripples can indicate past and possible future ground movements. Surface deposits and obstructions can provide information on previous site use and possible man-made conditions to be overcome. Water ponding, streams or dry ditches can suggest areas where sub-soil conditions and strengths differ from surrounding areas. Additional site investigation should be implemented at these locations to confirm the nature and extent of variations.

Useful information can be collected from records, for instance, surface features from Ordnance Survey (OS) maps, sub-soil details from geological maps, below-ground services from statutory authorities, mining operations from British Coal/the Mineral Valuers Office, sub-soil conditions/substructure constructions from local authority building surveyors offices.

A site visit however, is essential to gain a real appreciation of actual conditions. A walk around the site and observations of the adjacent areas can give a better feel for the extent and nature of sub-soil checks which should be implemented. The type of developments on adjoining areas and how the constructions have performed, can also give valuable information when it comes to consideration of development proposals for the site.

Some features can be more easily seen from the air and aerial surveying techniques are useful for identifying fault lines, outcropping strata and buried features. Archaeological sites and mine shafts have been located from air surveys when ground level checks have proved inconclusive.

The following sections highlight points to look out for and the implications of various features. Some features can be the product of a combination of the causes discussed and it should be appreciated that the following comments are intended for general guidance only.

4.2 IMPLICATIONS FROM SURFACE OBSERVATIONS

An initial site inspection will reveal the obvious physical features occurring or influencing the site. The main points will fall into the following categories:

(1) Changes in level, ground slopes and movement.
(2) Mounds, depressions and disturbed ground.
(3) Past or current activities – mining, quarrying, filling, buildings.
(4) Vegetation.
(5) Surface ponding or watercourses.

These points are discussed individually although overlaps are inevitable.

4.2.1 Changes in level, ground slopes and movements

Abrupt changes in level may mean discontinuity of sub-soil conditions which will require localized investigations. The angle of slopes for stability varies and is dependent upon soil type. Granular soils such as sands have a natural angle of repose of approximately 30° while hard rocks can be vertical or overhanging. Rock strength is wide-ranging and for sedimentary rocks can vary greatly depending on the angle of the force to the bedding plane (see Fig. 4.1).

Cross-section (a) in Fig. 4.1 is less stable than cross-section (b), relative to the direction of the applied load. This is due to the weakness in the bedding plane to shear forces. Percolating water in the upper layers of the rocks shown in section (a) are also more exposed to frost and weathering than those in section (b).

Cohesive soils are influenced more by factors such as moisture content, which can vary and affect slope stability.

Steep slopes on site therefore, can be an indication of likely soil strength but it should be noted however that strong vegetation can stabilize slopes which would fail at such steep angles if unprotected. Conversely, removal of such vegetation can cause instability. Observations in less protected or wet areas will often reveal signs of movement in the form of embankment slippage or surface ripples (see Fig. 4.2), which would indicate that long-term stability of the slope is questionable.

Other signs of movements relate to shrinkage cracking, hillside creep or heave affecting surface coverings of pathways or fissures in the landscape. These often indicate clay

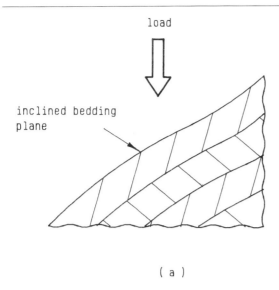

(a)

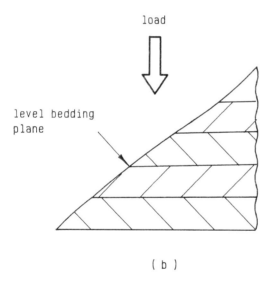

(b)

Fig. 4.1 Angle of force to bedding plane.

sub-strata sensitive to moisture changes or creep. Other surface movements can be the result of settlements from mining activities or brine extraction (see Chapter 6).

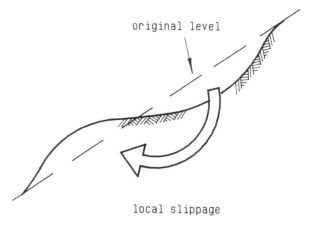

Fig. 4.2 Ground slippage.

4.2.2 Mounds, depressions and disturbed ground

Surface depressions may be produced as a result of the formation of swallow-holes which occur in chalk sub-soils. Water flowing through chalk strata forms voids in the chalk. These features are called swallow-holes, pipes and sink holes and tend to be related to topography. Steep slopes and drainage channels create concentrations of water flow which form voids in chalk. Some chalk areas contain numerous pipes with loose fills of clay, sand and flint debris (see Fig. 4.3). Typical diameters vary from 2 metres to 10 metres or more.

The pipes tend to be conical or cylindrical in shape and can be disturbed by adjacent constructions or changes to water drainage and collapse of adjacent voids. The installation of soak-away drains is commonly responsible for subsidence activity in chalk especially where the drain is in a cover of sand above the chalk. The conditions which develop in these debris filled pipes and voids is similar to that of fill containers discussed in Chapter 7 and reference should be made to that chapter for further information.

Surface observations, geological maps and historical records of the area should be used to provide information on the likelihood of collapse from these conditions. From this information the requirements and details of the ground investigation can be decided.

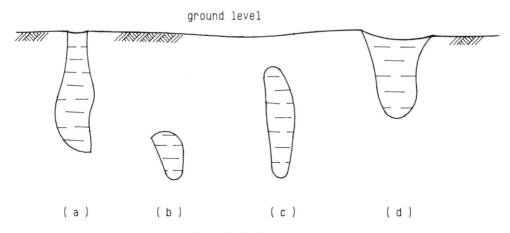

(a)   (b)   (c)   (d)

Fig. 4.3 Swallow-holes.

(1) Additional boreholes may reveal hidden pipes such as in Fig. 4.3 (b) which can prove expensive on a *blind* basis.

(2) It is far simpler to dig trial pits in the area of a proposed building, and these will reveal types (a), (c) and (d) in Fig. 4.3 which have the greatest effects on shallow footings.

Type (b) in Fig. 4.3 can then be catered for by designing reinforced footings to span over a notional future *soft spot* without the hit and miss expense of numerous boreholes.

Surface depressions can also be the result of bomb-holes or fallen trees. In wooded countryside recognition of bomb-holes and other ground disturbances are a little more obvious than in inner-city sites (see Fig. 4.4). On inner-city sites the disturbed contours have often been re-levelled but not in rural areas. The exploding bomb deposits disturbed ground on the circumference of the depression (bomb-hole), whereas the hole left by a fallen large tree causes disturbed ground to one side of the depression.

On developed sites it is more likely that the hole would have been levelled with fill material. The bomb-hole therefore, is likely to have a different fill in the top to that of the disturbed ground in the bottom and recognition of the true depth of the disturbed ground below the fill is not easy. This is due to the lower layers of the fill being very similar to the virgin ground (see Fig. 4.5).

Recognition of the location of such backfilled depressions relies upon observation of type and colour of vegetation, followed up with trial holes. A common surface filling of these depressions was fire ash with the depression being used as a convenient container for fire ash waste in built-up areas. The result of this activity has tended to favour vegetation in the form of nettles and other growths which do not rely upon a high-quality soil. Nettle growth has helped locate many backfilled holes and depressions during a site walkabout.

4.2.3 Past or current activities

MINING

The most obvious signs of past mining activities are the foundation and superstructures of winding gear and buildings along with dirt tracks and mounds of excavated debris.

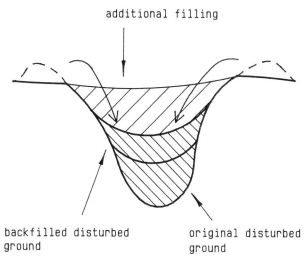

Fig. 4.5 Backfilled depression.

Many derelict mines however, have been levelled to the ground and topsoil has been imported and deposited over the original site. The only evidence for these situations at surface level is likely to be a variation in vegetation colour and vigour of growth. It is most important therefore, to make the relevant enquiries and desk studies recommended in Chapter 6.

Most mines will still have quantities of excavated material such as shale and poor-quality coal scattered around the area, which can be recognized from ground contours, vegetation and exposed shaley deposits. The implications of such observations are that it is possible and probable that mining has taken place and that addits, shafts, bell-pits, shallow or deep workings may exist below the site. The danger of subsidence, methane gas, combustion, collapse of shafts and underground tunnels should be investigated (see Chapter 6).

QUARRYING AND FILLING

Rock quarries are often partly or totally unfilled, and therefore the steep sides and access ramps are obvious remnants of past activities (see Fig. 4.6).

Sand and gravel quarries often penetrate below the water-table and lakes develop, of which many are now used for

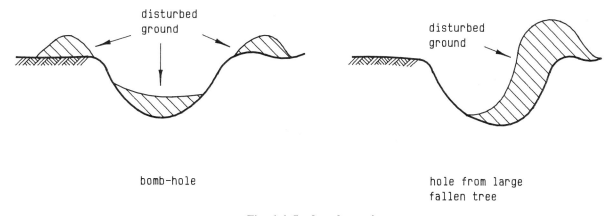

Fig. 4.4 Surface depressions.

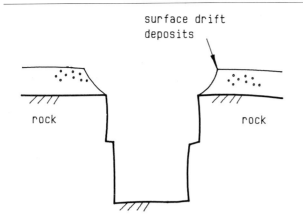

surface drift
deposits

rock   rock

Fig. 4.6 Rock quarry.

water sports. Other quarries however, have been used for tipping of refuse and rubbish and in some cases are totally filled back to the original ground level. Topographical evidence in such cases can often be seen in surface subsidence and cracking around the quarry/fill interface or in vegetation variations along a clear line at the edge of the filling. Also quarry waste or fill materials may litter the surface.

TIPPING AND FILL

Site mounds and hillocks are features usually indicative of possible surface tipping. In most cases on inner-city sites the fill will vary from soils and building rubble to household refuse. On inner-city sites old basements are often filled with demolition rubble, and this can be seen when the topsoil is scraped off the surface.

SHAFTS, WELLS AND CULVERTS

Surface identification of filled underground shafts is very difficult but vegetation variation may give some indication. In some areas shafts and wells may be visible at ground level, for example, mine shafts are sometimes kept open for groundwater observations or ventilation purposes, and these may have been extended upwards using brickwork or other forms of masonry. In the case of mine shafts it is often necessary to cap the shaft and isolate (sterilize) an area around the shaft from development. Wells, like shafts, can be filled, grouted and capped to prevent settlement (see Chapter 6).

Culverts and field drains are less obvious at ground level, but their presence should always be suspected on low-lying sites, boggy ground, or sites with names such as Marsh Lane, or Spring Fields, etc. The identification of the line of existing culverts usually relies on historical records for location, unless the outfall is on or near the site. Others may only be uncovered by excavation since no records exist of many of the old stone culverts used for drainage. The need to alter foundation designs locally over such culverts or to divert them means that their position can be critical to development and it is important to locate these lines as accurately as possible, or at the very least to include for possible diversion within the contract.

ADJOINING BUILDINGS AND RETAINING WALLS

Existing buildings and structures on or local to the site can be affected by future site development. It is therefore very important to record the locations and types of buildings for consideration in conjunction with the soil reports and other topographical observations. Existing foundations and retaining walls undermined by the reduced level dig of the new development will require underpinning. Moisture changes from dewatering in sensitive soils affect the loading capacity of adjoining sites, which in turn causes settlement. Buildings on soils sensitive to changes in stress from overlapping bulbs of pressure also restrict and affect the proposals for new developments.

Prior to the commencement of building works or development of the site it is advisable to prepare detailed records of the condition of existing structures so that any deterioration or damage can be identified.

The sensitivity of the occupants of any buildings to construction noise and vibration must not be overlooked. Hospitals, housing and offices are more sensitive than factories and warehouses, and the use of construction methods such as heavy sheet piling may be precluded, or at least the inclusion in the contract requirements of silent or vibrationless techniques may be required.

4.2.4 Vegetation

Vegetation growth relies upon plant food and an appropriate water supply. Different plants prefer varying soil conditions, some prefer acidic conditions, others alkaline conditions. Some like water-logged ground and others well drained soils. Disturbance of the ground changes soil drainage, imported fill changes the food and acidity from that of the surrounding ground. The results at surface level can vary from a total change of plant species, to colour and vigour variation in similar species. Reed grasses and willow trees prefer wet conditions, nettles will grow in ash and filled areas where other plants will not. Grasses often become yellow in poor food areas or dry soils. Aerial photography can identify even less obvious variations than ground level inspections, and such topographical observations can save endless abortive trial pit excavation by homing in on a likely location of a filled area.

4.2.5 Surface ponding or watercourses

Surface ponding, rivers and streams reveal valuable information relating to soil conditions and likely water-table levels. Eroding banks of streams and rivers reveal more direct information on the actual soils at these levels. The drainage of soil after rainstorms combined with stream embankment observation can create a picture of the ground conditions below.

For instance, free draining sub-soils such as sand and gravel will only pond if a clay or partly impervious topsoil overlies these sub-soil conditions or the water-table is very high.

EXAMPLE

One of the practice's engineers, having lived in the area of a particular site since childhood, remembered that when it was, in former times, a farmer's field there was often a pond in wet weather at one corner of the field.

The development in question was multi-phased and spread over many years, and the site investigation for the whole area had been carried out before phase 1. Phase 8 was partially over the area of the former pond, now landscaped as part of the developed works.

Because of this memory, a reluctant client was persuaded to pay for a further site investigation. This revealed considerable depths of peat below the former pond area, which the original site investigation had missed.

4.3 EFFECTS ON DEVELOPMENT ARISING FROM TOPOGRAPHICAL FEATURES

Following (or during) the site inspection the implications of the features observed can be considered with regard to development proposals. Dependent upon the stage of development plans it may be possible to introduce modifications to the buildings which will overcome difficulties or problems on the site. It may even be possible to exploit particular features, such as trees, ponds, or changes in level, by incorporating them into the overall development plan. The main points for consideration which arise from the topographical features are discussed below.

4.3.1 Sloping sites

The location of a building on a sloping site can be very important for both cost and function. Exploiting close contours where a change in level is required while keeping level areas to more widely spread contours can be effective to achieve an economical solution. However, it is necessary to consider this in the overall context of the development and other factors may have significant cost implications. Building on a sloping site will usually involve a cut-and-fill operation or retaining wall constructions or *stepped* foundations.

In the case of cut-and-fill, the level to be adopted for the ground floor slab of the development depends on a number of criteria. The main criteria is the level to suit the function of the building, and secondly, achieving an economical cut-and-fill operation. The economics are influenced by the ease of excavation of the sub-strata, the suitability of the removed material for re-use as filling, the cost of retention of the cut-and-fill faces and the implications on the building services and infrastructure. Where excavation of the materials is relatively simple and the material easy to compact, the cut and-fill can be balanced to avoid either importing of fill materials or removal of them from site (see Fig. 4.7).

Figure 4.7 shows a gently sloping site with a gravel sub-strata, in this case the materials can be balanced and re-used. In other cases where excavated material is not

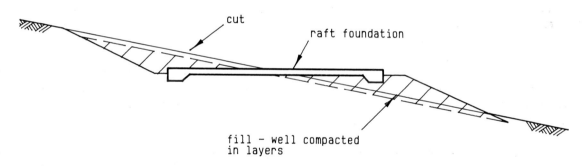

Fig. 4.7 Cut-and-fill — graded cut slope.

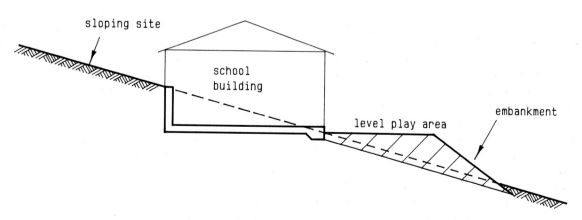

Fig. 4.8 Cut-and-fill — retained cut slope.

suitable for re-use, a balance between cutting and filling may not be achieved. From an economic point of view each site must be dealt with on its own merits, depending upon retention conditions, local disposal of excavated material, and availability of imported hardcore. The retention of cut-and-fill depends very much on the size of the site relative to the building. Retaining walls are much more expensive to construct than the cost of regrading the sub-soil materials, provided sufficient area is available to allow shallow regraded slopes to be achieved. Figure 4.8 shows an example of a school building constructed by the authors' practice.

Retention of materials by semi-basement walls/foundations is discussed in Chapter 15 but a possible conflict with mining requirements is mentioned here.

In mining areas a conflict can develop between the need to resist lateral pressure from the retained earth and the need to prevent and control lateral strains from the ground

due to mining activities. The details of basements and semi-basements in mining foundations become complex, and in many cases it is almost impossible to resolve the conflicting forces. In such areas double sided basements are best avoided and one sided retention only is preferable (see Fig. 4.9).

In the examples already considered, the cut-and-fill solution and retaining condition provide a level floor slab through the building. Changes in ground floor do occur and can be desirable in some buildings and in these situations changes or steps in the foundation can occur as shown in Fig. 4.10. Steps in the foundations can also occur in conjunction with level ground slabs, as shown in Fig. 4.11.

If *steps* in the foundation produce a significant change in the bearing strata then the introduction of joints through the building and foundation should be considered to avoid problems of differential settlement. From a buildability

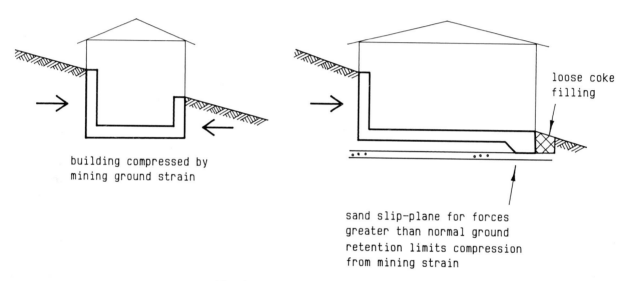

Fig. 4.9 Basement/retaining treatments.

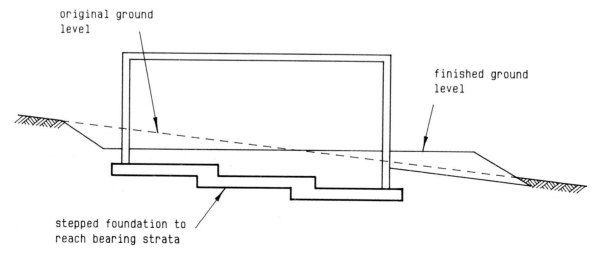

Fig. 4.10 Cut-and-fill — level slab/stepped foundation.

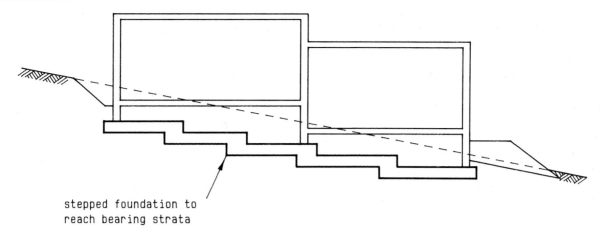

stepped foundation to
reach bearing strata

Fig. 4.11 Cut-and-fill − stepped slab and foundation.

point of view mass concrete step foundations are simpler to construct than reinforced concrete and therefore unless reinforcement is essential it should be avoided. The soffit of the foundation is often stepped to limit the tendency to slide. This requirement however, does not usually apply to ground beams between piles and piers, since they do not generally bear upon the ground and can be cast with a sloping soffit. It should be noted however, that where a foundation design combines the use of piles and partial ground support then the level of the underside of ground beams may be critical. The design of step foundations requires a balance to be struck to achieve buildability, economy and structural integrity.

Fundamentally the steps in the foundation should be placed as far apart as is practical. Where the ground slope is reasonably consistent the steps should be spaced with a regular going and rise to suit the dimensions of the super-structure construction, for example, if a brick masonry superstructure is to be built from the foundation the step going and rise should be based upon brick coursing and horizontal brickwork dimensions. Step positions should be set to avoid any intersecting foundations.

4.3.2 Slope stability

In addition to points already discussed a further important consideration when building on a sloping site is the long-term stability of the slope itself. Stability of cohesive soils on sloping sites requires detailed investigation to gather all the relevant sub-soil parameters to carry out slip circle analysis. Topography observations sometimes locate stability problems, such as a distorted fence line up the slope or ripples in the surface. Where the balance between adequate stability and inadequate resistance to the disturbing movement is triggered by the development, it is sometimes possible to transfer the vertical loads to a suitable lower level by the use of piling. It is important however, to prevent the transfer of loads via friction to the upper zone. It is also necessary to check stability of the lower levels where the load has been transferred.

When detailing foundations in cohesive soils of sloping sites it should be appreciated that certain locations are particularly sensitive to weathering and frost damage, such as the down slope edge of the foundation. In addition level foundations cut into a sloping site can reveal that the sub-strata at the reduced level at the back of the site is totally different to that at the front of the site. The variation can be such that a completely different strata with different consolidation, moisture content and bearing capacity are encountered. In addition the effect of cutting a level foundation into a sloping site is to reduce the length of the slip circle for shear resistance (see Fig. 4.12).

For sands and gravels the influence of loading tends to compact the granular materials to a denser consistency which improves the frictional resistance. However, when a level foundation is cut into the site, the surface area of slip and the mass resisting slip is much reduced in a manner similar to that of clay (see Fig. 4.12). Stability of sand slopes can therefore be sensitive particularly when shear resistance is critical. In addition the danger of surface erosion from water and wind demands a protective apron, particularly at the down slope edge of the foundation, to avoid undermining, especially where the sub-strata is fine sand.

As already mentioned the stability of rock slopes depends greatly on the angle of the bedding planes and water percolation through the rock. It is essential that changes in direction, folding, tilting and weathering are all recorded when slope stability is in question, so that measures such as rock anchors or other face protection methods can be assessed.

A further factor affecting stability is surcharge loading; the stability of a slope is generally adversely affected by loading placed upon it. In particular large point loads placed at shallow levels should be avoided. The design of the foundation for sloping sites should therefore reflect in the superstructure design to ensure the ground condition influences are considered within the general solution. Smaller point loads and uniformly distributed loads are often more easily dealt with, and these loads can be further distributed by the designer using rafts, step foundations or piles.

Soils are at their most vulnerable to the initiation of a collapse mechanism when level changes and embankments

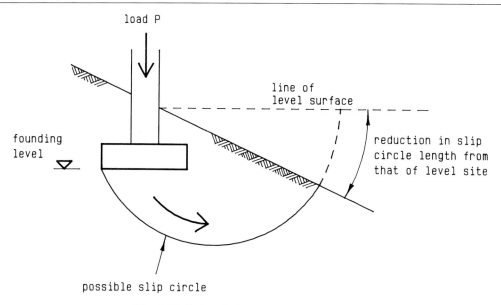

Fig. 4.12 Reduced slip circle on sloping site.

occur. Slip circle failure can be critical in embankments particularly when loaded by new constructions. (A typical failure mechanism is indicated in Fig. 4.13.)

It is important in such situations to ensure, by analysis, that the soil strength is adequate for the worst mechanism (i.e. that which gives the lowest safety factor). The analysis should be based upon thorough soil testing including the effects of moisture movement and drainage.

While the long-term stability of the slope is critical, further consideration must be given to the temporary conditions which may develop during the construction stage. Temporary works for foundation construction are not covered by this book other than where relevant to buildability. It is therefore only intended to deal with weakening effects in order to guide the engineer towards a buildable and economical solution.

On sloping sites it is generally less weakening to slope stability to excavate at right angles to the contours (see Fig. 4.14), provided that the normal timbering requirements for trench stability are in place. The reason for this is that excavations parallel to the contours tend to shorten the length of the slip circle resisting failure (see Fig. 4.15).

This weakening must also be taken into account in the permanent solution if such excavations are necessary. It is not only the length of the slip circle which is critical, for certain rock formations, the strength of the sub-strata is

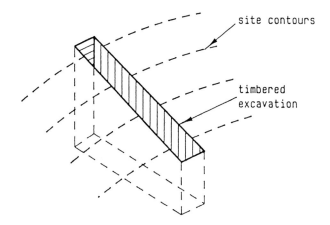

Fig. 4.14 Trench excavations on sloping site – at right angles.

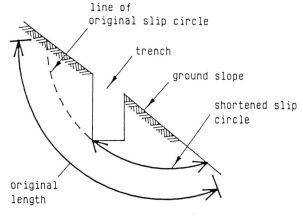

Fig. 4.15 Trench excavations on sloping site – parallel.

different relative to the direction of the bedding plane. In such cases a number of failure alternatives may need consideration, i.e. short failure lines through the strong axis and longer failure lines through the weaker axis.

Loose soils or soils containing silt veins or shrinkage cracks tend to become particularly unstable at changes in

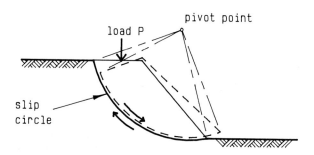

Fig. 4.13 Slip circle – embankment effect.

direction of trench excavations. These stability problems however, tend to be temporary and only apply during construction, the engineer should therefore be aware of them since they may influence the choice of foundation.

Similarly, stiff clays and strong rock formations can be criss-crossed with thin veins of silt through which moisture has percolated for many years. These veins form lines of weakness which when orientated critically to the excavation can be crucial to stability (see Fig. 4.16), especially temporarily during heavy rainfall.

Lenses of weaker material, cracking or faulting can be crucial to stability and it is therefore important to record sand lenses, silt veins, shrinkage cracking or faults in the soil investigation.

Slip circles are a wide subject and are often a civil engineering problem, not one for the structure/foundation designer, and so it is not proposed to labour the subject any further in this chapter. An example is included in Chapter 15 where the subject of basement wall/foundations is also discussed.

4.3.3 Groundwater

SANDS

The effect of water on sub-soil performance is such that water levels and water flow can be crucial to foundation design. Granular soils such as sands are weakened by water to the extent of halving the bearing capacity of stable sands. Running sand can develop due to the movement of fine sands and silts as a result of water flows which can cause the collapse of trenches and loss of material from below adjoining areas. These movements are not only crucial to the proposed excavation but also to the settlement of adjoining properties.

This is particularly the case when dewatering of sub-soil is implemented, since prevention of the loss of fines from the soil structure is crucial in preventing settlements of adjoining buildings. Soils which are not too fine (less than 0.006 mm) and relatively free draining are suitable for dewatering. A further option for some soils is to allow the groundwater to stabilize the trench and to concrete the foundations through the water with a tremie, but this only applies to soils which

do not run during excavation nor soften from contact with water in the trenches. Typically this would consist of firm to stiff wet clay or silt, the subject matter of the next section.

CLAYS

The effect of water content changes in clay soils is to produce ground heave or shrinkage (the best indicator to the likely shrinkage and heave characteristics of a clay-strata is historical performance, followed by PI (Plasticity Index)). The effect of trees on shrinkable soils through long dry summers has created a slight over-reaction to the shrinkage characteristics of soils and has caused the use of excessively deep foundations.

For example, on a site in the north of England and so not noted for shrinkable clays, mature trees were being removed and new ones planted following site development. The PI results were borderline, as small numbers of samples often are. The young engineer in charge decided wisely upon trench fill footings of the order of 2 m in depth. But additionally, due to concern about swelling and lateral movement of the clay, he also included compressible material down the sides of the trench fill. Such material is not only an unnecessary cost in itself in a non-shrinkable area, but is also labour intensive to secure in position. This negates the advantage of trench fill which is based upon simply digging a trench and filling it with concrete.

A common sense approach is therefore needed, based upon realistic data for the site location. Past performance of the soil along with information and observations from trial holes noting desiccation and fissures is most important. When collecting samples, it is crucial to maintain the moisture content by sealing each in a container immediately upon excavation. Tests on samples are valuable in assessing the likely characteristics of the soil, however observations on soil behaviour through periods of drought and in locations of low-rise buildings close to large trees are essential for a realistic assessment of likely sensitivity. Where similar buildings have been unscathed by cracking over previous years, the clay is unlikely to be sensitive.

Structures of high mass and small plan shape tend to move as one when subjected to swelling and shrinkage, these buildings give little away with regard to the characteristics of the soil. Smaller structures with sudden changes in mass stiffness or large plan shapes are much more sensitive. Cracking and movements tend to be more common on these types of buildings. It is therefore these types of buildings in particular which the engineer should inspect during the initial site visit.

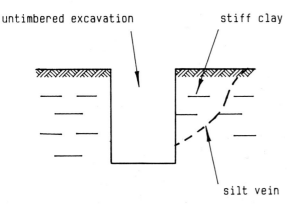

untimbered excavation   stiff clay

silt vein

Fig. 4.16 Trench excavation – stability.

PLASTICITY INDEX

The main basis for judgement of shrinkable soils tends to be the results of Plasticity Index tests (PI). The test for PI of soils is relatively crude but generally reasonably effective. However, the relationship between PI and shrinkage is not conclusive nor totally reliable. It is therefore essential that the engineer uses skill and judgement in gathering all the relevant information before making a decision on the combined results and not on the PI results in isolation.

Table 4.1

(a) Clay shrinkage potential

Plasticity Index (%)	Clay fraction (%)	Shrinkage potential
>35	>95	very high
22–48	60–96	high
12–32	30–60	medium
<18	<30	low

(b) Shrinkage potential of some common clays

Clay type	Plasticity Index (%)	Clay fraction (%)	Shrinkage potential
London	28	65	medium/high
London	52	60	high
Weald	43	62	high
Kimmeridge	53	67	high/very high
Boulder	32	—	medium
Oxford	41	56	high
Reading	72	—	very high
Gault	60	59	very high
Gault	68	69	very high
Lower Lias	31	—	medium
Clay silt	11	19	low

A guide to PI relative to shrinkage is shown in Table 4.1. Moisture changes occur in soils for many reasons. Some common causes are:

(1) Seasonal weather changes.
(2) Extraction of water by trees and other vegetation.
(3) Drainage.
(4) Flooding.
(5) Water-table fluctuations.
(6) Dewatering.

Many of the above causes can be altered or controlled and can therefore be influenced by the design and the actual development undertaken.

The effect of seasonal weather changes can be minimized by preventing soils from drying out in hot weather. Concrete pavings and skirts around buildings help prevent drying out and frost damage to perimeter foundations. The removal of trees early enough allows the balance of moisture to return to the ground prior to commencement of substructure construction. Avoiding development near existing trees can prevent moisture changes from this source affecting foundations.

Major changes in drainage of sensitive soils should be avoided if possible. If land drainage is essential to a development it should be carried out sufficiently in advance to allow a new equilibrium of moisture content to be reached. The control of groundwater on site during construction by pumping or dewatering should take into account the effects on sensitive soils. It is advisable to avoid prolonged pumping which may promote shrinkage of sensitive strata.

The effect of groundwater on the foundations of a development can also be wide ranging. Watercourses above ground can be observed and measured during surveys and are generally recorded on maps and other records. The

extent of underground watercourses is less obvious and often less well recorded. Above ground the watercourses can vary greatly depending on seasonal changes and rainfall. It is therefore not only important to gather information from maps, surveys and other records but also to contact river authorities and others responsible for watercourses who may have valuable information on flow, flooding, pollution, culverts, drainage and other local knowledge.

Watercourses, tidal effects and water fluctuations can cause:

(1) Structure flotation.
(2) Basement flooding.
(3) Reduced effective pressure.
(4) Erosion and scouring of sub-strata and foundations.

All these effects must be considered in the design of foundations.

The diversion of streams through culverts from within to around a development will create backfilled areas to be overcome within the foundation design. In addition the new watercourse container must ensure that no water continues on the old familiar route. The diversion must be total and permanent with no danger of erosion or leakage which may eventually affect foundations.

4.3.4 Settlement

Further considerations for substructure and superstructure performance relate to potential settlement problems.

SWALLOW-HOLES

Where ground investigation indicates only a remote risk of collapse from swallow-holes, normal foundations are appropriate. If an investigation reveals some risk of void migration and relatively small-diameter collapse, raft foundations designed to span such collapses are appropriate. In addition any voids found during investigation can be grouted to prevent collapse. Where the risk is greater, a substantial development would require a system of grouting voids and/or sleeve piling. The danger of piles shearing off during sub-soil collapse must not be overlooked and steel sleeving can increase robustness as well as allowing slip to occur. In addition to foundation precautions it is most important that no soakaways are installed within a risk zone and any drainage should be carried clear of the site before discharge.

MINING

Mining operations are also a major factor affecting potential settlements on a site. Inspection of the site may have revealed evidence of past workings and discussions with local people may reveal further information, some not well remembered but nevertheless useful. In the authors' opinion it is always worth taking the time to speak to locals passing the site, especially the older ones; information can be inaccurate, but often provides a clue for further investigation.

If mining has taken place in the area it is possible that a shaft may be located on the site. British Coal and other mining authorities keep reasonable records of most shafts. Other shafts, wells and old culverts are less well recorded, but previous advice on gathering information applies. It is inevitable however, that some shafts, culverts and wells will only be discovered during development excavations. It is most important to approach recorded information with caution since some records are based upon poor memory, others on poor quality measurement, some on mistaken mapping, but some are reliable. When excavating for a shaft in a recorded location it is important not to assume too quickly that recorded information is wrong simply because excavation has not revealed the existence of a shaft. It is equally important not to rely totally on the recorded information. If the shaft is not visible or obvious at ground level the recorded location should be marked on site and a surface scrape down to virgin substrata should be made. This scrape should pass over the shaft location into the surrounding ground, to observe any evidence of disturbed or filled ground which would indicate any previous excavation or filling. Should this scrape reveal only virgin ground throughout the area, then the area can be extended and widened until the actual shaft is found by the evidence of fill material. The fill can then be excavated to locate and check that this is in fact the shaft or past mine workings. A similar approach should be adopted for wells and culverts as for shafts.

The treatment of wells and shafts requires safety precautions to be observed during excavation, since collapse can be sudden and devastating. British Coal has special precautions to be observed when dealing with mine shafts and similar precautions should be adopted when dealing with wells. Backfill in shafts and wells tends to settle and void migration progresses towards the top of the aperture. In deep shafts the suction forces produced by sudden collapse of the fills can suck down a funnel of material from the

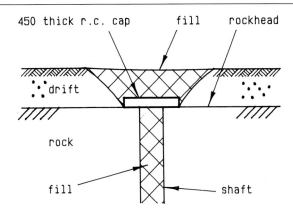

Fig. 4.17 Mine shaft filling and capping.

surrounding ground surface, taking machinery and equipment with it. To reduce this risk to an acceptable level for development, the fill should have any voids grouted and shafts or wells should be sealed with a reinforced concrete cap at rockhead (see Fig. 4.17).

4.4 SUMMARY

It can be appreciated from the points covered in this chapter that a visual inspection together with checks on recorded information is a vital point of the site appraisal and a prerequisite to trial pits, borehole investigation and sub-soil testing. The information obtained from each of these sources has implications on each of the other investigations which are carried out. Also the interpretation of information from each source is influenced by the inevitable overlap with other sources, and the engineer should bear this in mind at all times during the site appraisal/investigations.

CHAPTER 5

CONTAMINATED AND DERELICT SITES

5.1 INTRODUCTION

The general shortage of *good* sites for development purposes and the decline of industrial works in some areas has led to old, abandoned, industrial sites and waste dumps being considered for redevelopment. Engineers are being asked to investigate sites for building development which may have previously been considered unsuitable.

In some locations it may be political decisions which have provided the incentive to develop and reclaim an abandoned industrial site, particularly in regenerating the inner-cities, even though reclamation costs may be high. The redevelopment of such sites has the advantage of bringing back into use derelict land, thus providing local facilities, improving the urban landscape and avoiding the need to use scarce green field sites. The *total* cost to society may, overall, be lower since the infrastructure (roads, transport services, schools, hospitals, etc.) generally already exists.

It is inevitable that these 'less desirable' sites will require some special consideration and treatment to ensure satisfactory long-term use. The extent of this special consideration and treatment will be dependent upon the previous use of the site and the proposed end use. The problems and treatments associated with mineral extraction beneath the site is covered in Chapter 6, and the purpose of this chapter is to deal with problems arising from other forms of dereliction and contamination produced as a result of earlier above-ground operations. The range of dereliction and contamination is extensive. Typical examples include old gas, power or sewage works, landfill sites, abandoned iron and steel works, scrap yards, chemical works, and household refuse tips.

Derelict buildings or demolition materials on site may include asbestos, or other hazardous materials, and special precautions are necessary to remove these materials safely. Table 5.1 lists some commonly encountered contaminants and gives details of sites where they are likely to occur and the principal hazards they produce.

Researchers of many disciplines have been actively looking at the problems but little work appears to have been done on the solutions. Ultimately it is the engineer who has the responsibility to determine the best and safest method of reclamation and use.

Some industrial countries (such as Germany, Holland and the USA) have suffered severe problems from unsuitable reclamation and inadequate site treatment resulting in high incidences of illness being reported. It is therefore essential that the engineer should be aware of the difficulties of adequate assessment and treatment of contaminated sites. Besides the safety of eventual users or residents of the sites, construction workers are particularly vulnerable to any harmful chemicals that may be present. Construction workers can face the risk of explosion from methane, being overcome by fumes, and health risks from exposure to carcinogens and other toxic chemicals.

There are well-known precautions that workers should take such as the wearing of protective clothing and respirators. Making sure that precautions are observed at all times is difficult. It should also be appreciated that workers can be at risk in the act of gathering the samples from site from which the appropriate precautions can be determined. The risks involved in the development of derelict sites is made greater by the fact that most abandoned sites and contaminates are considered to be a harmless nuisance and a false sense of security develops which causes a careless approach to investigation and treatment.

5.1.1 State of the art

Compared with soil mechanics, geology and foundation design the risks and treatment of contaminated sites is in its infancy. The available information on acceptable thresholds, short-term and long-term risk, site solutions and treatments and code guidance is constantly being updated, added to and revised. It is therefore essential that in conjunction with the information provided in this book the engineer reads, digests and applies the latest guidance and information.

Currently, legislation is being developed and evolved with regard to a proposed Register of Contaminated Land. This is as a result of the provisions made under Section 143 of the Environmental Protection Act 1990 and other previous Acts of Parliament.

The Register of Contaminated Land proposed is a register, compiled by local authorities, of land within their area that has been used or is currently being used by an operation of a contaminative nature, for example, existing chemical installations or previous sites of steel works now redundant.

The Register will be compiled from archive sources of information currently within the possession of the local authority planning and environmental health departments and it is anticipated that it will be in force in the near future.

Table 5.1 Contaminant type and hazard (reproduced with permission from ICRCL Guidance Note 59/83, 2nd edn, 1987, Table 1)

Type of contaminant	Likely to occur on	Principal hazards
Toxic metals e.g. cadmium, lead, arsenic, mercury Other metals e.g. copper, nickel, zinc	Metal mines, iron and steel works, foundries, smelters; electroplating, anodizing and galvanizing works; engineering works, e.g. shipbuilding; scrap yards and shipbreaking sites	Harmful to health of humans or animals if ingested directly or indirectly. May restrict or prevent the growth of plants.
Combustible substances e.g. coal and coke dust	Gas works, power stations, railway land	Underground fires
Flammable gases e.g. methane	Landfill sites, filled dock basins	Explosions within or beneath buildings
Aggressive substances e.g. sulphates, chlorides, acids	*Made ground* including slags from blast furnaces	Chemical attack on building materials e.g. concrete foundations
Oily and tarry substances, phenols	Chemical works, refineries, by-products plants, tar distilleries	Contamination of water supplies by deterioration of service mains
Asbestos	Industrial buildings; waste disposal sites	Dangerous if inhaled

The main purposes of the Register will be broadly as follows:

(1) To identify areas of land subject to contaminative usage.
(2) As a means of controlling future development in respect of approving planning applications made to the local authorities. For example, before approval is given the local authority will have to satisfy itself that the intended development is suitable for a particular site.
(3) As a reference source of information to be used by other statutory, commercial and private organizations in establishing the source of migrating contamination. It is envisaged that the main user will be the National Rivers Authority in preventing pollution of watercourses.

With reference to site investigations, the Register will be open to any enquiry to confirm if the area of land is included in the Register and consequently if it is classed as contaminated. Future site investigations commissioned by developers will still remain confidential between consultants and developers. However, when the developer seeks approval from the local authority planning department the results of any site investigation may be requested for inclusion in the Register. Evidently this may have a consequential effect on land values.

With regard to treatment of contaminated sites identified by the Register, legislation will continue to be developed in consultation with interested organisations. However, currently the local authorities do have powers under Section 79–82 of the Environmental Protection Act 1990 to investigate statutory nuisances such as contaminated sites and to take action where appropriate or necessary.

In addition to dealing with the technical aspects of the problems arising from contamination of a site there are also the implications of the legal and financial responsibilities. The legislation will put the onus for dealing with contamination problems upon the owner or site purchaser. The engineer should therefore consider the effects of advice he may be giving to a prospective land purchaser since significant financial implications may arise in complying with the legislation. This chapter should therefore be used for guidance on principles rather than definitive criteria and solution.

5.1.2 Contamination implications
A site is considered to be contaminated if it contains chemical, physical or biological agents that may cause a nuisance, danger or health risk either during the development and construction stages, or in the longer term to end users of the site.

The word *contamination* tends to suggest hazardous conditions and is perhaps an emotive word often creating overreaction by the public and engineers alike. In some cases the site contamination may be no more than redundant shallow foundations of a previously demolished development which can be dealt with relatively economically and simply, and, too, presents no hazards to construction operations or end use. Chemical contamination of sub-soil can have occurred but at such a low level as not to create hazardous conditions. It should be appreciated however, that the same chemical material can be safe in certain conditions but hazardous in others, such as chemicals affected by water in a high or variable water-table.

The problems to be considered arise from the remains of previous site use or building operations which have left behind foundations or filled areas which will produce obstructions to new construction. Similarly industrial or chemical processes carried out on the site may have produced waste products which have been left on site in the form of unstable fills, obstructions or toxic material in the substrata. It must be remembered that the industrial revolution started in earnest over 150 years ago and while many companies may have long since ceased to exist their legacy of dumped waste and toxic by-products can remain active within the sub-soil.

To appreciate the range of contaminated sites and the implications of dealing with the subsequent problems which arise, it is perhaps easier to deal with these under two headings:

(1) *Physical obstructions* − non-toxic or hazardous, i.e., redundant foundations and services − (section 5.2).
(2) *Chemical and toxic contaminants*, i.e., risk to humans, animals, plants or building materials − (section 5.3).

5.2 REDUNDANT FOUNDATIONS AND SERVICES

On its simplest level consider a new housing development on an abandoned area of previously demolished houses. It is likely that these original houses were constructed off shallow masonry spread footings and some houses have cellars under part of the dwellings, the mains services, gas, water, electricity and sewers being located in the roads and pavements.

Demolition of the original houses would normally be carried out down to ground level only, with the footings left in place and the cellars backfilled with demolition material of doubtful quality inadequately compacted to support new construction. The new layout of houses and roads will almost inevitably be arranged such that the new houses straddle lines of demolished houses and roads. Consideration of these constraints are necessary in the structural design of new works above and below ground level.

5.2.1 Identification

To identify a site containing redundant foundations and services an indication of the extent of the problems can be determined by examination of record information. The Ordnance Survey maps provide reliable historical information on old street and housing layouts and information on old or redundant services can be obtained from the respective authorities. Previous treatment of redundant services can also be investigated and confirmed. On completion of an inspection of the record information it is useful to produce an overall layout drawing to show all relevant information relating to below ground obstructions as follows:

(1) Location of buildings and processes carried out.
(2) Existing foundations, plant bases, chimney bases, water towers, foundations, etc.
(3) Areas of tipping and filling.
(4) Lines of disused drainage and services.
(5) Extent of disturbed ground and boundary of virgin soils.

If from this information it is apparent that the contami-nation is likely to be obstructive rather than hazardous, the trial holes and boreholes can proceed to check the reliability of the desk-top study by checking a number of key obstructions, dimensions, directions and material information.

By overlaying drawings of record information and site investigations with proposed layouts, these can be used to build up detailed information for use in determining both foundation and superstructure proposals. From this process costings can be carried out more realistically on alternative treatments or constructions to achieve the most economical foundation.

A similar exercise should be carried out for the development over a derelict industrial area. Obstructions from old plant and large buildings may pose more extensive contamination problems and the selection of an economical foundation treatment will require a more detailed desk study of record information.

5.2.2 Sampling and testing

The main investigations relating to redundant foundations and services involve trial hole excavation to locate and record the actual positions and extent of obstructions which may affect the development. It is important to remember that normal soil samples and testing will be required to establish soil strength, settlement criteria, and soil consistency. This may be required for both the disturbed and virgin soils on site. In addition a check should be made on soil and water samples for naturally occurring chemicals which can affect building materials (e.g. sulphates). These items are discussed in section 5.3.

5.2.3 Site treatment

(1) If the old foundations are at shallow depth they can be removed (*grubbed up*) where they create obstructions to new foundations. Removal of these footings or breaking them down to a minimum of 500 mm below surface level would be advisable under floor slabs to avoid *hard spots* which can cause cracking in the ground floor slabs.

(2) In areas of old cellars, the removal of unsuitable backfill and replacement with properly compacted hardcore material can support new floating floor slabs. Alternatively suspended slabs may be adopted to avoid replacement of cellars fills, but it should be remembered that any material in the fills which may rot or deteriorate and cause a health hazard will require removal.

(3) If filled cellars are extensive and fill stable then treatment of the fill with ground improvement methods (see vibro/dynamic consolidation in Chapter 8) can be made and the tops of obstructions removed to accommodate raft or reinforced footings.

(4) The installation of joints in sub- and superstructures to reduce the effective size of units can produce a more economical foundation solution. Joints located at changes in ground conditions or over obstructions can also produce a more economical solution than designing a structure to straddle problem areas.

(5) Where large or irregular obstructions occur then partial or total removal may prove necessary. After removal of obstructions the use of a pile and beam or pier and

beam foundation can be used to reach suitable bearing strata. Suspended ground floor constructions can also be used to avoid the need for load-bearing backfill. Pile installation can be seriously impaired by obstructions in the ground and this factor should be investigated if pile foundations are under consideration. An alternative treatment is to import material to overlay the obstructions and provide a cushion or blanket to accommodate a raft and suitably designed superstructures.

(6) Old tanks, chambers and voids should be checked and cleared of chemical contamination, and be filled if not removed. Breaking through the tops of chambers will enable hardcore backfilling to be placed or alternatively p.f.a./cement injection treatments can be used.

5.3 CHEMICAL AND TOXIC CONTAMINANTS

5.3.1 Human and animal risk

The risk to humans and animals from toxic contaminants occurs through ingestion or contact, inhalation of fumes, dust or gases and explosion or combustion. Children tend to be more sensitive than adults and more exposed because of careless habits both dietary and behavioural. Edible plants can absorb metal in quantities dangerous to humans and animals. Grazing animals ingest appreciable quantities of soil, which can be direct ingestions of soil, contaminated drinking water or plant food. Skin contact may lead to absorption, chronic skin effects or acute skin irritation.

If chemical contamination is judged particularly toxic on a site, then it should be regraded to control run-off, water mains should be protected, gas and water migration controlled, dust suppressed, the site cleared of visible contamination and warning signs and perimeter fences erected. The local authority, water, gas, electricity, police and fire services should be informed, as necessary, and contact names left for emergency information. The extent and nature of contamination must be fully determined and suitable treatment devised. As previously stated the treatment of sites contaminated with toxic materials is still in its infancy but engineers are likely to meet such sites in increasing numbers in the future.

In addition to contamination by man there are other naturally produced gases which can be a danger to health. Radon is a radioactive gas naturally produced which needs special equipment to detect it, since it is colourless and has no smell or taste. Radon comes from the radioactive decay of radium which in turn originates from the decay of uranium. Small quantities of uranium are found in all soils and rocks. It is particularly prevalent in granite. Exposure to radon increases the risk of lung cancer.

Other hazardous gases can be present on derelict sites. For instance landfill sites can produce methane and carbon dioxide from the decay and chemical breakdown of the fill materials. Chemical reaction within the ground can also produce hazardous gases. The characteristics and effects of some gases which may be present on derelict sites is given in Table 5.2.

A further hazard is the combustion or the presence of potentially combustible materials below ground. Underground combustion has occurred in colliery waste materials where exothermic reactions have contributed to self-ignition. The burning of combustible materials underground leaves voids which may collapse later and result in settlements of surrounding sub-soils. During the combustion process, gases will be produced resulting in volume changes. The gas production creates hazardous conditions for buildings located on the site, its occupants and for construction personnel.

5.3.2 Plant risk

The toxic effect of harmful substances on plant life is not directly a technical problem for building works but affects gardens, landscaping, play areas, etc., and can be damaging to humans and animals (see section 5.2.1). The effect on plant life can cause a build-up of toxic substances at ground level due to the annual die back of plant life. The effect on landscaping can be unsightly and expensive to rectify. The engineer in his topographical surveys may find the damage to plants (e.g. a poor yellowed weak growth) a warning to possible toxic site conditions and a cautious approach should be adopted.

5.3.3 Risk to buildings

The composition of materials used below ground level in the construction of the foundations and services to buildings is such that deterioration can occur if contact with corrosive conditions occur. The deterioration of materials above ground level can be seen and monitored and engineers are familiar with the need to consider special precautions in extreme exposure conditions. Durability of concrete for example, has been scrutinized in the recent past and this has resulted in increased cement contents and concrete cover to reinforcement, to ensure satisfactory longer term performance of individual elements and the building as a whole. It can be appreciated therefore how much more critical are the foundations, which cannot be readily seen or monitored, yet support the total structure.

The durability of the sub-structure is obviously an important factor and appropriate British Standards give recommendations for timber, steel, concrete and masonry used below ground level. The following section is intended to give additional guidance and background when dealing with corrosive soils.

In the main it is the presence and movement of groundwater which activates attack, by carrying the corrosive material in solution into contact with the foundation. Once in contact with the surface of a foundation or drawn into it by capillary action, then a chemical reaction can develop.

It is not unusual during site investigation operations to confirm a water-table below the level of proposed foundations which consequently rises due to seasonal and long-term variations. Leakage from drains and services and local rain water run-off may carry chemicals in solution. Finally, it should be appreciated that corrosive conditions can occur on virgin sites and not only on derelict sites.

Some chemicals which can cause deterioration of the building materials used below ground are listed below and protection against deterioration is discussed in section 5.4.

Table 5.2 Characteristics and effects of hazardous gases (Leach & Goodger, *Building on Derelict Land*, CIRIA SP 78 (1991))

Gas	Characteristics	Effect	Special features
Methane	colourless odourless lighter than air	non-toxic asphyxiant	flammable limits 5–15% in air can explode in confined spaces toxic to vegetation due to deoxygenation of root zone
Carbon dioxide	colourless odourless denser than air	toxic asphyxiant	can build up in pits and excavations corrosive in solution to metals and concrete comparatively readily soluble
Hydrogen sulphide	colourless 'rotten egg' smell denser than air	highly toxic	flammable explosive limits 4.3–4.5% in air causes olfactory fatigue (loss of smell) at 20 p.p.m. toxic limits reached without odour warning soluble in water and solvents toxic to plants
Hydrogen	colourless odourless lighter than air	non-toxic asphyxiant	highly flammable explosive limits 4–7.5% in air
Carbon monoxide	colourless odourless	highly toxic	flammable limits 12–75% in air product of incomplete combustion
Sulphur dioxide	colourless pungent smell	respiratory irritation toxic	corrosive in solution
Hydrogen cyanide	colourless faint 'almond' smell	highly toxic	highly flammable highly explosive
Fuel gases	colourless 'petrol' smell	non-toxic but narcotic	flammable/explosive may cause anoxaemia at concentrations above 30% in air
Organic vapours (e.g. benzene)	colourless 'paint' smell	carcinogenic toxic narcotic	flammable/explosive can cause dizziness after short exposure has high vapour pressure

(1) *Sulphates*. Solutions of sulphates can attack the hardened cement in concrete and mortar. Sulphates occur mainly in strata of London Clay, Lower Lias, Oxford Clay, Kimmeridge Clay and Keuper Marl. The most abundant salts are calcium sulphate, magnesium sulphate and sodium sulphate. Sulphuric acid and sulphates in acid solution do not occur very often but may be found near marshy ground or colliery tips where the soils contain pyrites which are being slowly oxidized. As well as occurring naturally, sulphates are sometimes present in fill materials such as ash or shale.

Water movements passing through soils containing sulphates dissolve the salts which can then be carried in solution into contact with concrete or masonry elements. Water movements can be vertical or horizontal depending on site geology and seasonal variations. The chemical reaction between sulphates and cement occurs and deterioration of the structure ensues. Further supplies

of contaminated water lead to a continuing deterioration of the element until complete breakdown and failure can occur (see Table 5.10).

(2) *Phenols*. Aqueous solutions of phenols can attack plastics in the ground. Phenols are a group of chemical compounds which are derived from benzene (i.e. oil based). The common name for phenol is carbolic acid. Phenols come in many forms and concentrations, therefore it is difficult to give a clearly defined nature for these materials. They generally attack plastic and rubber based products and can have a detrimental effect on concrete if they are in a concentrated form. Phenols give off powerful fumes which can be dangerous to man in confined areas and are phytotoxic (i.e. they kill plants). Phenols have diffused through plastic water mains and tainted the water supply, without damaging the pipes.

Unlike some forms of contamination, phenols are

degradable and will disperse in the long term if the source of contamination is removed.

(3) *Metallic contaminants*. The risk of corrosion through electrolytic action can be increased by metallic contamination in the ground. Metallic pipework in underground services, sheet piling and other metal components in contact with the soil may become locally corroded by electrolytic reaction with dissimilar metals present in the ground.

(4) *Gases*. The presence of gases or combustible materials in the ground may not directly affect the building materials used in construction, however the effects on construction workers and future occupants requires consideration. Explosive risk or settlement effects on the building should also be considered.

5.3.4 Toxic contamination – site identification

In order that the engineer can identify potentially toxic contaminated ground he must be aware of the likely sources of the toxic materials and consider this during the study of record information of the site. Reference to historical data may give some indication of previous site usage, and when this usage can be linked with operations which produced or used potentially toxic materials then the site testing should be extended to check for the type or types of contamination. The *normal* site investigation approach and techniques are discussed in detail in Chapter 3, the following is therefore intended only to supplement that information.

A general site inspection can provide an indication of problems and variations within the site, for example, unusual odours, discoloured soil surfaces or water can be an indication of contamination. The type of vegetation or lack of it can suggest potential ground contamination. Deep rooted trees and heavy vegetation would not indicate high toxic levels whereas a more barren or yellowed appearance might suggest problems.

To assist the engineer in the identification of potentially contaminated sites, Table 5.3 gives details of the various industries, sites and contaminants.

In addition to manufacturing processes depositing chemical materials in the ground, fall-out from air-borne pollution can also cause contamination. Therefore land adjacent to former factories can become contaminated. Recent examples encountered by the authors' practice are:

- Heavy metal contamination around a redundant foundry,
- Fluoride contamination adjacent to a clay works,
- Asbestos contamination of disused railway sidings, attributed to the braking systems of the rolling stock.

Groundwater movements, in addition to mobilizing and activating toxic materials, can produce toxic solutions and gases. For example, decomposition of refuse material produces methane which can also occur naturally in peat bogs, etc. In addition to such gas emissions, other emissions due to spontaneous combustion of landfill sites, colliery spoil, coke and coal storage areas can occur and prove hazardous, as can gas leakage from abandoned services and mine workings.

A contaminated site does not necessarily mean that the site is unsafe or unusable; the application of sound engineering principles can solve the problems.

5.3.5 Contaminant investigation

Following an inspection of record information it is essential to produce an overall layout drawing on which can be shown relevant information with respect to potential contamination including the following:

(1) Location of buildings and the processes carried out.
(2) Areas of tipping or filling.
(3) Lines of old watercourses and vegetation.
(4) Existing/disused drainage runs.

From this information it is usually possible to zone the site into areas of high and low risk and then to produce an investigation borehole layout. Sampling from boreholes, rather than trial pits, is advisable in the first instance since trial pit excavations can expose site investigation personnel to toxic materials. The samples from boreholes can be handled more safely at this stage. On a recent investigation, in an area of an old tannery, checks were made for anthrax spores. The site personnel wore protective clothing and breathing apparatus during the work in these areas and the boreholes were sealed upon completion of sampling.

The on-site investigation should be arranged in stages so that information obtained is used to direct and determine further works. Initially a grid of boreholes, spaced at 30–40 m centres, should be used to give an indication of sub-soil contamination.

Analysis of the initial results will identify further areas for borehole investigation, such as closing down the centres as appropriate to 3–5 m if necessary. The information from tests from samples taken from the boreholes should then be used to produce layout drawings showing the extent, degree and levels of any contamination.

These layouts should be colour coded to indicate the degree of contamination and contours can show the extent of contaminated material at various levels below the site surface. Separate drawings can be used to show differing contaminants if necessary. The test information should be related to the original site layout of buildings, filled areas and watercourses, to identify any anomalous or inconsistent results.

5.3.6 Sampling and testing

It is recommended that samples should be analysed for contaminants with some portions of the samples being stored for further examination, if necessary. The principle hazards and minimum analysis to be included in investigations are shown in Tables 5.4 and 5.5.

Samples taken from sites of potential historical contamination should be tested and analysed as shown in Table 5.5.

The engineer should update these requirements on a regular basis to keep abreast of the latest information. Advice can be obtained from the specialist testing authorities.

Since water movements are usually contributory to problems arising from contaminated ground, monitoring of

Table 5.3 Potential contaminated sites related to past usage (BSI, DD 175: 1988)

Important note   This table should not be taken to mean that other types of site need not be investigated nor to mean that other contaminants are absent (see table footnote below).

Industry	Examples of sites	Likely contaminants
Chemicals	Acid/alkali works Dyeworks Fertilizers and pesticides Pharmaceuticals Paint works Wood treatment plants	Acids; alkalis; metals; solvents (e.g. toluene, benzene); phenols, specialized organic compounds
Petrochemicals	Oil refineries Tank farms Fuel storage depots Tar distilleries	Hydrocarbons; phenols; acids; alkalis and asbestos
Metals	Iron and steel works Foundries, smelters Electroplating, anodizing and galvanizing works Engineering works Shipbuilding/shipbreaking Scrap reduction plants	Metals, especially Fe, Cu, Ni, Cr, Zn, Cd and Pb; asbestos
Energy	Gas works Power stations	Combustible substances (e.g. coal and coke dust); phenols; cyanides; sulphur compounds; asbestos
Transport	Garages, vehicle builders and maintenance workshops Railway depots	Combustible substances; hydrocarbons; asbestos
Mineral extraction Land restoration (including waste disposal sites)	Mines and spoil heaps Pits and quarries Filled sites	Metals (e.g. Cu, Zn, Pb); gases (e.g. methane); leachates
Water supply and sewage treatment	Water works Sewage treatment plants	Metals (in sludges) Micro-organisms
Miscellaneous	Docks, wharfs and quays Tanneries Rubber works Military land	Metals; organic compounds; methane; toxic, flammable or explosive substances; micro-organisms

Ubiquitous contaminants include hydrocarbons, polychlorinated byphenyls (PCBs), asbestos, sulphates and many metals used in paint pigments or coatings. These may be present on almost any site.

groundwater levels and sample analysis over a period of time is invaluable in determining final site treatments. Piezometers should be installed as soon the investigations commence, and, if possible, left in place until development starts.

Scrap metals and other visible evidence of metals in the ground can be found by trial holes and inspection. Heavy metals in dust or in solution should be detected from tests on soil and water samples taken from boreholes during the ground investigation.

The previous use of the site and smells are the best guide to potential problems from gases on the site, but it should be remembered that some dangerous gases are odourless (see Table 5.1). Gases can also migrate from adjacent sites so the surrounding areas must be researched to check for old tips or quarries. Gases can be present in pockets on site as a

Table 5.4 Principal hazards and contaminants (based on Table 2 in ICRCL 59/83, 2nd edn, 1987)

Important note Consideration should be given to the inclusion of other contaminants if the site history has identified former uses likely to have introduced them.

Hazard (see Note 1)	Typical end uses where hazard may exist	Contaminants
Direct ingestion of contaminated soil by children	Domestic gardens, recreational and amenity areas	total arsenic total cadmium total lead free cyanide polycyclic aromatic hydrocarbons phenols sulphate
Uptake of contaminants by crop plants (see Note 2)	Domestic gardens, allotments and agricultural land	total cadmium (see Note 3) total lead (see Note 3)
Phytotoxicity (see Notes 2 and 3)	Any uses where plants are to be grown	total copper total nickel total zinc
Attack on building materials and services (see Note 2)	Housing developments Commercial and industrial buildings	sulphate sulphide chloride oily and tarry substances phenols mineral oils ammonium ion
Fire and explosion	Any uses involving the construction of buildings	methane sulphur potentially combustible materials (e.g. coal dust, oil, tar, pitch)
Contact with contaminants during demolition, clearance, and construction work	Hazard mainly short-term (to site workers and investigators)	polycyclic aromatic hydrocarbons phenols oily and tarry substances asbestos radioactive materials
Contamination of ground and surface water (see Note 2)	Any uses where possible pollution of water may occur	phenols cyanide sulphate soluble metals

Note 1. The hazards listed are not mutually exclusive. Combinations of several hazards may need consideration.

Note 2. The soil pH should be measured as it affects the importance of these hazards.

Note 3. Uptake of harmful or phytotoxic metals by plants depends on which chemical forms of these elements are present in the soil. It may therefore be necessary to determine the particular forms, if the total concentrations present indicate that there exists a possible risk.

result of the accumulation of gas produced over a long period of time or as a direct result of the current conditions which are continuing to produce gas. It is necessary to check for the presence of gas and then monitor the levels. Gases may be sampled in the atmosphere or in trial pits and boreholes. Below-ground presences and production of gases should be monitored over a period of time from sampling tubes sealed in backfilled boreholes or probes in voids in the ground.

The testing for radon is generally carried out over a three-

Table 5.5 Contaminant testing

Previous site usage	Analysis
Metal mines Iron and steel works Lead works Foundries and smelters Electroplating and galvanizing works Scrap yards Shipbreaking/shipbuilding yards	Toxic metals e.g. cadmium, lead, arsenic, mercury Other metals e.g. copper, nickel, zinc
Chemical works Refineries By-product plants	– chromium – lead, zinc and cadmium – arsenic – other extractable material – mercury – selenium – nickel – boron – oils – hydrocarbons
Tar and turpentine distillers Gas works	– phenols and tar residues – free and complex cyanides – thiocyanates – sulphur – sulphides
Refuse and landfill areas Filled dock basins	– methane gas – carbon dioxide
Tanneries and hide merchants	– chromium – sulphide – arsenic – biological contamination i.e. anthrax – other extractable material
All solid samples	pH; volatile matter; sulphate content; asbestos
All water samples	pH; sulphate; lead; zinc; cadmium; chromium; BOD (Biochemical Oxygen Demand); COD (Chemical Oxygen Demand); phenols; chlorides; iron; total dissolved solids; volatile total solids

month period using detectors supplied by the National Radiological Protection Board (NRPB). A survey by the Board has revealed the highest risk areas occur in parts of Cornwall and Devon and new properties in these most affected areas are being built to guidelines for low radon levels recommended by the NRPB. It is advisable to obtain the current information from the NRPB as the details are subject to continual updating.

Underground combustion is difficult to detect. The slow smouldering of materials can occur undetected over many years with little or no evidence at surface level. Tests on samples to determine calorific values of fill materials can identify potentially combustible materials.

The actual testing methods are undertaken by specialists under laboratory conditions in accordance with the relevant technical notes and standards. The analytical procedures should be agreed at the outset and the testing programme should be under constant review as the results are made available.

At present there are no official or statutory limits for toxic substances found on contaminated land, but from experience the concentration values found in any samples should be compared with local natural background concentrations. However, the Inter-Department Committee on the Redevelopment of Contaminated Land (*ICRCL Guidance Note 59/83*, second edition, July 1987) gives an indirect method for assessing the findings of the site investigations. This method is based upon *threshold* and *action* trigger values. The trigger values define three possible concentration zones for each contaminant. Further guidance for information on derivation and use of trigger values is given in ICRCL 59/83.

Zone 1   Contaminants found in low concentrations.
Zone 2   Contaminants found in concentrations above the threshold value and below the action value.
Zone 3   Contaminants found in concentrations equal to and above the action value.

In zone 1 the contaminants are considered to be in such low concentrations that the risk of hazardous conditions developing is low and no treatments are necessary.

In zone 2 the concentrations of contaminants while higher than the threshold value do not automatically need special treatments. In this range it is necessary to consider the end use of the site and decide if any precautions are required, the decision is to be based upon informed judgement.

In zone 3 the concentrations of contaminants are considered undesirable or unacceptable and some action is necessary – ranging from minor remedial treatments to changing the proposed use of the site.

To determine trigger values the ICRCL recommend that the contaminants are divided into three categories:

Category A.   Contaminants which may present a hazard in very low concentration e.g. methane and asbestos. Their threshold value is zero i.e. any measurable concentration requires action to be considered or taken.
Category B.   Contaminants which in a given concentration can produce a measurable effect on a *target* e.g. sulphate attack on building materials, phenol and organic compounds contaminating water supplies.
Category C.   Contaminants for which there is no *dose–effect* relationship between concentrations in the soil and the effects have been determined experimentally. Most of the contaminants which can affect man's health fall in this category but at present there is insufficient data to specify precise trigger values for these contaminants.

Tables 5.6 and 5.7 give guidance on trigger values. It should be appreciated that these values are based upon professional judgement taking account of currently available information. They are only applicable when used in accordance with the conditions and notes specified. The values are

Table 5.6 Tentative 'trigger concentrations' for inorganic contaminants (reproduced with permission from ICRCL Guidance Note 59/83, 2nd edn, 1987, Table 3)

Conditions
1. This table is invalid if reproduced without the conditions and footnotes.
2. All values are for concentrations determined on 'spot' samples based on an adequate site investigation carried out prior to development. They do not apply to analysis of averaged, bulked or composited samples, nor to sites which have already been developed. All proposed values are tentative.
3. The lower values in Group A are similar to the limits for metal content of sewage sludge applied to agricultural land. The values in Group B are those above which phytoxicity is possible.
4. If all sample values are below the threshold concentrations then the site may be regarded as uncontaminated as far as the hazards from these contaminants are concerned and development may proceed. Above these concentrations, remedial action may be needed, especially if the contamination is still continuing. Above the action concentration, remedial action will be required or the form of development changed.

Contaminants	Planned uses	Trigger concentrations (mg/kg air-dried soil)	
		threshold	action
Group A: *Contaminants which may pose hazards to health*			
Arsenic	Domestic gardens, allotments	10	*
	Parks, playing fields, open space.	40	*
Cadmium	Domestic gardens, allotments	3	*
	Parks, playing fields, open space.	15	*
Chromium (hexavalent)[1]	Domestic gardens, allotments		*
	Parks, playing fields, open space.	25	*
Chromium (total)	Domestic gardens, allotments	600	*
	Parks, playing fields, open space.	1000	*
Lead	Domestic gardens, allotments	500	*
	Parks, playing fields, open space.	2000	*
Mercury	Domestic gardens, allotments	1	*
	Parks, playing fields, open space.	20	*
Selenium	Domestic gardens, allotments	3	*
	Parks, playing fields, open spaces	6	*
Group B: *Contaminants which are phytotoxic but not normally hazards to health*			
Boron (water-soluble)[3]	Any uses where plants are to be grown[2,6].	3	*
Copper[4,5]	Any uses where plants are to be grown[2,6].	130	*
Nickel[4,5]	Any uses where plants are to be grown[2,6].	70	*
Zinc[4,5]	Any uses where plants are to be grown[2,6].	300	*

Notes:
* Action concentrations will be specified in the next edition of ICRCL 59/83.
1. Soluble hexavalent chromium extracted by 0.1M HCl at 37°C; solution adjusted to pH 1.0 if alkaline substances present.
2. The soil pH value is assumed to be about 6.5 and should be maintained at this value. If the pH falls, the toxic effects and the uptake of these elements will be increased.
3. Determined by standard ADAS method (soluble in hot water).
4. Total concentration (extractable by $HNO_3/HClO_4$).
5. The phytotoxic effects of copper, nickel and zinc may be additive. The trigger values given here are those applicable to the 'worst-case': phytotoxic effects may occur at these concentrations in acid, sandy soils. In neutral or alkaline soils phytotoxic effects are unlikely at these concentrations.
6. Grass is more resistant to phytotoxic effects than are most other plants and its growth may not be adversely affected at these concentrations.

Table 5.7 Tentative 'trigger concentrations' for contamination associated with former coal carbonization sites (reproduced with permission from ICRCL Guidance Note 59/83, 2nd edn, 1987, Table 4)

Conditions
1. This table is invalid if reproduced without the conditions and footnotes.
2. All values are for concentrations determined on 'spot' samples based on an adequate site investigation carried out prior to development. They do not apply to analysis of averaged, bulked or composited samples, nor to sites which have already been developed.
3. Many of these values are preliminary and will require regular updating. They should not be applied without reference to the current edition of the report 'Problems Arising from the Development of Gas Works and Similar Sites[1].
4. If all sample values are below the threshold concentrations then the site may be regarded as uncontaminated as far as the hazards from these contaminants are concerned, and development may proceed. Above these concentrations, remedial action may be needed, especially if the contamination is still continuing. Above the action concentrations, remedial action will be required or the form of development changed.

Contaminants	Proposed uses	Trigger concentrations (mg/kg air-dried soil)	
		threshold	action
Polyaromatic hydrocarbons[1,2]	Domestic gardens, allotments, play areas.	50	500
	Landscaped areas, buildings, hard cover.	1000	10 000
Phenols	Domestic gardens, allotments.	5	200
	Landscaped areas, buildings, hard cover.	5	1000
Free cyanide	Domestic gardens, allotments, landscaped areas.	25	500
	Buildings, hard cover.	100	500
Complex cyanides	Domestic gardens, allotments.	250	1000
	Lanscaped areas.	250	5 000
	Buildings, hard cover.	250	NL
Thiocyanate[2]	All proposed uses.	50	NL
Sulphate	Domestic gardens, allotments, landscaped areas.	2000	10 000
	Buildings[3].	2000[3]	50 000[3]
	Hard cover.	2000	NL
Sulphide	All proposed uses.	250	1000
Sulphur	All proposed uses.	5000	20 000
Acidity (pH less than)	Domestic gardens, allotments, landscaped areas.	pH 5	pH 3
	Buildings, hard cover.	NL	NL

Notes:
NL: No limit set as the contaminant does not pose a particular hazard for this use.
1. Used here as a marker for coal tar, for analytical reasons. See 'Problems Arising from the Redevelopment of Gasworks and Similar Sites' Annex A1[1].
2. See 'Problems Arising from the Redevelopment of Gasworks and Similar Sites' for details of analytical methods[1].
3. See also BRE Digest 250: Concrete in sulphate-bearing soils and groundwater[4].

only applicable after an adequate site investigation and do not apply to sites already developed.

To assist in assessing the significance of each contaminant Tables 5.8 and 5.9 have been included. Table 5.8 gives the Greater London Council guidelines for the classification of contaminated soil and Table 5.9 gives the Netherlands standards for contaminated soils. As previously noted the assessment and treatment of contaminated sites are deter-

mined on an individual basis and these tables are intended to help the engineer determine the appropriate treatment.

5.3.7 Site treatment

Since each site must be treated individually, it is not possible at the present time to give definitive and detailed advice on the treatment of a contaminated site. However, this section is intended to give the engineer broad guidance and advice

Table 5.8 Greater London Council guidelines for classification of contaminated soils

Parameter	Typical values for uncontaminated soils	Slight contamination	Contaminated	Heavy contamination	Unusually heavy contamination
pH (acid)	6–7	5–6	4–5	2–4	(less than 2)
pH (alkaline)	7–8	8–9	9–10	10–12	12
Antimony	0–30	30–50	50–100	100–500	500
Arsenic	0–30	30–50	50–100	100–500	500
Cadmium	0–1	1–3	3–10	10–50	50
Chromium	0–100	100–200	200–500	500–2500	2500
Copper (available)	0–100	100–200	200–500	500–2500	2500
Lead	0–500	500–1000	1000–2000	2000–1.0%	1.0%
Lead (available)	0–200	200–500	500–1000	1000–5000	5000
Mercury	0–1	1–3	3–10	10–50	50
Nickel (available)	0–20	20–50	50–200	200–1000	1000
Zinc (available)	0–250	250–500	500–1000	1000–5000	5000
Zinc (equivalent)	0–250	250–500	500–2000	2000–1.0%	1.0%
Boron (available)	0–2	2–5	5–50	50–250	250
Selenium	0–1	1–3	3–10	10–50	50
Barium	0–500	500–1000	1000–2000	2000–1.0%	1.0%
Beryllium	0–5	5–10	10–20	20–50	50
Manganese	0–500	500–1000	1000–2000	2000–1.0%	1.0%
Vanadium	0–100	100–200	200–500	500–2500	2500
Magnesium	0–500	500–1000	1000–2000	2000–1.0%	1.0%
Sulphate	0–2000	2000–5000	5000–1.0%	1.0%–5.0%	5.0%
Sulphur (free)	0–100	100–500	500–1000	1000–5000	5000
Sulphide	0–10	10–20	20–100	100–500	500
Cyanide (free)	0–1	1–5	5–50	50–100	100
Cyanide (total)	0–5	5–25	25–250	250–500	500
Ferricyanide	0–100	100–500	500–1000	1000–5000	5000
Thiocyanate	0–10	10–50	50–100	100–500	2500
Coal tar	0–500	500–1000	1000–2000	2000–1.0%	1.0%
Phenol	0–2	2–5	5–50	50–250	250
Toluene extract	0–5000	5000–1.0%	1.0%–5.0%	5.0%–25.0%	25.0%
Cyclohexane extract	0–2000	2000–5000	5000–2.0%	2.0%–10.0%	10.0%

Note: All numerical values expressed in milligrams per kilogram (mg/kg), which equates to parts per million (ppm), unless otherwise stated.

with respect to the various checks and processes to be considered.

The choice of a method for treating a contaminated site is mainly dependent upon the end use of the site. Cost will be a major factor but the most cost-effective solution may not be obvious so cost checks will be necessary to determine the solution.

The following is therefore intended to give an indication of possible options and points for consideration:

(1) The first process is to consider the site investigation results and implications with the client's proposals. While the client's requirements must be met, alternative layouts may be acceptable which can avoid areas of contamination and achieve satisfactory and more economical solutions than the original proposals.

(2) The principal construction options for dealing with a contaminated site are:
(a) Relocate the development.
(b) Remove contaminated material.
(c) Cover contaminated material with a suitable isolating *blanket*.
(d) A combination of (a), (b) and (c).
(e) Make use of contaminants, i.e. use methane for heating.

If the risk is too high the development may have to be relocated. Localized contamination is usually dealt with by removal of the material, whereas more extensive contamination is dealt with by a *blanket* cover, plus containment within the site boundaries. When considering the health aspect, higher concentrations of metal contamination of

Table 5.9 Netherlands standards for soil contaminants

Element/compound	Concentration in soil (mg/kg dry weight)		
	A^a	B^a	C^a
Arsenic	20	30	50
Barium	200	400	2000
Cadmium	1	5	20
Chromium	100	250	800
Cobalt	20	50	300
Copper	50	100	500
Lead	50	150	600
Mercury	0.5	2	10
Molybdenum	10	40	200
Nickel	50	100	500
Tin	20	50	300
Zinc	200	500	3000
Bromine (total)	20	50	300
Fluorine (total)	200	400	2000
Sulphur (total)	2	20	200
Cyanide (free)	1	10	100
Cyanide (complex)	5	50	500
Benzene	0.01	0.5	5
Ethyl benzene	0.05	5	50
Toluene	0.05	3	30
Xylene	0.5	5	50
Phenols	0.02	1	10
Naphthalene	0.1	5	50
Anthracene	0.1	10	100
Phenanthrene	0.1	10	100
Fluoranthrene	0.1	10	100
Pyrene	0.1	10	100
Benz(a)pyrene	0.05	1	10
Chlorinated aliphatics	0.1	5	50
Chlorobenzenes	0.05	1	10
Chlorophenols	0.01	0.05	5
PCB (total)	0.05	1	10
Organic chlorinated pesticides	0.1	0.5	5
Pesticides (total)	0.1	2	20
Tetrahydrofuran	0.1	4	40
Pyridine	0.1	2	20
Tetrahydrothiophene	0.1	5	50
Cyclohexanone	0.1	6	60
Styrene	0.1	5	50
Fuel	20	100	800
Mineral oil	100	1000	5000

[a] Below level A the soil is regarded as unpolluted.
    Above level A a preliminary site investigation is required.
    Above level B further investigation is required.
    Above level C remove soil or clean-up to level A values.

Table 5.10 Requirements for concrete exposed to sulphate attack (BS 8110: Part 1: 1985, Table 6.1)

Concentrations of sulphates expressed as SO_3				Type of cement	Dense, fully compacted concrete made with 20 mm nominal maximum size aggregates complying with BS 882 or BS 1047	
Class	In soil		In ground water		Cement* content not less than	Free water/ cement* ratio not more than
	Total SO_3 (%)	SO_3 in 2:1 water:soil extract (g/L)	(g/L)		(kg/m³)	
1	less than 0.2	less than 1.0	less than 0.3	All cements listed in 6.1.2.1 BS 12 cements combined with p.f.a.[†] BS 12 cements combined with g.g.b.f.s.[†]	300 —	— —
2	0.2 to 0.5	1.0 to 1.9	0.3 to 1.2	All cements listed in 6.1.2.1 BS 12 cements combined with p.f.a. BS 12 cements combined with g.g.b.f.s.	330	0.50
				BS 12 cements combined with minimum 25% or maximum 40% p.f.a.[‡] BS 12 cements combined with minimum 70% or maximum 90% g.g.b.f.s.	310	0.55
				BS 4027 cements (SRPC) BS 4248 cements (SSC)	280	0.55
3	0.5 to 1.0	1.9 to 3.1	1.2 to 2.5	BS 12 cements combined with minimum 25% or maximum 40% p.f.a.[†] BS 12 cements combined with minimum 70% or maximum 90% g.g.b.f.s.	380	0.45
				BS 4027 cements (SRPC) BS 4248 cements (SSC)	330	0.50
4	1.0 to 2.0	2.1 to 5.6	2.5 to 5.0	BS 4027 cements (SRPC) BS 4248 cements (SSC)	370	0.45
5	over 2	over 5.6	over 5.0	BS 4027 cements (SRPC) and BS 4248 cements (SSC) with adequate protective coating (see 6.2.3.3)	370	0.45

* Inclusive of p.f.a. and g.g.b.f.s. content.

[†] For reinforced concrete see 3.3.5: for plain concrete see 6.2.4.2.

[‡] Values expressed as percentages by mass of total content of cement, p.f.a. and g.g.b.f.s.

Note 1. See 6.2.4.3 for adjustments to the mix proportions.

Note 2. Within the limits given in this table, the use of p.f.a. or g.g.b.f.s. in combination with sulphate-resisting Portland cement (SRPC) will not give lower sulphate resistance than combinations with cements to BS 12.

Note 3. If much of the sulphate is present as low solubility calcium sulphate, analysis on the basis of a 2:1 water extract may permit a lower site classification than that obtained from the extraction of total SO_3. Reference should be made to BRE Current Paper 2/79 for methods of analysis, and to *BRE Digests 250* and *222* for interpretation in relation to natural soils and fills, respectively. (Note: *BRE Digest 250* has now been replaced by *BRE Digest 363*.)

All references within this table refer to the original document.

sub-soil can be made acceptable by covering the area with hardstanding.

When considering removal of contaminated materials it should be remembered that handling the materials on site involves special safety precautions including protective clothing for operatives. Removal from site must be carried out in covered and sometimes sealed transport and the material taken to licensed tips. It will be appreciated that significant costs can arise from such items.

The use of a *blanket* to cover the contaminated materials plus measures to contain the contaminants is an alternative to removal. The *blanket* is intended to seal the contaminated materials at a suitable depth to ensure safety of future users of the site. The *blanket* is designed to suit the actual site conditions and a satisfactory design is dependent upon sufficient detailed sub-soil data to enable accurate predictions of potential groundwater movements. The design and selection of suitable materials for the *blanket* is based on preventing percolation of contaminants to the surface. The *blanket* should incorporate a drainage and monitoring system to ensure its satisfactory performance. The use of a blanket cover must be considered in conjunction with sub-structure proposals since the integrity of the *seal* must be maintained.

Where contamination is of a natural source, as in the case of radon, removal is impractical and other precautions must be adopted. The basic principles involved are firstly to prevent radon entering the building through floors, walls and service entry locations. Secondly the radon-laden air from the ground trapped below the floor can be extracted by mechanical means. The details of how to do this can be obtained from the latest literature produced for the Department of the Environment (DoE). It is important that engineers work to the most recent updated guidance on contamination in general and radon in particular since the advice is likely to be under constant review.

5.4 FOUNDATION PROTECTION

In order to ensure the satisfactory performance of sub-structures it is necessary to know:

(1) Which elements can cause deterioration of the materials used in the foundations.
(2) How and where the elements occur.
(3) How attack develops and its effect.
(4) What checks and tests should be carried out to identify corrosive elements during site investigation works.
(5) What precautions can be taken to prevent attack.

With an appreciation and an understanding of the above points it is possible to incorporate the necessary design and construction of the foundations thus ensuring satisfactory performance and economy of the sub-structures.

(1) *Sulphates*. If sulphates cannot be prevented from reaching the structure the size of concrete members and quality of concrete requires careful consideration.

Fully compacted concrete of low permeability is essential in resisting sulphate attack. Massive forms of construction will deteriorate less quickly than thin or small sections. The rate of attack can increase if moisture can be lost by evaporation or leakage from any part of the concrete surfaces and replenishment can occur from other parts. Ground slabs and retaining walls are therefore more vulnerable than foundations and piles.

The quantity and type of cement used are also significant in resisting sulphate attack and Table 5.10 gives the appropriate recommendations considered. The severity of the potential sulphate attack is based upon results from site tests (see also BS 8110).

(2) *Phenols*. The use of plastic pipework to carry services should be avoided if phenols are present (or suspected). Alternative materials or protective coatings are necessary if plastic or rubber materials are used in areas contaminated by phenols. Increased sizes of concrete members, and additional depths of concrete cover to reinforcement should be considered if phenol concentrations are very high.

(3) *Metals*. To prevent metallic corrosion through electrolytic action the following alternatives should be considered:
   (a) Do not use metals in below-ground construction.
   (b) Remove the metallic contaminants from the critical areas of the ground.
   (c) Use protective coatings and layers for all metals in below-ground locations.

An evaluation of each option can be undertaken and the most suitable treatment adopted.

5.5 EXAMPLE

While it is not practical for a book of this length to give fully detailed examples it is hoped that the following will give a picture of some of the methods used in producing a solution to particular problems. It should be appreciated that dealing with an actual project can be complex, as ground conditions can be extremely variable. Engineering judgement must be exercised in these situations with consideration of the various contaminations and future site usages.

This example considers a derelict inner-city site which is to be reclaimed and redeveloped. The site is located in an area of mixed domestic and industrial development, where the industrial operations have declined.

The reclaimed site is to accommodate light industrial or domestic developments but no specific layouts were produced at the time the reclamation was planned.

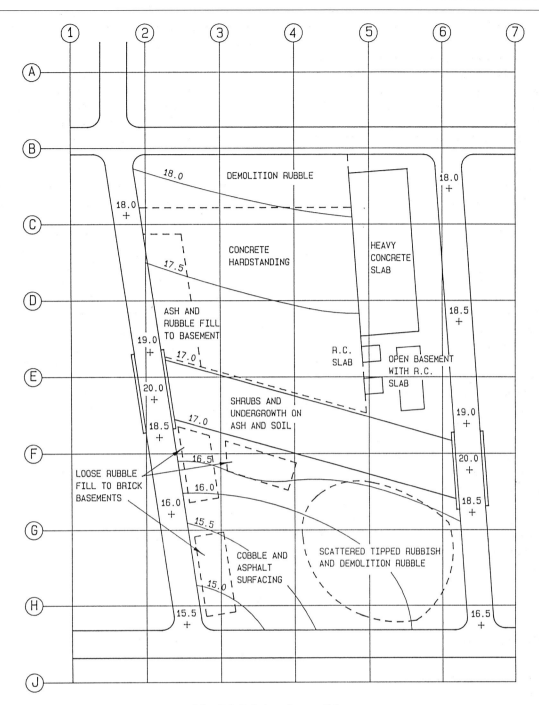

Fig. 5.1 Existing site conditions.

Figure 5.1 shows the site as existing and indicates the visible surface observations noted during an initial *walk around* survey. Demolition operations to ground level have been undertaken and filled basements are apparent. Demolition rubble is mounded at the south-east corner, and much of the site is covered by hardstanding of concrete, cobbles or asphalt. There are no visible signs of chemical contamination, the shrubs and undergrowth are of a healthy appearance and no discoloured ponding water or odours are apparent. The site levels fall from east to west and the site is generally below the level of the surrounding roads.

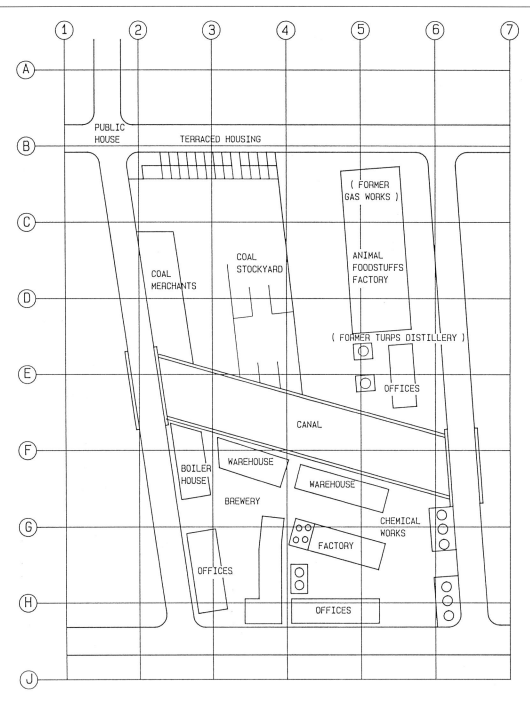

Fig. 5.2 Historical site usage.

Figure 5.2 shows the site layout indicating the past usage. This information was obtained during a desk-top study of historical information, i.e. Ordnance Survey maps, street plans, etc. as outlined in section 5.2.1. Checks on the previous usage of the adjacent sites are necessary to determine potential air-borne contamination or gas seepage. No such problems were identified for this site.

In addition to the historical checks, the desk-top study can collect information on sub-soil conditions. Geological maps and checks with the local authority engineers and building control officers provide useful information. In this case boulder clay underlaid the site and hence a reasonable bearing strata for strip and pad foundations was anticipated at shallow depth. A tracing paper copy of the layout can be overlaid on Fig. 5.2 to identify any anomalies with the physical observations on site. In this instance there was good correlation although the rubble mounded on the south-east corner was covering an area of possible chemical contamination.

Based upon the above information the site investigation works can be determined to confirm the nature and extent of underground obstruction and contaminants.

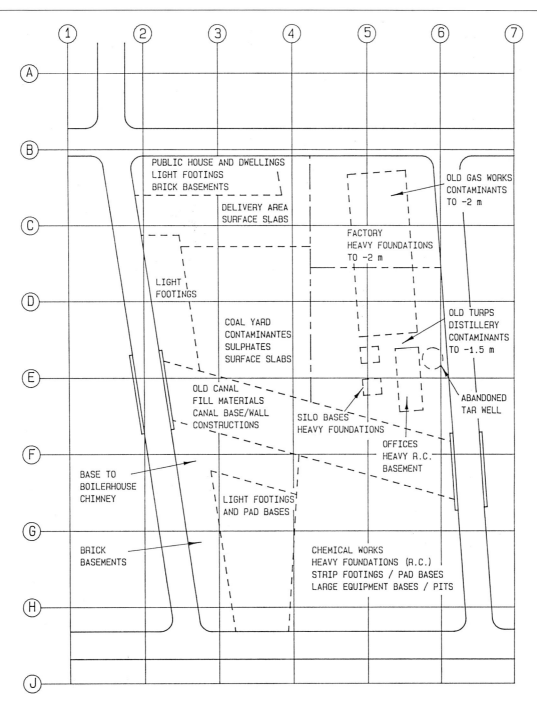

Fig. 5.3 Contamination layout.

Figure 5.3 shows the site layout indicating likely problems and is used as the basis for planning the detailed site investigations. If redevelopment proposals are available the implications of the existing conditions can be considered at this stage and any modifications to layouts investigated.

Based upon the information shown on Figs 5.1, 5.2 and 5.3 it is possible, at minimal cost, to gain a reasonable picture of the main problems of site reclamation and redevelopment. This information can be presented to a developer even before site acquisition.

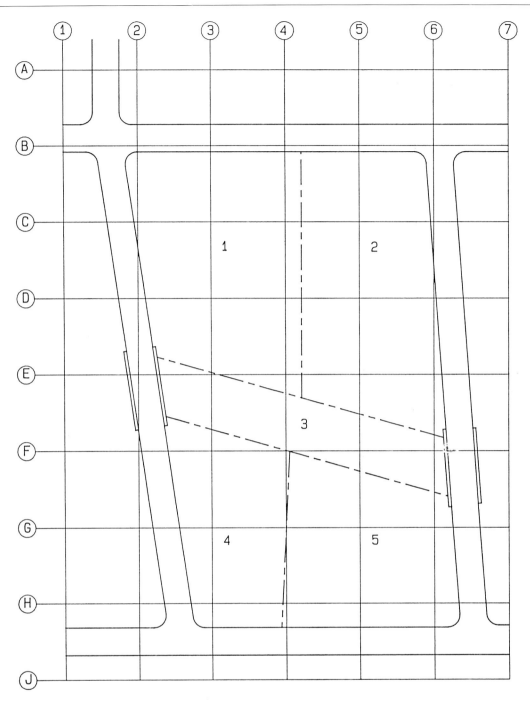

Fig. 5.4 Site zones.

Figure 5.4 shows the site layout divided into zones based upon the information shown on Fig. 5.3. The zoning is determined in this case on the degree of severity of buried obstructions, contaminants and relative levels.

SITE INVESTIGATIONS

Boreholes were located on a 40 m grid to the full site area arranged to provide a minimum of three boreholes in each zone.

Samples were taken from the boreholes to determine soil strength (see Chapter 3 for further details) and for contaminants as follows.

Contaminant testing
All water samples analysed for:

- pH
- sulphate
- lead, zinc and cadmium
- chromium
- BOD and/or COD
- phenols
- chloride
- iron
- total dissolved solids
- volatile total solids

Zone 1   10 boreholes − wide spectrum test in 2 boreholes plus, in coal yard area, tests on calorific value of samples.

Zone 2   10 boreholes.
Gas works and tar and turpentine distillers, tests for:

- phenols and tar residues
- free and complex cyanides
- thiocyanates
- sulphur
- sulphides

Zone 3   3 boreholes − wide spectrum tests in 1 borehole.

Zone 4   6 boreholes − wide spectrum test in 2 boreholes.

Zone 5   6 boreholes.
Chemical works, tests for:

- chromium
- lead, zinc and cadmium
- arsenic
- ether extractable material
- mercury
- selenium
- nickel
- boron

Contaminant test results were compared with trigger values prior to trial pit excavations. Where trigger values were exceeded, further boreholes were sunk and additional testing used to determine the extent, area and depth of contamination. In zones 2 and 5 closer spaced bores at 10 m centres were used to confirm contamination type, level, depth and area affected.

In addition to the boreholes a series of trial pits was excavated to give sub-soil profiles and to provide samples for testing. Precautions to protect operatives in areas where trigger values were exceeded were implemented and excavation works limited. Trial pits were used to expose existing sub-structure constructions. The trial pits were also arranged on an approximate grid of 40 m with additional trenching to expose the lines of redundant foundations.

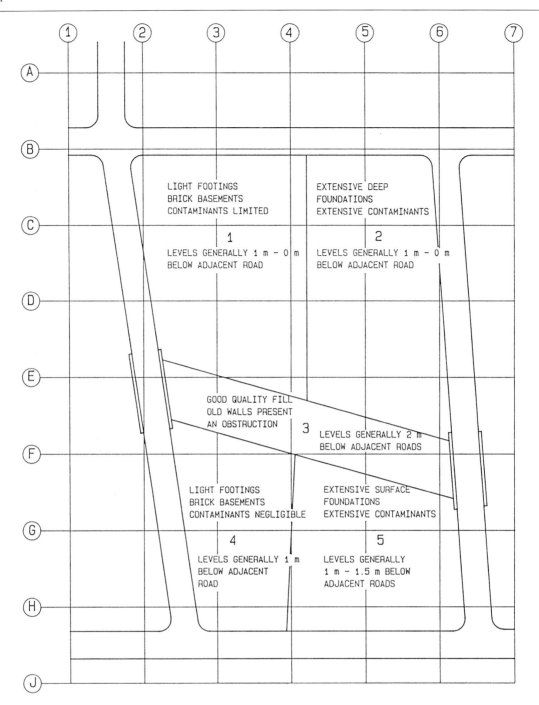

Fig. 5.5 Summary of zones/contaminant/obstruction findings.

Figure 5.5 shows the site layout divided into zones and a summary of the obstruction/contamination findings. The relationship of the adjacent levels are also indicated. The relative levels to surrounding areas are significant in this example since the raising of ground levels would be advantageous for both reclamation and redevelopment of the site. A sufficient depth of imported fill laid across the site would enable the majority of the underground obstruction to be left in position and for a new sub-structure to be designed to *ride* over using raft type foundations with the fill material acting as a cushion over the hard spots. In addition the fill can be used to isolate contaminants at an appropriate depth below surface levels.

A depth of 2−3 m above obstructions and contaminants was considered sufficient to achieve the required cushion. A barrier layer above the contaminants and below imported fill was considered adequate to contain contaminants to prevent migration upwards.

Possible alternative treatments would be to

(1) Remove obstructions or contaminants where they affect development proposals.
(2) Design new development around underground problems.
(3) Restrict development, leaving sterile areas where the treatment cost is prohibitive.

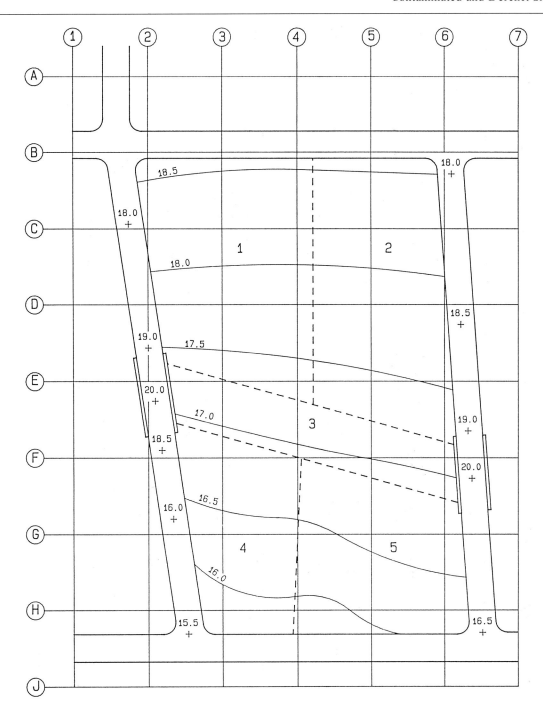

Fig. 5.6 Zoning and contouring of site.

Figure 5.6 shows the reclamation treatment which was adopted; it can be summarized as follows:

Zone 1 The light footings were grubbed up and localized lowering of existing levels undertaken to accommodate a minimum of 1 m of imported fill, with contaminants sealed in place.

Zone 2 Localized lowering of existing levels were implemented to accommodate a minimum of 1 m of imported fill. Local pockets of phenol were removed to licensed tips. The remaining contaminants were left in place and the site surface sealed with a barrier of clay and plastic membrane.

Zone 3 The obstructions were buried under 2.5 m of imported fill. There were no contaminant problems in this area.

Zone 4 The existing basements were filled with compacted hardcore of crushed brick, recycled on site. The light footings were grubbed up and site levels raised with imported fill.

Zone 5 The localized pockets of heavy metal were removed to licensed tips and the remaining contaminants left in place. The site surface was sealed with a combination of clay and polythene barrier to isolate the remaining contaminants. Site levels were raised and the underground obstructions buried under the imported fill.

Alternative fill materials were considered and ranged from demolition hardcore, MOT graded hardcore and dredged sand. Treatment of the fills could also vary (1) compacted in layers, (2) limited compaction and subsequent ground improvement using vibro, dynamic consolidation or preloading.

The sand option compacted in layers provided the most effective cost- and programme-efficient method. Continuous uninterrupted supplies of a consistent material provided an excellent formation for domestic or light industrial units using raft foundation constructions. The road and service installation could also be accommodated without problems or cost penalties. The solution adopted provided an evenly contoured site, with general levels slightly above the adjacent roads. The surface of the sand was treated to avoid erosion during redevelopment works on the site.

The subsequent development was a housing estate of one- and two-storey dwellings built in small terraces on raft foundations. The sub-structure and infrastructure were successfully constructed on the site without any problems arising from the reclamation works.

5.6 REFERENCES AND FURTHER INFORMATION

1. Wilson, D.C. & Stevens, C. (1988) *Problems Arising from the Redevelopment of Gas Works and Similar Sites*, 2nd edn. HMSO, London.
2. Building Research Establishment (1981) BRE Digest 250: *Concrete in Sulphate-bearing Soils and Groundwater*. BRE Watford, UK. Now replaced by BRE Digest 363, July 1991.

Further advice can be obtained from ICRCL and CIRIA, who produce detailed guidance on the testing and treatment of contaminated land.

CHAPTER 6

MINING AND OTHER SUBSIDENCE

6.1 INTRODUCTION

Subsidence due to mining of coal and other materials and the extraction of other minerals by pumping (i.e. brine pumping) can cause more severe stressing of structures than that caused by differential settlement. In addition to the vertical settlement there can also be horizontal movement of the supporting soil causing strain and stress, both in tension and compression, which can be transferred to the foundations with serious results to the superstructure.

The authorities responsible for the underground workings, i.e. British Coal and the Brine Authority, will usually give advice on the magnitude and distribution of the likely movement due to past, present and proposed future workings. The structure and its foundations must be designed to be robust enough or sufficiently flexible to safely withstand the effects of movements. It is even more important than normal that the foundation and superstructure design should be closely linked under the supervision of one engineer responsible for the project.

Long buildings should be broken up into shorter lengths by jointing and extensive buildings of larger plan area should be divided into smaller, independent units. When relatively shallow mining has ceased it may be possible to backfill the workings (known as *stowing*) or to take the foundations below them. In some cases the extra cost of providing strong, rigid foundations for houses and other small, lightly loaded buildings may far exceed the cost of repairing the possible minor cracking of such buildings with traditional or possibly semi-flexible foundations. Where this is inadvisable or uneconomic the superstructure design can be amended to resist safely the effects of subsidence (see also section 1.6). Typical examples are:

(1) The use of three-pinned arches in lieu of rigid portals.
(2) Simple supports instead of fixed end supports.
(3) Articulated structures and foundations (see section 6.10.7).
(4) Adding reinforcement to superstructure walls so that they act as deep, stiff beams.
(5) Groups of buildings should be kept separate, and isolated. If connections are unavoidable, such as covered corridors, concrete paths, they should form flexible links.
(6) Shallow raft foundations, particularly for low-rise buildings, can form the best resistance to tension and compression strains in the supporting ground.

Further examples are given in section 6.10.1.

There is considerable scope for engineering ingenuity and skill and sites should not be rejected, out of hand, because they are liable to subsidence.

The most common and widespread cause of subsidence is that due to coal mining and an understanding of this cause is not only helpful to the designer but it also facilitates the understanding of other forms of subsidence.

6.2 MECHANICS OF MINING SUBSIDENCE

The determination of the magnitude and rate of settlement is a complex and specialized topic and advice should be obtained from British Coal's specialist subsidence engineers. Nevertheless it is advisable that the designer of the foundation and the superstructure understands the ground behaviour due to mining subsidence since this will assist his fuller understanding of the subsidence engineer's report and help him to anticipate the effects on his project.

When coal, or other minerals, are extracted a sub-surface cavity is formed and the surrounding strata will *flow* into the cavity (see Fig. 6.1).

The action is, of course, three-dimensional and not merely two-dimensional, as shown for convenience in Fig. 6.1. The

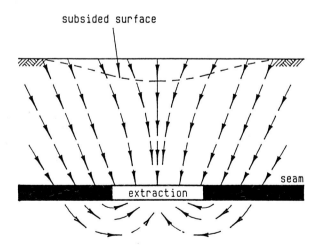

Fig. 6.1 Substrata *flow* into cavity.

resulting surface subsidence forms a trough- or saucer-like depression covering a much wider area than the extracted area. It will be appreciated that the wider the extraction or *width of workings*, *W*, then the wider will be the subsided surface (known as the *zone of influence*). Similarly the shallower the depth of seam from the surface, *H*, then the greater will be the magnitude of the maximum subsidence, *S*. (These points can be considered in some detail by examination of Fig. 6.2.)

The *angle of draw* is the angle between the lines from the edge of the worked area normal to the seam, and that to the point of zero subsidence at ground level. The angle has been found, by experience, to be $30° \pm 5°$ in most types of ground. The strain in the ground at the surface above the workings tends to be in compression and in tension above the area subtended by the angle of draw. The *horizontal* displacement of the surface tends to be zero at the point above the centre of the worked area rising to a maximum at the edge of the worked area then falling to zero again at the edge of the angle of draw (see Fig. 6.3).

The angle of draw, the magnitude of settlement, displacement and strain will depend on such factors as the thickness

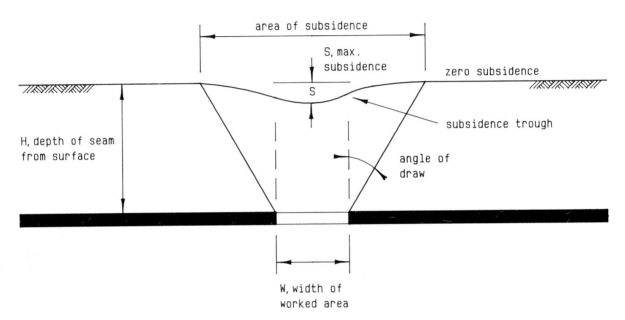

Fig. 6.2 Zone of influence.

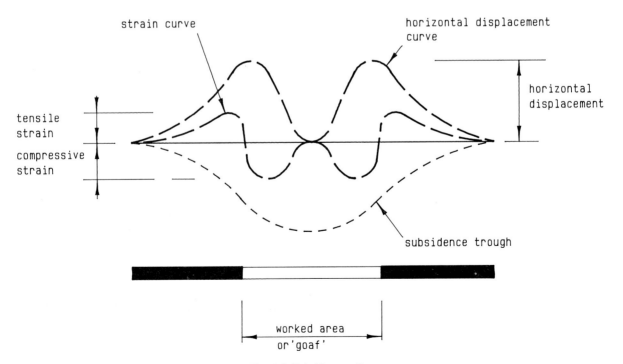

Fig. 6.3 Subsidence effect.

of the worked seam, its depth below the surface, the type of overburden, etc.

As can be seen from Figs 6.2 and 6.3, the ground (and the foundations resting on it) will be subject to vertical settlement − and to horizontal displacement strains in tension and compression. The effect on buildings is shown in Fig. 6.4.

For the sake of clarity a *static* case has been considered, i.e., the results of one part of the seam having been worked. But as the seam working is advanced then so too will the subsidence advance in the form of a subsidence *wave*. The ground strains, too, with the advance can change from tension to compression (see Fig. 6.5).

The magnitude of the vertical and horizontal displacement is significantly increased by the method of mining (it is also affected by the number of seams worked one above the other).

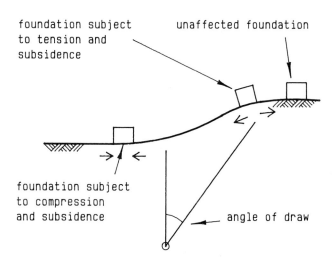

Fig. 6.4 Subsidence effect on buildings.

6.3 METHODS OF MINING

6.3.1 Longwall workings

The modern method of mining is to advance continuously on a wide face 200–300 m long. The roof near the face is supported on temporary supports (usually *walking* hydraulic jacks) as it advances and as the face advances so do the supports. The overlying strata either breaks through along the back edge of the supports or is partly supported by stowed material. At each end of the longwall face roadways are maintained for access of labour and plant, ventilation and the removal of the coal (see Fig. 6.6).

The subsidence waves advance in line with the longwall and at roughly the same speed as extraction. Though most of the subsidence is transmitted fairly rapidly to the surface as the overlying strata collapse into the worked seam (dependent on the overburden and other factors), the total subsidence may take up to two years to complete.

During subsidence the collapsed overlying strata bulks so that the surface subsidence is less than the thickness of the extracted seam.

The maximum subsidence likely to occur can be up to 80% of the coal seam thickness, when the width of the face exceeds $1.4 \times$ depth of seam − a common occurrence in longwall mining. The use of stowage reduces the surface subsidence but since stowage is an additional cost item in mining it is not often employed.

The horizontal strains can be as high as 0.008 for shallow workings but are generally 0.002. *Shallow* workings are defined as either less than 30 m below ground level or where the depth of the overburden is less than $10 \times$ seam thickness. The magnitude of the slope of the subsidence wave, or tilt of the ground surface, can be as high as 1 in 50 over shallow workings and the slope decreases as the depth to the working increases.

6.3.2 Pillar and stall workings (partial extraction methods)

During the 15th and 16th centuries methods of partial extraction started to be used which left pillars of unworked coal to support the roof (see Fig. 6.7). At most only half the

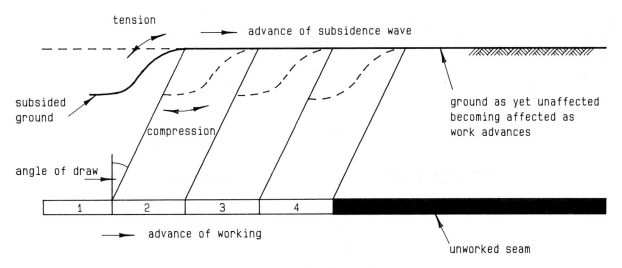

Fig. 6.5 Subsidence wave.

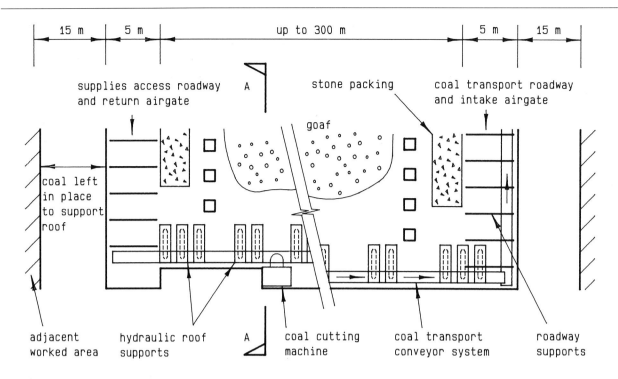

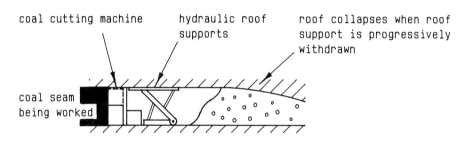

section A — A

Fig. 6.6 Longwall extraction.

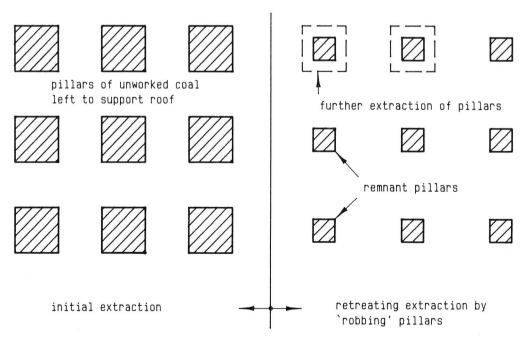

Fig. 6.7 Partial extraction — pillar and stall.

coal was extracted during the advance of the seam but when the limits of the seam were reached the miners, as they retreated from the workings, cut into the pillars thus much reducing them in cross-sectional area and, in some cases, totally removing the pillars.

In some mining areas the pillars formed a continuous wall with roadways between – and again, on retreating, the walls were *robbed* of coal. In either method the pillars could collapse due to overloading, or punch (shear) through the roof or floor, or weather away. Though such mines have long been abandoned, and thus ground settlement is likely to be complete, there can be problems today particularly where the depth of overburden is shallow and especially if it is of weak, friable strata. With roof collapse there is a risk that the cavity may *migrate* to the surface, i.e., continuous collapse of the overlying soil until the results of the cavity *backfill* reach the surface. The increased pressure on the pillar remnants due to a new building relative to the overburden pressure could be enough to cause them to collapse. Typical safe and less safe conditions are shown in Fig. 6.8.

Though detailed records of such workings were rarely kept, British Coal have a vast amount of information on such coal mines. Unfortunately, it cannot be guaranteed that all are known. It is advisable to check with boreholes, particularly when coal seams are at shallow depth and overlain by poor material.

Probably, nowadays, less than 5% of coal is mined by such methods in developed countries though the method is still used to win gypsum, limestone and ironstone.

6.3.3 'Bell-pits'

This form of mining evolved in about the 13th century, and was also used by unemployed miners in the strikes and economic depression of the 1920s and 1930s. Shafts generally 1.0–1.2 m in diameter were sunk to the level of the coal seam which could be up to 12 m below ground. The shafts were then widened at seam level to extract the coal until the area became too large to prevent roof collapse and the pit was abandoned.

The shafts can be as close as 10 m and where ironstone has been worked by this method the shafts can be as close as 5 m. (See Fig. 6.9.)

Evidence of bell-pit workings can be revealed by the cones of mine waste or ground depressions along the outcrop of main seams. Geophysical methods, such as seismic analyses, infra-red photography, etc., to detect the presence of pits are not always successful and it has been found more reliable to trench suspect areas (see section 6.4.1). The foundation designer's problem is that bell-pits were rarely properly backfilled and they were left to collapse leaving voids and the overburden with low load-bearing capacity.

6.4 ASSOCIATED AND OTHER WORKINGS

6.4.1 Abandoned mine shafts and adits

An increasing problem is the detection of abandoned mine shafts and adits which are often in an unstable condition. A mine shaft provides a circular vertical access to the coal whereas an adit is more usually square and inclined and often follows the dip of the seam. Though British Coal have records of over 100 000 shafts there are still many unrecorded. It is estimated, for example, that in Derbyshire alone there are over 50 000 lead mine shafts. Many of the shafts have not been properly backfilled, if at all. Many have been capped or plugged by using felled trees to form a scaffold at a depth not far below ground level on to which fill was tipped up to ground level. The fill material is often found to be unsatisfactory and consisting of refuse, degradable material, old tubs and the like.

There may be records of such shafts on early OS maps. Evidence of depressions should be checked or they may show up on aerial survey photographs. Drilling for shafts is not as successful in detection as trenching with excavators. Since the filled shafts can be unstable, regard must be paid to safety measures for the personnel and plant.

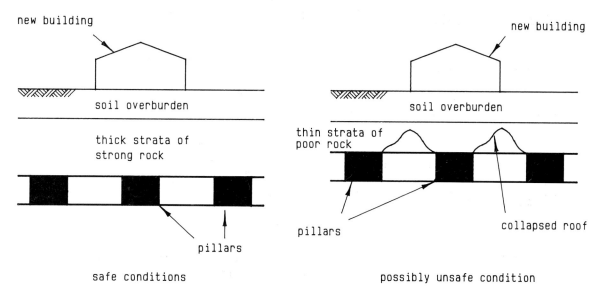

Fig. 6.8 Effect of overburden/rockhead strength.

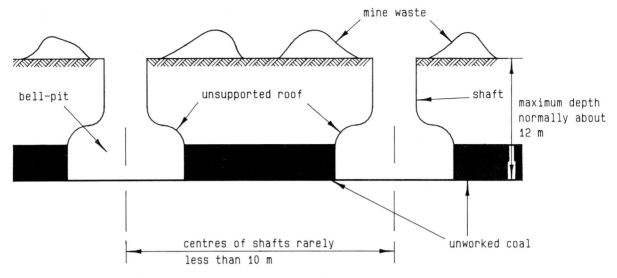

Fig. 6.9 Bell-pits.

Where a coal seam outcropped at the surface the main access would have been by adits. When abandoned, like shafts, they were often inadequately backfilled.

6.4.2 Fireclay and other clays
High-grade fireclay, for use in boiler lining and brick kilns, is still mined, as is red tile clay for floor tiles. The common mining method is the pillar and stall techniques.

6.4.3 Iron Ores
Some mining of iron ore is still continuing, on a relatively minor scale, usually by the pillar and stall method.

6.4.4 Other metals
Tin and copper have been mined in Cornwall and lead and zinc have been mined in the mountainous area of the Lake and Peak Districts, North Wales and in the northern Pennines. The minerals usually occur as vein deposits so that the workings are relatively narrow and localized.

6.4.5 Limestone
Limestone is mined when quarrying is not feasible and an alternative economic supply is not available. Again the pillar and stall method is a common technique and rarely gives rise to foundation design problems. (Should there be any doubt then the soundness of the pillars should be investigated.)

6.4.6 Salt
Salt extraction by brine pumping (brine being a mixture of salt and water) is common in Cheshire and also occurs to a limited extent in Lancashire, Yorkshire, Shropshire and other counties. Uncontrolled, or *wild* pumping, was stopped in 1930 because of the serious subsidence caused and some subsidence is still not complete. The extraction is now controlled by limiting the size of the cavities formed.

6.4.7 Chalk
Bell-pits have been sunk to extract flints in Berkshire, Hampshire, Hertfordshire, Kent, Norfolk and Suffolk.

Unrecorded workings for chalk have caused problems with crown holes migrating to the surface because of spalling or solution of the mine gallery roofs.

6.5 FAULTING

Mining areas, particularly coal, are often faulted and subsidence is sharp and sudden along the outcrop of the fault. It is therefore advisable to locate a building away from a fault.

6.6 NATURAL AND OTHER CAVITIES

6.6.1 *Dissolving* rock
Cavities can occur in sedimentary rocks due to sub-surface erosion caused by the movement of groundwater. Probably the most common cavities are those in limestone and chalk deposits and in salty strata (see section 6.4). There have been problems in wind–blown deposits, loess is the most common, but these are rare in northern Europe. The main foundation problem in this country is the formation of swallow-holes in chalk and limestone (see Fig. 4.3 in Chapter 4).

Chemicals in the deposit, chlorides and carbonates, can dissolve in water (known as *evaporites*) and are transported by underground springs. Over a period of time large caves, cavities and potholes are formed. Generally, there is sufficient rock and depth of overburden remaining so as not to cause foundation design problems. Advice should however, be sought from the Geological Survey in areas affected by such action.

6.6.2 *Dissolving* soils

A very common cause of foundation failures, particularly in housing, resulting in cracks in the walls, is the washing away of the supporting soil due to cracked water mains and sewers. The authors' consultancy has discovered this problem in numerous cases and now investigates, as a first step in a structural survey, the possibility of such leakage. The cure can be relatively simple – stop the leaks, replace the washed-out soil with compacted sandy gravel or underpin and repair any structural damage.

6.7 TREATMENT OF ABANDONED SHALLOW WORKINGS

6.7.1 Introduction
In the majority of cases it is not economic to treat abandoned workings, except for old shafts, but to design the foundations and superstructure to withstand or accept subsidence. However, consideration of treatment should not be dismissed out of hand, since there are cases when it is worthwhile to carry out remedial measures. The cost of such treatment should be compared to the alternatives of amending foundation and superstructure design.

The main treatment methods are:

(1) Excavate down to working and backfill — this is only feasible for very shallow workings, i.e., less than 5 m down for buildings.
(2) Partial grouting to improve bearing capacity or limit void migration.
(3) Full grouting of workings.

6.7.2 Excavate and backfill
This method can be used, and justified, when the cost is lower than grouting; the cost valuation of the site is reduced sufficiently because of the problem (to make the solution cost effective) and the alternative of adjustments to a normal foundation and superstructure are more expensive.

The backfilling of approved material must be compacted in the manner specified. The HMSO specification (Reference 1) gives detailed guidance on backfilling and compaction.

6.7.3 Partial and full grouting
Partial grouting tends to be limited to pillar and stall workings up to 20 m below ground level. Grout mixes are commonly of p.f.a. and cement in ratios varying from 12 : 1 to 20 : 1 with a crushing strength at 28 days of 1 MN/m^2 and a water content not exceeding 40% of the weight of the solids. Generally the grouting pressure should not exceed 10 kN/m^2 per metre depth. The grout is pumped through a grid of drill holes at between 3−6 m centres. Further details of grouting are given in References 2−6.

Grouting is a specialist operation and it is strongly advised that only experienced and competent contractors are invited to tender.

6.8 TREATMENT OF ABANDONED SHAFTS

The treatment of shafts, bell-pits, swallow-holes and the like by capping or other means is far more common than the treatment of shallow workings. The reason is obvious — shafts extend to ground level and must therefore either be avoided by relocating the proposed building or treated. British Coal prefer relocation, even if the shafts are securely plugged and capped, and suggest leaving a safety zone as shown in Fig. 6.10. It has also been suggested that the

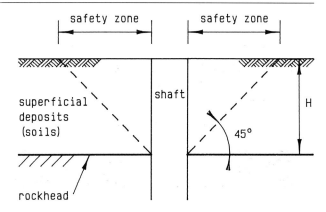

Fig. 6.10 Mine shaft − safety zone.

safety zone should equal H, up to a maximum of 30 m, or 2 × H, up to a maximum of 15 m depth of overburden.

Most old shafts were only partly filled, with the fill supported on a platform (often timber) just below ground level or within the depth of the superficial deposits.

Deterioration of the platform and/or shaft lining eventually leads to a collapse of the fill and the authors' consultancy has frequently been asked to advise in cases of sudden collapse.

Even shafts that appear to be filled are prone to collapse because the fill may migrate into the workings.

6.8.1 Capping
Reinforced concrete capping is the common method of covering a shaft (see Fig. 6.11). When buildings must be sited over capped shafts the dimensions of the cap will exceed those shown in Fig. 6.11 to ensure that the bearing capacity of the soil supporting the cap is not exceeded. The depth of the cap will also be likely to exceed that shown in Fig. 6.11 to accommodate bending and shear stress. Ideally the cap should be securely founded at rockhead and the shaft voids grouted.

6.9 EFFECT OF MINING METHOD AND METHOD OF TREATMENT

6.9.1 Introduction
Extra precautions in foundation design in mining areas are not always necessary. Foundations may be designed in the normal way when:

(1) Partial (see section 6.7.3) or complete consolidation, or other treatments have been successfully carried out.
(2) Where subsidence over old workings is complete and no new workings are envisaged.
(3) Where geotechnical surveys prove that there is a strong, thick overburden which will not subside.

Where these conditions do not occur the method of mine working must be considered.

6.9.2 Bell workings
If the extent and treatment cannot be guaranteed then:

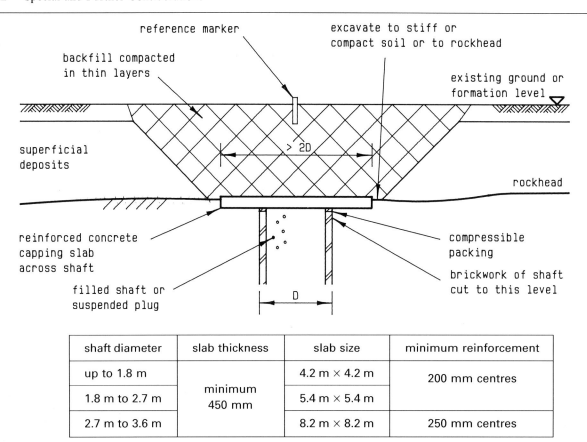

shaft diameter	slab thickness	slab size	minimum reinforcement
up to 1.8 m	minimum 450 mm	4.2 m × 4.2 m	200 mm centres
1.8 m to 2.7 m		5.4 m × 5.4 m	
2.7 m to 3.6 m		8.2 m × 8.2 m	250 mm centres

Notes
1 Reinforcement – use a minimum of 40 mm diameter in both directions at top and bottom of cap.
2 Cap must be a minimum of 3 m below any proposed adjacent building formation level or 1 m below ground level and be not less than 2D in width.
3 Cap should be founded on rockhead, if possible, or alternatively on a grouted base.
4 Vent pipes may be incorporated if the slab is placed on fill.
5 Extra precautions may be required where circumstances dictate that a building must be constructed over a shaft.

Fig. 6.11 Mine shaft – capping/filling.

(1) The proposed buildings should be relocated, or
(2) The foundations, such as piling or the use of deep basements, should be founded below the level of probable working for multi-storey buildings. Bell-pit workings are normally shallow so that such foundations can be economically feasible. Piling should not be used, however, where mining subsidence is still active (see section 6.10.6).
(3) For low-rise buildings stiff, strong foundations (i.e., doubly reinforced r.c. beams or, preferably, doubly reinforced two-way spanning rafts) should be provided (see Fig. 6.12). (This is in addition to the provision of structural movement joints discussed in section 6.10.1(3).) Typical worked examples are given in Chapter 13.

(*Note* Similar techniques can be used over swallow-holes and shafts.)

6.9.3 Pillar and stall
Due to long-term sub-surface erosion and weathering of the pillars, *punching* through the floor or roof, or increase of loading from new structures, can cause the pillars to collapse. The result, at ground level, is similar to bell-pit action in that loss of ground support can be sudden, unpredictable and localized in area. The foundation design should be similar to that described in the preceding section and Chapter 13.

In both bell-pits and pillar and stall working the associated ground movements are vertical, erratic and localized, and the use of reinforced rafted structures is usually the solution. Low-rise buildings are the worst affected – terraced housing of load-bearing brick walls, a brittle material, founded on unreinforced footings can be seriously damaged. Multi-storey structures with deep r.c. shear walls providing the main structural support are often more capable of resisting the effects of ground movements by spanning or canti-levering over the subsidence depression.

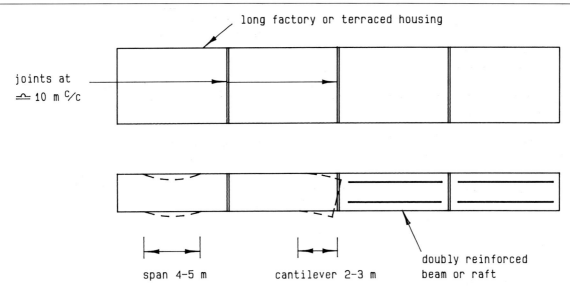

Fig. 6.12 Jointed structure/reinforced foundation.

CROWN HOLES

Where the overburden of the worked coal is weak, or a pillar fails, the roof can collapse into the workings and form in the first instance, a void above the workings. This void will *migrate* to ground level forming a depression known as a crown hole. The depth of the crown hole will not be equal to the depth of the seam since bulking of the soil collapsing into the seam will take place (see Fig. 6.13). Where crown holes are evident a check should be made for others and the likelihood of them occurring. The crown hole can be consolidated by grout injection or the foundation designed to bridge over it. Beams or rafts are commonly designed to *span* over possible 3 m diameter holes or cantilever 2 m.

6.9.4 Longwall workings

Most coal is now extracted by longwall working and the associated ground movements can be predicted with reasonable accuracy by subsidence engineers. The predictions are based on a relatively vast amount of empirical data and procedures which have been developed from continuous study of records and observations.

There is, as yet, no reliable scientific procedure for determining the magnitude, rate and form of subsidence.

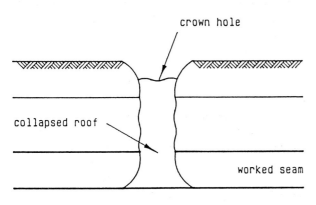

Fig. 6.13 Crown hole.

The area of ground surface affected by longwall mining is relatively large (whereas bell-pit and pillar and stall tend to be localized). The rate of subsidence depends on the rate of extraction and tends to be rapid at first, often 90% or more occurring within weeks of extraction, and then slows down and the residual subsidence can take two years and sometimes more to complete. The magnitude of subsidence will depend on the depth of worked seam and the depths and type of overburden. Advice on position, rate and magnitude of the subsidence can be obtained from the mining subsidence engineer. Fuller details are given in References 2, 3 and 5.

The subsidence wave advances in front of the working face causing, first, tension stresses in the ground, resulting in most damage to structures, followed by tilting and finally relatively short-term compression strain in the ground.

Foundations for superstructures for proposed buildings in potential subsidence areas, due to longwall workings, should be designed as rafts (see Chapter 13).

6.9.5 Rafts founded over longwall workings

Rafts, in addition to the normal pressures and stress, can be subject to ground strain in subsidence areas and it is advisable to understand the effect of such strain.

The structural strains in the raft are caused by *drag* from frictional forces generated by movement or strain in the supporting ground which can cause lengthening or shortening of the raft. A simplified explanation of the phenomena is shown in Fig. 6.14.

The figure shows that the raft is being dragged apart by the ground movement. The drag force is proportional to the weight of the structure and the frictional resistance (or coefficient of friction) at the ground–raft interface. The coefficient of friction between a concrete raft and sand slip-plane can be taken as 0.66 and the frictional force is usually taken as (weight of structure)/2 (see Design Example 4 in Chapter 13).

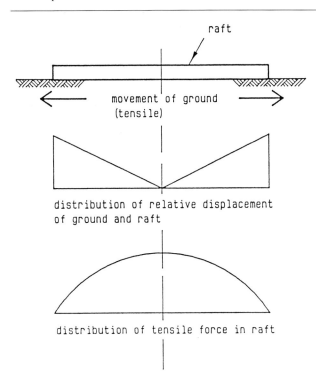

distribution of relative displacement
of ground and raft

distribution of tensile force in raft

Fig. 6.14 Effect of ground strain on raft.

6.10 DESIGN PRINCIPLES AND PRECAUTIONS IN LONGWALL MINING SUBSIDENCE AREAS

6.10.1 Introduction

The interdependence of the foundation and superstructure is an important design consideration in mining subsidence areas to produce a balanced and integrated structure.

Some advice on designing the building to cater for movement is given in the following sections, and the following points should be considered:

(1) Flexible superstructure structures with simply supported spans are preferable on flexible foundations, alternatively, stiff superstructures jointed to form small units can be accommodated on stiff foundations.

(2) Larger buildings should be jointed into smaller adjacent components with the joint extending also through their foundations which should comprise shallow, smooth-soffited rafts laid on a 150 mm thick layer of compacted sand which acts as a slip-plane to isolate the raft from the tensile and compressive ground strains.

(3) Large structures should be subdivided into smaller independent units by gaps or flexible joints, at least 50 mm wide, through superstructure, foundations, services and finishes. The necessary expansion joints in a superstructure can be used for such jointing.

(4) Avoid whenever possible the use of basements. Where these must be used the external walls should be protected from ground strains by the provision of a 100 mm thick expanded polystyrene layer, or similar, and the underside of the basement slab provided with a sand slip-plane (see 2 above).

(5) Connections between *pinned* structural members

should have adequate tensile strength to ensure that differential movement does not lead to progressive collapse.

(6) Masonry arches should be avoided.

(7) Brittle finishes should be avoided – use plasterboard and dry linings in lieu of plaster; avoid high-strength brittle mortars; use boundary fences and not walls, etc.

(8) *Wrap around* corner windows, projecting window bays, rigid concrete paving cast immediately against the external walls and similar potential problem areas should be avoided.

(9) Buildings should not be located over faults or in the area affected by the fault. Where this is not possible, is impractical or causes excessive extra costs the area of the building should be thoroughly jointed to isolate it from the rest of the structure and the foundation should be provided with extra stiffness. Subsidence and ground strains tend to concentrate along faults and can thus relieve, to some extent, the surrounding ground from disturbance.

(10) Retaining walls likely to be affected by ground strains should be free-standing and not be structurally integral with the superstructure.

(11) Excessive downstand beams and other projections below the smooth rafts should be avoided.

(12) Roofs should be provided with ample falls and where reversal of tilt is possible alternative outlets should be provided.

(13) Load-bearing masonry structures should not use brittle mortars; reinforcement may be added and planes of weakness, i.e., lining up of doors with windows over, should be avoided or the implications considered. Masonry structures have been successfully prestressed to resist tensile stresses due to subsidence.

(14) Consider using jacking points, where necessary, to re-level the building.

6.10.2 Rafts and strips for low-rise, lightly loaded buildings

Details of rafts for low-rise, lightly loaded buildings such as houses, single-storey clinics, primary schools and similar are shown in Fig. 6.15. Additional details are provided and explained in Chapter 13.

Where such rafts are expensive relative to the lower cost of housing repairs in areas of minor subsidence then British

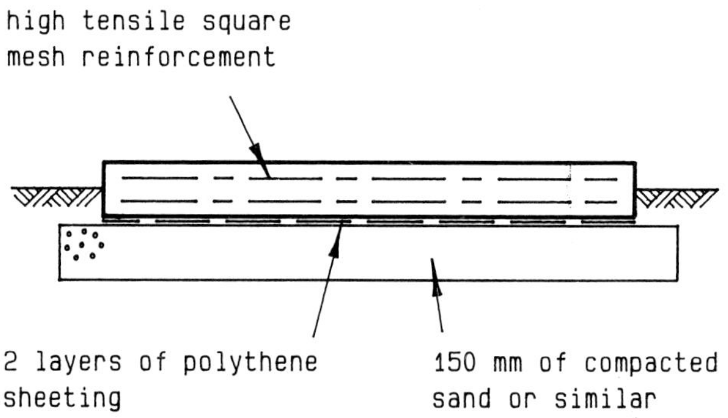

high tensile square
mesh reinforcement

2 layers of polythene   150 mm of compacted
sheeting   sand or similar

Fig. 6.15 Raft detail – low-rise/lightly loaded buildings.

Coal recommend strip footings, with some reinforcement, founded on a sand slip-plane with a compressible filler at the vertical ends of foundations to allow for longitudinal movement of the ground.

6.10.3 Rafts for multi-storey structures or heavy industrial buildings

Cellular rafts can be more economic than very thick, solid rafts and if basements are necessary and their use unavoidable they can in some cases be used as cellular rafts. The raft in a basement should be founded on a similar slip-plane as housing (see Fig. 6.16 (a)) and basement walls should, as described earlier, be externally clad with expanded polystyrene to absorb compressive strains in the ground (see Fig. 6.16 (b) and Chapters 9 and 13).

6.10.4 Jacking points

If it is vital to re-level the structure due to permanent tilting of the ground (a fairly rare occurrence) then the placing of hydraulic jacks or jacking points in the foundation under the walls and columns should be provided.

6.10.5 Service ducts

Where possible service ducts should be incorporated in the cells of cellular rafts. On some lightly loaded low-rise structures such as schools and hospital wards, requiring extensive services, consideration should be given to the use of suspended floors above the structural foundation raft with the void thus formed used to accommodate services.

Where this is expensive or impractical the service duct may have to protrude below the invert of the raft. It should then be designed as a box beam founded on a horizontal slip plane, clad externally with polystyrene, separated from the raft invert by three layers of felt or similar separator and with the raft more heavily reinforced over to prevent it breaking its back, in hogging, over the duct.

6.10.6 Piling

Piling should be avoided if at all possible since the horizontal ground movements may either shear through the piles or transfer excessive tension into the beam or slab over at the pile head.

Piles may be used over longwall workings when subsidence is complete and the overburden is too weak to support a raft and where there is strong rock below the worked-out seam. The piles should be taken below the seam, be of precast concrete or tubular steel filled with concrete and designed to withstand not only the structural load but also any possible downdrag.

A smaller number of large-diameter piles are preferable to a large number of small-diameter piles since their ratio of surface area to cross-sectional area is lower and thus reduces the effect of downdrag. Drilling the pile hole and sleeving it before inserting the pile could almost eliminate downdrag effects. To reduce transfer of stress, due to horizontal movement of the pile head, it may be worth considering topping the pile with two layers of unmortared slate or neoprene bearing pads and then capping with an oversized pile cap.

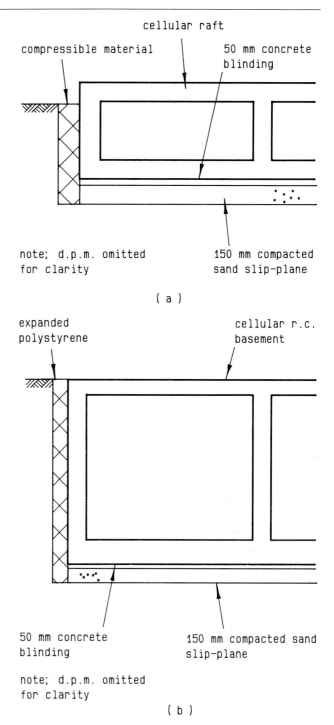

Fig. 6.16 Raft detail − multi-storey/heavy industrial buildings.

The employment of such a method of piling usually costs more than the use of a cellular raft so it is relatively uncommon. Furthermore, piling can disturb other previously stable mine workings and set off further subsidence.

6.10.7 *Articulated* foundation

Articulated or *three point support* has been used in a number of European countries. The foundation consists of three pads which support short, low-height columns resting on steel balls or other pinned joints. A beam and slab connect

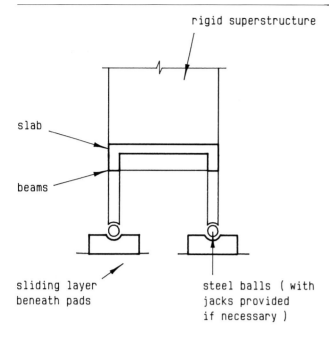

slab

beams

sliding layer
beneath pads

steel balls (with
jacks provided
if necessary)

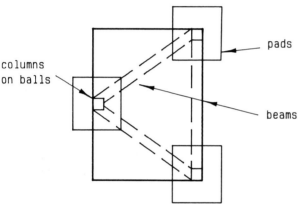

columns
on balls

pads

beams

Fig. 6.17 Articulated foundation.

and rest on top of the columns and form the base of the superstructure (see Fig. 6.17).

The *tripod* of pads will tilt as the subsidence wave passes but they will remain in the same plane. The superstructure will tilt but not suffer the effects of differential settlement or subsidence.

6.11 SUPERSTRUCTURES

6.11.1 Introduction

The superstructure, like the foundation, should be either completely flexible or completely rigid. Mixtures of the two techniques can lead to problems since partial strengthening may actually increase the damage due to ground subsidence and movement. It can sometimes help if rigid superstructures can slide on a slipping membrane, i.e., d.p.c. over the foundation. In other cases where the rigidity of the superstructure can enhance the stiffness of the foundation by structurally integral action then the connection between the superstructure and the foundation should be *fixed*.

6.11.2 Rigid superstructures

SINGLE-STOREY STRUCTURES
The authors' consultancy has designed a number of rigid single-storey structures which have successfully withstood two decades of mining subsidence (where other structures in the close vicinity have either collapsed or been severely damaged). A typical example, outlined in Example 2 of section 1.6, was the conversion of an r.c. column and beam frame into a Vierendeel girder.

MULTI-STOREY STRUCTURES
Where the structure has a relatively large number of walls, as in tall blocks of flats built in in situ concrete, then the walls act as deep, stiff beams and can easily cope with spanning or cantilevering over subsided ground areas. A large number of 14+ storey blocks of flats of plan dimensions of the order 25 m × 15 m built off 2 m thick r.c. rafts have successfully withstood the effects of subsidence.

6.11.3 Flexible superstructures

When it proves an economical alternative (and it often does), a flexible building with articulated joints, i.e., pins, will tolerate ground movement by readjusting its shape (see Fig. 6.18).

Simple surface foundations, capable of riding over the subsidence wave, are not only often adequate but are generally less expensive than a rigid foundation.

Care however must be taken in the readjustment of shape, in that the cladding is adequately overlapped to prevent water ingress and that internal finishes are flexible and not rigid or brittle. It is therefore advisable that other members of the design-and-build team are informed of the engineering decision so that the building design and construction implications can be appreciated and accommodated by all the various disciplines.

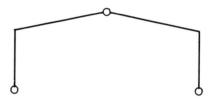

before ground movement

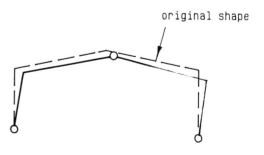

original shape

after ground movement

Fig. 6.18 Pin-jointed superstructure.

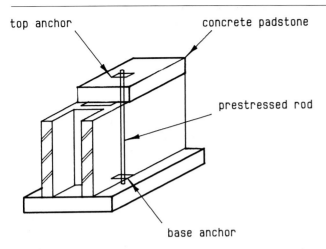

masonry diaphragm wall

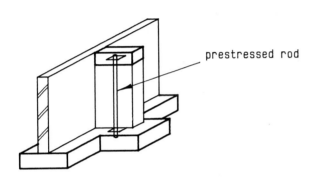

masonry fin wall

Fig. 6.19 Prestressed masonry.

SINGLE-STOREY STRUCTURES

The most common form of flexible superstructure is the three pinned arch shown in Fig. 6.18 and the authors' consultancy has designed a large number of such structures. They have been constructed in structural steel, precast concrete and glulam timber and used for industrial buildings, schools, churches and other buildings.

Over the past few years, where masonry has proved an economical alternative for tall single-storey structures, the authors have used prestressed, free-standing masonry diaphragms and fins (see Fig. 6.19). The roof sits, simply supported, and can be tied down to resist wind uplift on such walls.

MULTI-STOREY STRUCTURES

One of the earliest, best known and widely used techniques is the CLASP system (Consortium of Local Authorities Special Programme). This like the other techniques discussed is founded on a thin, flexible r.c. raft with the coefficient of friction between the raft and supporting soil reduced by a slip-plane of sand covered by a polythene membrane.

The frame is pin jointed and provided with diagonal bracing between columns to resist horizontal forces. The bracing incorporates springs to permit the steel frame to *lozenge* in any direction (see Fig. 6.20). All external cladding, internal finishes and their fixings are designed so that movement can take place without distortion or cracking. The floors and roofs act as stiffening diaphragms but have flexible fixing to the frame.

A number of timber framed housing and four-storey maisonette systems have been developed and, with care to eliminate damp penetration, have performed satisfactorily.

6.12 MONITORING

Whenever possible the performance of the structure should be monitored and the information gained passed to British

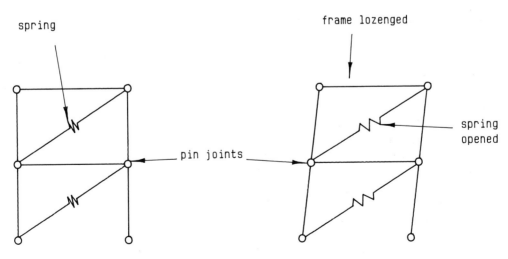

Fig. 6.20 Pin jointed/spring braced superstructure.

Coal and the Building Research Establishment (BRE). This is particularly important for innovative, or non-standard, design. The more records and information on ground, foundation and superstructure interaction acquired then the more efficient can become future design.

6.13 REFERENCES

1. HMSO (1986) *Specification for Highway Works*, Parts 1–7.
2. Institution of Civil Engineers (1977) *Ground Subsidence*. ICE, London.
3. CIRIA (1984) *Construction over abandoned mine workings*.
4. British Standards Institution (1986) *Code of practice for foundations*. BS 8004, BSI, London.
5. British Coal (1975) *Subsidence Engineers' Handbook*.
6. Department of the Environment (DoE) (1976) *Reclamation of Derelict Land: Procedure for Locating Abandoned Mine Shafts*.

CHAPTER 7

FILL

7.1 FILLED SITES

7.1.1 Introduction

The main body of written information on soil mechanics and foundations deals with virgin ground and civil engineering solutions. The majority of site problems, on the other hand, relate to fill materials and structural building foundations. Fill materials encountered during soil investigation are often contaminated so that there is an overlap between this and other chapters.

Some publications suggest that the use of landfill sites is a recent problem. However, the authors have been dealing with such sites since the late 1950s and these sites have for a long time been 'bread and butter jobs' for many structural engineering practices. Filled sites are at present being reclaimed in greater numbers and developed more economically than in the early days and there is a greater awareness of the hazards and dangers of contamination and gas emissions. The treatment of hazards and gas emissions must be considered in relation to carrying out the site investigation operations and it is necessary to advise on measures to ensure satisfactory long-term development. This chapter however concentrates on the structural aspects of filled sites and refers only briefly to these hazards which are covered more fully in Chapter 5.

7.1.2 Movement and settlement

The word *container* as used in this chapter is defined as the periphery surface of virgin ground within which the fill is contained. Some important characteristics which relate to movement and settlement of fill within a *container* are listed below:

(1) The outer conditions surrounding the fill material i.e. the shape, strength, surface roughness and sub-surface condition of the fill container.
(2) Sediment, water or chemical deposit at or near the surface or within the container.
(3) The properties of the fill material i.e. consistency, density, strength, decay characteristics, gas emissions, moisture content, void ratio and chemical content.
(4) The history of placing and more recent disturbances.
(5) The direction, location and orientation of proposed structures and loadings.
(6) Test results from the fill material.

From an evaluation of the above characteristics for a particular site the designer can decide on the actions and design requirements for the site including the necessary treatment of the fill. Advice can also be given suggesting any re-orientation or revision to the location of the proposed works to minimize potential problems. To assist the evaluation, the implications of these characteristics will be considered under the following headings:

- The container surface
- The container edges
- The container base
- The container sub-strata
- Water
- The fill material
- Fill investigations
- Settlement predictions: (1) fill alone (2) combined effects
- The development and its services – treatment and solutions.

7.2 THE CONTAINER

7.2.1 The container surface

The periphery conditions at the interface between the fill and the virgin ground are most important, i.e. the container shape, edge condition and base condition affect the behaviour of the fill material within it, and Fig. 7.1 indicates some typical sectional examples of surface shape and resulting fill cross-sections.

It can be seen that the container shape will affect the resulting fill settlement, since the depth of fill can vary considerably across the site and the width of fill can also vary with depth of step positions (see Fig. 7.1 (e)).

It can also be seen from Fig. 7.1 (e) that, should consolidation of the fill occur, voids would develop below the overhanging steps. The importance of the variation of the plan shape with depth is indicated in Fig. 7.2 since migration of the fill layers downward into new cross-sections with reduced or increased plan area affects the final settlement profile and magnitude. Since settlement results from a total volume change, these effects can result in differential settlements at surface level, which are also indicated in Fig. 7.2.

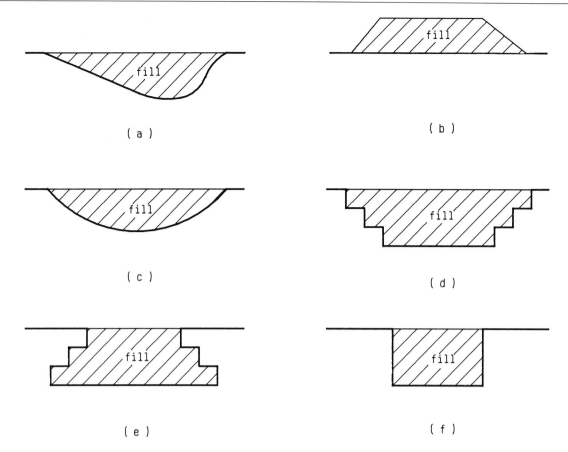

Fig. 7.1 Fill cross-section.

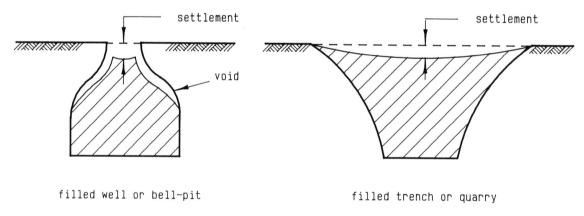

filled well or bell-pit   filled trench or quarry

Fig. 7.2 Fill consolidation.

7.2.2 The container edges

The restraint at the edges of a container can delay or reduce locally the total settlement. Restraint at the container edges can be due to frictional drag or mechanical keying against the face of the container (see Fig. 7.3).

Sudden settlement can occur in these restrained zones when either:

(1) Consolidation causes the voids to migrate upwards to the surface level, or
(2) Lubrication of the container face by water or other liquid reduces the friction or erodes the fill.

Backfilled open-cast quarries in rock areas are particularly vulnerable to such settlements around the edges of the quarry, especially when damaged land drains, etc., have been left discharging down the quarry face. The quarry itself can also act as a sump collecting surface water and ground-water. Water is not the only possible seepage however, since other contaminants may also discharge into old open-cast workings around the edges of the container (see Fig. 7.4).

7.2.3 The container base

Many fill containers have been left derelict prior to filling

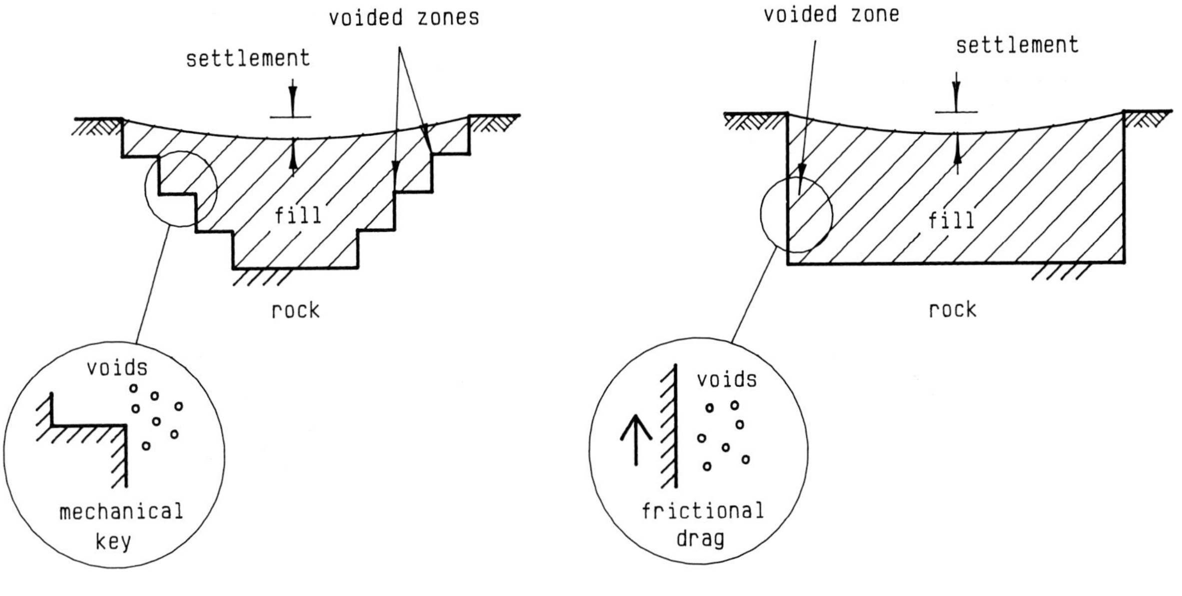

voided zones

settlement

fill

rock

voids

mechanical
key

voided zone

settlement

fill

rock

voids

frictional
drag

Fig. 7.3 Fill edge restraint.

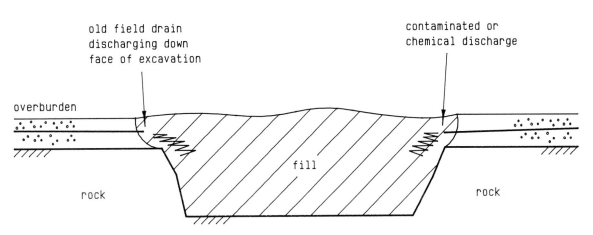

old field drain
discharging down
face of excavation

contaminated or
chemical discharge

overburden

fill

rock

rock

Fig. 7.4 Filled quarry.

and sediments and waste products often litter the base. Soft silts, decaying vegetation, old car bodies, etc., are often at the base of the more recent fill materials and create excessive settlement, sometimes over very long periods. Standing water at the base of the container is a common condition prior to filling and often these sites have been filled without any engineering supervision or control. There was rarely an intention to develop at the time of filling and the designer should expect uncontrolled tipping of waste materials. Figure 7.5 indicates some of the conditions which commonly exist.

In addition, the edges and sometimes the base of the container may be undermined by tunnels and other remains of previous shallow mine workings which pre-date open-cast activities above them. The container base itself may have hazards local to its surface or a short distance below, which have resulted from previous underground workings. These hazards may have structural implications and/or be the source of possible gas emission.

7.2.4 The container sub-strata

The peripheral and internal differential movements of the fill are significantly affected by the container sub-strata. A yielding container sub-strata may allow a relatively uniform settlement to occur between the container and its contents. A container of sand backfilled with similar material to a similar compaction and density (see Fig. 7.6) will settle in a similar manner to that of the virgin sand. In this case, loading of the varying depths of fill will produce a varying settlement due to the fill consolidation but this will be partly compensated by the yielding of the sub-strata below the container.

At the other extreme a similar shaped container in rock, backfilled with sand, could result in critical differential settlements (see Fig. 7.7).

The non-yielding surface of the container in Fig. 7.7 will be reflected at surface level due to the greater consolidation over the deeper fill areas. This example illustrates the need to assess the container and its sub-strata, along with the fill

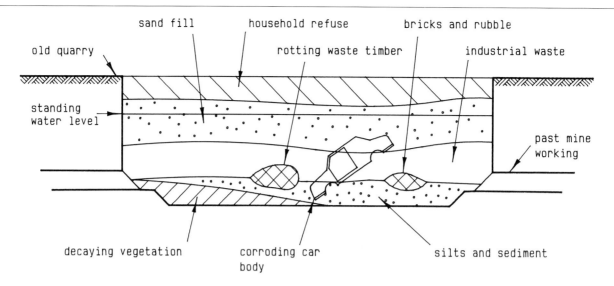

Fig. 7.5 Uncontrolled filling.

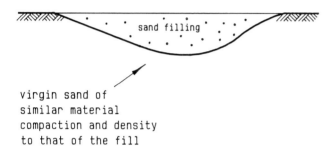

Fig. 7.6 Sand fill on sand sub-strata.

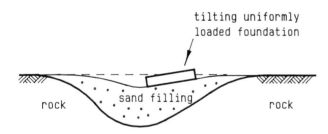

Fig. 7.7 Sand fill on rock sub-strata.

material, in arriving at settlement predictions. It also highlights the detrimental effect which strong, unyielding containers can have in relation to differential settlements.

7.3 WATER

Water and water movement can have an effect on many different situations and materials, some of which are briefly discussed below.

7.3.1 Effect of water on combustion

The introduction of water into combustible fills can, under some circumstances, increase the likelihood of combustion by carrying oxygen into the fill material. In other circum-

stances the water may decrease the chances of combustion due to the cooling effect of the passing water (see Chapter 5 for more information).

7.3.2 Effect of water on chemical solutions

Water rising to the surface can carry chemical solutions against surfaces of concrete and other materials used in the construction (see Chapter 5 for more information).

7.3.3 Water lubrication

Movement and variation in the water-table can effect the removal of fines from fill materials and can also remove and/ or lubricate the face of materials down the edges of the container as previously mentioned.

7.3.4 Water inundation

The strength and settlement of the fill material can be greatly affected by submergence in water and serious collapse settlement of loose and unsaturated fill material can occur on inundation with water.

7.3.5 Organic decay

Water can accelerate decay of organic material deposited within the fill.

7.3.6 Information from water

Tests on water samples taken from the boreholes can provide valuable information on contaminants and soluble chemicals. Variation in standing water levels between boreholes can be an indication that impervious layers may occur between the positions.

7.4 THE FILL MATERIAL

7.4.1 Introduction

Various factors affect the structural performance of fills and these include the type, quality, density and consistency. Table 7.1 indicates these qualitative classifications of fills.

Table 7.1 Qualitative classification of fills

Classification	Description
Nature of material	Chemical composition Organic content Combustibility Homogeneity
Particle size distribution	Coarse soils, less than 35% finer than 0.06 mm; fine soils, more than 35% finer than 0.06 mm (BS 5930: 1981)
Degree of compaction	Largely a function of method of placement: thin layers and heavy compaction − high relative density; high lifts and no compaction − low relative density; end tipped into water − particularly loose condition. Fine grained material transported in suspension and left to settle out produces fill with high moisture content and low undrained shear strength, e.g. silted up abandoned dock or tailings lagoon.
Depth	Boundary of filled area Changes in depth
Age	Time that has elapsed since placement: if a fill contains domestic refuse, the age of the tipped material may be particularly significant, since the content of domestic refuse has changed considerably over the years; during the last 40 years the ash content has decreased while the paper and rag content has increased; the proportion of metal and glass in domestic refuse has also increased during this period; it may be that more recent refuse will be a much poorer foundation material than older refuse not only because there has been a shorter time for settlement to occur, but also because the content of material which can corrode or decompose is greater
Water-table	Does one exist within the fill? Do fluctuations in level occur? After opencast mining a water-table may slowly re-establish itself in the fill

These properties affect considerably the amount of total and differential settlement which can be expected within the life of any proposed development. For example, consider two similar containers as shown in Fig. 7.8.

Container A is filled on a cleaned surface with compacted layers of consistent granular material to a consolidated granular mass, similar to that of the surround ground of the container. Container B is filled by end tipping various waste materials into the sedimentary deposits (uncleaned) of the disused depression.

The design of foundations for Container A can be carried out using normal criteria for design, similar to that of the surrounding virgin soils. For example, shallow foundations for low-rise structures would be suitable. The design of foundations for Container B would present greater difficulties and settlements could prove impossible to predict to any

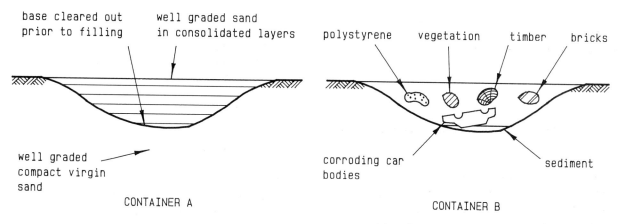

Fig. 7.8 Factors affecting differential settlement.

degree of accuracy. Large long-term settlement would result from decaying vegetation and from corroding car bodies if traditional shallow foundations were constructed in the fill, therefore either piling or vibro-techniques would be more suitable (see Chapters 8 and 14).

The most common and widespread conditions relate to derelict and abandoned sites in the inner-cities where buildings have been demolished into old basements and depressions. There are numerous sites with basements and cellars filled with brick rubble, timber, steel joists, etc. Old sewers and similar abandoned services remain and many have collapsed. Old basement walls and foundations form hard spots and cause obstruction to piling and vibro-compaction operations. Often the sub-strata has consolidated from previous buildings and the container edge is of brick construction. Many such sites are best dealt with by piling or vibro-techniques and the hard spots reduced in level. This is just one type of filled site; the problems and solutions on fill are varied and numerous and reference should be made in particular to Chapters 9 and 10 on foundation types and solutions, Chapter 5 on contaminants and Chapter 8 on ground improvement techniques.

7.5 FILL INVESTIGATIONS

7.5.1 Special requirements
In order to assess the likely settlement on fill, the site investigation generally needs to be more detailed than that for virgin sites. Some reasons for this are:

(1) Possible gas emissions and other health hazards are more likely and need to be revealed early.
(2) Knowledge on the boundary conditions of the site and/or its container need to be revealed.
(3) Fill tends to be more varied and the consistency, strength and organic content of the fill needs to be determined.
(4) Obstructions are more likely in fill and therefore the effect of obstructions, etc., old basement walls, cellars, and abandoned sewers needs to be assessed.

The numerous and varied methods of filling such sites make the task of investigation seem daunting. A few simple procedures, however, if adopted, provide a systematic approach which can prevent excessive expenditure caused by incorrect sequencing of the soil investigation.

7.5.2 Suggested procedures
The suggested procedures are as follows:

(1) A desk-top study and historical review of the site and its surroundings should be carried out. A preliminary investigation into the likelihood of gas and chemical waste should be made and its effect on the approach to the site investigation should be assessed. Information relating to the placing of the fill, the previous use of the site, the possibility of contamination, etc., should be noted.
    If the desk-top study indicates the likelihood of hazardous conditions, then the recommendations for contaminated sites, detailed in Chapter 5, should be followed. If the desk-top study reveals the likelihood of fills without a health hazard then a site *walkabout* should be undertaken followed by the procedures indicated below.
(2) A simple trial hole investigation, using an excavator, should be carried out. The purpose of these trial holes is to provide a general feel of the site and the conditions likely to be encountered by a more detailed investigation. The holes are inspected from surface level, thus preventing the need for timbering and other expensive works associated with deeper trial holes. These holes will reveal the likely strength consistency, organic content and an indication of the boundary conditions to the experienced eye.

At this stage a decision can be made as to whether or not to spend money on a detailed analysis of the fill. For example, if the fill is unlikely to contain health hazards or chemicals but appears unsuitable for load-bearing pressures, then the more detailed testing and recording of information will be carried out for soils below fill level.

Alternatively for fills which appear to have suitable load-bearing capacity, detailed testing and borehole logging of the materials will be required. Therefore from the initial trial holes the extent of detailed soil investigation requirements and testing procedures can be established.

For example, fills of inconsistent material properties such as soft clays intermingled with topsoils and pockets of organic material are likely to be unsuitable for load-bearing purposes. However, consistent firm granular fills free from organic content are likely to prove suitable for surface spread foundations. A detailed investigation of the fill for the latter case would prove cost effective in foundation economy.

The ground investigation should commence with trial pits starting on a wide grid and closing the investigation to a more suitably spaced grid relative to the consistency of the information revealed. Notes should be taken on the nature of the fill (its composition, variability, moisture content, organic content, etc.). Based upon the information revealed suspected boundaries or possible edges of the fill should be excavated through at right angles with deep trenched trial holes to reveal the shape of the edge conditions. All relevant information should be recorded. A grid of boreholes should then if necessary be driven and changes in material, sample locations, in situ tests and water ingress should be recorded.

Standpipe piezometers can be sealed into selected boreholes if required to monitor water levels. Grid levelling stations can be established if needed and related back to a fixed datum to monitor settlement and differential movement due to own weight or test loading.

7.6 SETTLEMENT PREDICTIONS

7.6.1 Settlement: fill only
The settlement of fill sites results from a number of applied compressive forces. The major forces are those from:

(1) The own weight of the fill.
(2) The weight of the proposed structure.
(3) Water movement or inundation.
(4) Preloading.

The most significant force in deep fill is generally that resulting from its own weight and this will often be the principle cause of long-term settlement. In loose unsaturated fills the designer should consider the hazard of inundation which may cause collapse settlement. In normal granular fills the majority of settlement due to own weight occurs as the fills are placed but in many fills this can leave significant creep settlement to occur from constant effective stress and moisture movement. For many fills the rate of creep decreases relatively quickly with time and, when plotted against the logarithm of time elapsed since deposition, produces an approximately linear relationship. This linear relationship approximation for prediction of settlements, however, can be unreliable and applies only where conditions in the fill remain unaltered. Some typical values for the percentage vertical compression of the fill that occurs during a log cycle of time after the placing of the fill are shown in Fig. 7.9. The designer must apply his experience and judgement in the use of such graphs.

The consolidation of cohesive fills is much slower than granular fills and when haphazardly intermixed with other materials can make time-related settlements impossible to predict. When fine material is placed under water however, a soft, cohesive, low permeability fill is formed and the resulting settlement is controlled by a consolidation process as water is squeezed out from the voids of the fill. The process of consolidation occurs as excess porewater pressure dissipates slowly from the fill (see Chapter 2). In some such fills disturbance can cause liquefaction to occur. However, in general if such fills are consistent, the settlement can be predicted to some degree of accuracy by applying normal soil mechanics theories in relation to consolidation.

The compressibility of fill materials varies widely, depending on the particle size distribution, the moisture content, the existing stress level, the void ratio and the stress increases likely from the proposed development. In general it is the limitations of differential settlement which will determine the design bearing stresses for the proposed structures and not the bearing capacity of the fill. Assuming that settlement is the limitation that applies to the fill, simple calculations can be based on compressibility parameters which are related to one-dimensional compression and the BRE have proposed some typical values of constrained modulus for small increments of vertical stress and for a number of different fill types. Table 7.2 indicates constrained modulus for an initial stress in the region of $30 \, kN/m^2$ for increases in stress of approximately $100 \, kN/m^2$.

These constrained moduli are only applicable to small increments in vertical stress for limited conditions and therefore are very restricted in their applications. In addition they do not give the more critical information which relates to differential settlement. The differential movements should be calculated based upon observed variations in the fill material plus an allowance for some additional variation from that observed.

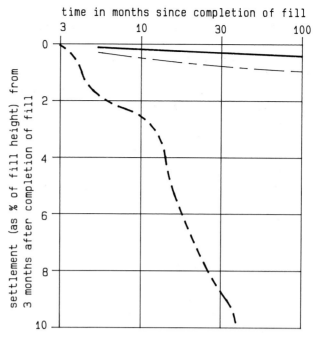

Fig. 7.9 Settlement rates of different types of fill (vertical compression plotted against log_{10} time). (Reproduced from Building Research Establishment *Digest 274*, Table 1, by permission of the Controller of HMSO, Crown copyright.)

It should be noted that the values given in Table 7.2 for the creep compression rate parameter α are applicable to settlement under self-weight in various fills. However, these can also be used in general for settlements resulting from applied loads. In such calculations zero time should be related to the application of the load, not the placing of the fill. It should also be appreciated however, that the value for domestic refuse is different under the two different types of loading and the values given relate mainly to the decay and decomposition of organic matter. Provided that the creep due to the decay is taken into account in the assessment of the own weight loading then the increase in stress due to the building will not be a function of this organic decay. A value of 1% is suggested by the BRE for this condition.

The designer must use his judgement in the use of such predictions and needs to consider that the word *fill* is the only soil description used in soil mechanics which embraces such a wide variety of materials. The present excellent research on fills is nevertheless only touching on the edges of the subject. In such circumstances it is evident that much more work will be required before real predictions can be made and mathematical calculations become relatively accurate. In the meantime engineering judgement based on

Table 7.2 (Reproduced from British Research Establishment
Digest 274, **Table 2, by permission of the Controller of**
HMSO, Crown copyright.)

(a) Creep compression rate parameter α

Fill type	Typical values of α (%)
Well compacted sandstone rockfill	0.2
Uncompacted opencast mining backfill	0.5–1.0
Domestic refuse	2.0–10.0

(b) Compressibility of fills

Fill type	Compressibility	Typical values of constrained modulus (kN/m²)
Dense well-graded sand and gravel	very low	40 000
Dense well-graded sandstone rockfill	low	15 000
Loose well-graded sand and gravel	medium	4 000
Old urban fill	medium	4 000
Uncompacted stiff clay fill above water-table	medium	4 000
Loose well-graded sandstone rockfill	high	2 000
Poorly compacted colliery spoil	high	2 000
Old domestic refuse	high	1 000–2 000
Recent domestic refuse	very high	—

experience combined with the present research knowledge is the only reliable method.

The following design examples are given as guidance only and should not be relied upon alone in predicting the movement likely to occur in practice.

EXAMPLE 1

Calculate the settlement prediction for the first ten years after placing of a fill 12 m thick, consisting of loose compacted colliery spoil. The fill has an α value of 1% for a $\log_\alpha$ cycle of time between one and ten years for a one-dimensional compression.

$$\text{Compressive settlement after ten years} = \frac{1 \times 12 \times 10}{100} = 120\,\text{mm}$$

EXAMPLE 2

If a development is proposed which will on average increase the vertical stress in the upper fill of the site used in Example 1 by 50 kN/m², determine the predicted approximate increase in total compression strain in the top 1 m thick layer.

From Table 7.2, the approximate constrained modulus for this fill would be in the region of 2000 kN/m².

The constrained modulus is also equal to the increase in

vertical stress divided by the increase in vertical strain,

$$\text{i.e. constrained modulus} = \frac{\Delta\sigma}{\Delta\varepsilon}$$

In this case
$$2000 = \frac{50}{\Delta\varepsilon}$$

therefore
$$\Delta\varepsilon = \frac{50}{2000} = \frac{1}{40}$$

settlement
$$\rho 1 = H_1\,\Delta\varepsilon = 1\,\text{m} \times \frac{1}{40} = 25\,\text{mm}$$

which is the settlement in the top 1 m due to the increased load.

In order to calculate the total increase in settlement the stress increase in each layer would need to be calculated, splitting the depth of fill into suitable layers relative to the thick layers and the settlement for each layer calculated as indicated above. The total settlement would be the summation of the individual settlements of all these layers.

In order to calculate differential settlements it is necessary to make comparisons between two locations on the site. It is often assumed that for small sites the two greatest differences in conditions in the site investigation data could exist between these two locations. These would be the loosest/poorest material with the highest stress compared with the firmest material under the lowest stress. The difference in movement between these two positions would give some guidance as to the differential settlement that may be expected.

7.6.2 Settlement: combined effects

The limiting effect of differential settlement has to be predicted from combining all the various movements, causes and time relationships. Since even on virgin sites, this is not a precise science it should be appreciated that the prediction of settlement of fills is even less accurate. The designer therefore must allow himself margins of safety relative to the nature and extent of detailed information he has obtained from the site, its loading and sub-strata conditions. This safety margin should also relate to the designer's experience, since unfortunately the prediction of settlements is as much an art based upon experience as it is a science. To build up the analysis of the time related differential settlements, the designer should determine approximate values of movement related to various times within the development. A typical diagrammatic representation of the total settlement for two locations within a container, for a particular point in time, are shown in Fig. 7.10.

To arrive at this accumulation of settlement the designer must consider for each location the magnitude of settlement and time relationship caused by the following load conditions:

(1) The own weight of the fill.
(2) Consolidation of the container sub-strata.
(3) Creep consolidation.
(4) The effects of the proposed development.
(5) Decay, corrosion, etc.

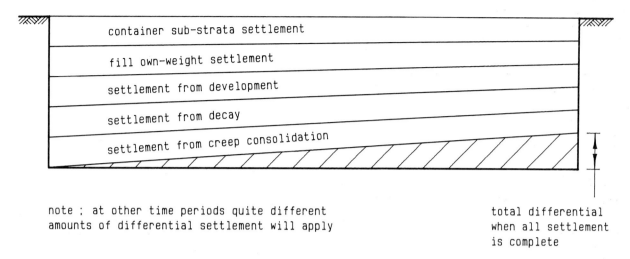

container sub-strata settlement

fill own-weight settlement

settlement from development

settlement from decay

settlement from creep consolidation

note : at other time periods quite different
amounts of differential settlement will apply

total differential
when all settlement
is complete

Fig. 7.10 Combined effects of differential settlement.

From this information the critical time point which gives the maximum differential and total settlements from combining the effects under (1) to (5) can be determined.

In order to select the critical conditions the designer need not carry out detailed calculations but can summate these mentally prior to making a decision. It is on the broad consideration of these differential settlements that a decision is made as to whether to pursue reliance on the fill as a load-bearing strata, use the fill in a stabilized state or to transfer the load to the underlying strata. Should it be decided to place reliance on the fill, then a detailed analysis for that condition should be made. Any decision to transfer the load to the lower strata should take into account the likely negative skin friction on piles or other structures used. It may be that on certain sites a combination of these conditions will be selected to suit different locations (see Fig. 7.11). The filled quarry in this example has a number of

problems which are solved in varying ways for different positions on the site.

For example, it is proposed to consolidate the last phase of the work i.e., area 'A' by preloading since time is available prior to construction commencing on this portion of the site. Vibro-compaction is proposed for area 'B' i.e., the portion of the site restrained by the rock face against normal consolidation. The remaining central area 'C' requires no treatment since the fill has consolidated naturally over a long period of time.

7.7 THE DEVELOPMENT AND ITS SERVICES

7.7.1 Sensitivity

When developing fill sites it is necessary to give careful consideration to the sensitivity of the building and its foun-

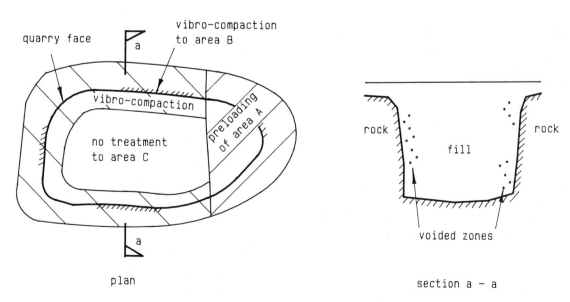

quarry face

vibro-compaction
to area B

a

vibro-compaction

preloading of area A

no treatment
to area C

a

plan

rock   rock

fill

voided zones

section a – a

Fig. 7.11 Typical proposed solution for filled quarry.

dations. For example, the tall tower on a narrow foundation constructed in the soil straddling the edge of a young, end tipped, clay filled, rock quarry, shown in Fig. 7.12, would be particularly sensitive to movement.

It would be necessary therefore to consider:

(1) A special treatment for the fill.
(2) A wider foundation.
(3) Transferring the load to the bedrock sides of the container.
(4) Transferring the load to the container base.
(5) Relocation of the development.

In general, the most suitable solution is to relocate the structure and avoid straddling the edge of a quarry. When relocation is not an acceptable option then piling or vibro-compaction can be considered. The engineer must however satisfy himself that migrating voids at the quarry edge will not shear off or damage the piles as settlement takes place. The load from the settling fill must also be taken into account in the pile loads by assuming negative skin friction on the piles through the fill layer (see Chapter 14).

With long blocks of buildings in such locations, jointing into rectangular units of smaller dimension should be adopted to minimize the stresses and differential settlement in any one block. Where significant differential settlements are expected, services into a building or its foundation should be constructed to absorb the movements by the use of flexible joints or telescopic connections at ground/foundation interface (see Fig. 7.13 which indicates a pile foundation on a fill site).

Where the design of the services and foundation is carried out by different engineers it is essential that the foundation engineer communicates to the services engineer the need to accommodate the differential movements between the settling ground below the development and the limited movement of the foundation through which the services pass.

In low-lying areas filling may be required to achieve sufficient elevation above sea level to prevent flooding. The sub-soils of such low-lying areas often contain silts, peats

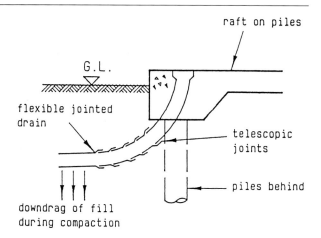

Fig. 7.13 Sensitivity — flexible service connections.

and other soft virgin strata which are prone to excessive settlement. In these low-lying areas the designer should, if adopting piling, allow for the effects of negative skin friction or downdrag on the piles as the stratum settles. Piling for drainage runs should be avoided in the peaty or silty areas when overburden filling is to be used for elevating the level, since these drains would need to act as beams supporting the overburden and are likely to fracture under such loading. The most successful method to adopt for such drainage is to predict the differential settlements likely to occur over the site and provide flexible jointed pipework at drainage and foundation interface to cater for the differential plus a tolerance. The falls should be suitably improved to allow for the settlement to occur without affecting the run-off from the drainage system (see Fig. 7.14).

7.7.2 Treatment and solutions

In the soil investigation for any building foundation it is never practical to reveal and test all the sub-strata, therefore information on soils is limited and assumptions are made. In the case of fill sites, the information tends to be less reliable than that for virgin sites despite a more thorough investigation. It is inevitable therefore that the designer will have

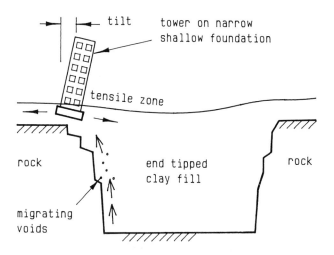

Fig. 7.12 Sensitivity — building over edge of quarry.

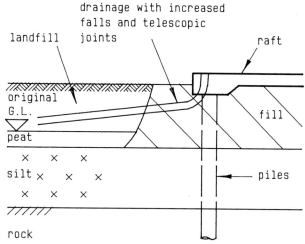

Fig. 7.14 Sensitivity — increased drainage falls.

to make a judgement on the most economic solution for the development. The solution must therefore embrace and accommodate the likely variables and will be based upon experience.

The ground may be partly used untreated where small differential movements are expected but in more critical areas, dynamic consolidation, vibro-compaction or piling may be adopted. On some larger sites, savings on foundations may be made by relocation of critical buildings to the better ground (see Chapters 8 and 14 for further information).

On sites where vibro-compaction is necessary on part of the site, other less critical areas where vibro-compaction would not normally have been used may prove economically viable to develop using this technique. This is because the economics of the process improve once the plant has been established on site. Other ground improvement methods which may be considered are the use of hardcore blankets, preloading, improved drainage, and water inundation (see Chapter 8 on ground improvements).

The use of pin-jointed frameworks rather than fixed joints, the sub-division of long buildings by jointing, the use of flexible joints in services and telescopic joints at interfaces (where large differentials are likely) are methods of absorbing these variables. Monitoring of preloading and large-scale load tests give a feel for the possible accuracy of such judgements. However, even these observations tend to be short-term when related to the normal life expectancy of a development. Therefore allowances for changes in moisture content and other variables have to be extrapolated from the observed conditions in order to give settlement predictions.

7.7.3 New filling for development
Two basic methods of filling sites for future development are:

(1) The use of carefully selected materials placed under controlled conditions to a density suitable to ensure an adequate founding material for the proposed development.
(2) The use of a suitably cheap inert material, end-tipped with a view to using either time or future compaction to obtain a suitable foundation strata.

In method (1) the chosen material is generally placed in layers, each layer being consolidated with a vibrating roller. In method (2), if time is available, the material is allowed to consolidate over a very long period of time, but for quicker results deep vibro-compaction or dynamic consolidation methods can be adopted.

The method adopted will generally depend upon the economic consideration based upon time, money, available materials, laying and compacting costs, etc. It is cost effective and preferable (for development sites) to remove any water and sedimentation, including topsoil, from the base of the container prior to filling. This process will reduce the settlement and unpredictable behaviour of these lower soft deposits which are prone to consolidation and decay. When filling on a soft sub-strata, the introduction of base layers of geotextile fabric placed prior to filling may help the filling process. Filling for future development should preferably be compacted in layers. When cohesive and granular materials are used it is best to interspace the layers to form efficient and short drainage paths to speed up settlement/consolidation of the cohesive materials and create a more uniform cross-section.

The first layers should generally be a granular material to help drainage particularly since, in some cases, the sub-strata itself may be cohesive and in others the sediment in the bottom of the container may not have been fully removed.

When selecting suitable fills, granular materials are preferable to cohesive fills from a settlement point of view. Materials such as crushed rock, gravel or course sands are free-draining and consolidate more easily than clays. In all cases the materials used must not be contaminated in such a way as to present an environmental or health hazard. When filling under controlled conditions for a suitable founding strata, the edge conditions relating to mechanical keying and friction may require more compaction and control (see section 7.2).

Special specification clauses and supervision may be necessary for these locations to ensure satisfactory compaction to overcome the edge restraints and consolidate out the voids from the fill alongside the quarry or container face.

7.8 CASE EXAMPLES

7.8.1 Introduction
The following examples are given in broad outline only to clarify the approach and general solutions to particular problems. The actual projects involved a mass of information, drawings and reports which had to be digested, sifted and summarized in order to arrive at a clear and practical approach to design. The authors recognize that often the most difficult step for inexperienced designers is recognizing, from such a mass, what the real problems are, but they advise that this will come from the application of experience and logic.

7.8.2 Example 1: Movement of existing building on fill
This first example highlights the unpredictable nature of fill when a wide range of materials are involved on one site. The example should help to broaden the designer's outlook when trying to solve the wide ranging problems resulting from developing on fills and remind the designer that the word *fill* discloses nothing about the fill material other than that it has not been deposited naturally.

The site is an inner-city fill site where houses had been constructed on raft foundations some 35 years previously. The foundations had performed successfully with no apparent defects until the last six months of that period. At that time a problem developed simultaneously in three blocks of semi-detached properties and revealed itself when these properties began to settle differentially and crack internally and externally. The movement seemed surprisingly rapid after such a long period of dormancy.

The cracking developed quickly and caused serious concern and distress to the occupants. The gap between the semi-detached properties was closed off by a garden wall and the access gate to the rear. The initial investigation was limited to shallow trial holes around the edges and centre of the raft and deeper boreholes in the accessible areas between the properties. A desk-top study was carried out relating to the history of the area and the site. The information gathered from these investigations revealed that the fill was generally ash, that the material was very old and generally compact but that gaps existed between the underside of the raft and fill material in various positions. The general area was part of a zone from which water had been pumped for commercial use over a period of hundreds of years.

Ancient wooden water pipes had been uncovered in excavations around the area which pre-dated the recorded pumping. A level survey revealed that excessive differential settlement was occurring to the raft foundation.

After studying all the information gathered, the engineering conclusion was that the differential settlement could not be fully and satisfactorily explained by the information so far gathered. It was therefore decided that further, more detailed borehole information was necessary and this was obtained by demolishing some of the separating garden walls to allow access for a rig into the rear gardens.

Boreholes were driven and details of the materials encountered were recorded. The bores revealed large voids at 3 m below surface level which had formed in remnants of large pockets of salt within the fill. It was considered that these pockets of salt had dissolved over a long period of time and that the cavities in the compact fill had been able to reach quite large sizes, prior to collapsing below the foundations. It was these collapsing cavities which were now causing the excessive differential settlements.

The solution adopted was to grout the voids and construct a stiffer raft by underpinning operations. The new rafts were designed to span and cantilever over any similar depressions resulting from future dissolved salt voids. A depression of 2.5 m diameter was used in the design. (See Fig. 7.15).

The solution was implemented and no further problems have been reported.

7.8.3 Example 2: New development on existing colliery fill

It was proposed to develop a derelict area in a mining town. The sub-soil consisted of colliery fill overlying clay above shallow mine workings. Three recorded mine shafts existed close to the site.

The fills in the area varied from 1–8 m thick and from loose to firm colliery waste (see typical borehole in Fig. 7.16). Generally the shallow mine workings were found to be collapsed, however the possibility of some migration of voids did exist. The development consisted of units of two-storey domestic premises and infrastructure. The soil

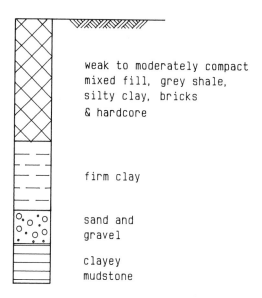

weak to moderately compact mixed fill, grey shale, silty clay, bricks & hardcore

firm clay

sand and gravel

clayey mudstone

Fig. 7.16 Borehole log.

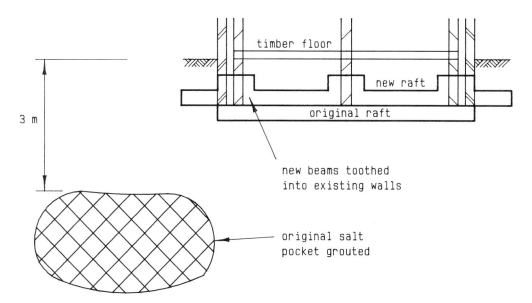

timber floor

new raft

original raft

3 m

new beams toothed into existing walls

original salt pocket grouted

Fig. 7.15 Example of foundation treatment over settlement of fill.

investigation was carried out using trial holes and boreholes. From the details of the fill material it was evident that the ground could not be relied upon as a load-bearing strata without treatment. Piling was not considered a suitable foundation option due to the mine workings at the lower levels (see Chapter 6). Serious consideration was given to the use of vibro-compaction which would also help to *chase* down the fills and tamp out small migrating voids that may have been approaching the surface. It was necessary to check that none of the mine shafts were located in the area of the development and a desk-top study was undertaken. This study revealed that the three recorded shafts had been located outside of this site and had been treated. Other possibly untreated shafts were identified from records but were located a long way from the area of the site and did not affect proposals on this site.

Trial holes were excavated and inspected by the designer and by the vibro-compaction contractor and it was decided that the dry process would be suitable to improve the fill materials inspected. In addition to designing the foundation to bear on the treated ground in the normal way, the design incorporated a raft solution which could span and cantilever over any depressions created as a result of the migration from voids which may have remained below the level of the vibro treatment and which could not be incorporated into the ground improvement.

Generally, site works went very smoothly. However, one small area of the site was found to be at variance from the materials uncovered in the ground investigation. In this area very soft, loose, fine colliery waste in a waterlogged condition was discovered and the particular dry vibro process was ineffective in this material. The plant being used had a side-fed poker and a trial probe showed that as the vibrating poker was pushed in the ground the fill materials closed in around the poker preventing access for the stone. As the poker was withdrawn, the fill material squeezed back into the hole and deposited stone which simply plugged the top of the holes at high level. In this material, it was evident that either a bottom-fed poker or the wet process must be used to achieve satisfactory treatment. In this case the side-fed poker was used with the wet process to achieve successful installation of probes.

7.8.4 Example 3: New development on new filling

This site was part of a much larger development of domestic properties. The main part of the development was successfully founded on traditional simple shallow foundations on firm sand. Part of the site however, was low lying and sloped down to a strip of peaty deposits overlying silty sands. The developer (who had not consulted an engineer) began excavations for simple strip footings and uncovered peaty deposits overlying silty sands which became running sand during foundation excavations. Figure 7.17 shows a typical section through this location and indicates the typical layout of the semi-detached dwellings.

To determine the extent and nature of the problem a series of trial holes were dug along the run of properties and revealed that the soft deposits existed for the full length of the semi-detached dwellings. The peat material was removed and excavations carried down to the silty layers and trial layers of hardcore filling were installed. The trials revealed that the fine virgin soils in the base of the excavation came through any thin (i.e. 150 mm thick) layers of hardcore. A medium layer (i.e. 235 mm thick) produced a wave ahead of the compacting machine. It was realized however, that the problem was one of slow dissipation of porewater pressure from the silty materials during stress changes from the compaction plant. It was decided therefore to spread a layer of 450 mm thick well graded hardcore using a tracked vehicle prior to the introduction of the compaction plant and this was carried out followed by subsequent thinner layers compacted in long strips similar to road construction. This formed a stiff, hardcore blanket on top of the silty materials upon which a flexible raft foundation could be constructed. This was one of the early uses of the blanket raft, described in Chapter 9 and indicated in Figure 7.18.

The raft was designed to span over a nominal diameter depression to take into account the likely differential settlement that may have occurred due to the soft silty nature of the underlying silts. The possible depression diameter chosen from predictions of settlement was 2.5 m. The scheme proved very successful and is now more than 20 years old. This form of construction has been repeated successfully for similar circumstances on many subsequent occasions.

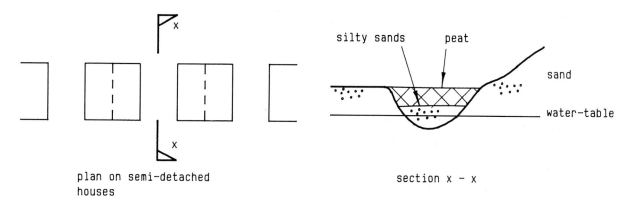

Fig. 7.17 Example of blanket raft – ground conditions.

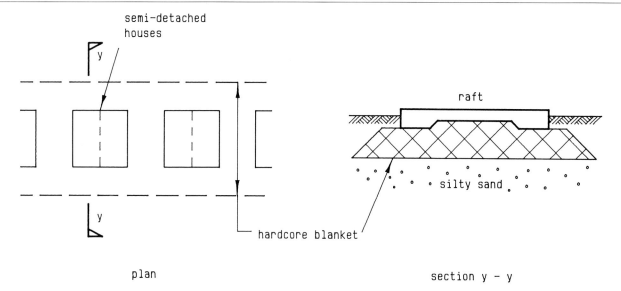

semi-detached
houses

plan

section y - y

raft

silty sand

hardcore blanket

Fig. 7.18 Example of blanket raft — foundation solution.

7.8.5 Example 4: New developments on existing preloaded fill

Fill sites are rarely simple to investigate and solve for development. This case, however, consisted of a strip of filled land which was left after the removal of a 6 m high disused railway embankment. The proposal was to construct a row of semi-detached domestic properties along the centre-line of the original railway embankment. The embankment had existed for some 60 years and was being removed prior to purchase of the site. The remaining fill was approximately 2 m deep and consisted of consolidated layers of clay inter-layered with sand. The clay layers were 450 mm thick and the sand approximately 150 mm thick, similar to the construction of the full height of the original embankment. The removed embankment could now be classed as removed overburden. Calculations indicated that overburden stresses prior to removal of the embankment exceeded the loading stresses from the proposed development and that future heave would be minimal. It was proposed therefore to use a nominal crust raft (see Chapter 13) on a thin bed of hard-core (see Fig. 7.19).

The raft was constructed by excavating the perimeter thickening and casting the 250 mm thick mass concrete blinding. The reinforced edge strip was then poured to the underside of the perimeter brickwork; the mass concrete strip supporting the reinforcement on spacers off the blind-ing. The zig-zag continuity bar was pushed into the wet concrete of the edge strips and the strip allowed to cure. Three courses of masonry were then constructed as shown in Fig. 7.20 and the polystyrene cavity fill inserted to form the vertical shutter face to the slab. The blinded hardcore was constructed below the slab and the slab reinforcement caged up from it. The slab was poured between the poly-styrene using the top of the brickwork as the shutter tamp support.

By constructing in this sequence it was possible to cast the whole of the raft without the use of shuttering and the solution proved economical and structurally successful.

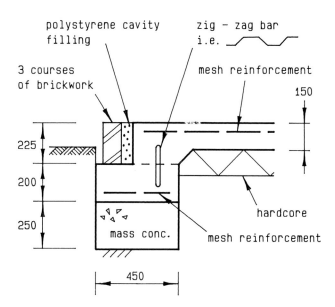

polystyrene cavity filling

zig - zag bar
i.e. ⌐_⌐

3 courses
of brickwork

mesh reinforcement

150

225

200

250

mass conc.

hardcore

mesh reinforcement

450

Fig. 7.19 Section through edge thickening.

7.8.6 Example 5: New development on existing backfilled quarry (purchase of coal rights)

The site consisted of a disused opencast quarry which had been filled 30 years previously with sandy clay material. The quarry was underlain by existing coal seams and was close to a fault line. The sandy clay fill, though quite old, varied in consistency and density in the upper layers. In addition, the NCB were extracting a coal seam which was approaching the quarry and due to pass under it some time after com-pletion of the properties. The predicted subsidence from this seam was likely to be erratic and substantial. In order to minimize the possible effects of the future mine workings under the site, an approach was made to the NCB to check the feasibility of purchasing the coal rights for the seams below the site. Under normal circumstances coal value would be such that this approach would receive little con-

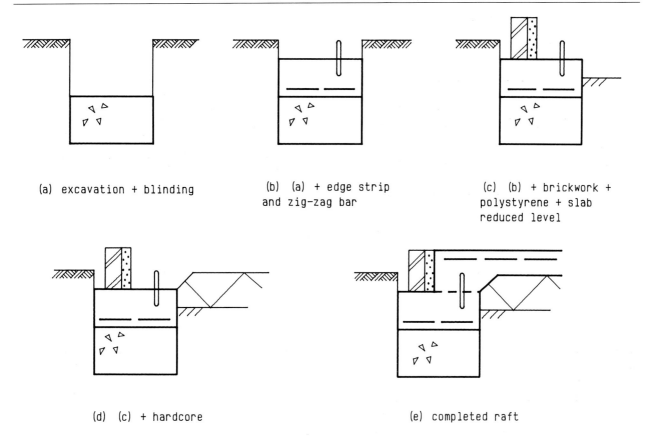

(a) excavation + blinding

(b) (a) + edge strip and zig-zag bar

(c) (b) + brickwork + polystyrene + slab reduced level

(d) (c) + hardcore

(e) completed raft

Fig. 7.20 Construction sequence.

sideration. However, in this case, the possibility was feasible, since the seam was due to run out after passing under the site owing to the fault line, also the quality of coal approaching the fault was deteriorating. The NCB therefore agreed to sell the coal rights for a nominal sum. The sum was less than the cost of increasing the foundation strength to deal with the potential settlements condition. The effect of purchasing the coal rights minimized the risk and inconvenience to the property owners that would be caused by damage to services. The compaction of the fill at the edges of the quarry was investigated by long trenching excavations and deep boreholes and was found to be consistently consolidated below the level of 4 m. The upper fill was found to be suitable for compaction by the use of the dry vibro process. Due to the cohesive nature of the fill the foundation was designed to span between the probe positions (see Chapter 8 for further information). The foundation solution in this case therefore embraced:

(1) Purchasing the coal rights,
(2) Vibro-compacting the upper layers,
(3) Picking up and redirecting all incoming drainage including field drains from the edges of the quarry,
(4) Using a lightweight downstand raft designed to span between the vibro hard spots and to absorb differential settlements resulting from consolidation effects below the vibro.

The solution proved economic and successful.

7.8.7 Example 6: Development on new fill (prevention of flooding)

It was proposed to use a low-lying peaty farmland site for a large housing development on the edge of an existing town. In order to prevent flooding, the site needed to be lifted by approximately 1 m. The existing sub-soil consisted of 1 m of peat overlying 16 m of soft silt with bands of silty clay, overlying hard marl. The total depth to firm strata was in the region of 17 m. From inspection it was clear that large settlements would result from any additional load at surface level. A detailed soil investigation and settlement analysis revealed that, even assuming that the peat layer was removed, the fill site would settle under its own weight, within the life of the development, by some 250 mm in the poor areas and approximately 50% of this in the better areas. A feasibility study was undertaken based upon boreholes from the surrounding areas. Using experienced engineering judgement of the conditions, the study assumed that the peat would be removed from all road, hardstanding and service run areas for a width which allowed a 45° dispersion through the replacement filling. These areas would be left after backfilling for a period of three months before excavating through and constructing the drainage and other services. The areas of gardens and housing would be filled on top of the existing peat and the houses and garages would be piled through this construction. The roads and hardstandings would be lifted 300 mm above the required minimum levels for flooding and the gardens 450 mm above to allow for consolidation. All services would be provided with flexible joints and extra falls to maintain flow after differential consolidation of the sub-strata. At all locations where services passed from one condition into another, i.e., passing from service trench into house foundation, they would have telescopic joints and enter the foundation in a vertical direction.

A brief analysis indicated that a development based upon these assumptions was feasible. The scheme was therefore progressed using a detailed analysis of the differential movements between roads, service trenches and houses as the basis for detailing. The details of service junctions were prepared making allowance for a safety margin for inaccuracy of the analysis, i.e., extra differential movement was allowed for in the details to that estimated from the analysis.

The site was developed over 20 years ago and has proved both economic and successful. The economics of the site were made attractive to the developer by the low cost of purchase of a very large site in a very good location.

CHAPTER 8

GROUND IMPROVEMENT METHODS

8.1 INTRODUCTION

The treatment of weak or loose soils to improve their load-bearing capacity and reduce their potential settlement characteristics has proved to be cost effective in achieving an economical substructure solution to many developments. The treatment is known as *ground improvement* and there are various methods available.

The main problem associated with providing foundations which perform satisfactorily on poor ground is the effect of differential settlements. The main object of ground improvement therefore is to achieve a reduction and more uniform ground settlement due to the applied loads thus reducing differential movements to within acceptable limits. Settlements are usually caused by the vertical load delivered by the building, and its foundation, which can result in consolidation, compaction and shear strain of the soils. In addition the rates of settlement are closely related to soil drainage. Ground improvement therefore aims to consolidate and compact the soil and improve its shear resistance and make its drainage characteristics more uniform. This reduces the magnitude of differential settlement under loading and improves the load-bearing capacity of treated soil.

It is not proposed to deal with temporary strengthening of the soil, such as dewatering, freezing, etc., since these are mainly construction aids in development. Long-term ground improvement treatments include:

(1) Mechanical methods
   (a) surface rolling
   (b) vibro-stabilization   } *which directly aids consolidation.*
   (c) dynamic consolidation
(2) The installation of drainage systems, *which accelerate consolidation.*
(3) Preloading, *which directly aids consolidation.*
(4) Grout injection, *which improves soil strength and reduces settlement.*

8.2 SURFACE ROLLING

8.2.1 Introduction
Surface rolling of imported granular materials or hardcore in preparation for floor slabs and road construction is common practice. In such cases materials can be selected with the aim of obtaining an appropriate type and grade to suit compaction and economy. These materials are generally granular and well graded, this is not to say that other less ideal materials could not be compacted to a suitable strength, but since labour and haulage costs are high it is a sensible approach to be selective.

There is little doubt that loose dry well graded granular materials are easier to compact than wet clay, however, the range in between contains many suitable materials which may achieve the desired result if the compaction method is varied and appropriate for the conditions. Existing substrata on sites vary and while vibro-compaction, dynamic consolidation and in some cases piling may be needed for foundation success, there are other sites with shallow depths of loose material which can be satisfactorily improved by surface compaction.

8.2.2 Method
Dry loose granular materials are generally compacted by specified compaction plant until no further movement occurs at surface, or to a specified number of passes of a roller or tamper. These soils tend to be predictable and compaction requirements can be assessed from the soil investigation data. For materials containing more fines however, pore-water pressure dissipates more slowly under stress and compaction is hindered. In such cases imported hardcore is laid to form a blanket over the surface prior to compaction. This blanket tends to crust up the surface and prolong the induced stress therefore allowing more time and more even dissipation of water (see Fig. 8.1).

It should be noted that unless the porewater has time to escape under compaction, then compaction will not be fully achieved. For example, a vibrating roller inducing excessive stress into a soil where water cannot escape, simply transfers the stress to the water particles and a quagmire results. It is therefore better to increase the stressed period and reduce the level of the stress as shown in Fig. 8.1. This allows compaction to occur more gently and slowly and is more suitable in soft, damp or wet materials. The hardcore blanket also allows the porewater to escape vertically into the hardcore layer rather than being trapped beneath the roller surface. In such materials re-distribution of stress to encourage more even settlement can be achieved by additional crusting up of the surface with hardcore layers and the adoption of a blanket raft (see Chapters 9, 10 and 13).

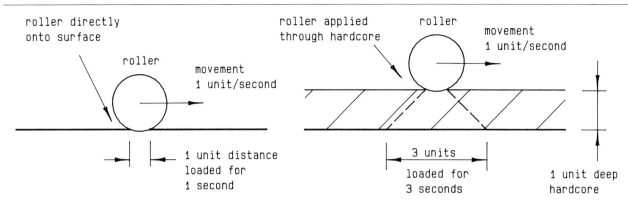

Fig. 8.1 Surface rolling.

Trial runs on finer soils are recommended at the commencement of work on site to enable the finalization of the best combination of the required thickness of hardcore layer, number of layers, types of hardcore, number and weight of roller passes, etc. to achieve the total thickness. The materials to be compacted, the bearing capacity to be achieved and the variety of schemes are numerous and wide ranging. The correct solution is therefore reliant upon a judgement using plant information, soil investigation and test results combined with a great deal of experience. To assist the engineer the following tabulated information on plant and hardcore is provided as a guide for successful use. This method of compaction is an economic alternative applicable to many sites particularly for low-rise building foundations (see Table 8.1).

8.2.3 Soil suitability and variation

The most suitable sites for treatment by surface rolling are those where compaction of a loose well graded granular material of shallow depth is required. In such cases a small number of passes with a specified vibrating roller until no further movement of the surface is apparent can achieve a much improved bearing capacity with reduced total and differential settlement. The roller will search out the softer areas and the requirement to continue to vibrate roll until no further movement occurs will concentrate the large number of passes into the most needed locations. The ideal sites with such sub-strata are few and far between, however, fine sands, silty sands, silty sandy gravels, demolition rubble and other mainly granular mixtures can be treated by this method.

8.2.4 Site monitoring

Monitoring of compaction quality is mainly visual and consists of inspection of the material as it is being compacted. This provides extensive information to the experienced eye. The movement of sub-strata below the roller, the permanence of downward movement, horizontal movement or wave action ahead of the roller all reveal information relating to the success of the compaction operation (see Fig. 8.2).

Table 8.1 Hardcore grading and compaction

Hardcore material should be composed of granular material and shall be free from clay, silt, soil, timber, vegetable matter and any other deleterious material and shall not deteriorate in the presence of water. The material shall be well graded and lie within the grading envelope below:		Hardcore material should be placed and spread evenly. Spreading should be concurrent with placing and compaction carried out using a vibrating roller as noted below:	
BS sieve size	**Percentage by weight passing**	**Category of roller (mass per metre width of roll)**	**Number of passes for layers not exceeding 150 mm thick**
75.0 mm	90–100	Below 1300 kg	not suitable
37.5 mm	80–90	Over 1300 kg up to 1800 kg	16
10.0 mm	40–70	Over 1800 kg up to 2300 kg	6
5.0 mm	25–45	Over 2300 kg up to 2900 kg	5
600 µm	10–20	Over 2900 kg up to 3600 kg	5
75 µm	0–10	Over 3600 kg up to 4300 kg	4
		Over 4300 kg up to 5000 kg	4
Note: This grading falls between the recommendations of the Department of Transport for sub-base Type 1 and Type 2 gradings.		Over 5000 kg	3
		Compaction should be completed as soon as possible after material has been spread	

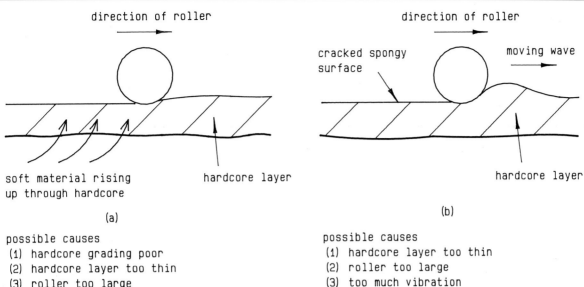

Fig. 8.2 Monitoring surface rolling.

Materials with poor permeability, high moisture content and weak interaction will be spongy i.e. depress as the roller passes over and rise again behind the roller. With such material there is also a chance of a forward moving wave and a general liquid-like behaviour if overloaded by the vibrating equipment. It is useless to use heavy equipment on these soils. Soils have a limit to the reaction they can supply to the passing roller and soft weak soils with a high moisture content are particularly critical (see Fig. 8.3).

The reaction of the soil can be increased by steady compaction which allows time for water to dissipate and the gradual build up of soil strength by squeezing out water and voids. The initial trial runs are essential to check and adjust the number of passes and the thickness of hardcore layers to achieve compaction of the total hardcore thickness (see Fig. 8.2).

As the work progresses experienced supervision is generally all that is required, however, if compaction is to be tested then plate tests (as used for vibro-stabilization) can be used and/or density tests can also be carried out. Density tests are only suitable where the loose density and compacted density requirements can be assessed i.e. when consistent materials are being improved. The density test involves the removal of loose samples from the soils to be compacted, volume and weight measurements are taken of the loose materials. The material is then compacted and samples re-taken, measuring volume and weight of the sample. From these tests, combined with laboratory tests the void ratio of loose and compacted materials can be obtained and a specified requirement checked against actual site compaction.

8.3 VIBRO-STABILIZATION

8.3.1 Introduction

The process of ground improvements using vibration techniques was originally developed in Germany in the 1930s,

(a) problem

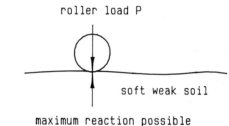

maximum reaction possible from soil less tnan P therefore ground disturbance occurs with no compaction

(b) solution

gradual crusting up of surface, first rolling after 250 mm layer placed, roller load can increase on upper layers

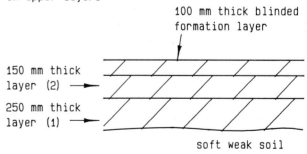

Fig. 8.3 Surface rolling – crusting surface.

further development continued in the USA and West Germany after the Second World War. The method originally involved the compaction of deep layers of loose sand using a vibrating poker inserted into the ground. The vibration reduces the void ratio and increases the compaction of the sand and thus its density, strength and settlement resistance. Although originally intended for treatment of loose sands, developments in the last 15–20 years have been increasingly to strengthen fills and cohesive soils by forming within them stone columns known as granular compaction points or gravel piles.

Vibro-stabilization techniques have been used in this country for over 30 years, and the applications have ranged from the treatment of soft organic and alluvial deposits to sands, gravels and rubble fills, the materials treated being extended as the methods of compaction improved. The structures supported on stabilized ground vary from single- to multi-storey buildings and from infrastructure to industrial plant installations such as storage tanks.

The installation of stone columns is carried out using a vibrating poker, whose features and components are shown in Fig. 8.4. The poker is suspended from a crane that is usually crawler mounted which assists movement on difficult site surfaces. The poker contains eccentric weights rotated by an electric or hydraulic motor to create vibrations in a horizontal plane. Relatively low frequencies are used to achieve the required compaction of the surrounding soils and stone columns. A jetting medium is used when forming the stone columns, the medium being either water or compressed air depending upon the nature of the ground being treated.

Sets of jets which carry the jetting fluid are located on the vibrator. The lower set of jets at the probe's tip aids penetration, the upper sets, discharging above the vibration, help removal of unwanted material during penetration and aid compaction. The choice of the jetting fluid will be dependent upon the nature of the ground and the position of the water-table. If the water-table is within the depth of the stone columns, it is necessary to use water as the jetting fluid i.e. the *wet* system, since the use of compressed air would result in air bubbling through the water-table causing the sides of the hole to collapse, thus preventing the formation of the stone columns. It may also be necessary to use water to aid the penetration of the probe where the water liquefies soft deposits, thus permitting the penetration of the poker. The amount of water used in the *wet* process is high and this may lead to problems during the construction of the

foundations, particularly on non-porous grounds, unless adequate drainage or pumping facilities are provided. In addition, the introduction of large quantities of water into the ground can, in some soils, have a detrimental effect on the bearing capacity of the ground. When water is used in the forming of the stone columns the treatment is referred to as *vibro-flotation*.

Generally the dry process is used in ground made up of mainly granular material of coarse grained particles such as sands, gravels, brick and demolition fill. It is, however, necessary to prevent the sides of the hole being drawn in by the reduction in pressure as the poker is withdrawn, and this is achieved by compressed air being passed through the poker and on certain rigs by bottom-feed skips for the delivery of stone to the poker point.

8.3.2 Working surfaces

The provision of a working surface, usually a hardcore bed, is often required to avoid difficulties of movement of the machine between compaction points. Also, suitable surface gradients must be provided to enable the machine to move across an uneven site. The actual treatment is such that the compaction process cannot be achieved to surface level. The top 600 mm depth is usually considered to be untreated and foundations are generally located at a minimum of 600 mm below the working surface. However, for most contracts an agreement can be reached to remove less than 600 mm of *untreated* material, i.e. say 300–450 mm, and to compact the remainder with a surface roller prior to construction (see section 8.2 on surface rolling) particularly where raft foundations or floor slabs are involved.

8.3.3 Method

The poker is first vibrated into the soil under its own weight, assisted by air or water jetting, to the required depth for treatment. At this point the bottom jets are closed and jetting takes place through the top of the vibration unit. The compaction process of the soil is achieved by gradually withdrawing the poker in predetermined steps – usually 300–600 mm – with the compaction process held for between one and two minutes in each step. In granular materials a cone shaped depression tends to develop at ground level around the poker indicating that compaction is occurring. Well graded backfill material is constantly fed into the space formed around the poker, or through the shaft to a bottom feed as it is withdrawn, so that a column or pile is formed and the compaction of the soil is completed. The process can achieve compaction in the soil around the poker ranging from 1.2–3 m in diameter (see Fig. 8.5).

Various terms are used to describe different vibro-stabilization methods. While vibro-stabilization is the general term used to cover ground improvement or stabilization using a vibrating poker to achieve deep compaction, there are three differing techniques, as follows:

(1) *Vibro-compaction.* This method is used to compact granular (non-cohesive) soils, and may employ stone columns in finer grained materials. The method may be wet or dry to suit groundwater conditions.

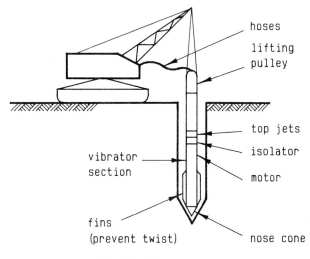

hoses

lifting
pulley

top jets

isolator

motor

vibrator
section

fins
(prevent twist)

nose cone

Fig. 8.4 Vibro rig.

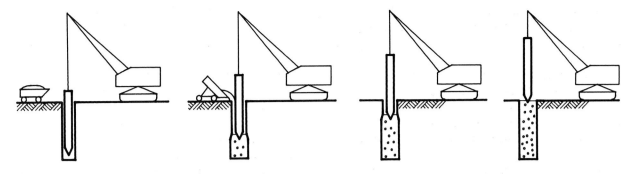

Fig. 8.5 Vibro method.

(2) *Vibro-displacement.* This method is used to improve cohesive materials, employing the dry process to form compacted stone columns within the clays.

(3) *Vibro-replacement.* This method is used to improve soft cohesive materials. Using the wet process, disturbed materials are washed away and replaced by compacted stone columns − alternatively in some soils the dry process using a bottom-feed method via a hopper and supply tube direct to the toe of the vibrator is adopted (see Fig. 8.6).

8.3.4 Vibro-compaction

The concept of compacting deep layers of loose granular materials beyond the range of surface vibrations by vibro-stabilization methods is based upon the response of the material to the mechanical vibrations set up in the soil. The mechanical vibrations destroy the inter-granular friction within the soil and the particles rearrange themselves under gravitational forces into a more dense state. Since the process of rearrangement occurs in an unconstrained and unstressed state it is therefore permanent. During the compacting process the initial void ratio and compressibility of the granular soils are greatly reduced while the frictional resistance and modulus of deformation are increased. It can be appreciated that while this process can be satisfactory for granular materials, cohesive materials require further con-

sideration, therefore the materials most suited to improvement by vibro-compaction range from medium-to-fine gravel to fine uniform sand (see Fig. 8.7).

The lower limit of treatments is determined by the silt and clay sized particles and organic matter. A high fines content reduces the permeability of the soil and dampens the vibrations thus reducing the degree of compaction possible making the process inefficient or uneconomic. However, the authors feel that the range of vibration frequency on present rigs could be extended to embrace the natural sensitive frequencies of a greater range of soils.

It is at present considered that vibro-compaction can be applied to soils containing up to 10% fines or permeability greater than 10^{-6} m/s. However, the process is moving towards a wider range of materials. The upper limit of material suitable for treatment is governed by the ability of the vibro-float to penetrate coarse granular materials such as cobbles. In order to achieve good penetration the material should be loose and well graded and include a complete range of particle sizes; however, suitable compositions for relatively low bearing capacities are achievable in more poorly graded fills. If the rate of penetration of the poker is reduced by the increase in particle size and density, the compaction process becomes less successful, less practical and less economic.

The vibrating principle of compaction is not effective for

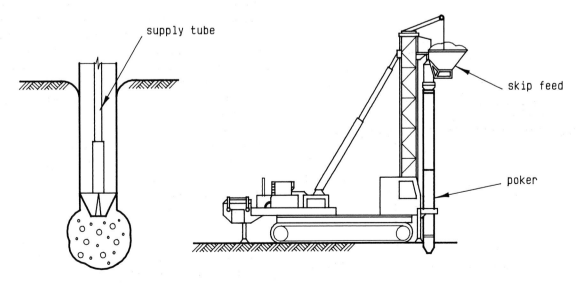

Fig. 8.6 Bottom-fed vibro method.

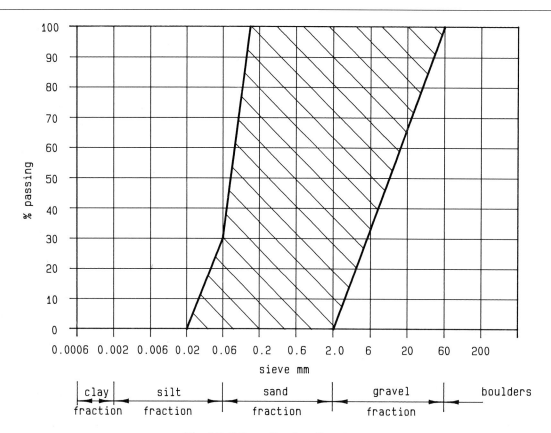

Fig. 8.7 Soil grading for vibro treatment.

clays and some silts since the cohesion between particles is not overcome by vibrational forces. Improvements in cohesive soils can be achieved by vibro-displacement or vibro-compaction to install stone columns within the cohesive materials.

A further application of the method is the use of the stone columns to both stabilize the sub-strata and speed up settlement by shortening the drainage paths. For example, stone columns have been used in areas of soft silt overlying gravel where the site level requires lifting to an acceptable minimum height above sea level to prevent flooding (see Fig. 8.8

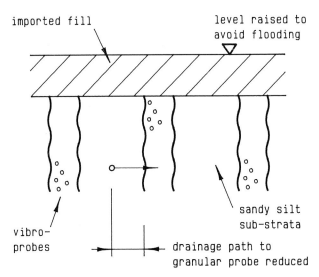

Fig. 8.8 Soil drainage using vibro.

which shows the sub-strata and shortened drainage paths).

By using stone columns and leaving the site preloaded (see section 8.5) for a period of time prior to constructing the foundations, the differential settlement can be brought within acceptable limits in many situations without the need for a more expensive piling solution.

8.3.5 Vibro-displacement

In the vibro-displacement process the penetration of the vibrator into partially saturated soils results in shear failure of the soil which is displaced readily forming a cylindrically compacted zone. When used in soft-to-firm clays this material usually has sufficient cohesion to maintain a stable hole when the process is used. The vibro-float is removed from the hole and selected granular material used to backfill the hole in stages of about 1 m. The vibro-float is returned to displace the granular fill into the surrounding clay material. The process is repeated until a compacted stone column is formed. The individual stone columns are usually in the order of 600–900 mm in diameter and can achieve bearing capacities of between 100–200 kN.

8.3.6 Vibro-replacement

The vibro-replacement method employs the wet process or bottom-feed dry process and is generally applied to the softer more sensitive clays, saturated silts and alluvial or estuarine soils with an undrained shear strength of less than 20 kN/m². In the wet process the poker penetrates the soil using the water jets to cut an oversized hole, to the required depth of treatment. The vibro-float remains in the hole while the selected granular backfill is placed (the vibro-float being withdrawn under the surging action as the stone compaction point is formed). The continuous flow of a large volume of water is used to keep the hole free of lumps of soft clay or silt materials while the stone backfill is placed. The expansion of the stone column is halted by the passive pressure of the surrounding materials. The stone columns formed by this method tend to be fairly constant in diameter although localized increased diameters can occur where softer layers are encountered. The diameters of the stone columns are usually in the order of 900–1100 mm.

The stone columns act as drainage paths formed in the cohesive material which improves the dissipation of excess porewater pressures resulting from the applied structural loads, thus improving the load-carrying and settlement characteristics of the soil. The surrounding cohesive ground provides lateral restraint to the stone columns, thus maintaining their bearing capacity. As load is applied to the columns during the construction of the structure over, the columns will tend to dilate, thus displacing the surrounded ground and increasing its density and, thus, its bearing capacity and it also creates a more uniform ground bearing pressure. This dilation of the stone columns is associated with a reduction in length of the column, thus causing limited settlement to occur. The uniformity of this settlement will depend on the stiffness and uniformity of loading from the structure over and the consistency of the material providing lateral restraint to the columns. If the supported ground is reasonably homogeneous and the loading evenly distributed then the settlement that takes place will be reasonably uniform.

Any problems caused by the settlement would be limited to junctions with existing buildings or at changes in types of foundations. It would always be prudent to make provision for differential settlement at such locations and, indeed, the complete separation of the structure would be wise.

Since the bearing capacity of the probes in cohesive materials is dependent upon the restraint offered to the probe by the surrounding ground, it is not possible to use this system in very soft clays with inadequate restraint. It should be noted that the introduction of the stone columns acts as drainage for the excess porewater in the clay, thus as the drainage of the clay takes place an increase in its strength is achieved so offering more restraint to the stone columns.

8.3.7 Summary of vibro-stabilization
To sum up, stone or gravel columns are generally used in areas of soft sub-strata or fill where sufficient upgrading of the bearing capacity or reduction in differential settlement can be achieved by one of the applications mentioned. In such situations the stone/gravel column is usually much cheaper and in some situations much more suitable than the concrete pile alternative. For example, the gravel column has a particular advantage in mining areas where the use of concrete piles could result in the foundation developing unacceptable ground strains and the piles could shear off during subsidence, due to the brittleness of concrete. The gravel column can be used incorporating a slip-plane

between the top of the pile and the underside of the foundation in the normal manner. (See Chapters 6 and 9 and the section on sandwich rafts in Chapter 13.)

8.3.8 Design considerations – granular soils
The improvement in bearing capacity of granular materials by deep compaction methods is related to both the depth of treatment and to the spacing of compaction points which increases the density of the material. The increased density results in an increase in the bearing capacity and reduction of differential settlements. The spacing and locations of compaction points are designed to improve a uniform zone of increased density beneath foundations at relatively shallow depth. The difference in soil conditions at actual compaction point locations and mid-way between points is therefore not considered significant.

8.3.9 Design considerations – cohesive soils
In cohesive materials the improvement method cannot be considered to perform in a similar manner as in granular soils. In the short term it is the actual compaction point which carries the majority of the construction loads while the surrounding clay maintains the stone column diameter at an increased porewater pressure. In the long term the porewater pressure will dissipate, the lateral resistance of the clay will reduce and the compaction becomes more uniform. It is therefore necessary for the foundation to be capable of performing over the range of support conditions which are time related. The period of time for these changes to take place is difficult to predict and depends upon the type of clay, its consolidation and its drainage capabilities.

The compacted stone column can, when used in soft clays, be considered as a grid of flexible piles which partly transmit construction loadings to deeper bearing strata. The authors recommend that foundations on clay soils should be designed in the short term to totally or partly span between compaction points and the engineer must use his discretion and the soils information to determine the extent of spanning to be assumed. If in doubt total spanning should be adopted (see Fig. 8.9).

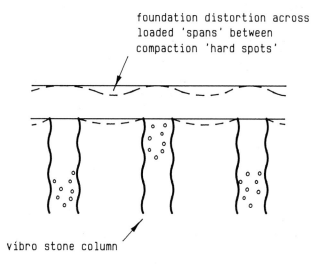

foundation distortion across loaded 'spans' between compaction 'hard spots'

vibro stone column

Fig. 8.9 Foundation performance on vibro.

There are a number of specialist companies which have considerable experience in vibro-stabilization treatments and they will assist in the design of a suitable treatment for the site conditions and allowable bearing pressure required. An evaluation of the site investigation and particularly the bore-hole information will enable a treatment type, depth and spacing to be determined, but it is essential that trial pit excavations are also carried out to supplement the borehole data.

If the wet process is selected, consideration should be given to an adequate supply of water and ease of effluent discharge from the site surface from the vibro-process. In some cases the provision of storage tanks may be necessary to collect water supplied *off-peak* (overnight) to enable the treatment process to continue efficiently during the day. The disposal of effluent water, if not considered and planned for, can become a major problem. Water with soil materials in suspension cannot be taken direct into drainage systems without treatment. Settling lagoons can be used for this process but the size, location and cost must not be overlooked.

The effect of the ground treatment on adjacent buildings and services should also be considered. There are a number of cases where the vibration during the installation of compaction points has caused distress to nearby existing foundations and services. The distance between existing constructions and proposed lines of compaction points will depend on factors such as the position and form of the existing construction, ground conditions, depth of treatment and the vibro method to be used. The excavation of *relief trenches* between compaction points and adjacent services or structures can be used to reduce the distance or effects on existing works. In spite of these warnings it should be appreciated that the frequency of vibrations are designed to affect the soils and not the buildings and the method can often be used relatively close to most structures. The inspection and recording of the condition of adjacent structures is of course advisable prior to commencement of new works alongside existing constructions.

8.3.10 Testing
The effect of vibro-stabilization treatments in cohesive and granular soils are different and this should be taken into account when testing the effectiveness of the treatment and design of the sub-structure. Plate load tests are the usual method of testing the design and workmanship of the treatment. It should be remembered that plate load tests have a limited pressure bulb and are usually carried out over a short time-span. They do, however, appear to give a reasonable guide to the quality of work. A more extensive test method is to carry out zone tests which cover a larger area and give a more accurate prediction of the performance of the treatment. On cohesive soils the tests should, where practical, be extended to cover as long a period as possible to permit dissipation of excess porewater and allow maximum settlements to occur. This method of testing is expensive and time-consuming and is rarely justified on smaller contracts where the cost of such testing may outweigh the

cost advantages of vibro-stabilization. In these situations quality control during construction, plate load tests and experienced engineering judgement is required to ensure suitable treatment.

Briefly the plate load test consists of applying a load to a small steel plate and measuring settlement and recovery during loading and unloading. The plate is usually 600 mm in diameter laid on a sand bed at a minimum of 600 mm below the level at which treatment is undertaken. The plate is lightly preloaded to achieve bedding of the plate on the sand and the test loading is applied in increments up to working load and on to 1.5, 2 or 3 times the working load depending upon the test basis. The settlements are recorded during loading and unloading. The test results are compared with predetermined acceptable values for settlement at working load and maximum test load. The test loading is usually applied by kentledge using the machine which carried out the vibro work; this generally limits the maximum test load to approximately 12 tonnes.

Figure 8.10 shows some typical examples of vibro-stabilization methods, their use and selected foundation type.

8.3.11 Vibro-concrete
A recent development of ground improvement using vibro-stabilization methods is the use of concrete columns in place of the stone ones. This method can be used in ground conditions where stone columns would not work because the surrounding soils are very soft. The integrity of the stone column is lost if the surrounding soil is highly compressible and the stone pushes into the soft material. The vibro-concrete method can be used in sub-soil conditions that could not be treated successfully using stone compaction points. In soft ground conditions the poker is penetrated to firmer ground below and the concrete column formed from the *bottom-feed poker*. A toe is established in the firmer ground and the concrete pumped down the poker and out of the bottom. The poker is withdrawn as the concrete is pumped to form the concrete column which gives ground improvement in a zone that can be treated and maintains cohesion of the column in the compressible area. This method can be used as an alternative to a piling solution. Vibro-concrete columns usually range between 500 mm and 800 mm in diameter and maximum economic lengths of 12–15 m are reported by the specialists.

8.4 DYNAMIC CONSOLIDATION

8.4.1 Introduction
Dynamic consolidation is the term given to describe a ground improvement treatment which is achieved by repeated surface tamping using a heavy weight which is dropped on the ground surface. The weights (or tamper) are usually between 10–20 tonnes, although much higher weights have been used. The tamper is dropped from a height of between 10–20 m although heights up to 40 m have been used (see Fig. 8.11).

GROUND CONDITIONS		DEVELOPMENT	FOUNDATION SOLUTION	VIBRO TREATMENT
	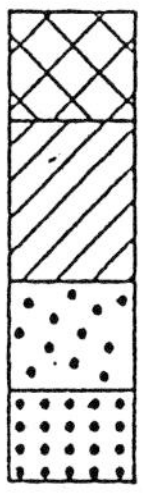 **1.75 m** Demolition fill **1.75−2.35 m** Compact fill (Mainly sub-soil) **2.35−3.2 m** Compact red sand **3.2−3.6 m** Hard red sandstone (trial pit dry)	Two- and three-storey housing of traditional construction	Traditional strip footings on vibro-improved ground Vibro treatment on load-bearing wall lines	Dry process adopted Probes at **1.5 m** centres on centreline of load-bearing walls Probes carried through fill to sand layer Depth of treatment **2.5 m** Allowable bearing pressure **150 kN/m²**
	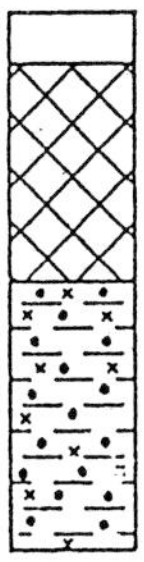 **0−0.1 m** Topsoil **0.1−2.4 m** Soft to firm brown and grey sandy silty clay with ash and bricks **2.4−6.0 m** Firm to stiff dark brown slightly sandy to sandy silty clay Becoming stiffer with depth (borehole dry)	Five-storey residential building Load-bearing masonry construction with suspended concrete floor slab (including ground floor)	Traditional strip footings on vibro-improved ground Vibro treatment on load-bearing wall lines Footings 0.7 to 1.20 m wide reinforced with two layers of B785 mesh	Dry process adopted Two lines of probes at **0.95** to **1.5 m** staggered centres on centreline of load-bearing walls Probes carried through fill to clay Depth of treatment **3 m** Allowable bearing pressure **150 kN/m²**
	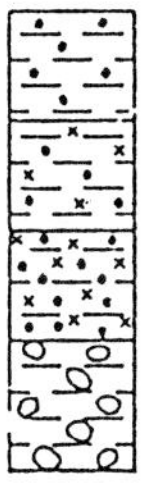 **0−1.0 m** Sandy clay probable fill **1.0−2.2 m** Firm, sandy, silty clay **2.2−3.8 m** Soft very sandy silty clay **3.8−6.0 m** Stiff boulder clay	Tall single-storey factory/warehouse Steel portal frame with steel sheeting and dado masonry	Pad bases beneath columns, with masonry walls on strip footings between bases Vibro-improved ground beneath foundations and ground slab	Dry (bottom-feed) process adopted Probes on **1.5 m** grid under pad bases (**2.8 m** square pad on nine probes) Probes at **1.6 m** centres on centreline of footings Probes at **2.0 m** grid beneath slab area Depth of treatment **4 m** Allowable bearing pressure: **100 kN/m²** to pads/strips; **25 kN/m²** to slabs
	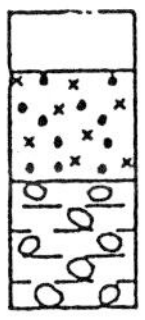 **0−0.15 m** Topsoil **0.15−2.4 m** Loose saturated silty sand **2.4−6.0 m** Firm to stiff boulder clay	Two-storey institutional building, part load-bearing masonry part r.c. frame	Pad bases to columns, strip footings to load-bearing walls Vibro-improved ground beneath foundations and ground slab	Wet process adopted Probes on **1.5 m** grid under pad bases (**2.0 m** square base on four probes) Probes at **1.5 m** centres on centreline of footings Depth of treatment **2.5 m** Allowable bearing pressure; **150 kN/m²** to pads/strips; **25 kN/m²** to slabs
	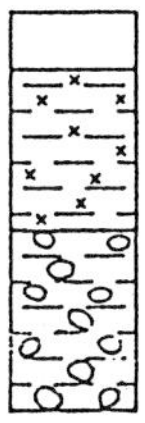 **0−0.3 m** Topsoil and sub-soil **0.3−2.7 m** Soft to very soft bands of clay and silts saturated **2.7−6.0 m** Firm to stiff boulder clay	Tall single-storey load-bearing masonry sports hall	Wide strip footings on vibro-improved ground 1.5 m wide footing reinforced with C785 mesh	Dry (bottom-feed) process adopted two lines probes at **1.25 m** staggered centres on centreline of load-bearing walls Probes at **1.8 m** staggered centres under slab Depth of treatment **2.8 m** Allowable bearing pressure: **150 kN/m²** to footings; **25 kN/m²** to slab *Note* Following testing programme the treatment centres reduced to **0.75 m** in localized area of very soft ground to achieve settlement test criteria
	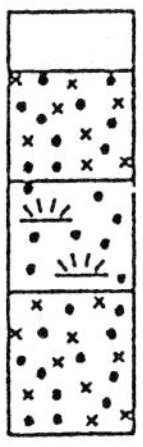 **0−0.2 m** Topsoil **0.2−1.8 m** Loose brown fine silty sand **1.8−2.2** Loose moist dark brown peaty sand **2.2−9.5** Greyish brown fine silty sand	Two-storey teaching block, load-bearing masonry construction	Crust raft on vibro-improved ground Raft slab incorporated internal thickening under load-bearing wall lines	Wet process adopted" Probes at **1.7 m** centres on centreline of raft edge and internal thickenings Probes at **2.5 m** grid under floor areas Depth of treatment **4.8 m** Allowable bearing pressure **110 kN/m²** "This project was undertaken in late 1970s before bottom-feed dry vibro-treatment was available (it is considered that the dry bottom-feed method would have proved effective in this case)

Fig. 8.10 Examples of vibro treatment/foundation solutions.

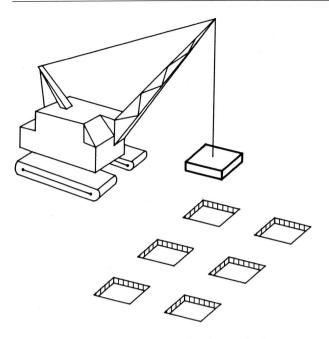

Fig. 8.11 Dynamic consolidation method.

8.4.2 Method

The tamper is dropped on a grid pattern over the whole site area. The process is repeated two to five times until the consolidation required is achieved. The number of repetitions or *passes* is dependent upon soil conditions. The times between successive passes are related to soil permeability.

The treatment is designed to suit the type of development and sub-soil conditions which determine the tamper weight and height dropped, the number of passes and the phasing of passes. On-site testing and monitoring is necessary to ensure satisfactory compaction of the soils. The test results are compared with tests carried out prior to site compaction and allow the changes of soil characteristics to be monitored after each pass. Dynamic consolidation has been used for a wide range of structures and loading conditions on different types of ground including fills where depths in excess of 10 m have been treated.

8.4.3 Usage

Dynamic consolidation has been shown to be cost-effective for the treatment of relatively large areas where mobilization costs can be absorbed more easily. Unlike vibro-stabilization which is a localized treatment − under lines of foundation for example − dynamic consolidation is a treatment of the site area as a whole. The effects on surrounding structures, foundations and services must be considered and usually a clear zone of about 30 m is required to avoid disturbance/damage from vibrations or flying debris.

The major use of dynamic consolidation in this country has been to compact loose fills on large open sites which have been shown to respond well to treatment. The voids within the fills have gradually closed by repeated tamping at the surface. Most saturated materials can be improved by dynamic consolidation but as permeability reduces the treat-

ment becomes less effective. In order to enable continuous operation of the plant and achieve economic treatment, site areas should generally be greater than 10 000 m², although on very permeable sites economical operations have been achieved on smaller areas.

8.4.4 Site checks

Site checks of soil characteristics are usually carried out by monitoring:

(1) Porewater pressure − using a piezometer.
(2) Accurate levelling to measure enforced settlements and ground heave.
(3) Modulus of deformation and limit pressures using pressure metres in boreholes.

In comparison with vibro-compaction the system is clumsy and crude and is economically applicable to a limited number of developments. It is nevertheless the most economical system for some sites and should not be discounted.

8.5 PRELOADING

8.5.1 Introduction

In their natural state soils are consolidated by the effects of the own weight of their overburden. The weight of the materials removed during excavation works to accommodate the new foundations is taken into account on *normal* foundation designs, since this overburden has preloaded the soil thus improving the bearing capacity and settlement characteristics of the underlying soils. The removal of overburden by natural erosion may have occurred, or the passage of glaciers in past ice ages could have temporarily created overburden conditions. The beneficial effects of this preloading in these cases are realized much later.

The temporary preloading or prestressing of a site with a surcharge or prestress to improve soil conditions prior to construction works is therefore a logical approach. This course of action depends upon whether an appropriate improvement in ground condition can be achieved in a suitable time-scale and the preload materials and prestress anchorages are available at an economical cost. This approach has provided an economical foundation solution on many sites (see Fig. 8.12). This solution does however restrict site progress and operations.

8.5.2 Method

Consolidation of uncompacted granular materials and fills can be achieved fairly quickly with only a short duration of preloading, whereas cohesive soils or saturated materials may require a lengthy period of surcharge to achieve the required consolidation because of the longer drainage times. Improvement of drainage in saturated material may be employed to speed up the consolidation process, by the use of sand drains and sand wicks. It should be appreciated that it is not necessary to cover the whole site with the designed surcharge at the same time but to move the *surcharge* around the site − alternatively foundations can be cast and

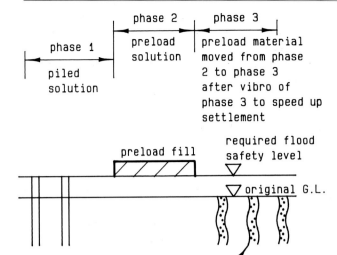

cross-section through large site

Fig. 8.12 Example of preloading.

prestressed using ground anchors to achieve significant settlement prior to construction of the building over (see Fig. 8.13).

This does, however, rely upon a suitable layer of anchorage material within a reasonable depth. It is not essential that the ground anchors be anchored into rock since many modern anchors have other methods of restraint into softer materials, and the engineer should consider all the alternatives.

8.5.3 Design of surcharge

In order to prepare surcharge designs it is necessary to determine the following:

(1) Soil properties from site investigation.
(2) Bearing capacity of the soil.
(3) Bearing pressures of the foundations.
(4) Total settlement.

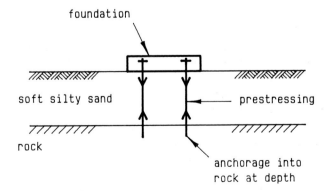

Fig. 8.13 Soil prestressing using ground anchor.

(5) Acceptable settlement.
(6) Lengths of time to achieve settlements.
(7) Weight of surcharge to provide bearing pressure required to produce the excessive portion of settlement.
(8) Increased weight factor (1.2−1.5).
(9) Length of time for surcharge.
(10) Method statement for surcharge operation.
(11) Monitoring procedure to ensure that design consolidation is achieved.

When considering the requirements for the treatment of fills it should be appreciated that settlements are not as easily calculated since settlements are caused not only by the weight of the buildings and their creep settlements, but also by decomposition of organic materials, collapse compression and liquefaction. It is therefore important to locate the various soil deposits and consider any additional movements from these criteria. In many cases soil deposits may, if non-uniform in thickness or quality or unsuitable in type, rule out the possibility of using surcharge alone to motivate the excessive differential settlements to a suitable limit. In other cases the time required for adequate consolidation by surcharge may not be available. In some such cases improvement of drainage may accelerate settlement and a combined surcharge with improved drainage may be adopted.

8.5.4 Installation of drainage systems

Time can be the limiting factor relating to allowable foundation design pressures on sub-soils when surcharge or prestress is used to limit final settlements. Settlement which can be induced prior to construction and application of finishes can help to limit the amount of critical differential settlements. Installation of drainage systems can accelerate settlements and even out differential time relationships. To assist drainage of imported fills, horizontal drainage can be inserted during laying in the form of blankets of granular material or matting. These layers can be put in at specified vertical centres to shorten the drainage path (see Fig. 8.14). For existing fills, vertical sand drains, wicks, vibro or other vertical drainage can be driven at specified centres to shorten the length of drainage and dissipate more quickly the porewater pressure (see Fig. 8.15).

These processes combined with preloading or prestressing can for certain conditions reduce post-development settlements to acceptable limits.

8.6 GROUT INJECTIONS

8.6.1 Introduction

The design of suitable grout for particular criteria is complex and specialized. The design depends on criteria such as the size of voids, access to them, knowledge of their location and risk of grout penetration into services or other restricted areas. These and other aspects affect the type of grout and method of placing. It is not intended to deal in depth with grout design but to provide only some basic background knowledge to enable the engineer to call upon grouting services when appropriate to solve foundation problems.

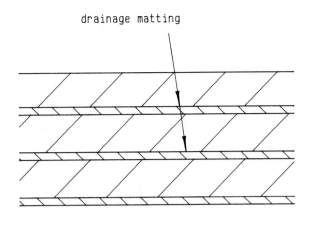

drainage matting

granular layers to
assist drainage

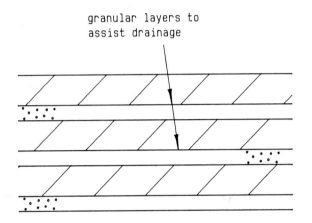

Fig. 8.14 Horizontal soil drainage.

For structural foundations some of the more common applications occur in filling voids in loose materials, swallow-holes, shallow mining, shafts and wells. Guidance on these applications will give a feel of the conditions under which grouting can prove a viable solution to sub-strata strengthening (see Fig. 8.16).

In all these applications it is important that the engineer realizes that since grout is designed to penetrate the most difficult location there is a danger that unwanted penetration into existing cracked sewers, service ducts, etc., within

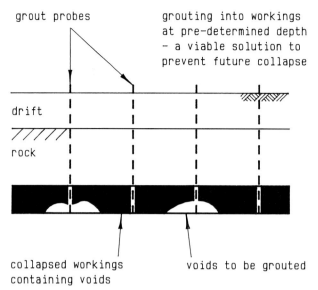

grout probes

grouting into workings
at pre-determined depth
– a viable solution to
prevent future collapse

drift

rock

collapsed workings
containing voids

voids to be grouted

Fig. 8.16 Grout injection method.

reach of the grouting zone may occur. These aspects must not be overlooked in the design of the system. In sensitive locations grouting may be totally unsuitable due to the risk of damage and difficulty in sealing around the zone of grout treatment.

8.6.2 Loose soils
In the majority of cases strengthening of loose soils is most successfully achieved by a vertical pressure and/or vibration rather than grout injection, particularly where the depth of loose material can be influenced by surface rolling or vibro-stabilization (see sections 8.2 and 8.3). There are situations, however, where deep loose materials exist which may also contain larger voids and in these cases grout injection is a more appropriate and safe solution. For example, where loose soils exist in shafts or old workings below rockhead the grout can search out migrating voids. It is the use of grout to fill these larger void pockets rather than the small well distributed loose material which is appropriate to grouting. For this reason grouting injection tends to be used for mine shafts, swallow-holes and shallow mine workings rather than for strengthening of weak soils.

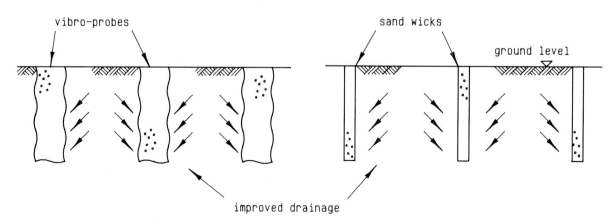

vibro-probes

sand wicks

ground level

improved drainage

Fig. 8.15 Vertical soil drainage.

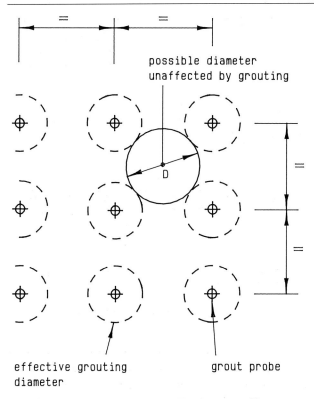

effective grouting
diameter

grout probe

Fig. 8.17 Grout injection grid/effective grout diameter.

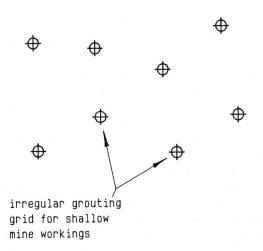

irregular grouting
grid for shallow
mine workings

Fig. 8.18 Grout injection irregular grid.

8.6.3 Swallow-holes

A brief explanation of the structure of swallow-holes and the risk of collapse of the loose soil filling is mentioned in Chapter 4. Where a desk study and soil investigation has revealed a real risk of swallow-hole collapse, the risk can be embraced within the design by either (a) allowing a predicted diameter collapse to be designed into the foundation solution ensuring suitable spanning capability or (b) where more severe conditions prevail, combining a grid of grouting holes to reduce the likely spanning requirements to a more practical diameter (see Fig. 8.17).

The depth of the drilling and grouting should be decided from the ground investigation and assessment of the depth of improved ground required to reduce the risk from collapse to an acceptable level.

8.6.4 Shallow mining

Voids from shallow mine workings can result in sudden collapse – see Chapter 6 for details relating to mine workings. The risk of collapse can be accommodated by designing the raft foundation to span and cantilever over the collapsed ground or the voids can be grouted. A further alternative is a combination of a spanning raft and grout injection.

Grouting on a regular grid however should be avoided in these situations since mining methods also tended to have a regular grid. The method should be to grout in a way similar to that used for swallow-holes, but on a non-regular grid (see Fig. 8.18). The depth of the holes should be related to mining records and the information obtained from drilling (see Chapter 6 on mining for further details).

8.6.5 Mine shafts, wells and bell-pits

Mine shafts and wells have been discussed in Chapter 4 and further information is also available on shafts in Chapter 6. It is essential that adequate safety precautions be taken in any treatment of these man-made conditions and reference should be made to Chapter 6. The difference in shape and void formation when the apertures have been previously filled can be seen in Fig. 8.19.

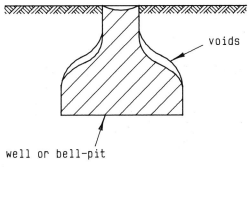

well or bell-pit

voids

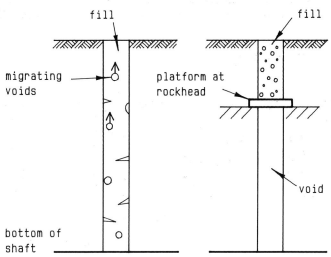

migrating
voids

platform at
rockhead

bottom of
shaft

void

Fig. 8.19 Void formation.

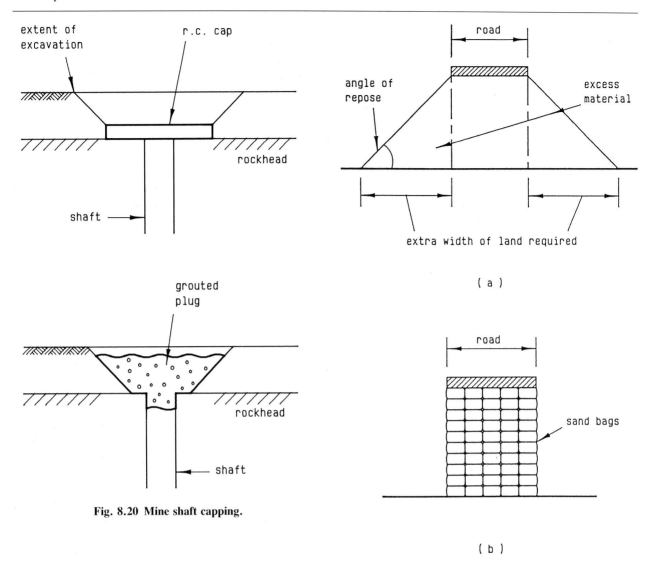

Fig. 8.20 Mine shaft capping.

Fig. 8.21 Reinforced earth.

From Fig. 8.19 it is apparent that knowledge on the shape and depth of the aperture is important to the likely location and type of voids formed. A desk-top study of the information available is therefore helpful to the drillers carrying out the investigative and grouting activities. The grouting is generally carried out by drilling a number of holes down the shaft for grout injecting and working the grout up from the lower levels of the aperture. On completion of grouting a cap or plug is generally inserted, at rockhead in the case of mining, or at ground level for shallow wells. The cap can be cast by excavating to the rockhead or appropriate level. Alternatively the cap can be formed using grout injection of the sub-soil over the shaft. (See Fig. 8.20.)

8.7 LIME/CEMENT STABILIZATION

Soil stabilization requires a high degree of uniform particle bonding. A review of grouting processes illustrates that penetration by cements in most soils is impractical, but injection that displaces soil fabric and mixing that destroys it, allows cement to be introduced into the soil. These

processes are at present mainly applicable to civil engineering use. Deep consistent ground improvement to achieve foundation bearing capacities is at present impractical and uneconomic for structural foundations.

8.8 REINFORCED EARTH

8.8.1 Introduction
The road embankment formed in the normal manner of compacted layers of granular material (see Fig. 8.21 (a)) requires a wide strip of land to be purchased and uses a large amount of material in the side banks. The embankment so formed can also restrict the design of road layouts at junctions and slip roads.

If the sand is contained in sacks and the resulting sand bags carefully placed and built up (see Fig. 8.21 (b)), a more economical use of material and width of land is achievable.

Sand, and other non-cohesive soil, is stronger in compression than it is in shear or tension. Concrete, too, is strong in compression and weak in tension, so steel reinforcement is added to compensate for the tensile and shear weakness. The principle of adding reinforcement to earth (i.e. the tensile strong sacks) is basically the same and the resulting *reinforced earth* – now a composite material – is stronger than the earth acting on its own.

In reinforced concrete, care is taken to ensure that the reinforcement is free of oil, loose rust, or other material which would reduce the bond (or friction) between the concrete and the reinforcement. The tensile stresses are transferred to the reinforcement by this *bond*. (A thickly oiled stainless steel bar would not be gripped by the concrete in a beam and when load is applied to the beam the bar would slip and carry little, if any, tensile force.) Similarly in reinforced earth sufficient friction must be developed between the reinforcement and the earth. The reinforcement is usually in a mesh, or net form, of steel or polymers. Steel is strong, stiff and more resistant to creep than polymers but can suffer from corrosion. In general polymers are not so strong and more liable to creep deformation than steel but can be more durable.

If reinforcement is substituted for sacks in the road embankment shown in Fig. 8.21 (b), then, while the embankment would be sound, there could be side surface erosion of the earth (see Fig. 8.22). This can be prevented by fixing facing units to the end of the reinforcement (the facing units are usually segmental and made of a durable material). Early facing units were of elliptical steel sheeting but now precast concrete cruciform and other facing sections are more common.

The combination of selected granular fill, reinforcement and facing units comprise reinforced earth fill.

The most common application of reinforced earth has been in retaining walls and embankments. By comparison to walls there has been far less application, and little research and development, of reinforced earth for rafts or other types of foundations. This is probably because the ground around buildings can usually incorporate a simple embankment and, to date, reinforced earth is not usually cost competitive with ground improvement techniques such as vibro-stabilization combined with a banked arrangement. However, construction costs fluctuate and the designer may consider costing alternative preliminary designs using a reinforced earth solution.

8.8.2 Foundation applications

In addition to the more common use of earth reinforcement in embankments and road construction, benefit can be gained from increased tensile resistance of soils for foundations. For example, reinforcement can improve soil behaviour during surface rolling for soil strengthening. The behaviour of the soil can be improved by restraint from side spread by the introduction of horizontal reinforcement between the layers of material.

This is particularly useful where *crusting up* of the surface is required (see Fig. 8.23). A reinforcement mesh which also improves drainage is even more beneficial for soils of poor permeability and some modern meshes perform this dual role. Products are constantly being improved and the engineer should check the latest developments. The basic engineering principles however do not change and improved tensile strength and drainage with long-term durability and performance are major factors affecting foundation designs.

8.8.3 Patents

It is claimed that the innovation of reinforced earth was made by Henri Vidal, a French architect, in the 1960s. He found that mounds of dry sand could stand at a steeper angle when horizontal layers of pine needles were incorporated in the sand. He undertook the early research and development and subsequently patented the technique. *Reinforced Earth* is the trademark of the Reinforced Earth Company Limited which is Henri Vidal's exclusive licensee in the UK.

Two other systems, of limited applications in the public sector, have been developed in the UK after the Department of Transport paid Henri Vidal a lump sum under license agreement in 1980. The systems are basically the same (i.e. selected fill, reinforcement and facing unit). One system – the York system used by the DoE – uses facing units of lightweight glass-reinforced cement cast in the form of hexagon-based pyramids. The other system – the Websol system – does not use steel reinforcement and employs 'Paraweb' as a substitute. Paraweb, made by ICI (Imperial Chemical Industries), consists of synthetic fibres encased in a polythene sheath. Facing units of precast concrete, T-shaped in front elevation, are used with this system. Since

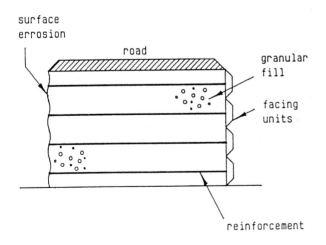

Fig. 8.22 Facing to reinforced earth.

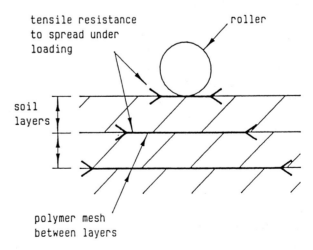

Fig. 8.23 Reinforced crust.

there have been a number of lawsuits (and some are pending) it is advisable to check that there is no infringement of copyright in further developments.

8.8.4 Research and development
There has been with reinforced earth (as there was in the early days of soil mechanics) a veritable flood of laboratory research – most of it into the investigation of failure modes and soil reinforcement bond. However, as with early soil mechanics research, the results, while of value in understanding behaviour, are of somewhat debatable value to the designer. This is mainly due to the difficulty of simulating construction site conditions in a laboratory. Researchers are well aware of the difficulty and there is likely to be an increase in in situ testing to provide adequate empirical design data.

PART 3

FOUNDATION TYPES: SELECTION AND DESIGN

CHAPTER 9

FOUNDATION TYPES

9.1 INTRODUCTION

This chapter describes the various types of foundations in general terms, and defines the different functions, materials employed and how they behave. The foundation types are discussed in four groups i.e.:

- Strip and pad foundations,
- Surface spread foundations,
- Piled foundations,
- Miscellaneous elements and forms.

9.2 FOUNDATION TYPES

The design of foundations involves the use of many different combinations of structural elements and foundation types which in turn vary to perform a wide variety of functions. It is therefore not surprising that the foundation scene has grown into a jumble of rather poorly defined elements and forms. In addition to providing guidance on the elements and forms available, this chapter suggests a more clearly defined terminology in an attempt to help clarify the issue. It is possible therefore that even the experienced engineer may at first find some of the terms unfamiliar. However, the authors have found that with use, the terms prove to be of great assistance. Since this chapter covers modern developments in foundation design this has resulted in the introduction of further new terms. Wherever existing terms clearly define the structural element or foundation form they have been retained, but more vague definitions such as 'rigid raft', etc. have been deliberately omitted, since such terms are in danger of misinterpretation and cover a widely varying group of foundations.

In addition to the design of the foundations to support the applied loads, without excessive settlement and distortion, there is a need to resist or prevent the effects of frost-heave and/or shrinkage and swelling of sub-strata. The many different loads and conditions demand different solutions, however the foundation types can generally be defined and the main types are described in this chapter. Various references are made in the text to relative costs. Appendix M should be consulted for more detailed cost guidance.

9.3 GROUP ONE – STRIP AND PAD FOUNDATIONS

Strip footings and pad bases are used to deliver and spread superstructure loads over a suitable area at foundation (formation) level. The foundation is required to be stiff enough to distribute the loadings onto the sub-strata in a more uniform manner.

9.3.1 Strip footings

Strip footings are used under relatively uniform point loads or line loads. The main structural function of the strip is to disperse the concentration of load sideways into an increased width of sub-strata in order to reduce the bearing stress and settlement to an acceptable limit. A cross-section through an unreinforced concrete strip footing showing the assumed dispersion of load is shown in Fig. 9.1.

A further major structural function is to redistribute the loads in the longitudinal direction where the loading is non-uniform or where the sub-strata resistance is variable (see Fig. 9.2). The width of the strip is usually decided by calculating the width required to limit the bearing stress and choosing the nearest excavator bucket size up from that dimension. From a construction point of view, the strip depth is used as a means of levelling out irregularities in the trench bottom and the width has to absorb the excavation tolerances which would be unacceptable for the setting out of walls etc.

There are a number of different types of strips which include masonry strips; concrete strips – plain and reinforced; trench fill – concrete and stone; reinforced beam strips – rectangular and inverted T, and these are described in the following sections.

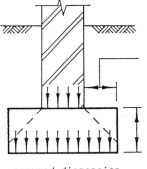

footing projection from wall to allow tolerance for building wall – usually 100 mm minimum

thickness of footing determined by dispersion line passing through side of footing

assumed dispersion approximately 45°

Fig. 9.1 Typical strip footing.

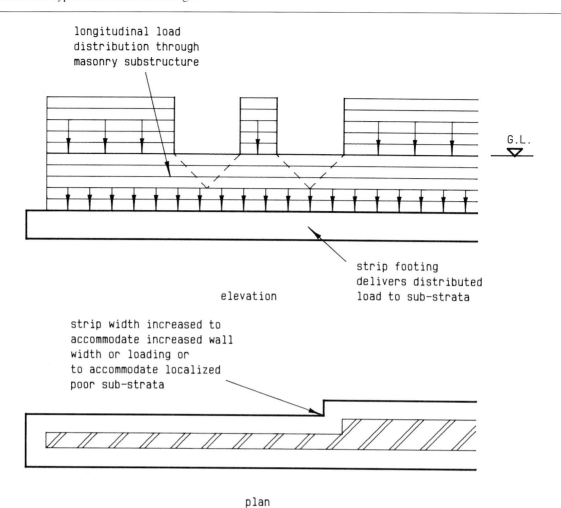

longitudinal load
distribution through
masonry substructure

G.L.

elevation

strip footing
delivers distributed
load to sub-strata

strip width increased to
accommodate increased wall
width or loading or
to accommodate localized
poor sub-strata

plan

Fig. 9.2 Strip footing load spread distribution.

9.3.2 Masonry strips

Masonry strips are rarely used these days, however they can be adopted where good quality sub-strata exists and the raw materials for masonry construction are cheap and abundant. The wall is increased in width by corbelling out the masonry to achieve the required overall foundation width as shown in Fig. 9.3.

It should be noted that it can be important, particularly when using masonry strips in clay or silt sub-strata, to bed the masonry units in mortar and to completely fill all joints. The reason for filling the joints is mainly to prevent the strip footing acting as a field drain with the water flowing along the surface of the formation level and through the open joints of masonry. The authors have found clear evidence of induced settlement due to softening of the clay surface below *dry* random rubble strips (*dry* random rubble being a term for dry stacking without mortar and not dry meaning no moisture).

9.3.3 Concrete strips – plain and reinforced

The concrete strip footing replaced the corbelled masonry in more recent constructions. In plain (unreinforced) strip footings the thickness is determined by the requirement for the line of dispersion to pass through the side of the footing as shown in Fig. 9.1. The width of the trench must also allow

working space for the bricklayers to build the masonry off the footing.

The profile of the reinforced concrete strip is similar to the unreinforced strip except that it can generally be made thinner in relation to its projections since it no longer relies upon an approximate 45° line of load dispersion. The strip is often reinforced with a fabric or lattice reinforcement. The longitudinal bars are the main bars selected to suit the longitudinal bending expected on the strip and the cross bars designed to cater for the cantilever action on the projections (see Fig. 9.4).

9.3.4 Concrete trench fill

Concrete trench fill consists of a mass concrete strip cast into the open trench making use of the trench sides as a shutter (see Fig. 9.5).

Concrete trench fill is often used where strip loads are required to be transferred to relatively shallow depths through soft material which is capable of standing up, without extra support, for at least a period adequate to cater for the construction sequence to be adopted. The trench fill can embrace requirements for heavy loads going down to rock or light loads on soft sub-strata (see Fig. 9.6).

The requirement for working space within the trench for bricklayers is not a factor in determining the width of

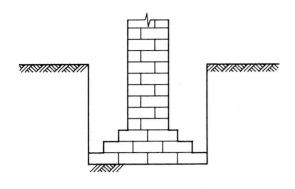

corbelled brick
footing

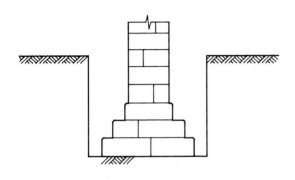

corbelled stone
footing

Fig. 9.3 Masonry strip footing.

excavation with this method. Pouring concrete to within 150 mm of ground level overcomes this consideration.

9.3.5 Stone trench fill
Stone trench fill consists of stone deposited into the open trench excavation and compacted in layers. It is particularly useful in areas where poor quality sands, sandy silts, etc., exist. The material immediately below the topsoil is often suitable for the general floor slab loading but not for the more heavily loaded external and internal strip loadings. Suitable sub-strata for the strip loads often exists at a shallow depth and stone trench fill can be used down to these levels (see Fig. 9.7).

9.3.6 Rectangular beam strips
Rectangular beam strips consist of rectangular reinforced ground beams which are designed to be of sufficient width to reduce the bearing pressures on the sub-strata to an acceptable value. The beam is required to be of sufficient cross-section to resist the induced bending moments and shear forces in the longitudinal direction. The beam is reinforced with either ladder reinforcement or caged reinforcement to suit the design conditions. Figure 9.8 shows a typical beam strip supporting point loads.

9.3.7 Inverted T beam strips
The inverted T beam strip fulfils the same function as the rectangular beam strip but the cross-section is modified to an inverted T so that the flanges reduce the contact pressure on the ground (see Fig. 9.9).

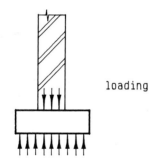

loading

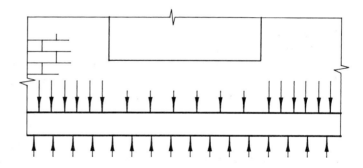

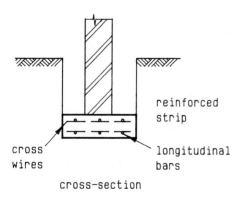

reinforced
strip

cross
wires

longitudinal
bars

cross-section

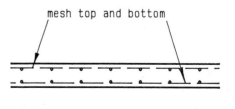

mesh top and bottom

longitudinal section

Fig. 9.4 Reinforced strip footing.

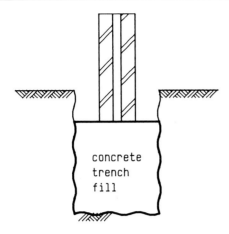

Fig. 9.5 Concrete trench fill.

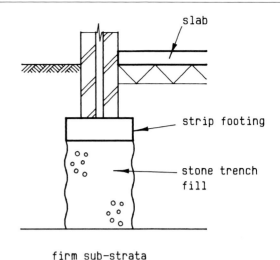

firm sub-strata

Fig. 9.7 Stone trench fill.

9.3.8 Pad bases

Pad bases are used under point loads from columns and piers. There are a number of different types, including mass and reinforced concrete both shallow and deep, which are described in the following sections.

9.3.9 Shallow mass concrete pads

Shallow mass pads consist of mass concrete pads supporting point loads from columns, piers, etc. (see Fig. 9.10).

They are used for varying conditions of sub-strata where suitable load-bearing soils exist at shallow depths below the effects of frost and general weathering. They are particularly economic where the side of the excavation can be used as a shutter and where a suitable depth of mass can be accommodated to disperse the load without the need for reinforcement. The general assumption for load dispersion is as

mentioned previously i.e. a 45° spread through the mass concrete (see the typical example shown in Fig. 9.11).

9.3.10 Shallow reinforced concrete pads

Reinforced concrete pads are similar to the mass concrete pads but for the same conditions can be thinner when reinforced with steel. The reduction in thickness is made possible by the introduction of reinforcement on the tensile face of the pad which increases the pad's resistance to bending moment (see Fig. 9.12).

9.3.11 Deep reinforced concrete pads

Deep reinforced concrete pads are similar in cross-section to the shallow reinforced pad but are constructed at depth in situations where the suitable sub-strata is not available at

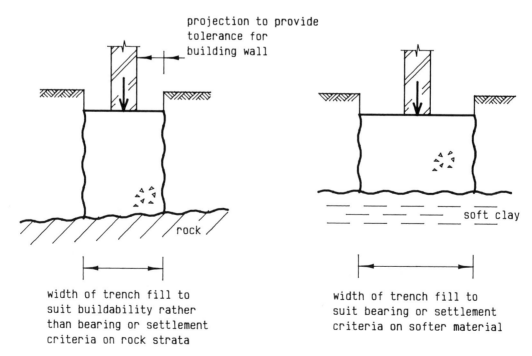

Fig. 9.6 Concrete trench fill.

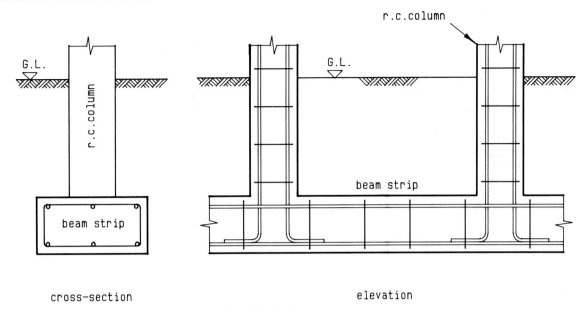

cross-section   elevation

Fig. 9.8 Typical beam strip.

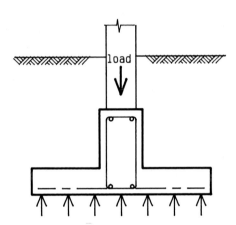

Fig. 9.9 Inverted beam strip.

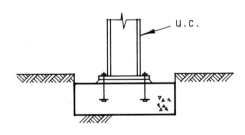

Fig. 9.10 Shallow mass concrete pad.

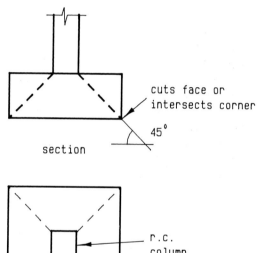

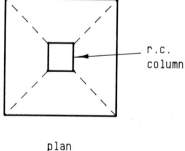

plan

Fig. 9.11 Load spread on mass concrete pad.

high level. Such pads are not often economic and the more competitive mass concrete or piles and caps are often used. However in some situations they can prove to be a suitable solution – see Fig. 9.13 which indicates a typical example of such a use.

9.3.12 Deep mass concrete pads
Deep mass pads consist of mass concrete pads cast with their soffit at depths in excess of 1.5–2 m. They are generally used where a suitable ground bearing strata is relatively deep and where the piling alternative is more expensive, i.e. a small number of pads are required or access for piling is difficult and expensive. Deep mass pads tend to be of two types, one being constructed up to high level using a basic cross-section and the other using a reduced and shuttered cross-section for the upper levels (see Fig. 9.14).

An alternative to concrete for the upper reduced cross-section is to construct a brick pier off the mass concrete pad (see Fig. 9.15). This solution has the advantage of avoiding the need for expensive shuttering and can result in an overall saving. If brickwork is adopted it is necessary that

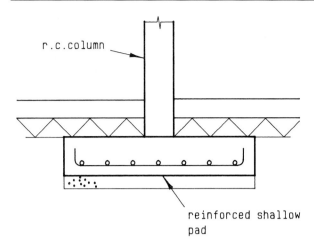

Fig. 9.12 Shallow reinforced concrete pad.

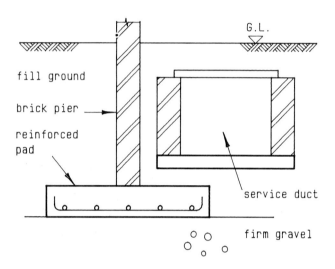

Fig. 9.13 Deep reinforced concrete pad.

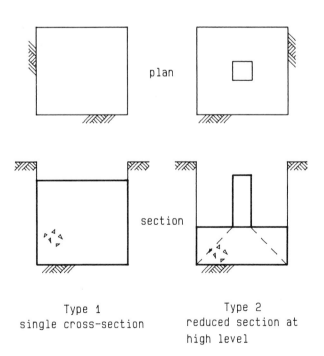

Type 1
single cross-section

Type 2
reduced section at
high level

Fig. 9.14 Deep mass concrete pad.

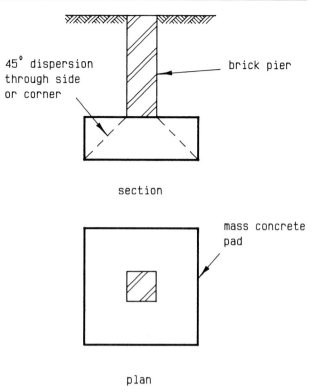

Fig. 9.15 Deep mass concrete pad with brick pier.

the pad size provides the necessary working space for the bricklayers to build the pier.

9.3.13 Balanced pad foundations

Balanced pad foundations are used where a number of loads are required to be supported on a single pad and where excessive variations in pressure could produce unacceptable differential movement. They consist of reinforced concrete pad bases designed for the critical design loading with the aim of keeping the differential ground stresses and hence settlements to an acceptable level. This requirement could be the result of a sensitive sub-strata and/or a sensitive superstructure over. There are a number of different types of balanced pad foundations which include rectangular, trapezoidal, holed and cantilever, and these are described in the following sections (see also Chapter 12).

9.3.14 Rectangular balanced pad foundations

A typical rectangular balanced foundation supporting two point loads from a sensitive structure that has only a small tolerance to accommodate differential settlement is shown in Fig. 9.16. The problem has been overcome by adjusting the cantilevered ends of the base to produce a constant ground bearing pressure for the load conditions.

9.3.15 Trapezoidal balanced pad foundations

The trapezoidal balanced foundation is used in similar circumstances to the rectangular balanced foundation. Adjusting the width of each end of the pad in relation to the load supported can produce a more economic solution. This is particularly useful where two point loads of different sizes

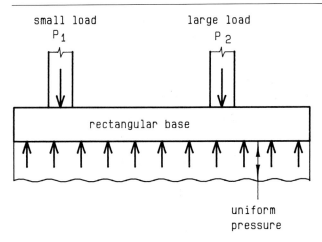

Fig. 9.16 Rectangular balanced pad foundation.

need to be supported and a relatively uniform bearing pressure is required (see Fig. 9.17).

It is also useful where adjustments by cantilever action are not possible, for example, where two different column loads on the edge of opposite building lines require support (see Fig. 9.18).

9.3.16 Holed balanced pad foundations

The holed balanced foundation is a pad type foundation supporting a number of loads and transferring the load to the bearing strata in a relatively uniform fashion. The allowable variation in bearing pressure and differential settlement is again determined from the ground conditions and sensitivity of the superstructure. The resultant load and its position are determined for the critical load case. While with the rectangular base the balancing is done by varying the cantilever and with the trapezoidal base by varying the end dimensions, in this case the balancing is done by

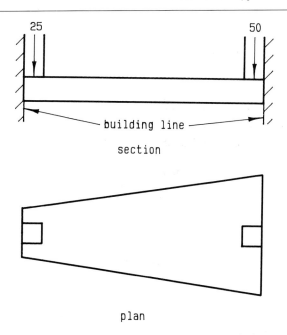

Fig. 9.18 Trapezoidal balanced pad foundation.

forming a hole in the base positioned so as to move the centroid of the base to coincide with that of the resultant load (see Fig. 9.19).

9.3.17 Cantilever balanced pad foundations

The cantilever balanced foundation consists of a ground beam picking up loading from the superstructure and cantilevering out over a pad foundation with the pads designed, theoretically, to have uniform bearing stress (see Fig. 9.20).

The need for a cantilever arrangement can be produced by restrictions from adjacent buildings or existing services (see Fig. 9.21).

9.4 GROUP TWO – SURFACE SPREAD FOUNDATIONS

Surface spread foundations consist mainly of rafts and are generally used where the normal ground bearing sub-strata is relatively poor and the depth to suitable load-bearing soils is excessive or the load-carrying capacity of the soil deteriorates with depth. Surface spread foundations are therefore employed to distribute the superstructure/substructure loads over a large area of the ground thus reducing the contact

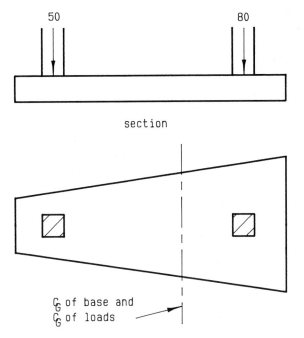

Fig. 9.17 Trapezoidal balanced pad foundation.

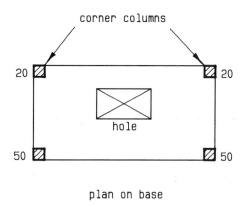

Fig. 9.19 Holed balanced pad foundation.

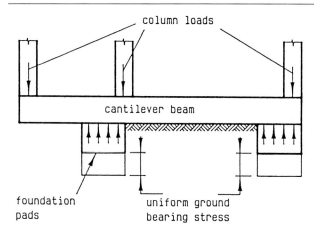

Fig. 9.20 Cantilever balanced foundation.

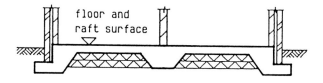

Fig. 9.22 Typical raft foundation.

bearing pressure. Since most structures also require a ground floor slab it is usually economic to incorporate it with the foundation into one structure/element. This can be done by making the upper surface of the raft foundation coincide with the top surface of the floor slab. A simple example is shown in Fig. 9.22.

Surface spread raft foundations are often adopted in areas of active mining as the best means of resisting excessive distortion, tensile and compressive forces, etc., resulting from the ground subsidence. These and other types of surface spread foundations are discussed in the following sections. It should be noted that rafts do not necessarily distribute the loads as a uniform contact pressure to the sub-strata, on the contrary, most rafts are relatively flexible foundations and will have higher contact pressure under loaded points and edge thickenings than below the main slab areas.

9.4.1 Nominal crust raft

A nominal crust raft is basically a ground-bearing reinforced concrete floor slab with nominal thickenings around the edges. Internal thickenings are sometimes incorporated in the raft (see Fig. 9.23).

The slab acts as a surface crust to the sub-strata thus evening out any small local differential settlement movements which could result from variations in imposed loading on the top of the slab and/or local variations in settlement characteristics of the sub-soil. The design is generally carried

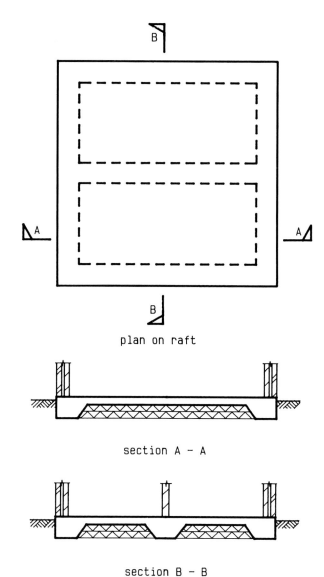

plan on raft

section A - A

section B - B

Fig. 9.23 Nominal crust raft.

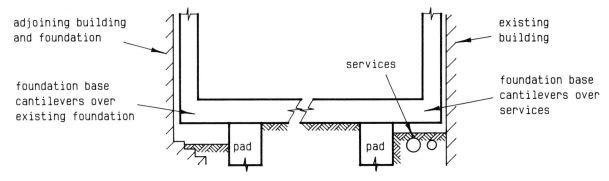

Fig. 9.21 Cantilever balanced foundation.

out either by sizing the raft from previous experience or by calculation based upon nominal assumptions.

9.4.2 Crust raft

The crust raft is a stiffer version of the nominal crust raft. The ground slab and thickening which form the crust are combined into a total raft design. Heavier loads on soil of low bearing capacity determine the size and depth of the thickenings. The thickness of the slab is dictated by the overall raft design which generally exceeds the nominal slab requirements.

9.4.3 Blanket raft

The blanket raft consists of a concrete crust raft constructed on a stone *blanket* which in turn is built up in layers off the reduced sub-strata level (see Fig. 9.24). The crust raft and *blanket* interact to support and span the loading over any localized soft spots or depressions. The main difference between this and the crust raft is the introduction of the stone blanket. This blanket effectively disperses any heavy point and edge loads or imbalance of load. Composite action between the crust raft and the stone blanket is the basis of the action and design of this foundation system.

9.4.4 Slip-plane raft

The slip-plane raft consists of a concrete raft constructed on a slip-plane layer, such as sand of known friction or shear resistance, which is located between the raft and the sub-strata. The slip-plane is constructed in sufficient thickness to ensure that a straight failure plane could occur under excessive longitudinal ground strain (see Fig. 9.25). The depth of penetration of the raft into the ground is kept to a minimum to avoid picking up loading from ground strains. However,

the depth below finished ground level must take account of potential frost heave.

9.4.5 Cellular raft

A cellular raft consists of an arrangement of two-way interlocking foundation beams with a ground bearing slab at the underside and a suspended slab at the top surface. The upper and lower slabs are usually incorporated within the beams to form I sections. The intersecting beams effectively break the large slab into two-way spanning continuous small panels (see Fig. 9.26).

The top slab is cast using precast soffits or other forms of permanent formwork such as lightweight infill blocks. These rafts are used on sites subject to severe mining activity or in areas of poor ground where large bending moments are to be resisted. They are also used in locations where a valuable increase in bearing capacity can be achieved by the removal of the overburden and where deep foundation beams are required.

9.4.6 Lidded cellular raft

The lidded cellular raft is very similar in profile to the cellular raft and is used in similar locations, i.e. severe mining conditions, areas of poor ground where the raft will be subjected to large bending moments, etc. The main difference however is the use of a lighter form of upper slab designed to be separate to the main foundation (see Fig. 9.27).

The detail at the seating of the upper floor depends upon the need for re-levelling and the possible number of times adjustments to line and level may be necessary.

9.4.7 Beam strip raft

The beam strip raft consists of (ground-bearing) downstand beams in two or more directions which support the heavy

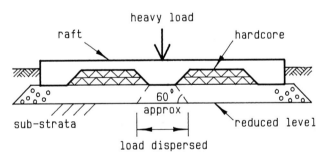

Fig. 9.24 Blanket raft.

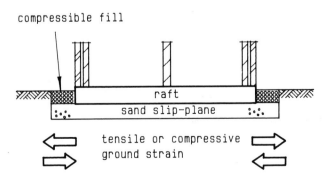

Fig. 9.25 Slip plane raft.

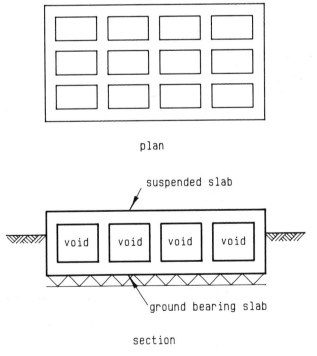

Fig. 9.26 Cellular raft.

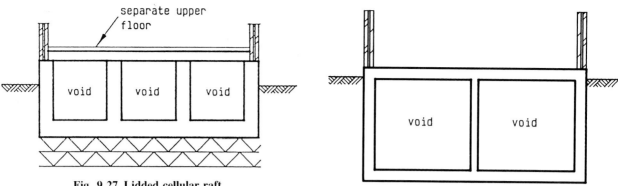

Fig. 9.27 Lidded cellular raft.

Fig. 9.29 Buoyancy raft.

uniform or point loads from the structure. The beams are tied together by a ground-bearing slab supported on the hardcored dumplings, i.e. the raised areas of hardcore protruding up between the beam lines (see Fig. 9.28).

This raft is mainly used in areas of either mining activity or soft alluvial deposits where a stiffened beam is required on the main load lines. The tying of the ground floor slab into the beams prevents lateral distortions of the beam and evens out any local differential settlements. This type of raft is more economic than the cellular form and is used where conditions are not as severe.

9.4.8 Buoyancy (or 'floating') raft

A buoyancy raft is similar to a cellular raft and is a deep raft with large voids. The main weight of removed earth is replaced with practically weightless voids of the raft (see Fig. 9.29). Basement accommodation can be provided in this form of construction. Basement slabs together with retaining walls form the raft.

It is used for heavily loaded structures in areas of low ground-bearing capacity.

9.4.9 Jacking raft

The jacking raft is used in areas where the expected subsidence would tilt or distort the structure to an unacceptable degree and where re-levelling of the raft produces an economic and viable foundation for the design conditions. The jacking raft is used in locations of excessive or unpredictable subsidence, for example, in areas subjected to brine or other mineral extraction. A typical jacking raft for a domestic property is shown in Fig. 9.30.

9.5 GROUP THREE – PILE FOUNDATIONS

9.5.1 Introduction

Piles are generally used as a means of transferring loads down through unsuitable bearing strata either by skin friction and end bearing or end bearing only into a firm layer at greater depth (see Fig. 9.31).

There are many different types of piles including concrete – in situ and precast; steel; timber; stone. The cross-section of the pile and the installation method vary

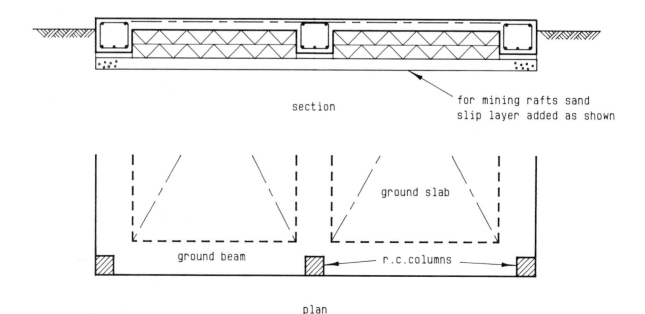

Fig. 9.28 Beam strip raft.

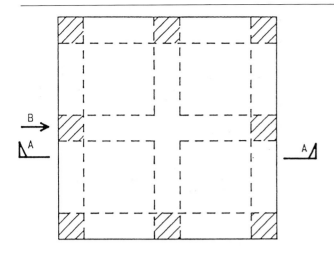

plan on raft

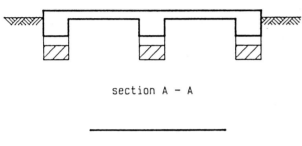

section A – A

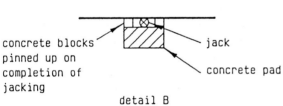

concrete blocks
pinned up on
completion of
jacking

jack

concrete pad

detail B

Fig. 9.30 Jacking raft.

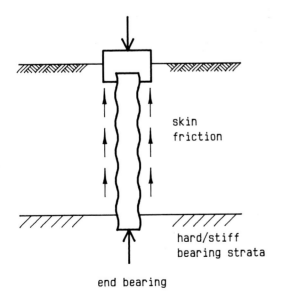

skin
friction

hard/stiff
bearing strata

end bearing

Fig. 9.31 Typical pile foundation.

significantly. In addition to transferring loads to greater depths below surface level the stone pile system (vibrostabilization) can be used to upgrade the bearing capacity of the sub-strata (see Fig. 9.32) – further information on this aspect can be found in Chapter 8.

The distribution of load through the sub-soil varies with the various types of pile and different installation methods. Some piles are suitable for sandy soils, others for clay soils or end bearing into rock. The aim however is generally the same and that is to provide an economic means of support for the foundation and its loads. The various pile types and/ or systems have advantages and disadvantages which make each pile more suitable and competitive for particular situations and soil conditions. There is perhaps a danger of the designer, having selected a competitive system on the first piling job, making the assumption that it is also the appropriate pile system to use on future contracts and ignoring the fact that the competitive tender probably related as much to the site and ground conditions as it did to anything else.

There is, therefore, a need for designers to understand the various types of pile, their best application, and possible limitations, etc., in order to provide good engineering solutions for design purposes. The following sections which describe the various types will assist the engineer in his choice of suitable pile systems and applications.

9.5.2 Stone/gravel piles

The stone, or gravel, pile is mainly used as a means of strengthening a sub-strata by pushing into it a series of stone columns using vibration or jetting methods which compact the ground around the stone and replace the void created with a compacted stone column (see Fig. 9.33).

There are basically three main applications which require quite different design judgement and approaches to site testing, and these aspects are dealt with in detail in Chapter 8. In general terms, stone or gravel piles are used in areas of soft sub-strata, or fill, where sufficient upgrading of the bearing capacity or reduction in differential settlement can be achieved. In such situations the stone/gravel pile is usually much cheaper and in some situations much more suitable than the concrete pile alternative. For example, the gravel pile has a particular advantage in mining areas where the use of concrete piles could result in the foundation picking up unacceptable ground strains and/or possibly the

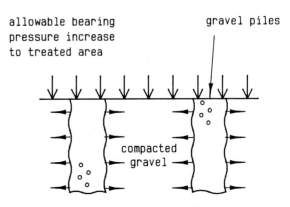

allowable bearing
pressure increase
to treated area

gravel piles

compacted
gravel

Fig. 9.32 Stone pile (vibro).

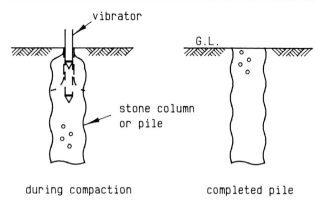

Fig. 9.33 Stone/gravel piles (vibro).

piles shearing off during subsidence due to the brittle form and limited capacity to resist horizontal ground strain. The gravel pile can be used incorporating a slip-plane between the top of the pile and the underside of the foundation in a manner similar to that described for the slip-plane raft in section 9.4.4 and Fig. 9.25.

9.5.3 Concrete piles

Concrete piles are generally used to transfer loads through an unsuitable bearing material to a deeper load-bearing strata. This is achieved either by skin friction and end bearing or end bearing alone (see Fig. 9.34).

There are many different types and systems of piles, however the main types are:

(1) Driven precast piles.
(2) Driven cast in situ piles.
(3) Bored piles.
(4) Augered piles.

These piles can also be divided into either displacement or replacement methods dependent on the system of driving,

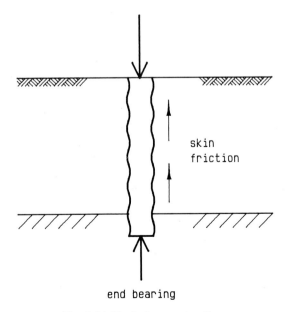

Fig. 9.34 Typical concrete pile.

i.e. either removal of material, termed *replacement*, or wedging apart of material, termed *displacement*. Typical examples of these types are shown in Fig. 9.35.

DRIVEN PRECAST PILES
Driven precast piles can be used in areas where the soils, through which the pile is to be driven, are relatively soft and unobstructed and where the length of pile required can be determined to a reasonable accuracy. The piles can be cast to any suitable cross-section, i.e. square, rectangular, circular, hexagonal, etc. The shape and protection to the point of the pile is determined from the end bearing requirements and driving conditions. The pile head and the reinforcement are designed to take account of the pile-driving impact loads. Some disadvantages of this method of piling are that the pile can be damaged in a location out of sight during driving and the pile can be displaced if it meets an obstruction such as a boulder in the ground. In addition the accuracy of the estimated length is only proved on site and short piles can be difficult to extend and long piles can prove to be expensive and wasteful. A further disadvantage is the relatively large rig required for driving and the need for hard-standings that are often required to provide a suitable surface for the pile-driving plant.

DRIVEN CAST IN SITU PILES
Driven cast in situ piles use steel, or precast concrete, driving tubes which are filled with in situ concrete after driving. Variations in pile lengths can be more easily accommodated using segmental liners. The piles can be cast accurately to the required length and the driving or liner tube can be driven in short lengths. In some cases the tube is left in position permanently and in other cases the tube is withdrawn and used to tamp the concrete by lifting and dropping the liner tube. In other situations the tube is withdrawn and vibrated as the concrete is poured and additional compaction achieved by impact to the surface of the wet concrete. Driven piles can therefore have a smooth or irregular side surface depending on the method of driving and this results in differing friction and mechanical keying to the surrounding soil which varies depending on the pile type and sub-soil conditions. Again large rigs are required for driving cast in situ piles and hardstanding requirements can prove expensive.

In piling systems where the liner tubes are withdrawn there is a danger that the tube can lift the upper portion of in situ concrete leaving a void a short distance below the surface or squeezing during withdrawal can cause necking. This can happen where the mix is not carefully controlled or where the liner tube is not withdrawn at a steady slow rate. Driven cast in situ piles, however, can prove to be economic for sands, gravels, soft silts and clays, particularly when large numbers of piles are required. For small numbers of piles the on-site cost can prove expensive. Driven precast piles and driven cast in situ piles can prove particularly suitable where groundwater or soft inclusions occur in the sub-strata.

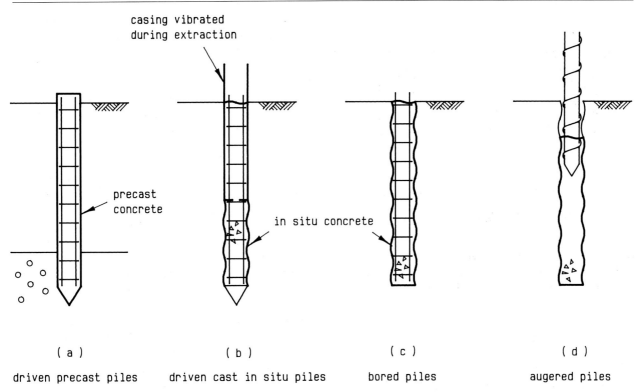

Fig. 9.35 Concrete pile types.

BORED PILES

The bored pile is usually formed by using a simple cable percussion rig. The soil is removed by shell and auger and the hole filled with in situ reinforced concrete as required (see Fig. 9.36).

For filled sites or soft clay sites overlying stiff clay or rock, small to medium bored piles often prove to be economic. The relatively small on-site cost of bored piles means that smaller sites can be piled more economically than they can using a driven piling system. The bored pile is not usually economic in granular soils where removal and disturbance of surrounding ground can cause excessive removal of soil and induce settlement in the surrounding area. During piling operations the hole can be lined with a casing which can be driven ahead of the bore to overcome difficulties caused by groundwater and soft sub-soil but sometimes difficulties of withdrawing the casing after casting can prove expensive.

AUGERED PILES

The augered pile is usually constructed by screwing a rotary auger into the ground. The material is either augered out in a similar manner to that of a carpenter's bit and the open hole filled with concrete or alternatively an auger with a hole down its centre is used and a cement grout injected under pressure down the hole during withdrawal of the auger. Augers can be used to drill large-diameter holes in a wide range of soils, the range having been extended by the use of bentonite slurry to assist the support of the sides of the hole in soft silts and clays. In addition, the large-diameter auger can be used with under-reaming tools to enlarge the end bearing base of the pile (see Fig. 9.37).

Probably one of the most successful auger methods is the use of the hollow tube auger in soft silt, etc., where water and squeezing of soft silts in the surrounding ground can cause necking problems for many other systems, i.e. squeezing in of the pile shaft due to side pressure. The use of the hollow auger and injected sand cement grout can produce good-quality piles in these soft and difficult conditions at competitive prices, particularly where large numbers of piles are involved. For small numbers of piles the on-site cost of the rig and grouting plant can prove to be prohibitively expensive.

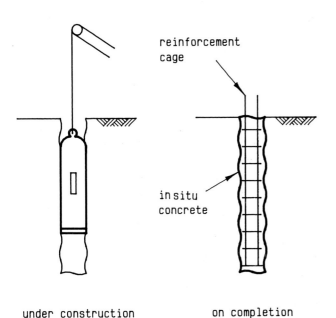

Fig. 9.36 Bored pile.

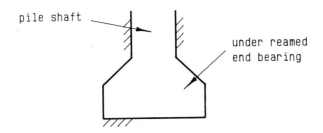

Fig. 9.37 Under-reamed augered pile.

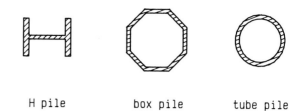

Fig. 9.39 Typical steel pile cross-sections.

9.5.4 Timber piles

Timber piles are suitable for temporary works and where kept permanently below the groundwater level they are suitable for permanent works. Timber piles have been used very successfully in marine environments. They are driven by percussion means similar to precast concrete piles, have good flexibility and resistance to shock and, if kept permanently wet or permanently dry, they can have a very long life.

There is some danger from attack by marine organisms below water, and from micro-fungal attack and wood-destroying insects when kept dry. However, careful selection of the species of timber and the use of preservatives can overcome most of the problems. In Victorian piles charred faces were used to prevent surface deterioration. To assist in the driving of the pile, steel hoops are often used around the head of the pile and steel shoes on the toe to prevent damage from impact forces in these locations and to ease the driving. There is some danger of undetected damage below ground level in a similar manner to that of the precast concrete pile. However jetting or pre-boring in difficult conditions can help in overcoming this problem. The pile can be in the form of trimmed tree trunks or shaped timber cross-sections (see Fig. 9.38).

Timber piles are usually in a range of 5–12 m long and if lengths in excess of this are required they can be spliced using specially designed steel connections. With the exception of jetties, temporary works and sewer supports, timber piles are not often used in general structural foundations and detailed information is, therefore, outside the scope of this book.

9.5.5 Steel piles

Steel piles, like timber, are driven by percussion means and have a variety of suitable cross-sections. In addition to the common sheet piles, the three main types are H sections, box piles and tube piles. Typical sections are shown in Fig. 9.39.

typical timber pile
cross-sections

Fig. 9.38 Typical timber pile cross-sections.

The main use of steel piles is for temporary works, retaining walls and marine structures. The problem of corrosion of the steel can be overcome by suitable protection. However, account should be taken of the abrasion during driving on the final performance of such protection/coatings. In addition to coating, however, increased metal thickness and cathodic protection may be appropriate for particular locations and conditions. However, detailed information on these aspects along with detailed information on steel piles and piling is outside the scope of this book due to their limited use in structural foundations.

9.5.6 Anchor piles

Anchor piles are piles used to resist uplift or inclined tensile forces in the surrounding ground. They are used as:

(1) Reaction piles for pile testing.
(2) Piles to resist uplift forces from flotation.
(3) Anchorage to react to cantilevered foundations, etc. (see Fig. 9.40).

The piles are designed as tension piles transferring their load to the ground by friction, by under-reaming or by bonding into unfractured rock (see Fig. 9.41). It is most important that allowance be made, when anchoring into rock, for the possible damage and shattering of the rock or pile surface during driving.

It is also important to give special consideration to the use of piles as anchor piles for testing if they are to be incorporated as working piles in the final scheme.

9.5.7 Anchor blocks

Anchor blocks are used in situations where anchorage is required horizontally or near the ground surface or at long distances from the foundation, to keep the anchorage outside of the zone of influence of active pressures, etc. Tie beams or rods are sometimes used to transfer the load to the anchor block (see Fig. 9.42).

The anchor block gains its resistance from the surrounding ground in the form of friction, weight of earth and passive pressure (see Fig. 9.43).

The anchor block can be a wedge shape or any other suitable shape most economic for the load and surrounding ground conditions. The designer must ensure that the friction assumed in design can be developed after construction and particular care should be taken to see that trench sides are undisturbed and that the interface is suitably constructed. In addition the movement required to generate passive resistance and friction resistance must be catered for in the design

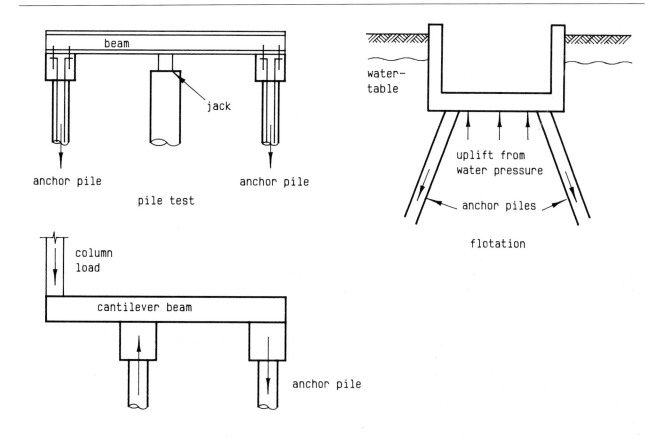

Fig. 9.40 Anchor piles.

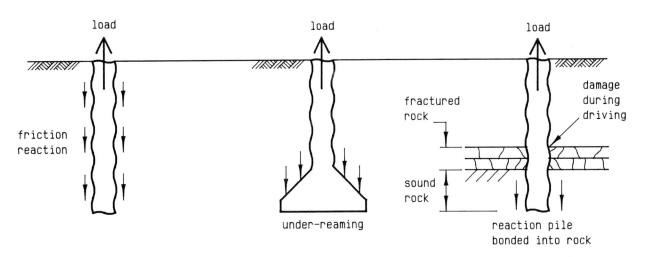

Fig. 9.41 Tension piles.

and detail to make sure that failure of other parts of the structure is not caused during the motivation of the reaction.

9.5.8 Pile caps and ground beams

Many piled foundations consist of a number of relatively small-diameter piles and they require a practical driving tolerance. Pile caps or capping beams are required to accommodate this tolerance and to pick up varying widths of superstructure elements. In some situations, where large-diameter piles are used, pile caps or beams are unnecessary.

However, this is the minority case. The caps consist of concrete pads or beams constructed at the head of the piles to provide the connection between the pile and superstructure (see Fig. 9.44).

In addition to transferring the vertical load from the superstructure to the foundation, there is often a need to provide lateral restraint to the tops of the piles, particularly where less than three piles are provided. For small low-rise developments (single- or two-storey construction) the amount of restraint required to the tops of the piles can be

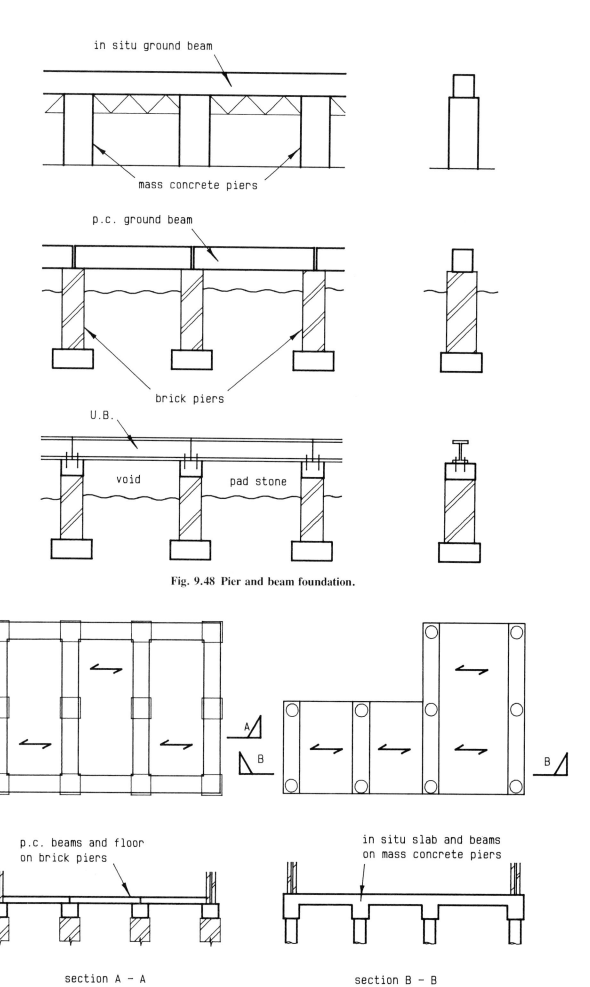

in situ ground beam

mass concrete piers

p.c. ground beam

brick piers

U.B.

void

pad stone

Fig. 9.48 Pier and beam foundation.

A

A

B

B

p.c. beams and floor
on brick piers

in situ slab and beams
on mass concrete piers

section A – A

section B – B

Fig. 9.49 Typical pile/pier and ground beam arrangement.

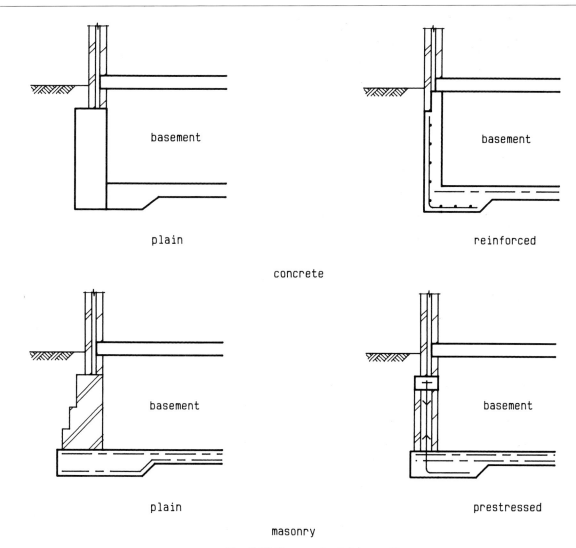

plain   reinforced

concrete

plain   prestressed

masonry

Fig. 9.50 Basement retaining walls.

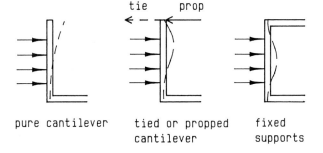

pure cantilever   tied or propped   fixed
               cantilever   supports

Fig. 9.51 Retaining walls — design approach.

reinforced or post-tensioned (see Fig. 9.53). The engineer should apply his skill and ability in arriving at the most suitable and economic form for each individual situation.

The design of retaining walls in relationship to foundations does mean that the normal design to retain earth can become secondary to or parallel to the overall foundation behaviour. For example, where the building is constructed on a raft foundation and the retaining wall becomes part of the raft, then continuity of raft stiffening ribs are most critical to the design and detail (see Fig. 9.54).

The location of settlement or other movement joints through foundations which embrace the retaining walls can be critical to or dictate the structural behaviour of the wall, for example, by effectively removing the prop/tying action of the upper floor slab of a change in level (see Fig. 9.55).

In mining areas the need to relieve horizontal ground stress by allowing the foundation to move relative to the sub-strata can conflict with the need to resist lateral loads in a retaining situation. On sloping sites this conflict can often be overcome by the detail shown in Fig. 9.56.

Where a basement is required on a flat mining site the conflict is more difficult and much greater forces have to be resisted by the building foundations (see Fig. 9.57).

9.6.5 Grillage foundations

Grillage foundations consist of a number of layers of beams usually laid at right angles to each other and used to disperse heavy point loads from the superstructure to an acceptable ground bearing stress (see Fig. 9.58).

Grillage bases are rarely economic these days for permanent foundations except for very heavy loads. However their prefabricated form can prove very useful for temporary works particularly where re-use of the foundations is required (see Fig. 9.59).

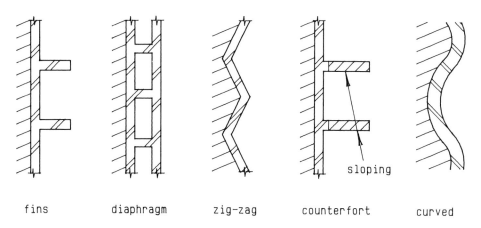

fins   diaphragm   zig-zag   counterfort   curved

Fig. 9.52 Masonry retaining walls/plan forms.

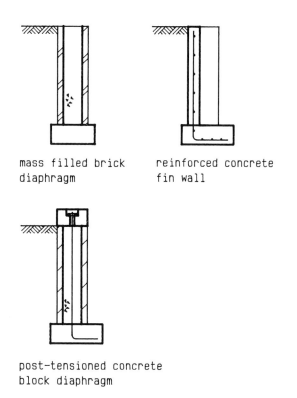

mass filled brick
diaphragm

reinforced concrete
fin wall

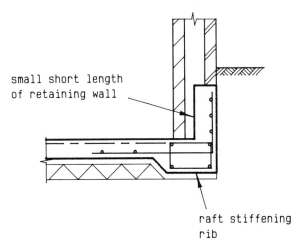

post-tensioned concrete
block diaphragm

Fig. 9.53 Masonry retaining walls – structural forms.

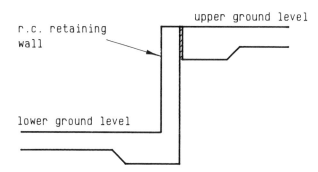

r.c. retaining
wall

upper ground level

lower ground level

Fig. 9.55 Retaining wall/movement joint.

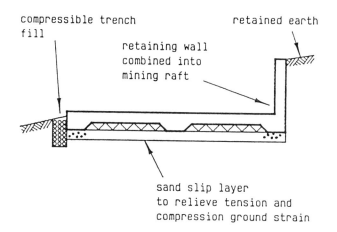

compressible trench
fill

retained earth

retaining wall
combined into
mining raft

sand slip layer
to relieve tension and
compression ground strain

Fig. 9.56 Mining raft slab/retaining wall.

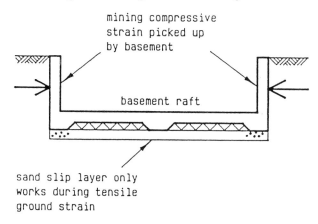

mining compressive
strain picked up
by basement

basement raft

sand slip layer only
works during tensile
ground strain

Fig. 9.57 Mining (basement) raft/retaining wall.

small short length
of retaining wall

raft stiffening
rib

Fig. 9.54 Retaining wall/raft slab.

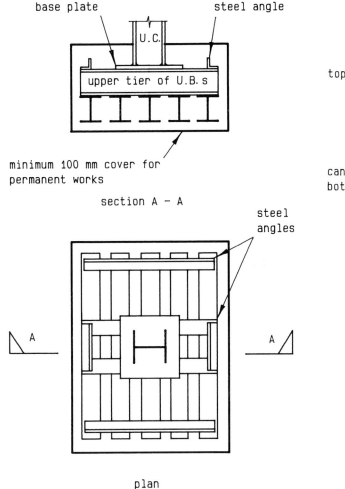

Fig. 9.58 Grillage foundation.

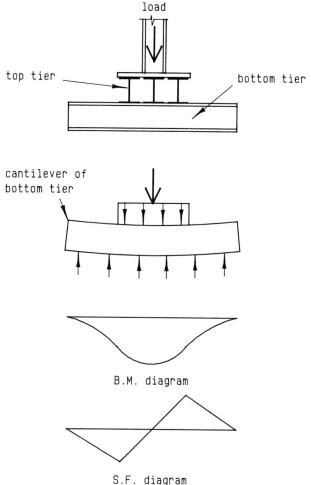

Fig. 9.60 Grillage foundation – bending and shear diagrams.

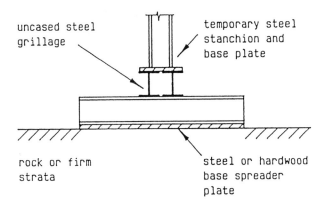

Fig. 9.59 Grillage foundation – temporary works.

The grillage beam can be in any material, the most usual being either steel, precast, concrete or timber. In some permanent situations, however, where unusual circumstances exist, such as an abundance of durable timber or the possible re-use of existing rolled steel sections, the grillage can prove both successful and economic. In permanent conditions durability becomes an important design factor and protection and/or the selection of suitable materials is a major part of the design. In the case of steel grillage below ground this is usually achieved by encasing the grillage in concrete. The concrete for average ground conditions would usually require to provide a minimum cover to the steel of 100 mm. In the case of timber grillages the selection of a suitable species of timber and/or suitable preservation protection is crucial to the design, in a similar way to that for timber piles.

The design of the grillage is carried out by calculating the loads and moments applied from the superstructure and determining the required base area using a suitable allowable ground bearing pressure for the condition involved. From this area, the number and size of each grillage layer can be decided. The layers are then designed to cantilever from the edge of the layer above which determines the beam sizes required to resist the applied bending moments and shear forces (see Fig. 9.60).

If the grillage is encased in concrete and the sequence and method of construction and loading is compatible with the design requirements, the composite action of the beam and concrete can be exploited.

CHAPTER 10

FOUNDATION SELECTION AND DESIGN PROCEDURES

SECTION A: FOUNDATION SELECTION

10.1 INTRODUCTION

This chapter describes selection of an appropriate foundation solution. Section A deals with the selection process by considering the type, nature and availability of information required, its collection and its validity when determining a suitable foundation. General guidance on the relationship of sub-soil conditions/suitable foundation/factors affecting choice are also given. In section B, the design calculation procedures are discussed.

Earlier chapters (Part 1) have covered the principles of design, soil mechanics, geology, and site investigations. Subsequent chapters (Part 2) have covered other factors and considerations which may affect the actual site that is to be developed. The different foundation types i.e. strips, rafts, piled, etc., were discussed in Chapter 9 and the actual design approach and calculation method for each type is covered in Chapters 11–15.

10.2 FOUNDATION SELECTION

The selection of the appropriate foundation solution is perhaps the most important part of the design process and most difficult to define. The engineer should not confuse structural calculation and analysis with design. Calculation usually involves analysing, from certain parameters, the forces and stresses involved in a particular structural element. Structural design is the process of exploiting engineering knowledge in an attempt to produce the most suitable and economic structure. The foundation selection is governed by many factors which include: sub-soil conditions, past site usage, adjacent construction, size/scale of development proposals, timescale/cost limitations. While this is not a comprehensive list it can be appreciated that the sub-soil or any one factor is only part of the overall equation.

The selection of the foundation type to be adopted to accommodate the various criteria is a design process which evolves. It is necessary to approach this process on a broad front taking account of all the relevant information and balancing the factors which can vary as decisions are made. Foundation design should therefore be carried out using a careful blend of geology, soil mechanics, theory of structures, design of materials, experience, engineering judgement, logic and down-to-earth engineering. The designer must understand that sampling and testing soils, while being

by no means an accurate science, is, when corrected by logic and practical experience, an excellent guide for use in the design of foundations. Foundation design is one of the most challenging aspects of engineering and no two foundation conditions are the same. The need for the engineer to see and feel the subject cannot be overemphasized.

Foundation design like other structural design requires a good sound basic approach in order to achieve a truly successful result. There is a tendency in engineering to use test information and theory in a rigid uncompromising way when preparing designs or, alternatively, where experience has shown such theories and tests to be suspect, to revert to an attitude that testing and theory is of no use in practical design. Both approaches are wrong, design should be the process of using all the available tools and information in an attempt to produce the most suitable solution. It should be appreciated that theory is the process of simplifying by assumption the things which actually happen in order to make it possible for a human brain to understand and analyse. Very few assumptions are correct and hence the errors produce a variation from reality. Part of the process of design is to understand such theories, remember the assumptions made and their likely effect on the answer, and to make due allowance for such errors in the choice of the solution.

The sampling and testing of sub-soil materials involves numerous practical site problems, such as disturbance of samples and errors in testing. If the test results are used in isolation such testing is often unreliable. Part of the design process is therefore to understand the practical difficulties, seek indications of unreliability in the results, assess the implication and magnitude of such errors, and make suitable allowance in the design.

The most demanding and exciting part of the engineering process must be approached next, that of making use of all this information in the design of foundations. However, the designer must first make sure that he has all the necessary information and has assessed its reliability.

10.3 INFORMATION COLLECTION/ASSESSMENT

Chapters 1 and 3 have discussed the assessment of information from ground investigation and the collection of details of the soil, groundwater, chemicals, etc., together with the need to inspect the site and surrounding buildings.

In order to assist in this procedure it is useful to use a check-list which can be monitored against actions to make sure that other important items have not been missed. This list can be updated and extended in the light of experience for particular types of jobs and group conditions. Below is a suggested initial check-list for general buildings in the UK.

CHECK-LIST 1 − INFORMATION REQUIRED REGARDING SITE
The following list is not presented in any significant order but as a reminder of the various points to consider.

(1) History of the site.
(2) Soil qualities.
(3) Water-table details.
(4) Chemical qualities (pH values, sulphates, combustion, swelling, ground contamination).
(5) Mining situation (coal, brine, clay, tin, lead, etc.).
(6) Access to site.
(7) Site contours and vegetation.
(8) Overhead and underground services.
(9) Existing tunnels, etc.
(10) Condition of existing buildings on and around the site.
(11) Foundations of adjoining buildings.
(12) Proposed superstructure requirements.
(13) Acceptable settlements and movements.
(14) Type of contractors likely to be employed.
(15) Availability of materials relative to the site location.

In addition to the collection of the information listed above concerning the actual site to accommodate the new development, it is also necessary to have a clear understanding of the client's requirements and criteria for the development proposals.

In parallel to site data collection the following points in check-list 2 should be established.

CHECK-LIST 2 − INFORMATION REGARDING SITE DEVELOPMENT
(1) Nature of the proposed development and phasing of works.
(2) Future development/extensions.
(3) Extent of any possible repositioning of building(s) within site area.
(4) Site features to be retained.
(5) The loads required to be supported.
(6) The amount of settlement/differential movement which can be tolerated.
(7) Any plant, equipment or chemicals likely to be used in the building.
(8) The need for any tanks, basement and/or underground services.

All these items can significantly affect structural consideration and foundation solutions. It is also important to check with the client which of his requirements are rigid and which are flexible in order to be able to make realistic recommendations and adjustments which could produce economies or improvements to the scheme.

10.4 GENERAL APPROACH TO CHOICE OF FOUNDATIONS

Having collected the information about the site including that noted in check-list 1 and obtained from the client answers to the queries including those noted in check-list 2, the foundation selection process can start. As discussed previously the design process evolves as the effects of the various constraints are dealt with or any problems are solved. Tables 10.1, 10.2 and 10.3 give descriptions of the basic sub-soil and site types and a general guide to suitable foundations. In broad terms these tables will assist in the selection of foundation type. The lists are by no means exhaustive nor is the selection ever as simple as these tables may suggest, however, they should prove very useful as a general guide.

Table 10.3 gives details of suitable foundation types to suit varying depths and strengths of bearing strata.

Table 10.1 gives details of foundations to account for varying sub-soil types ranging from rock to peat. The table gives comments on the effects of trees and shrubs on cohesive soils and gives notes on factors to be considered when selecting foundation type.

Table 10.2 gives details of suitable foundations to account for particular site conditions covering sloping, filled or affected by mining, old foundations, groundwater problems. The table gives notes on factors to be considered when selecting the foundation type.

While Tables 10.1−10.3 give a general guide to the foundation selection by considering the factors which can influence this choice and earlier chapters have highlighted and discussed these points, check-list 3 provides a further list of points for consideration during the foundation selection process.

CHECK-LIST 3 − POINTS TO CONSIDER WHEN ASSESSING SUB-SOIL CONDITIONS
(1) The extent of site investigations.
(2) The amount of information available prior to site investigations.
(3) The possibility of errors in the information received.
(4) The variability of the ground conditions.
(5) The inaccuracy of the soil mechanics.
(6) The effects of removal of the overburden.
(7) The effects of the groundwater.
(8) The seasonal effects of the groundwater levels.
(9) The effects of frost and seasonal weather changes.
(10) The effects of trees.
(11) The effects of the water-table on the depth at which various foundations will be considered.
(12) The effects of settlement.
(13) Variations of pressure with time.
(14) Variations of loading with time.

These factors will influence the bearing/settlement capacity of the sub-soil.

When evaluating test results or information from so-called specialists these data should be very carefully interpreted since the information on which their experience

Table 10.1 Foundation selection to suit sub-soil type

Sub-soil type	Suitable foundation	Factors to be considered
Group 1 Rock; hard sound chalk; sand and gravel, sand and gravel with little clay content, dense silty sand	Strips/Pads/Rafts	(1) Minimum depth to formation for protection against frost heave 450 mm for frost susceptible soils. (2) Weathered rock must be assessed on inspection. (3) Beware of swallow-holes in chalk. (4) Keep base of strip or trench above groundwater level where possible. (5) Sand slopes may be eroded by surface water − protect foundation by perimeter drainage. (6) Beware of running sand conditions.
Group 2 Uniform firm and stiff clays (a) where existing nearby vegetation is insignificant	Strips/Pads/Rafts	(1) Trench fill likely to be economic in this category. (2) Minimum depth to underside of foundation 900 mm. (3) When strip foundations are cast in desiccated clay in dry weather, they must be loaded with the structure before heavy rains return.
(b) where trees, hedges or shrubs exist close to the foundation position or are to be planted near the building at a later date	Concrete piles supporting reinforced ground beams and precast concrete floor units OR Concrete piles supporting a suspended reinforced in situ concrete slab OR Specially designed trench fill (possibly reinforced) in certain clay soils depending on position of foundation relative to trees OR Rafts	(1) Clay type and shrinkage potential, distance of trees from foundation and spread of roots dictate necessity or otherwise of piling. (2) Type and dimensions of pile depend on economic factors. (3) Where a suspended in situ concrete ground slab is used a void must be formed under it if laid in very dry weather over clay which is desiccated. (4) Where existing mature trees grow very close (e.g. within quarter of mature tree height) to the position in which piles will be installed. It might be prudent to design for sub-soil group 2(c). (5) Where trees have been or will be planted at a distance of at least one to two times the tree height from the foundation, a strip foundation may be suitable. (6) In marginal cases, i.e. with clay of low to medium shrinkage potential and in the perimeter zone of the tree root system, reinforced trench fill can be used.

Table 10.1 continues

Table 10.1 *Continued*

Sub-soil type	Suitable foundation	Factors to be considered
(c) Where trees and hedges are cut down from area of foundations shortly before construction	Reinforced concrete piles (in previous tree root zone) OR Strip foundations as in groups 2(a) and 2(b) (outside previous root zone) OR Rafts	(1) Piles must be tied adequately into ground beams or the suspended reinforced concrete slab. An adequate length of pile must be provided to resist clay heave force, and the top section of the pile possibly sleeved to reduce friction and uplift. (2) Special pile design may be required for clay slopes greater than 1 in 10 where soil creep may occur and it is necessary to design for lateral thrust and cantilever effects. (3) In marginal cases, i.e. with clay of low to medium shrinkage potential and in the perimeter zone of the tree root system, reinforced trench fill can be used.
Group 3 Soft clay, soft silty clay, soft sandy clay, soft silty sand	Wide strip footing if bearing capacity is sufficient and predicted settlement allowable OR Raft OR Piles to firmer strata below − for small projects consider pier and beam foundations to firm strata	(1) Strip footings should be reinforced depending on thickness and projection beyond wall face. (2) Service entries to building should be flexible.
Group 4 Peat	Concrete piles taken to firm strata below. For small projects, consider pad and beam foundations taken to firm strata below. Where no firm strata exist at a reasonable depth below ground level but there is a thick (3−4 m) hard surface crust of suitable bearing capacity, consider raft.	(1) Pile types used are bored cast in place with temporary casing; driven cast in place; and driven precast concrete. (2) Allow for peat consolidation drag on piles. (3) Where peat layer is at surface and shallow over firm strata, dig out and replace with compacted fill. Then use raft or reinforced wide-strip foundations depending on expected settlement. (4) Where raft is used, service entries should be flexible. Special high-grade concrete and protection may be necessary in some aggressive peat soils.

Ground improvements of sub-soil Groups 3 and 4 by vibro treatments can often be achieved and can be an effective and economical solution when used in conjunction with raft or strip foundations

Table 10.2 Foundation selection to suit varying site conditions

Site condition	Suitable foundation	Factors to be considered

Filled site	Concrete piles taken to firm strata below. For small projects consider beam and pier foundations taken to firm strata below. If specially selected and well compacted fill has been used consider
	(1) Raft or
	(2) Reinforced wide-strip footings
	(3) Strip/pad/raft on ground improved using vibro or dynamic consolidation depending on fill type
		(1) Allow for fill consolidation drag on piles, piers or deep trench fill taken down to firm strata below.
		(2) Proprietary deep vibro and dynamic compaction techniques can with advantage improve poor fill before construction of surface or shallow foundations.
		(3) If depth of poorly compacted and aggressive fill is small remove and replace with inert compacted fill, then use reinforced strip or raft foundations.
		(4) Deep trench fill taken down to a firm stratum may be economic if ground will stand with minimum support until concrete is placed.
		(5) Allow flexible service entries to building.
		(6) Avoid building a unit partly on fill and partly on natural ground.
		(7) Take precautionary measures against
		  (a) combustion on exposure to atmosphere,
		  (b) possible toxic wastes,
		  (c) production of methane gas.

Mining and other subsidence areas	Slip-plane raft	(1) Where a subsidence wave is expected, building should be carried on individual small rafts. Avoid long terrace blocks and L-shaped buildings. (2) In older mining areas, locate buildings to avoid old mining shafts and bell-pits. (3) In coal mining areas, consult British Coal in all cases. (4) Avoid piled foundations.
Sloping site	Foundations to suit normal factors and soil conditions, but designed for special effect of slope	(1) Strip foundations act as retaining walls at steps. With clay creep downhill, design and reinforce for horizontal forces on foundations. Provide good drainage behind retaining wall steps. (2) Foundations are deeper than normal, so keep load-bearing walls to a minimum. Keep long direction of building parallel to contours.

Table 10.2 continues

Table 10.2 *Continued*

Site condition	Suitable foundation	Factors to be considered
		(3) In addition to local effects of slope on foundations, consider total ground movement of slopes including stability of cohesionless soils, slip and sliding of cohesive soils. (4) Make full examination of all sloping sites inclined more than 1 in 10. (5) The presence of water can increase instability of slope. (6) Special pile design may be required for clay soil slopes greater than 1 in 10 where soil creep may occur and it is necessary to design for lateral thrust and cantilever effects.
Site containing old building foundations	Normal range of foundations. It is possible to use strips, piling, and pads but beware of varying depths of fill in old basements, causing differential settlement, and old walls projecting into fill over which slabs may break their back.	(1) Notes relating to 'filled site' apply. (2) Where possible, dig out badly placed or aggressive fill and replace with inert compacted material. (3) Remove old walls in filled basements, or use piers or piles carrying ground beams to span such projections. (4) Deep trench fill down to firm strata at original basement level may be economic. (5) Trench fill depths may vary greatly as old basement depth varies. Some formwork may be required in loose fill areas. (6) Remove old timber in demolition material − a source of dry rot infection.
Site with groundwater problems	Normal range of foundation types can be used. Consider piling through very loose saturated sand to denser stratum to provide support for raft or strip foundation at high level above groundwater. Consider use of proprietary vibro-replacement ground techniques to provide support for raft or strip foundation at high level above groundwater.	(1) In sand and gravel soil, keep foundation above groundwater level where possible. (2) Avoid forming steep cuttings in wet sand or silty soil. (3) Consider use of sub-surface *shelter* drains connected to surface water drains, and allow for resulting consolidation or loss of ground support. (4) Take precautions against lowering of groundwater level which may affect stability of existing structures.

Table 10.3 Foundation selection to suit bearing strata strength and depth

Sub-soil conditions	Suitable foundation
Condition 1 Suitable bearing strata within 1.5 m of ground surface	**Strips** **Pads** **Rafts** When loading on pads is relatively large and pad sizes tend to join up or the foundation needs to be **balanced** or connected then **continuous beam** foundations are appropriate. Strip foundations are usually considered the norm for these conditions but rafts can prove more economical in some cases.
Condition 2 Suitable bearing strata at 1.25 m and greater below ground surface	**Strips** **Pads** } on improved ground using vibro or dynamic **Rafts** } consolidation techniques
Condition 3 Suitable bearing strata at 1.5 m and greater below ground surface	As Condition 2 plus the following **Piles** and **ground beams** **Pier** and **ground beams** **Piles** and **raft**
Condition 4 Low bearing pressure for considerable depth	As Condition 2 plus the following **Buoyant rafts**
Condition 5 Low bearing pressure near surface	As Condition 2 plus the following **Rafts** Ground improvement using preloading to support reinforced strips on rafts

is based is generally limited by their specialized activities. For example, recommendations from one expert may clash with the requirements of another. It is therefore up to the engineer to gather the data and reassess in overall terms the reliability, relevance and practicality of both the information and recommendations being made. When selecting a foundation type it should also be appreciated that prices of materials and labour vary depending on the timing and location of the project. The size of the contract can have a significant effect on the economics of the solution. For example, the effects of fixed costs such as those for getting piling rigs on and off the site can be very small when spread over a large number of piles, on the other hand they can prove to be the major cost when a small number of piles are to be driven. It is also necessary to keep up-to-date with piling and ground improvement techniques to ensure that decisions made on cost and performance are current.

High costs can be generated by complex shuttering details to foundations. These costs can be reduced significantly if details are simplified, for example, concrete can be cast against hand packed hardcore in raft construction. Two-stage concrete pours for raft edges can use earth faces for *shuttering* to the first pour. Brickwork built off the raft edge can act as shuttering to the second pour (see Fig. 10.1).

When adopting this form of construction it is necessary to increase reinforcement cover against earth faces and to provide reinforcement connection between the first and

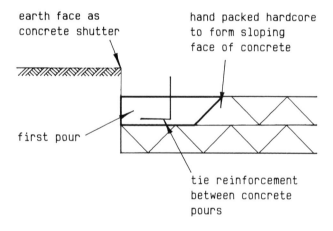

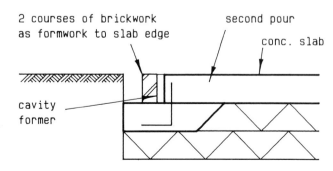

Fig. 10.1 Raft edge construction.

second concrete pours. An appreciation of construction methods and problems is also helpful in determining which foundation type to adopt. Pouring concrete under water using tremie techniques can suggest trench fill rather than strip and masonry (where it would be necessary to pump out the water to enable the masonry to be constructed). Deep excavations in waterlogged ground are best avoided and alternative foundation solutions using rafts should be considered. It is advisable to avoid deep strips, pier and beam and piled foundations where mining is a problem.

In addition to construction considerations affecting the foundation selection, basic decisions in the design process can be significant. Varying the shape, length or rigidity of the foundation can have a major influence on performance. The introduction of joints in the substructure and superstructure can be exploited in the foundation design and selection. Adopting a composite design for the foundation can also affect the type of foundation to be selected.

The importance of the above items are dealt with in detail in other chapters but they are included here as a useful reminder of the early part of the design process and to assist in the gathering of all relevant information.

10.5 QUESTIONING THE INFORMATION AND PROPOSALS

Having gathered the information together for the design, the engineer's first considerations should take account of the following:

(1) *Is the investigation sufficient to design a safe and economic foundation?*
   For example, a ground investigation that was undertaken which collected samples and arranged testing in the light of background information and which predicted a piled solution may have lacked detailed investigation of the upper strata – this would be necessary in order to consider a raft as an alternative (see Fig. 10.2).
   An alternative example would be a ground investigation based upon boreholes and sample collection/testing at

shallow depth envisaging surface spread foundations – this would not provide the information necessary to consider a piled solution.

If a piled foundation solution is subsequently found to be necessary, important sub-soil information would not be known.

In some cases the engineer is only called in after the initial site investigation has been completed and the alternative foundation solution is only appreciated at that stage, making further sub-soil investigation necessary in order to design the most economic foundation.

(2) *Is the initial proposed scheme appropriate for the ground conditions identified?*
   Could the proposals be modified without detriment to the successful functioning of the building and give significant savings on foundations or significant reductions in predicted differential settlement?
   For example, where piling is necessary for a single-storey building the economic span for ground beams, etc., often produces loads in the piles which do not fully exploit their load-carrying capacity. In such situations consideration should be given to changing the building form from single-storey to multi-storey (see Fig. 10.3).

(3) *What blend of superstructure and foundation should be employed?*
   For example, in active mining areas, the combination of superstructure and foundation can be very critical in accommodating movements and the whole structure must be carefully considered when taking account of subsidence (see Fig. 10.4).

(4) *Is the arrangement of the superstructure supports very critical to the foundation economy?*
   For example, the design of the superstructure should not be made completely independently of the foundation economy. In the same way the foundation economy should not be considered independently of the superstructure. A typical example of this kind of problem is the use of fixed feet on portal frames which often create greater additional costs on the foundation than they do savings on the superstructure (see Fig. 10.5).

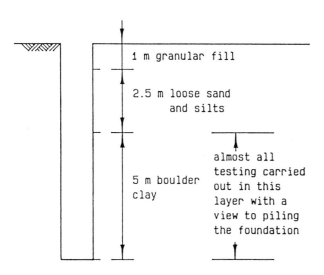

Fig. 10.2 Borehole log.

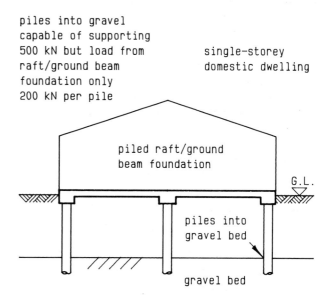

Fig. 10.3 Typical situation for low-rise construction.

note : pinned base portal frames can
be used in mining areas but care is
needed to ensure that the tie is able
to resist both the tie force and mining
stress and that the elongation of the tie
does not adversely affect the moments in
the frame

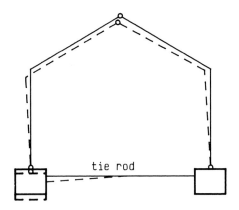

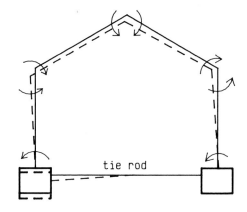

three pin arch subjected to
differential vertical settlement
of support
original ——————
final — — — — —
rotation takes place at pin
joints and arch relocates itself
above the foundation

fixed arch subjected to a similar
differential vertical settlement
of supports
rotation cannot take place without
inducing increased bending moments
as shown therefore increased
danger of failure

Note : walls and other building elements need to be detailed to accommodate
both mining movement and the associated distortion of the frame

Fig. 10.4 Effect of subsidence on pinned/fixed frame.

(5) *Is the proposed layout and jointing of the foundation exploiting engineering knowledge to provide the most economic foundation?*

For example, the choice of the lengths and jointing of continuous ground beams can have extreme effects on the foundation moments and forces and hence on the costs, as is shown by the following example.

Consider a series of six columns at 10 m centres, the four outer columns having a load of 500 kN and the two inner columns having a load of 250 kN (see Fig. 10.6).

Assume that a ground beam is positioned under these columns in one continuous length of 50 m. The total load on the beam is 2500 kN and it is symmetrical. Assuming a stiff beam, a uniform distributed pressure below the beam of 50 kN per metre run would result.

Referring to the diagrams shown in Fig. 10.6, it can be seen that the point of zero shear occurs at the mid-length of the beam and that the maximum resulting shear force is equal to 500 kN. Since the maximum bending moment is equal to the area of the shear force diagram to one side of the point, the maximum bending moment is as follows:

$$\text{Max. BM at mid-point} = -\frac{(500 \times 10)}{2} \times 2 - \frac{(250 \times 5)}{2}$$

$$= -5000 - 625 = -5625 \, \text{kNm}$$

It also follows that the bending moment diagram would be approximately that shown in Fig. 10.7.

If the resulting bending moments are considered, it can be seen that for its full length the beam is hogging and the resulting deflected shape would be of convex outline (see Fig. 10.8).

It is also apparent that much smaller bending moments and shear forces would result if a deflected shape similar to a normal continuous beam supporting a uniform load could be achieved (see Fig. 10.9).

If, therefore, we set out to achieve this and keep in mind that the fixed forces in this case are the column loads, then we must aim for a continuous uniformly loaded beam with reactions equal to the column loads.

If we now refer back to the original loads and consider them as reactions, we can then place upon them a beam

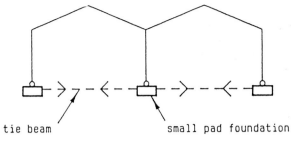

tie beam   small pad foundation

typical pinned base portal

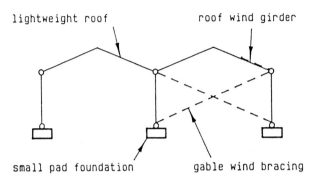

lightweight roof   roof wind girder

small pad foundation   gable wind bracing

typical beam and column
construction, pinned feet

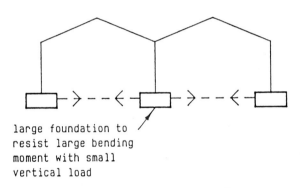

large foundation to
resist large bending
moment with small
vertical load

typical fixed base portal
for same building

Fig. 10.5 Arrangement and effect of superstructure support on foundation.

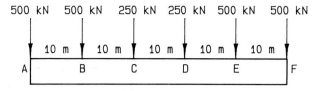

500 kN   500 kN   250 kN   250 kN   500 kN   500 kN

10 m   10 m   10 m   10 m   10 m

A   B   C   D   E   F

loading beam 1

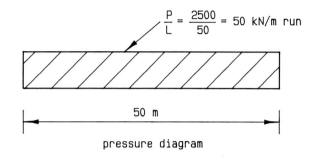

$$\frac{P}{L} = \frac{2500}{50} = 50 \text{ kN/m run}$$

50 m

pressure diagram

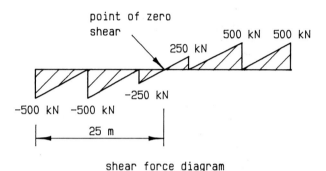

point of zero
shear

250 kN   500 kN   500 kN

-250 kN

-500 kN   -500 kN

25 m

shear force diagram

Fig. 10.6 Continuous foundation beam 1.

uniformly loaded with similar reactions (see Fig. 10.10). The beams in Fig. 10.10 have been chosen by ending the beam near the smaller loads and cantilevering out over the heavier loads.

If these beams are now adopted, assuming that ground conditions and site boundaries will allow this, the revised bending moments and shear forces can be assessed. The total load on each beam is now 1250 kN and the length of each beam is 25 m. Let the resultant load act at a distance x from the 250 kN load (see Fig. 10.11). Taking moments about this column's position:

$$x = \frac{(500 \times 10) + (500 \times 20)}{1250} = 12 \text{ m}$$

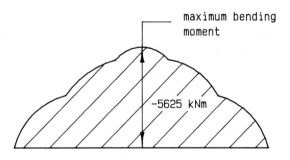

maximum bending
moment

-5625 kNm

Fig. 10.7 Bending moment diagram beam 1.

Fig. 10.8 Deflected shape beam 1.

Fig. 10.9 Desired deflected shape beam 1.

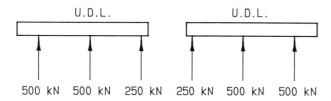

Fig. 10.10 Jointed foundation beam.

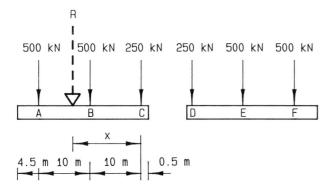

Fig. 10.11 Jointed foundation beam — resultant load location.

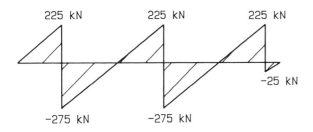

Fig. 10.12 Jointed foundation beam — shear force diagram.

Since a 0.5 m cantilever has been given to this end of the beam the resultant load acts at 12.5 m from each end and hence is symmetrical.

The resulting pressure, again assuming a stiff beam, is 50 kN per metre run, as for the previous beam. Referring to the shear force diagram for this beam (see Fig. 10.12), it can be seen that a number of zero shear points occur. Consider point A and again using the area of the shear force diagrams to obtain the bending moments:

$$\text{BM at A} = \frac{225 \times 4.5}{2} = 506.25\,\text{kNm}$$

$$\text{BM at B} = 506.25 - \frac{(275 \times 5.5)}{2} + \frac{(225 \times 4.5)}{2}$$

$$= 506.25 - 756.25 + 506.25 = 256.25\,\text{kNm}$$

$$\text{BM at C} = \frac{250 \times 0.5}{2} = 6.25\,\text{kNm}$$

Bending moments can similarly be obtained for each position on the beam, but by inspection the maximum value will be 506.25 kNm at A.

If the result on this beam is compared with that of the previous beam it will be found that the maximum bending moment is less than one-tenth of that of the earlier solution and the shear force has been reduced to approximately one-half, both emphasizing the importance of the *selection* of the foundation beam arrangement to be used.

The previous example illustrates the need for the engineer to use his basic knowledge of structures to exploit the conditions. It can be seen that it is not economic to have a continuous beam foundation which bends in either a hogging or dishing form under a number of loads unless site restrictions prevent alternative solutions. The aim should therefore be to achieve bending more in the form of a normal continuous beam being bent in alternative bays in each direction. To achieve this aim it is necessary to inspect the loads and to relate these to continuous members which would have similar reactions (see Fig. 10.13).

10.6 EXPLOITATION OF FOUNDATION STIFFNESS AND RESULTING GROUND PRESSURE

In addition to the basic knowledge of structural theory there is also a need to keep in mind the basic pressures which develop in various ground conditions and the effect of the foundation stiffness on these pressures (see Fig. 10.14).

Again armed with this knowledge the engineer should exploit the conditions to his advantage.

For example, the stiffness of a foundation should only be sufficient to distribute the applied load down to a suitable bearing capacity and accommodate the resulting settlement, since the bending moment developed would be much smaller on the flexible foundation than the rigid foundation. The flexible foundations will result in higher pressures directly under the load and when the stiffness and pressures are taken into account the resulting bending moments produced are much smaller.

10.7 CONCLUSIONS

As stated in the introduction to this section the foundation selection process is the most difficult to define. The checklists, guides and the examples should have given the reader a feel for the foundation selection and design process.

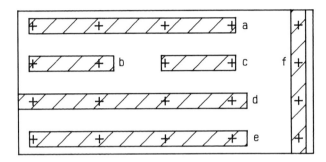

plan showing columns thus ☑
and load thus 200 in kN

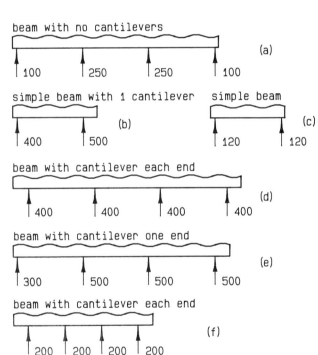

beam with no cantilevers

(a)

100   250   250   100

simple beam with 1 cantilever   simple beam

(b)   (c)

400   500   120   120

beam with cantilever each end

(d)

400   400   400   400

beam with cantilever one end

(e)

300   500   500   500

beam with cantilever each end

(f)

200 200 200 200

beam layout selected to give small bending
moments using knowledge of continuous beams
with similar reaction relationship to column
loads

Fig. 10.13 Beam loads.

Practical application of the design process is obviously
necessary to gain experience and confidence of foundation
design/selection which is one of the most challenging and
rewarding aspects of engineering.

rigid foundations   flexible foundations

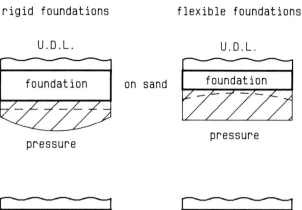

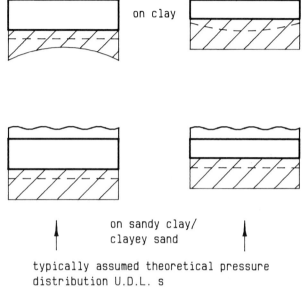

on sandy clay/
clayey sand

typically assumed theoretical pressure
distribution U.D.L. s
deflected shape shown thus   — — — — —

for point loading on rigid foundation pressure
distribution would be similar to those for
U.D.L. s, however for flexible foundation see
below

point loads

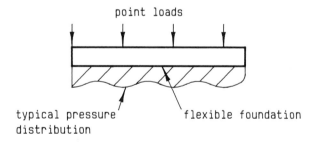

typical pressure   flexible foundation
distribution

**Fig. 10.14 Foundation stiffness and resulting ground
pressure.**

SECTION B: FOUNDATION DESIGN CALCULATION PROCEDURE

10.8 INTRODUCTION

The currently accepted hybrid approach to foundation design
in the UK means that while bearing pressures are checked
on a working stress basis, the foundation members are

designed using limit-state methods. This often leads to confusion in the design of foundations though with some forethought and methodology in the superstructure design it is a relatively straightforward matter to establish the foundation loads in a format which can be used for both parts of the design process.

In the vast majority of cases the design method can be simplified and this shall be looked at in detail first to establish the principle to be adopted. In addition a method for dealing with most foundation and loading types will also be introduced.

10.9 DEFINITION OF BEARING PRESSURES

The site investigation, laboratory analysis, established principles of soil mechanics, and most importantly the engineer's own judgement are used to assess the *allowable bearing pressure* which the soil can support – this assessment is covered in section A of Chapter 2.

This allowable bearing pressure is required to provide a sufficient factor of safety against failure in terms of bearing capacity (i.e. ultimate collapse failure), usually taken as 3, and against settlement (i.e. serviceability). This allowable bearing pressure is assessed in one of two forms:

(1) *Total allowable bearing pressure*. The maximum pressure which can be applied at the soil/foundation interface by the foundation and the loads acting upon it.
(2) *Net allowable bearing pressure*. The maximum increase in pressure which can be applied at the soil/foundation interface (i.e. the difference in pressure after the foundation is loaded compared with that in the soil before construction is started).

The difference between these two cases in its simplest form is shown in Fig. 10.15. It is common practice for simple foundations such as axially loaded pads and strips to be designed on the basis of checking the net allowable bearing pressure against the load from the superstructure ignoring the weight of the foundation. This is a valid method given that the weight of the foundation is typically of the same order of size as the weight of the soil it replaces ($24 \, \text{kN/m}^3$ for concrete compared with $20 \, \text{kN/m}^3$ for soil). Providing the surcharge remains the same the error involved is minimal compared with the inaccuracies of basic soil mechanics.

It is, however, fundamentally important that the total allowable bearing pressure is not confused with the net allowable bearing pressure. If, for example, a soil has a total allowable bearing pressure of $80 \, \text{kN/m}^2$ at a depth of $2 \, \text{m}$,

(a) before construction

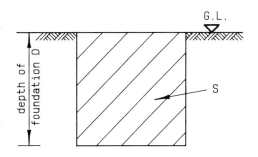

existing soil pressure
$s = \gamma D$

existing overburden load
$S = \gamma D A$

where γ = density of soil
A = area of foundation

(b) after construction

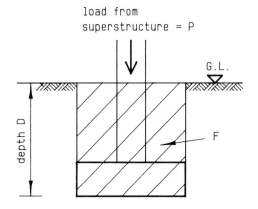

load from superstructure = P

final total load
$T = P + F$
where F is the load from the self weight of the buried foundation and its backfill

final total bearing pressure
$t = \dfrac{T}{A}$

$t = \dfrac{P + F}{A}$

$= p + f$

final net bearing pressure
$n = t - s$

$= \dfrac{P + F}{A} - \gamma D$

normally the density of the concrete foundation and backfill $\simeq$ density of soil and $s = \gamma D \simeq \dfrac{F}{A}$ and the net bearing pressure $n \simeq \dfrac{P}{A}$

Fig. 10.15 Definition of loads and pressures – simple case.

the load which the foundation can support is

$$P = (80 - 2\gamma)A$$

Taking γ as $20\,\text{kN/m}^3$ and where A is the area of the base in m^2

$$P = 40\,A\,\text{kN}$$

If, however, the total allowable bearing pressure was erroneously taken to be the net allowable bearing pressure, then the load which the foundation could support would be calculated as

$$P = 80\,A\,\text{kN}$$

Thus in this example the error is 100%!

When considering wind loading conditions the total allowable bearing pressures are increased by 25% in line with the factor used in permissible stress design codes used for structures prior to the introduction of the limit-state design codes.

It should not be forgotten that unless specific reference has been made to the contrary the allowable bearing pressure is usually based on the ultimate bearing capacity (typically with a factor of safety of 3) without an assessment of settlement as this will be dependent on the type, size and actual applied bearing pressure adopted in the design. It is therefore necessary for the engineer to make a separate assessment of the allowable bearing pressure in relation to settlement criteria and he should ensure that the site investigation provides him with the necessary information to make that assessment.

10.10 CALCULATION OF APPLIED BEARING PRESSURES

Figure 10.15 shows the typical example where both the loading is axial and there is no variation in ground level or surcharge. While this simple example will cover a large proportion of foundations constructed, the more general situation needs to be considered, firstly for calculating the total and net bearing pressures with variations in surcharge and/or ground levels and then for the effects of introducing asymmetrical loading.

While on good bearing soils modest surcharges and/or changes in ground levels will have little effect on the bearing capacity of the soils, in poor soil conditions or where the load changes are significant they can have a dramatic effect.

For a general case therefore the *net increase in load*, N, is given by the formula

$$N = (\text{total load after construction}) -$$
$$\quad (\text{total existing load})$$
$$= T - S$$

where T = total load after construction at underside of foundation

and S = existing load at underside of foundation
$$T = P + F$$

where P = load from superstructure

and F = load from foundation
$$= F_S + F_B \text{ (see Fig. 10.16)}$$

where F_S = final foundation surcharge load

and F_B = load from buried foundation and backfill
$$S = S_S + S_B \text{ (see Fig. 10.16)}$$

where S_S = existing surcharge load (taken as zero except where it has acted as a permanent load)

and S_B = load from existing overburden.

Therefore the net increase in load may be rewritten as

$$N = T - S$$
$$= P + F - S$$

and the net increase in soil pressure, for an axially loaded foundation, is given by

$$n = \frac{N}{A}$$
$$= \frac{T}{A} - \frac{S}{A}$$
$$= t - s$$

Alternatively,

$$n = \frac{P + F}{A} - \frac{S}{A}$$
$$= p + f - s$$

This is shown diagrammatically in Fig. 10.16.

It should be noted that where the soil level has been significantly reduced by a major regrading of the site or by construction of basements and the like, consideration should be given to the effects of heave particularly in clays or where there are artesian groundwater pressures.

It is almost always sufficiently accurate to take the weight of the new foundation and backfill as equal to the weight of soil displaced, i.e. $F_B \sim S_B$. Thus the equations for net increase in load and net increase in soil pressure simplify to:

$$N = P + F_S - S_S$$

and

$$n = \frac{P + F_S - S_S}{A}$$
$$= p + f_S - s_S$$

When the ground levels and surcharge pressures are only nominally changed, $F_S \sim S_S$, and so the formulae reduce to

$$N = P$$
$$n = \frac{P}{A}$$
$$= p$$

i.e. the net increase in soil load is equal to the load from the superstructure as mentioned previously.

In the examples above, the foundations have been axially loaded such that the total bearing pressure is given by

$$t = \frac{P + F}{A} \quad \text{(see Figs 10.15 and 10.16)}$$

While this is the most common situation, and it is clearly an efficient design principle to create a foundation which uses the maximum available bearing pressure over its entire

a) new ground level higher than existing

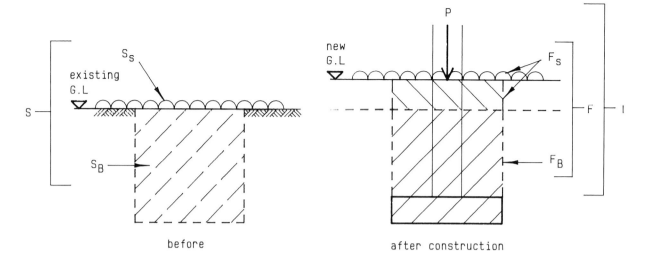

b) new ground level lower than existing

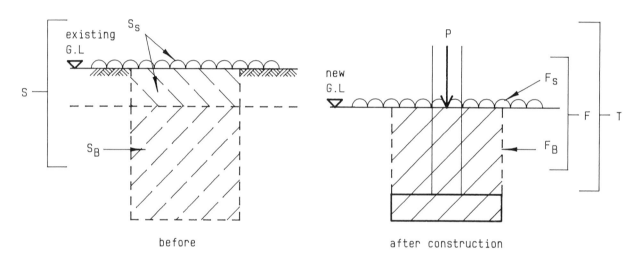

```
                 existing load = S
              total final load = T    = P + F
                      net load = T - S = P + F - S
```

notes :   1) the existing surcharge S_S must have been in place for sufficient time to be
           considered as a permanent load. If this is not the case the existing surcharge S_S
           should be ignored.
           2) If the net bearing pressure is negative then consideration should be given
           to the effects of heave, particularly on clay soils

Fig. 10.16 Definition of loads and pressures – general case.

base, there are many occasions when this is not practical and non-uniform foundation pressures have to be considered. This non-uniformity is typically caused by:

(1) The applied superstructure load P not being on the centroid of the foundation.

(2) The superstructure being fixed to the foundations such that moments are transferred into the foundation (e.g. fixed bases of rigid sway frames).

(3) The application of horizontal loads.

(4) Variations in relative loads on combined bases (e.g. bases carrying two or more columns).

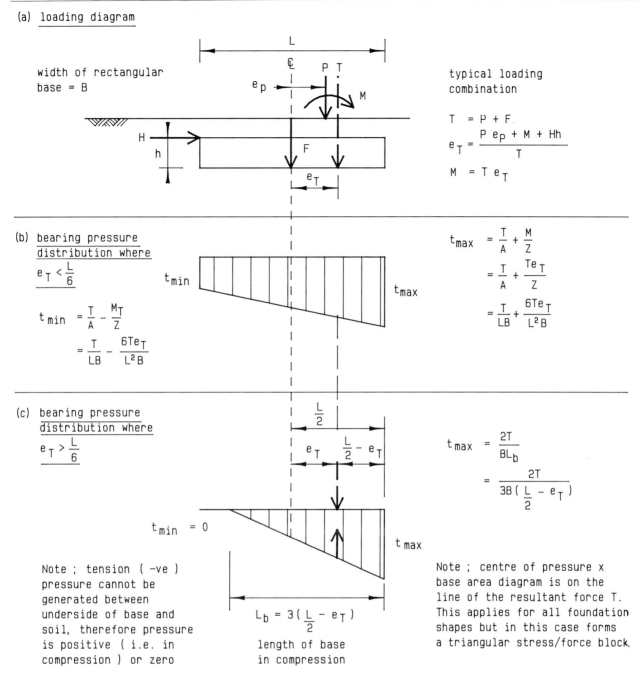

(a) loading diagram

width of rectangular
base = B

typical loading
combination

$$T = P + F$$

$$e_T = \frac{P\,e_p + M + Hh}{T}$$

$$M = T\,e_T$$

(b) bearing pressure
distribution where

$$e_T < \frac{L}{6}$$

$$t_{min} = \frac{T}{A} - \frac{M_T}{Z}$$

$$= \frac{T}{LB} - \frac{6Te_T}{L^2 B}$$

$$t_{max} = \frac{T}{A} + \frac{M}{Z}$$

$$= \frac{T}{A} + \frac{Te_T}{Z}$$

$$= \frac{T}{LB} + \frac{6Te_T}{L^2 B}$$

(c) bearing pressure
distribution where

$$e_T > \frac{L}{6}$$

$$t_{min} = 0$$

$$t_{max} = \frac{2T}{BL_b}$$

$$= \frac{2T}{3B\left(\dfrac{L}{2} - e_T\right)}$$

$$L_b = 3\left(\frac{L - e_T}{2}\right)$$

length of base
in compression

Note ; tension (–ve)
pressure cannot be
generated between
underside of base and
soil, therefore pressure
is positive (i.e. in
compression) or zero

Note ; centre of pressure x
base area diagram is on the
line of the resultant force T.
This applies for all foundation
shapes but in this case forms
a triangular stress/force block.

Fig. 10.17 Foundation in bending about single axis.

Thus in a general case the total pressure under a base with a small out-of-balance moment is

$$t = \frac{T}{A} \pm \frac{M_T}{Z} \quad \text{for single axis bending (see Fig. 10.17 (a) and (b)), and}$$

$$t = \frac{T}{A} \pm \frac{M_{Tx}}{Z_x} \pm \frac{M_{Ty}}{Z_y} \quad \text{for biaxial bending (see Fig. 10.18).}$$

The moment M_T is calculated by taking moments about the centroid at the *underside* of the foundation. In these cases it is usually beneficial to consider the *total bearing pressure* which allows for the balancing effect of the resultant force due to eccentric loads and/or applied moments.

As with simple beam design if

$$\frac{M_T}{Z} > \frac{T}{A} \qquad \text{or}$$

the pressure will be negative and tension, theoretically, will be developed. However, for most foundations it is impossible to reliably develop tension, and the foundation pressure is either compressive or zero.

For a simple rectangular foundation

$$\frac{M_T}{\left(\dfrac{BL^2}{6}\right)} > \frac{T}{BL}$$

$$\frac{6M_T}{L} > T$$

$$\frac{M_T}{T} > \frac{L}{6}$$

$$e_T > \frac{L}{6}$$

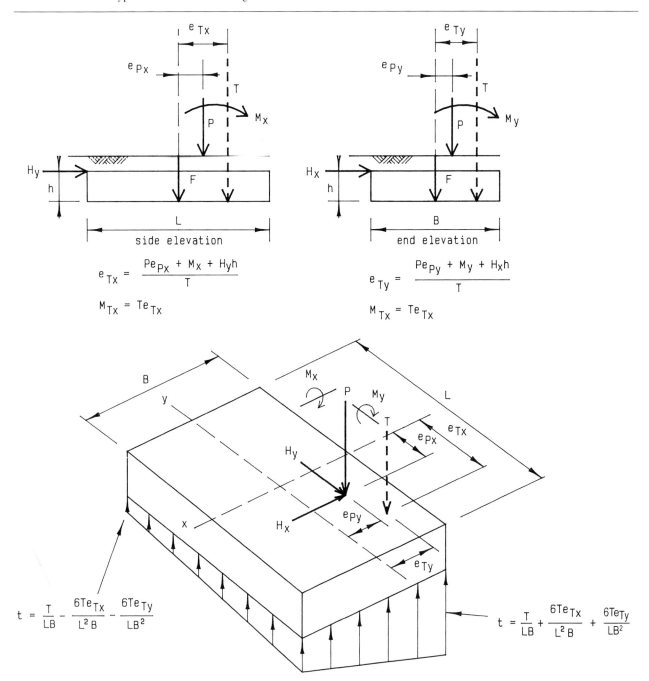

$$e_{Tx} = \frac{Pe_{Px} + M_x + H_yh}{T}$$

$$M_{Tx} = Te_{Tx}$$

$$e_{Ty} = \frac{Pe_{Py} + M_y + H_xh}{T}$$

$$M_{Tx} = Te_{Tx}$$

$$t = \frac{T}{LB} - \frac{6Te_{Tx}}{L^2B} - \frac{6Te_{Ty}}{LB^2}$$

$$t = \frac{T}{LB} + \frac{6Te_{Tx}}{L^2B} + \frac{6Te_{Ty}}{LB^2}$$

Fig. 10.18 Foundation in biaxial bending.

where e_T is the resulting eccentricity of the foundation. Therefore if e_T is less than $L/6$, the foundation will be fully in compression. This is known as the *middle third rule* which is illustrated in Design Example 6 in Chapter 11 (section 11.3.2).

Where e_T is greater than $L/6$, a triangular stress distribution is generated under part of the base and zero under the remainder, and the maximum bearing pressure is calculated using the *shortened base theory*, which, for a rectangular base is

$$t_{max} = \frac{2T}{3B\left(\dfrac{L}{2} - e_T\right)}$$

(see Fig. 10.17 (c)).

Again benefits can be made by considering the total

bearing pressure, thus utilizing the foundation loads which reduce the overturning and increase the effective length of the pressure diagram. Consideration should also be given to the positioning of the base so that the vertical loads P and F are used to counteract the effects of any moment or horizontal loads. In the example shown in Fig. 10.17, the load P should be to the left of the centreline such that the formula for calculating the total eccentricity becomes

$$e_T = \frac{-Pe_P + M + Hh}{T}$$

The ideal situation is that e_T should be zero or

$$e_P = \frac{M + Hh}{P}$$

While it is appropriate to compare the existing *load* with the new *load* on the ground when designing axially loaded foundations, in the more general case where the loads are eccentric, it is necessary to consider the allowable bearing *pressure* (net or total) with the applied foundation *pressure* (net or total) and it is recommended that pressures are compared rather than loads in all cases to maintain consistency and avoid confusion.

Eccentrically loaded rectangular pad or strip foundations are generally designed on the middle third rule where this applies. For other shapes and conditions a trial and error basis is adopted. A base size is selected and the resulting bearing pressures compared with the allowable; the base size is adjusted up or down and the calculations repeated until the maximum bearing pressure is close to the allowable. Experience will soon enable the engineer to make a fairly accurate first guess on the size of base required and reduce the number of iterations necessary.

10.11 STRUCTURAL DESIGN OF FOUNDATION MEMBERS

This section covers the design of the foundation elements in terms of the structural resistance to the applied forces, but does not cover durability factors which are catered for by reference to BS 8110 or BS 5950.

The vast majority of foundations are constructed from concrete, either plain or reinforced, precast or in situ, though a few foundations utilize masonry or steel grillage systems. Each of these materials are currently designed using limit-state design methods familiar to most practising engineers. The simplest to design are the mass concrete or plain masonry foundations which rely on natural load spread through the foundation to enable the point or line loads at the top of the foundation to be distributed out to the full area of the base of the foundation. The load spread is usually taken to occur along a 45° line such that the thickness at the base of the foundation should be no less than the maximum outstand between the edge of the column or wall applying the load to the foundation and the edge of the foundation (see Fig. 10.19). No other structural design is required for such foundations providing they are not required to span over soft spots. It should be remembered,

as with any structural element, that the worst case loading condition needs to be determined and the loading case which produces the highest column axial load may not be the one which creates the worst bearing pressure or elemental stresses. This is particularly so when considering foundations which are required to resist column base moments and/or the wind loads (it is frequently the case that the size of a base on a bracing line is determined by the minimal dead load and maximum wind uplift) or when designing balanced bases.

In the normal case the total unfactored column/wall load from the superstructure will be of the form

$$P = G + Q + P_W$$

where G = superstructure dead load (vertical)
Q = superstructure imposed load (vertical)
P_W = superstructure wind load (vertical component)
and the factored load from the superstructure will be

$$P_u = \gamma_G G + \gamma_Q Q + \gamma_w P_W$$

where γ_G, γ_Q, γ_w are the appropriate partial load factors for the case under consideration, which can be taken from Table 10.4.

The unfactored (characteristic) foundation load has previously been expressed (see Fig. 10.16) as

$$F = F_B + F_S$$

where F_B = load of the foundation and backfill
F_S = foundation surcharge load.
For ultimate limit-state calculations it should be rewritten as

$$F = F_G + F_Q$$

where F_G = foundation dead load ($= F_B$ + dead load component of F_S)
F_Q = foundation imposed load ($=$ imposed component of F_S)
and the factored load from the foundation will be

$$F_u = \gamma_G F_G + \gamma_Q F_Q$$

(γ_G and γ_Q can be taken from Table 10.4).
The total unfactored (characteristic) load is

$$T = (\text{superstructure load}) + (\text{foundation load})$$
$$T = P + F$$
$$= (G + Q + P_W) + (F_G + F_Q)$$

wall or column

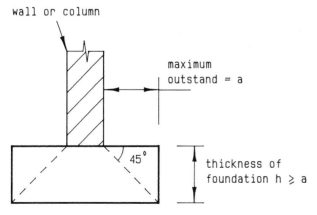

Fig. 10.19 Load spread in mass concrete foundation.

Table 10.4 Typical load cases for ultimate limit-state design of structural foundation members

Load case	Partial safety factor for loads		
	Dead γ_G	Imposed γ_Q	Wind γ_w
Dead + Imposed	1.4	1.6	—
Dead + Wind uplift	1.0	—	1.4
Dead + Wind downthrust	1.4	—	1.4
Dead + Imposed + Wind	1.2	1.2	1.2
Accidental loading	1.05	0.35[a]	0.35

[a] Except for areas of storage where $\gamma_Q = 1.0$

and the total factored load at the underside of the foundation is

$$T_u = (\gamma_G G + \gamma_Q Q + \gamma_W P_W) + (\gamma_G F_G + \gamma_Q F_Q)$$

or $$= \gamma_G(G + F_G) + \gamma_Q(Q + F_Q) + \gamma_W P_W$$

In simple cases where wind loads are not critical the calculations can be made simpler by using an overall combined partial load factor γ_P for the superstructure load such that

$$P_u = \gamma_P P$$

Frequently γ_P is taken conservatively as 1.5 (being half-way between $\gamma_G = 1.4$ and $\gamma_Q = 1.6$ for the dead + imposed case) on the basis that very few building structures support a total imposed load greater than the total dead load. Alternatively a closer assessment can be made on the ratio between dead and imposed loads and the value of γ_P obtained from Fig. 10.20.

Similarly combined partial safety factors γ_F and γ_T can be used for the foundation and total loads where

$$F_u = \gamma_F F$$
$$T_u = \gamma_T T$$

Again these may be obtained from Fig. 10.20. The use of these combined factors is illustrated in the design examples in Chapters 11–14.

Having calculated the factored loads it is then necessary to establish the factored foundation pressures, and to determine the resulting moments and shears in the foundation elements, which should be designed in accordance with the appropriate British Standard.

While the loads already utilized to establish that the allowable bearing pressure is not exceeded are unfactored service loads, the factored loads are required for the design of the members. Some discipline is therefore required when designing the superstructure to keep the dead, imposed and wind loads separate so that they can be easily extracted. This can be achieved either by recording the working load reactions separately so that the loads can be used directly in the determination of bearing pressure and factored up for the design of the elements, or by recording the factored

reactions separately so that the loads can be used directly for the design of the elements and factored down for the determination of bearing pressures. While there is no particular advantage in which way it is undertaken it is recommended that a consistent approach is adopted for each project to avoid errors.

10.12 GENERAL DESIGN METHOD

This section gives a general design method based on the previously described calculations which allows for a systematic design process.

PART 1: CALCULATION OF BEARING PRESSURES FOR CHECKING AGAINST ALLOWABLE BEARING PRESSURES

(1) Determine the relevant load cases to be considered using engineering judgement and guidance from the limit-state code appropriate to the foundation material. The load factors to be used can be taken from Table 10.4 which can be of further help in assessing critical load cases.

For *each* of the load cases the following procedures should be adopted.

(2) Calculate the superstructure characteristic (unfactored) load in terms of

$$P = G + Q + P_W$$
$$H = H_G + H_Q + H_W$$
and $$M = M_G + M_Q + M_W$$

where P = superstructure vertical load
H = superstructure horizontal load
M = superstructure moment
and the subscripts are
G = dead load
Q = imposed load
W = wind load.

If the base is subject to biaxial bending calculate H and M about both x and y axes to give H_x and M_x and H_y and M_y.

(3) Estimate the foundation size using the middle third rule where applicable, and calculate the foundation load

$$F = F_G + F_Q$$

where F_G = dead load from foundation
F_Q = imposed load from foundation.

(4) Calculate the total vertical load at the underside of the foundation

$$T = P + F$$

(5) Calculate the eccentricity of the total load

$$e_T = \frac{Pe_P + M + Hh}{T}$$

where h is the thickness of the base.

If the base is subject to biaxial bending, calculate e_{Tx} and e_{Ty} for the two axes from

$$e_{Tx} = \frac{Pe_{Px} + M_x + H_y h}{T}$$

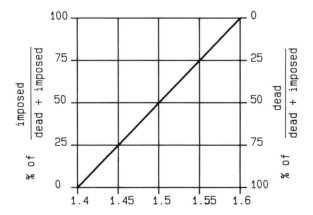

Fig. 10.20 Combined partial safety factors for dead + imposed loads.

$$e_{Ty} = \frac{Pe_{Py} + M_y + H_x h}{T}$$

Consider if economy could be gained by offsetting the base to cancel out or reduce this eccentricity, and recalculate as necessary.

(6) Assess which bearing pressure distribution is appropriate and calculate the total bearing pressure, t, in accordance with section 10.10 and Figs 10.21–10.23.

(a) Axial loading: i.e. uniform pressure $e_T \sim 0$

$$t = \frac{T}{A}$$

(b) Axial plus bending with base pressure wholly compressive:

(i) For single axis bending, the general equation is

$$t = \frac{T}{A} \pm \frac{Te_T}{Z}$$

which becomes

$$t = \frac{T}{A} \pm \frac{6Te_T}{BL^2}$$

for a rectangular base because

$$e_T \le \frac{L}{6}$$

(ii) For biaxial bending, the general equation is

$$t = \frac{T}{A} \pm \frac{Te_{Tx}}{Z_x} \pm \frac{Te_{Ty}}{Z_y}$$

which becomes

$$t = \frac{T}{A} \pm \frac{6Te_{Tx}}{BL^2} \pm \frac{6Te_{Ty}}{LB^2}$$

for a rectangular base because

$$\frac{e_{Tx}}{L} + \frac{e_{Ty}}{B} < \frac{1}{6}$$

(c) Axial plus bending with zero pressure under part of the base:

(i) For single axis bending, use the shortened base theory to establish the pressure diagram under the base such that the resultant is under the line of the total load T. From this calculate the maximum total bearing pressure. For a rectangular base, this becomes

$$t = \frac{2T}{3B\left(\dfrac{L}{2} - e_T\right)}$$

(ii) For biaxial bending, it is recommended that this situation should not be allowed to develop. Consider increasing the size of base or adjust its position relative to the column, to reduce e_{Tx} and/or e_{Ty} as appropriate.

(7) Check t against the total allowable bearing pressure, t_a

(*note*: the total allowable bearing pressure can be increased by 25% when resisting wind loads), or calculate the net bearing pressure from

$$n = t - s$$

where s is the existing soil pressure.

If n or t are greater than the corresponding allowable bearing pressure, n_a or t_a as appropriate, then adjust the base size and/or the column eccentricity and recalculate from section (3).

If n or t are less than the corresponding allowable bearing pressure then check settlements (see section A of Chapter 2) and move on to section (8) below. If settlements are not satisfactory, revise the base size and recalculate from section (3).

If n or t are very much less than the corresponding allowable bearing pressure then, for economy, reduce the base size and recalculate from section (3).

(8) If the base is of plain concrete, calculate the minimum depth where h_{min} = maximum distance from edge of column to edge of base (see Fig. 10.19).

(9) If the base is of reinforced concrete proceed to Part 2 below.

Foundation design complete.

PART 2: CALCULATION OF BEARING PRESSURES FOR DESIGN OF REINFORCED CONCRETE OR STEEL FOUNDATION ELEMENTS

Before progressing with the design of the reinforced concrete elements of the foundation the engineer must make an assessment as to whether it is necessary to make a full re-analysis of the bearing pressures in the manner described above but using factored loads, or whether sufficient accuracy can be achieved by taking the short cut of multiplying the bearing pressures by an overall factor γ_P, γ_F or γ_T as appropriate. In the vast majority of cases the short cut method is perfectly satisfactory but there are cases when it is not and the engineer must be careful! If in doubt he should use the full method. Typical cases where the short cut method *does not apply* include the following:

(1) Where wind loading forms a significant part of the foundation loading particularly where generating uplift.
(2) Where a live load applied to the structure will increase the horizontal load, H, and/or moment, M, without a proportional increase in the vertical load, P.
(3) Where there is partial zero pressure under the foundation (T is outside the middle third rule for rectangular foundations).

Short cut method

(1) Determine the ultimate total pressure distribution, t_u, under the base from

$$t_u = \gamma_T t$$

where γ_T = combined total load factor assessed with the aid of Fig. 10.20

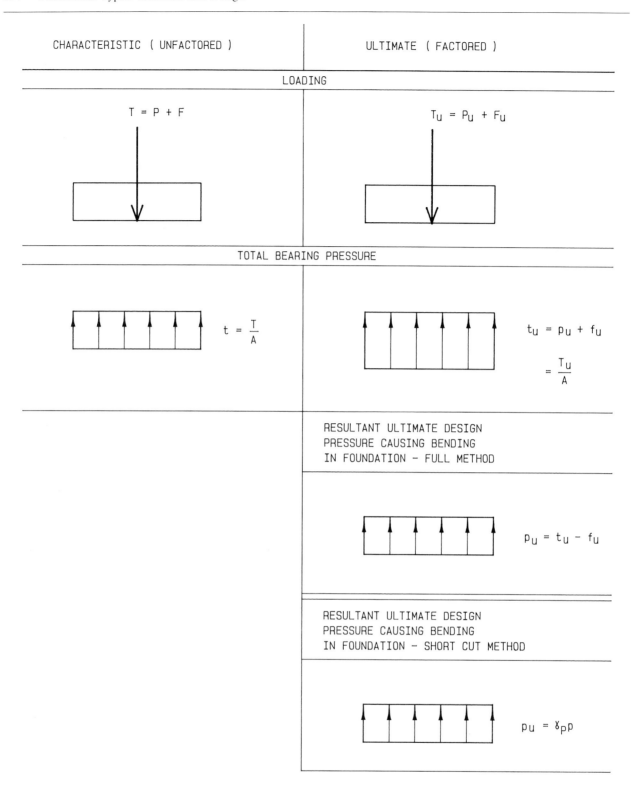

Fig. 10.21 Design of axially loaded foundation.

t = unfactored total stress distribution from Part 1, section (6) (a) or (b).

(2) Determine the ultimate foundation pressure distribution, f_u, under the base from

$$f_u = \gamma_F f$$

where γ_F = combined foundation load factor assessed with the aid of Fig. 10.20 (usually 1.4 unless

there is an imposed load element or uplift is being considered).

(3) Determine the resultant ultimate design pressure causing bending, p_u, from

$$p_u = t_u - f_u$$

(see Figs 10.21–10.23).

Note that in the case of *axially loaded foundations* this

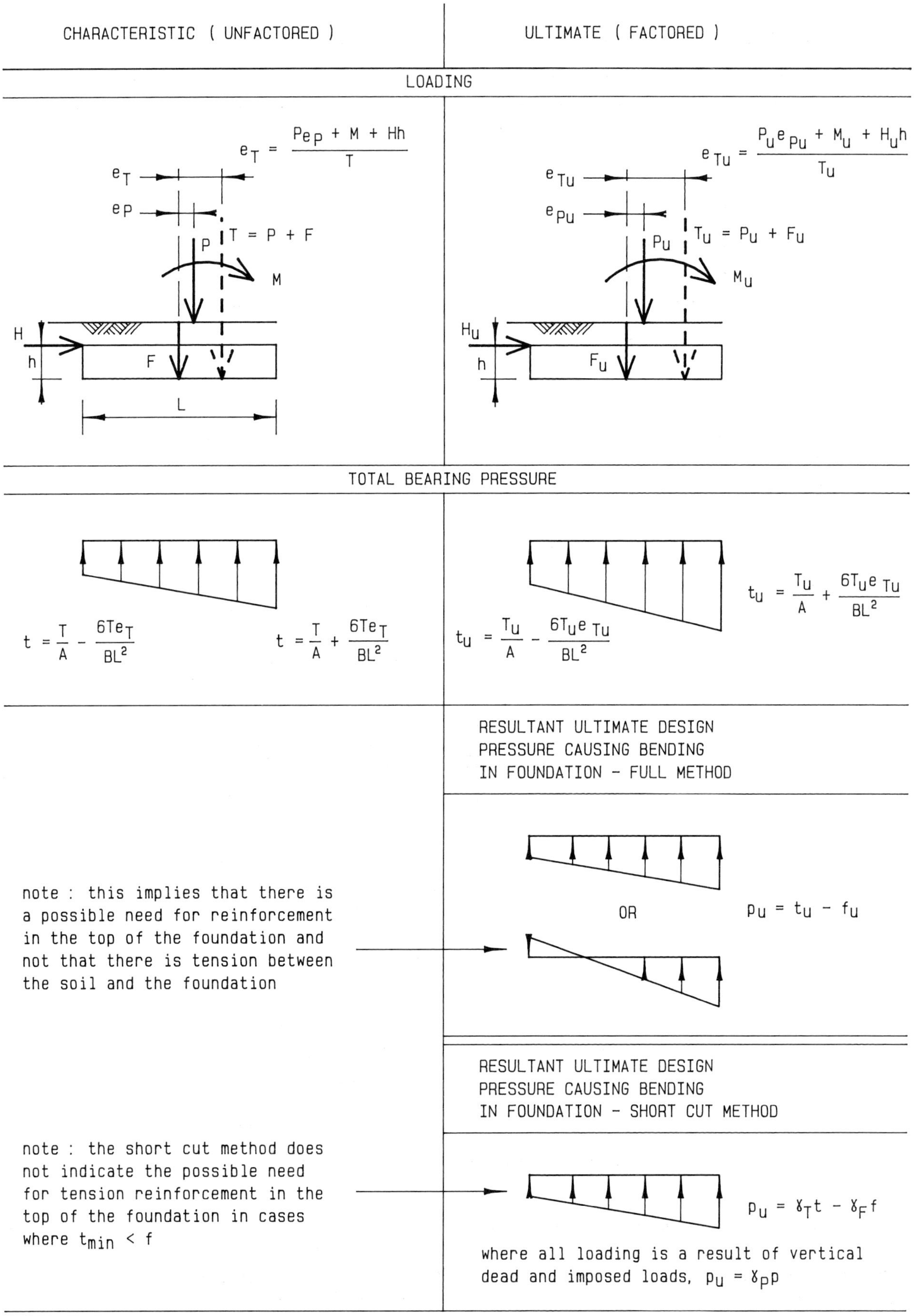
CHARACTERISTIC (UNFACTORED)
ULTIMATE (FACTORED)
LOADING
$e_T = \frac{Pe_P + M + Hh}{T}$
e_T
e_P
$T = P + F$
P
M
H
h
F
L
$e_{Tu} = \frac{P_u e_{Pu} + M_u + H_u h}{T_u}$
e_{Tu}
e_{Pu}
$T_u = P_u + F_u$
P_u
M_u
H_u
h
F_u
TOTAL BEARING PRESSURE
$t = \frac{T}{A} - \frac{6Te_T}{BL^2}$
$t = \frac{T}{A} + \frac{6Te_T}{BL^2}$
$t_u = \frac{T_u}{A} - \frac{6T_u e_{Tu}}{BL^2}$
$t_u = \frac{T_u}{A} + \frac{6T_u e_{Tu}}{BL^2}$
RESULTANT ULTIMATE DESIGN PRESSURE CAUSING BENDING IN FOUNDATION – FULL METHOD
note : this implies that there is a possible need for reinforcement in the top of the foundation and not that there is tension between the soil and the foundation
OR
$p_u = t_u - f_u$
RESULTANT ULTIMATE DESIGN PRESSURE CAUSING BENDING IN FOUNDATION – SHORT CUT METHOD
note : the short cut method does not indicate the possible need for tension reinforcement in the top of the foundation in cases where $t_{min} < f$
$p_u = \gamma_T t - \gamma_F f$
where all loading is a result of vertical dead and imposed loads, $p_u = \gamma_P p$

Fig. 10.22 Design of foundation in bending – base fully in compression.

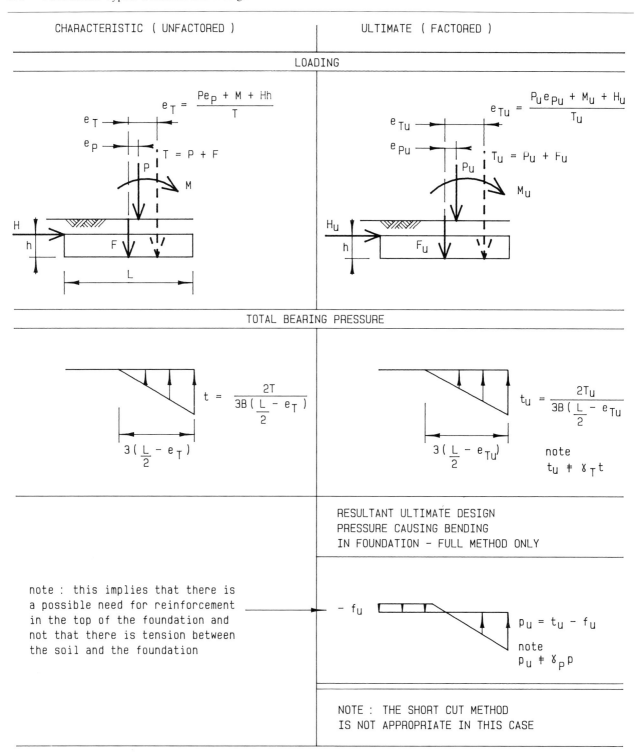

Fig. 10.23 Design of foundation in bending — zero pressure under part of base.

calculation can be reduced further by calculating p_u directly from the superstructure pressure

$$p_u = \gamma_P p$$

where γ_P = combined superstructure load factor assessed with the aid of Fig. 10.20

p = bearing pressure due to the superstructure. (see Figs 10.21 and 10.22).

(4) Having determined the resultant ultimate design pressure, p_u, this is used to determine the ultimate shear, bending moments and axial forces using accepted structural theory and to design those elements in accordance with the appropriate British Standards.

Foundation design complete.

Full method

(1) Calculate the ultimate superstructure loading in terms of

$$P_u = \gamma_G G + \gamma_Q Q + \gamma_W P_W$$
$$H_u = \gamma_G H_G + \gamma_Q H_Q + \gamma_W H_W$$
$$M_u = \gamma_G M_G + \gamma_Q M_Q + \gamma_W M_W$$

where γ_G, γ_Q and γ_W are the load factors appropriate to the load case under consideration (see Table 10.4).

(2) Calculate the ultimate foundation loading in terms of:

$$F_u = \gamma_G F_G + \gamma_Q F_Q$$

(3) Calculate the total ultimate load

$$T_u = P_u + F_u$$

(4) Calculate the total ultimate load eccentricity

$$e_{Tu} = \frac{P_u e_P + M_u + H_u h}{T_u}$$

or for biaxial bending

$$e_{Txu} = \frac{P_u e_{Px} + M_{xu} + H_{yu} h}{T_u}$$

and

$$e_{Tyu} = \frac{P_u e_{Py} + M_{yu} + H_{xu} h}{T_u}$$

(5) Calculate the ultimate bearing pressure distribution using the procedure in section 10.10 but with ultimate loads:

(a) Axial loading: i.e. uniform pressure $e_{Tu} \sim 0$

$$t_u = \frac{T_u}{A}$$

(b) Axial plus bending with base pressure wholly compressive:

(i) For single axis bending, the general equation is

$$t_u = \frac{T_u}{A} \pm \frac{T e_{Tu}}{Z}$$

which becomes

$$t_u = \frac{T_u}{A} \pm \frac{6 T_u e_{Tu}}{BL^2}$$

for a rectangular base because

$$e_{Tu} < \frac{L}{6}$$

(ii) For biaxial bending, the general equation is

$$t_u = \frac{T_u}{A} \pm \frac{T_u e_{Txu}}{Z_x} \pm \frac{T_u e_{Tyu}}{Z_y}$$

which becomes

$$t_u = \frac{T_u}{A} \pm \frac{6 T_u e_{Txu}}{BL^2} \pm \frac{6 T_u e_{Tyu}}{B^2 L}$$

for a rectangular base because

$$\frac{e_{Tx}}{L} + \frac{e_{Ty}}{B} < \frac{1}{6}$$

(c) Axial plus bending with zero pressure under part of the base, i.e. $e_{Tu} > \dfrac{L}{6}$:

(i) For single axis bending, for a rectangular base

$$t_u = \frac{2 T_u}{3B\left(\dfrac{L}{2} - e_{Tu}\right)}$$

(6) Calculate the resultant ultimate design pressure from

$$p_u = t_u - f_u$$

where f_u is the factored pressure due to the foundation construction and backfill defined by

$$f_u = \frac{F_u}{A}$$

(7) Having determined the resultant ultimate design pressure, p_u, this is used to determine the ultimate shear, bending moments and axial forces using accepted structural theory and to design those elements in accordance with the appropriate British Standards.

Foundation design complete.

Figures 10.21–10.23 show the various stress distributions graphically and clearly show the difference between working and ultimate loads and stresses, and the resultant ultimate pressure for foundation element design using both the full and the short cut methods.

CHAPTER 11

DESIGN OF PADS, STRIPS AND CONTINUOUS FOUNDATIONS

11.1 UNREINFORCED CONCRETE PADS AND STRIPS

11.1.1 Introduction

In general, shallow pads and strips are the economic foundation for most structures where ground conditions allow this solution.

The suitability of shallow strips and pads should be one of the first considerations for the engineer and their use tends to form the *normal* foundation criteria against which the extra over cost of *abnormal* foundations tends to be judged.

This does not mean however, that strips and pads should be used wherever possible since as they become deeper or more heavily reinforced the alternatives of vibro-compaction and/or piles becomes competitive (see section A of Chapter 10). However, at shallow depth, they are the economic alternative.

11.1.2 Trench fill

A brief description of trench fill strips is given in section 9.3.4. The design of such strips is relatively simple, and it is true to say that there is more design involved in making the decision to adopt such a foundation than in analysing and sizing the appropriate trench fill.

Trench fill is often used in an attempt to:

(1) Reduce the foundation width where brickwork below ground would need a wider footing to suit working space,
(2) Reduce the labour content of construction, and
(3) Speed up the construction of the footing, for example, in conditions where trench supports are not necessary for short periods but would be required if the trench were left open for a significant time.

The saving in excavation, labour, time and/or temporary works can in some situations be quite considerable. However, in loose ground the quantity of concrete used can become both difficult to predict and/or considerable in quantity particularly if trenches meet or cross at right angles.

Strips excavated through poor ground to reach a suitable bearing strata can prove troublesome due to instability of the trench sides, particularly at changes in direction of the strip (see Fig. 11.1). This can be overcome by using suitable trench supports. However, the problem can often be more economically assisted by good design.

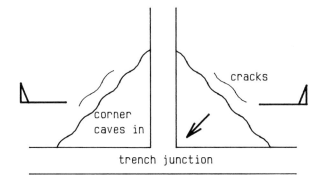

plan

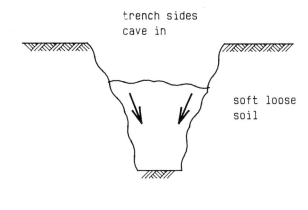

section

Fig. 11.1 Trench instability at change in direction.

For example, Fig. 11.2 shows two alternative designs for the same house foundations: in (A) the trenches would fail under much less critical conditions than the trenches in (B) since this scheme avoids trench direction changes and hence avoids the corner failure conditions of the trench sides.

A disadvantage in some situations is the tendency of the trench strips to pick up, via passive resistance, any longitudinal or lateral ground strains which may occur in the strata around the foundation. This can prove to be a major problem in active mining areas and in sub-strata sensitive to moisture changes such as shrinkable clays. In some situations

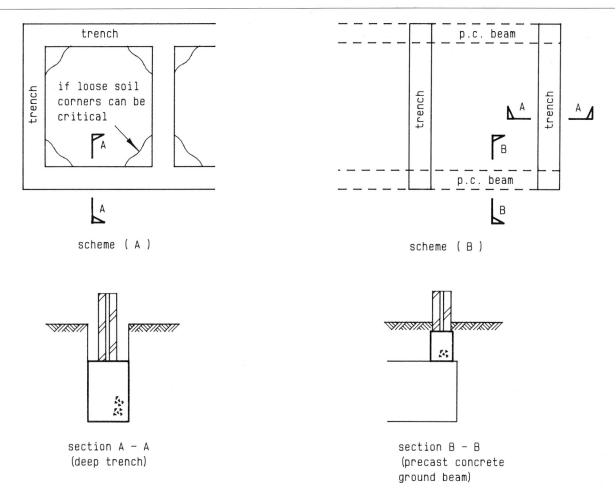

Fig. 11.2 Trench fill alternatives.

this problem can be overcome by the insertion of a compressible bat against the trench faces (see Fig. 11.3), but this must be considered for all directions and for conflicting requirements since passive resistance is often exploited in the superstructure and foundation design.

In addition the high level of the concrete can create problems for drainage and services entering the building if these are not pre-planned and catered for. The top surface should be low enough so as not to interfere with landscaping

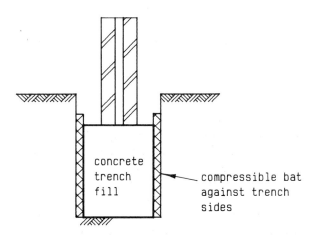

Fig. 11.3 Trench fill with compressible side formers.

and planting. In some situations concrete trench fill can create undesirable hard spots, and stone trench fill should be considered.

Stone trench fill used under the strip loads to transfer the loads to the lower sub-strata is more yielding than concrete trench fill which may produce excessive differential movement between the main strip load area and the general slab (see Fig. 11.4.)

In soft wet conditions, the soft materials at the surface of the trench bottom can be absorbed into the voids of a first layer of no fines stones blinded by a second layer of well graded stone. The second layer prevents the soft materials from oozing up through the hardcore. This can prove to be a clear advantage for difficult sites where the material is sensitive and wet and where good clean trench bottoms are difficult or impractical to achieve. By this method a stable trench fill can quickly and easily be achieved in relatively poor ground (see Fig. 11.5).

Compaction difficulties can be experienced in narrow trenches cut in dry or relatively stiff sub-strata where compaction of the fill at the edges is partly restricted by the frictional resistance of the trench sides. This tends to show itself in the concave surface of the compacted layer (see Fig. 11.6). However, this can be overcome by using suitably graded stone in relatively thin layers and by extra compaction at the edges of the trench.

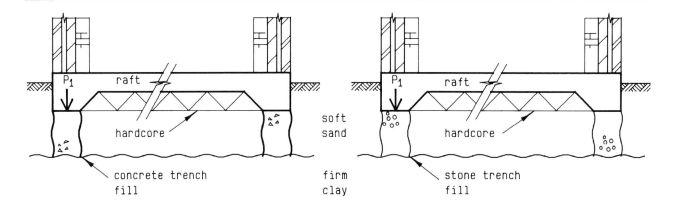

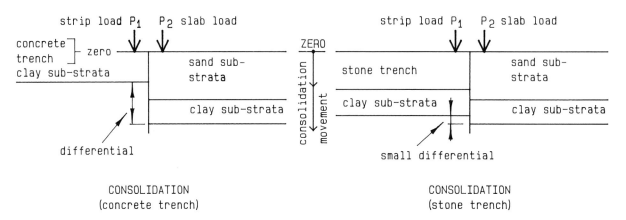

Fig. 11.4 Stone versus concrete trench fill.

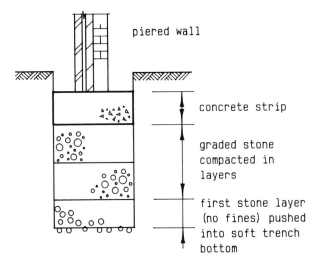

Fig. 11.5 Trench fill in poor ground.

Selection of suitably graded and shaped stone is particularly important, for example, single sized rounded stone will tend to compact automatically during filling in a similar way to say filling a trench with marbles. The marbles immediately fall into contact on more or less the maximum compaction due to the standard radius involved. However, in some locations it is important to avoid forming a *field drain* within the fill which may attract moving water, therefore well graded material is essential in these situations.

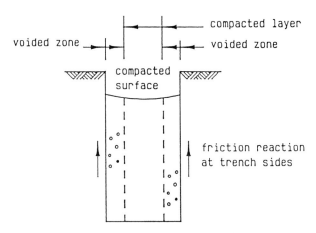

Fig. 11.6 Concave compacted surface.

11.1.3 Trench fill design decisions

A typical trench fill foundation is shown in Fig. 11.7 where (a) indicates a typical section, (b) shows the typical design forces, and (c) illustrates the possible externally applied ground strains which can prove critical. Such strains can and do cause serious damage to buildings and their finishes.

The considerations to be made therefore in the design decisions relate to:

(1) The depth and type of suitable ground-bearing strata relative to foundation loading.

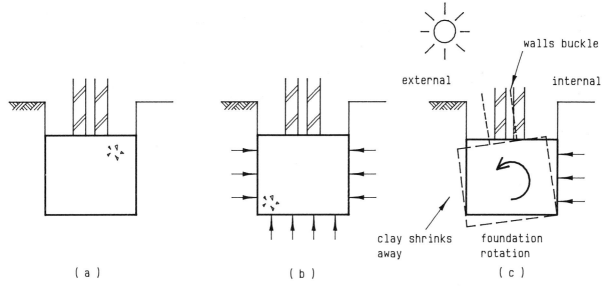

Fig. 11.7 Trench fill conditions.

(2) The likelihood of large horizontal ground strains due to moisture changes in the sub-strata, mining activity or frost.

(3) The economy of trench fill versus normal strip footings.

Considering (1), strip footings can prove economic for medium loads at shallow-to-medium depths on firm-to-stiff sub-strata, for example, $100-300\,kN/m^2$ at $600-1500\,mm$ deep on firm-to-stiff clay, firm-to-dense sand or firm-to-stiff sandy clay or clayey sand.

Considering (2), provided that the sub-strata is not a sensitive clay, i.e. not highly shrinkable, and provided that there is no likelihood of large ground strains from mining or other activities, then trench fill footings should be considered. Standard trench strips should generally be avoided where lateral forces may be picked up from below ground on the sides of the footing. A void or cushion can however be adopted to prevent the transfer of such forces where this is appropriate. Alternatively, the foundation and its superstructure may, in certain circumstances, be designed to resist the force transferred.

Considering (3), the economics of any potential solution vary with time and are dependent on:

(a) Material costs and availability,
(b) Excavation technology and machinery,
(c) Weather conditions and likely stability of trenches, and
(d) Manpower availability and other conditions relating to a particular site.

As a general rule providing that the trenches can remain stable for a suitable period to allow excavation and casting of concrete, and that conditions (1) and (2) are appropriate, then trench fill should be considered in any comparison exercises, particularly if the width of a normal strip footing would be dictated by the working space required for bricklaying below ground.

Under such conditions, the solution is likely to prove competitive both from a cost point of view and speed of construction. The speed of construction can be particularly important where trench stability is time related and where foundations are being constructed through a winter period. Under such conditions a mass concrete trench fill could be adopted. However, the use of stone or concrete is an economic decision based upon the ground conditions, the long- and short-term stability of the trenches, and the availability of the materials (see Table 10.2).

11.1.4 Sizing of the design

In the case of mass concrete trench fill the foundations can be sized using the assumptions that dispersion of load through the strip can be assumed to be at an angle of 45°. In the case of stone fill a dispersion of between 60° and 45° should be assumed depending on the size, quality, type, grading and shape of the stone being used.

In general a 60° dispersion would be a conservative assumption and is often adopted. Though the normal depth of trench fill tends to be more than adequate to provide the required dispersion, nevertheless there is a need to check to see that certain requirements are met. These requirements would be similar to those adopted for unreinforced strips. The sizing of unreinforced strips or mass concrete or stone trench fill for typical uniform axially loaded conditions i.e. the centreline of the resultant load is on the centreline of the strip footing, as shown in the cross-section in Fig. 11.8, would be as follows.

On the basis of Fig. 11.8 (a), assuming P equals the total load per unit run, B equals the required breadth of the strip, h equals the required depth of the strip, and n_a equals the allowable bearing pressure, then it follows that $B = P/n_a$, and for mass concrete, assuming a dispersion of 45°,

$$h = \frac{B - t_w}{2}$$

It also follows from Fig. 11.8 (b) that for a stone mass fill based upon an assumed dispersion of 60°:

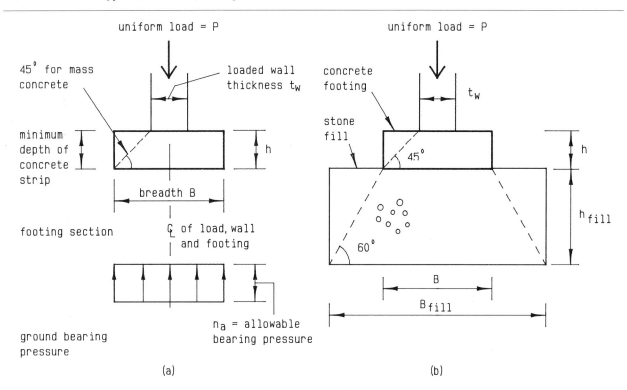

Fig. 11.8 Load dispersion through mass concrete strip.

$$B_{fill} = \frac{P}{n_a}$$

$$\text{and} \quad h_{fill} = \frac{(B_{fill} - B)\sqrt{3}}{2}$$

where B is based on the bearing capacity of the fill material but is not usually less than 450 mm.

On this basis the minimum breadth and depth of a continuous uniformly loaded strip can be easily sized. However, many strips are not uniformly loaded and, in the case of trench fill, the sizes to be adopted for economic reasons are often not those minimum sizes demanded by this part of the design. For the non-uniformly loaded strip for example, it may be necessary to consider the load dispersion in the longitudinal direction in order to arrive at the critical loading on the strip for determining the cross-section (see Fig. 11.9).

As can be seen from the figure the concentrated loads from the piers can be assumed to disperse this load at a similar dispersion angle to that assumed in the cross-section,

i.e. 45° for mass concrete trench fill and 60° for a stone trench fill. Using these assumptions in conjunction with an assumed depth and the critical loaded length, the depth can be determined by trial and error. Varying the assumed depth the length determined from this calculation can be used to arrive at the minimum cross-section as indicated previously. An alternative to this approach is to assume that the local loads on the strip are similar to a mass pad, and the section can be determined from the plan requirements to disperse the load (see the following example).

A typical mass pad load dispersion is shown in Fig. 11.10. The load P being axial can be supported on a pad with its centre of gravity on a similar axis to the centre of gravity of the load. This solution would produce a uniform bearing pressure. If we assume the area required is A, then

$$A = \frac{P}{n_a} = B \times L$$

where n_a = net allowable bearing pressure and B and L are the dimensions of the pressure area (see Fig. 11.10).

When deciding on the plan dimension it is preferable to select a similar relationship for B and L to that of the existing breadth and length of the pier. However in some situations an unrelated relationship has to be adopted. In this case the worst direction of dispersion should be used to determine the depth of the pad, e.g. the wider cantilever should be considered at the appropriate angles (see Fig. 11.11).

In the case of trench fill and deep mass pads the required depth would normally be well within the nominal depth of the selected strip to the extent that only the plan dimensions generally need to be calculated.

The design of the shallow mass pad is carried out by first

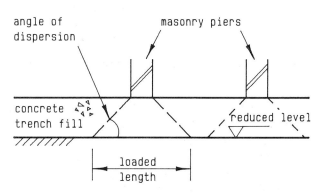

Fig. 11.9 Load dispersion from piers.

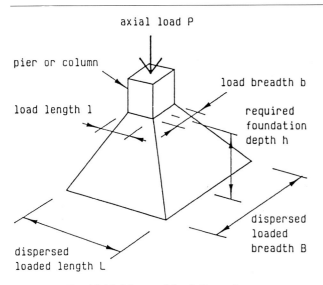

Fig. 11.10 Mass pad load dispersion.

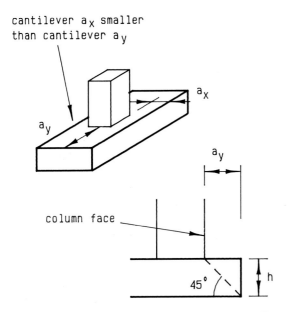

Fig. 11.11 Pad depth determined from maximum cantilever.

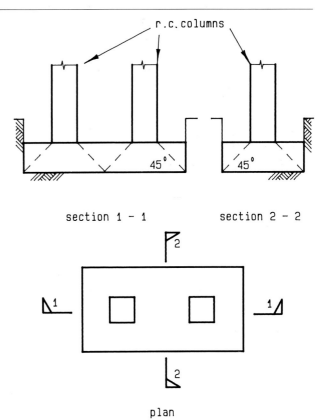

Fig. 11.12 Rectangular mass concrete pad/strip.

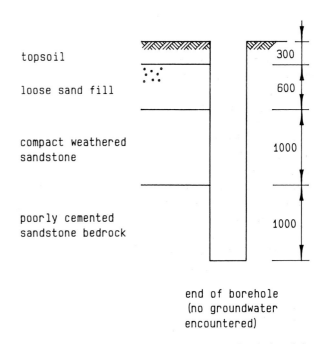

end of borehole
(no groundwater
encountered)

Fig. 11.13 Borehole log for Design Examples 1, 2 and 4.

determining the area of pad required from the estimated load and allowable bearing pressure and then determining the most economic shape to spread the load and give a minimum depth. For example, for the two point loads indicated in Fig. 11.12, a rectangular strip is the most appropriate with a minimum unreinforced concrete depth equal to the maximum projection.

The design of a typical trench fill strip and pad foundations are given in the following examples.

11.1.5 Design Example 1: Trench fill strip footing

The internal load-bearing wall for a four-storey office block is to be supported on a strip foundation. Borehole investigations produced the consistent soil profiles shown in Fig. 11.13.

Soil analysis shows that the sand fill is an unreliable bearing strata. The weathered sandstone has net allowable bearing pressures of $n_a = 400 \, \text{kN/m}^2$ for strip footings and

$n_a = 550 \, \text{kN/m}^2$ for pads, both with a maximum of 20 mm settlement. The sandstone bedrock has a net allowable pressure of $n_a = 2000 \, \text{kN/m}^2$ for pad foundations.

By inspection of the soil profile and analysis in Fig. 11.13, the strip will be founded in the compact weathered sandstone. The relatively even distribution of the loading will not lead to unacceptable differential settlements and, as the sides of the excavations do not collapse in the short-term, mass concrete trench fill footings have been selected as the most appropriate foundation type.

Loadings
The loadings from the four-storey structure have been calculated (as working loads) as follows.

		(kN/m run)
Dead load from floors and roof	=	137
Dead load from 215 mm thick load-bearing wall	=	55
Superstructure dead load, G =		192
Superstructure imposed load (from floors and roof), Q =		93
Net load = superstructure total load, $P = G + Q$ =		285 kN/m run

Size of base (normal method)
The foundation surcharge is considered small enough to be neglected. The minimum foundation width is given by

$$B = \frac{\text{superstructure load}}{\text{net allowable bearing pressure}}$$

$$= \frac{P}{n_a}$$

$$= \frac{285}{400}$$

$$= 0.71 \, \text{m}$$

In many instances this approximate method is satisfactory. Where the new foundation surcharge is large, or the allowable bearing pressure is low, the following method should be used.

Size of base (allowing for foundation surcharge)
Dead load from new surcharge

$$= 20 \, \text{kN/m}^3 \times 0.3 \, \text{m} \qquad = 6 \, \text{kN/m}^2$$

Imposed load from new surcharge

$$= 5 \, \text{kN/m}^2 \text{ distributed load} \qquad = 5 \, \text{kN/m}^2$$

New foundation surcharge $\qquad fs = 11 \, \text{kN/m}^2$

The weight of the new foundation is taken as approximately equal to the weight of soil displaced, and thus is excluded from the above loads.

From section 10.10, the net bearing pressure is

$$n = \frac{P}{A} + f_S - s_S$$

In this case the existing surcharge $s_S = 0$.

$$n = \frac{P}{A} + f_S$$

$$400 = \frac{285}{B \times 1} + 11$$

$$B = \frac{285}{(400 - 11)}$$

$$= 0.73 \, \text{m}$$

As may be seen, the *normal method* value of $B = 0.71$ m in this example is sufficiently accurate for all practical purposes.

Final selection of foundation width must take into account the width of the wall, together with an allowance for tolerance. It should also try to suit standard widths of excavator buckets which are in multiples of 150 mm, e.g. 450 mm, 600 mm, 750 mm, etc. In this case a width of $B = 750$ mm would be appropriate, as shown in Fig. 11.14.

Actual net bearing pressure (ignoring foundation surcharge)
The actual net bearing pressure beneath the strip footing may now be calculated, if required.

$$\text{Actual net bearing pressure, } n = \frac{\text{superstructure load}}{\text{foundation width}}$$

$$= \frac{P}{B}$$

$$= \frac{285}{0.75}$$

$$= 380 \, \text{kN/m}^2$$

In this example, n is nearly equal to the allowable pressure, n_a. If it were significantly less, settlements would be expected to be less than the 20 mm anticipated at a pressure of $n_a = 389 \, \text{kN/m}^2$.

11.1.6 Design Example 2: Deep mass concrete pad base
A steel-framed building is to be built on a site adjoining that in Design Example 1 (see section 11.1.5), where variable fill extends down to the level of the bedrock. A heavily loaded

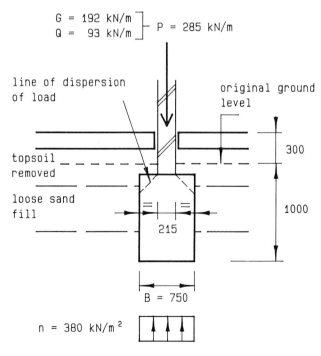

Fig. 11.14 Trench fill strip footing design example.

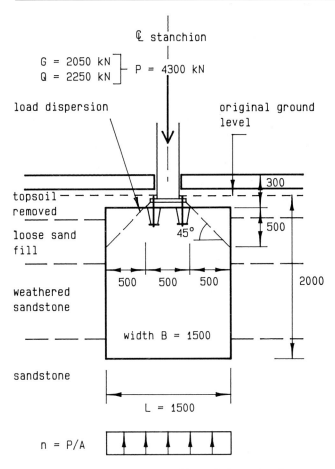

G = 2050 kN
Q = 2250 kN] P = 4300 kN

℄ stanchion

load dispersion

original ground level

topsoil removed

loose sand fill

weathered sandstone

sandstone

45°

300

500

2000

500   500   500

width B = 1500

L = 1500

n = P/A

Fig. 11.15 Mass concrete pad base design example.

stanchion, carrying axial load only, is to be supported on a pad foundation.

It has been decided to found the heavily loaded base in the sandstone bedrock, in order to minimize settlement. The base is to be constructed as a mass concrete pad.

Loadings
The superstructure working loads are as follows:

Superstructure dead load,	$G = 2050\,\text{kN}$
Superstructure imposed load,	$Q = 2250\,\text{kN}$
Net load = superstructure total load, $P = G + Q =$	$\underline{4300\,\text{kN}}$

Allowable bearing pressure
From Design Example 1, the sandstone has a net allowable bearing pressure of $n_a = 2000\,\text{kN/m}^2$.

Size of base
The foundation surcharge due to the groundbearing slab is small and can be neglected. Therefore,

$$\frac{\text{Minimum area of}}{\text{foundation}} = \frac{\text{superstructure load}}{\text{net allowable bearing pressure}}$$

$$A = \frac{P}{n_a}$$

$$= \frac{4300}{2000}$$

$$= 2.15\text{m}^2$$

$$= 1.47\,\text{m} \times 1.47\,\text{m}$$

Therefore a $1.5\,\text{m} \times 1.5\,\text{m}$ pad foundation will be used, as shown in Fig. 11.15.

The stanchion bases are set at a common depth of 300 mm below slab level, and the remaining depth of excavation down to the sandstone rock is taken up by the mass concrete base.

The minimum depth of base required, before it becomes necessary for reinforcement to be introduced, is 500 mm (see depth to angle of load dispersion in Fig. 11.15). Clearly a mass concrete base is adequate in this instance.

The choice between the full sized mass concrete pad and the stub column solution is determined from economic considerations.

The economic change-over point is where the cross-section required for groundbearing purposes becomes excessively wasteful in terms of the cost of concrete compared with the cost of introducing shuttering to form the smaller cross-section. Situations where this would apply are:

(1) Where the pads are very deep,
(2) Where the allowable groundbearing pressures are very low, or
(3) Where, due to the nature of the ground, a shuttered pad is required in any case.

The lower pad plan size is determined from the loading and the allowable groundbearing capacity. If the section is reduced at higher level, the size at the point where it is reduced is generally based upon a 45° dispersion of load through the mass concrete (see Fig. 11.16).

The upper pedestal cross-section is determined from the load, the allowable bearing stress below the base plate and the allowable compressive stress on the mass concrete in conjunction with a suitable practical and economic size for construction. For example, the size determined from stress considerations often needs to be rounded up to a larger practical mass concrete cross-section particularly where the mass pier is relatively tall (see Fig. 11.17).

11.1.7 Unreinforced concrete strips
The unreinforced strip footing requires slightly better ground conditions than trench fill to maintain trench stability during construction of the masonry over it.

The adoption of a thin strip means that the trenches tend to remain open longer during construction than in the case of the trench fill solution to allow bricklayers or masons to work from within the trenches. However, the overall cost of the work often proves less than for trench fill and on many sites proves to be easily achieved. It has the added advantage of more easily accommodating services but suffers similar disadvantages to trench fill in active mining areas. The choice between trench fill and concrete strips usually depends upon cost.

The width of the strip is generally the nearest suitable excavation standard bucket width to that of the design width required from the calculations. However, for deep strips the

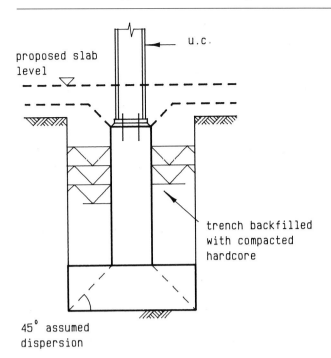

45° assumed
dispersion

Fig. 11.16 Stub column pad base.

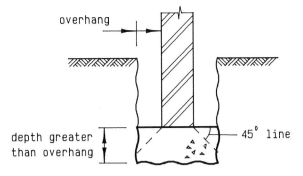

Fig. 11.18 Unreinforced strip.

11.2 REINFORCED CONCRETE PADS AND STRIPS

11.2.1 Introduction
A brief description of reinforced pads and strips is given in sections 9.3.1 and 9.3.3.

These pads are used in similar locations to those of the mass concrete pad, but where the reduction in cost of mass concrete exceeds the cost of the additional labour and materials. These extras would include providing the reinforcement and any extra shuttering, blinding, or working space which may prove necessary for the reinforced solution.

The plan size and shape is determined from the vertical load and allowable bearing stress in conjunction with any physical requirements. The depth and amount of reinforcement is determined from the resulting bending moments and shear force considerations (see Fig. 11.20) or from past experience. The experience basis is often used where reinforcement needs are related to variable ground for a familiar location and use or where there is a need to cater for a number of time-related variations in differential settlement.

11.2.2 Design decisions
The decision to reinforce a concrete foundation of this type usually follows the realization that the ground conditions are variable and/or deep trench fill is uneconomic.

working space required for bricklayers can determine the width required. The thickness is generally selected to be greater than the overhang (i.e. this is based upon a 45° dispersion of load through the mass concrete, see Fig. 11.18).

Where this guidance would give a thinner strip than that practical from a construction point of view, or that desirable from a performance requirement, a greater nominal thickness is used. Longitudinal bending considerations, particularly where the strip requires to be continuous below door openings etc. (see Fig. 11.19), is one of the situations which may demand a thicker strip than that given by the general 45° line. However, this would only apply if dispersion of load along this length of footing is required to reduce the bearing pressure.

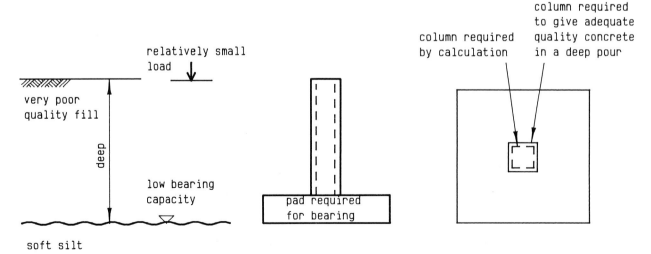

Fig. 11.17 Mass pier criteria.

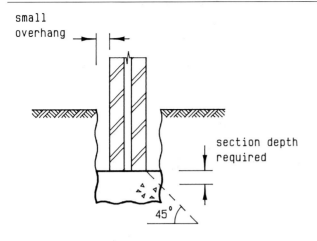

section

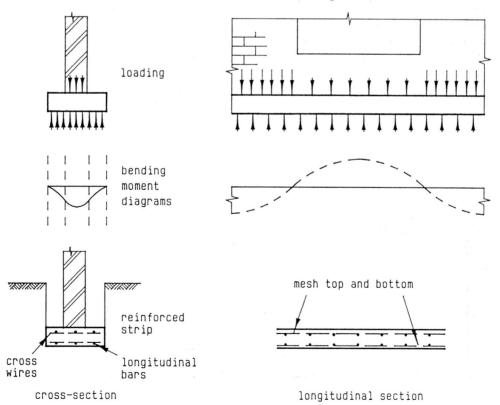

Fig. 11.19 Continuous strip through opening.

Reference to Table 10.2 and Chapter 10 on choice of foundation types will assist in this decision.

11.2.3 Sizing up of the design

The depth and width of the reinforced concrete strips are determined in a similar way to that adopted for unreinforced strips and trench fill. The depth to the underside of the footing is determined by the ground conditions and the level of suitable sub-strata, taking into account the need to be below the effect of any critical frost heave or swelling and shrinkage of sub-strata.

The fabric reinforced strip is used generally where there is both relatively poor ground and smallish loads or where some slight movements are expected from differential settlement or subsidence.

More heavily reinforced strips, using bars and not fabric, are used where ground conditions are more critical and/or loading more excessive (see Fig. 11.21.)

For particularly heavy loads and/or poor ground, beam strips are often used (see sections 9.3.6 and 9.3.7).

For axially loaded strip foundations, the breadth of the strip required is:

$$B = \frac{P}{n_a}$$

where P is the superstructure load/unit run and n_a is the net allowable bearing pressure.

The thickness of the foundation should be determined by designing for the cantilever action of the strip taking into account the bending, shear and bond stresses to be accommodated and allowing for the longitudinal moments and forces (see Fig. 11.20).

Fig. 11.20 Reinforced concrete strip design conditions.

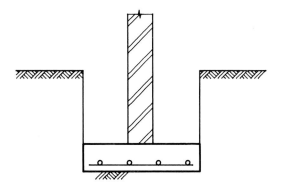

Fig. 11.21 Section through reinforced strip.

For strip footings a generous thickness for bending is necessary in order to maintain the shear and bond stresses within permitted limits and in order to produce an economic balance for the ratio of concrete to reinforcement. The detailed design of a reinforced concrete strip is covered in Design Example 3 which follows, but in general the calculated foundation thickness required for shear and bending compression is rounded up to the nearest 50 mm as the economic thickness for the strip foundation.

11.2.4 Design Example 3: Reinforced strip foundation
The load-bearing wall of a single-storey building is to be supported on a wide reinforced strip foundation.

A site investigation has revealed loose-to-medium granular soils from ground level to some considerable depth. The soil is variable with a safe bearing capacity ranging from $75-125 \, \text{kN/m}^2$. Also some soft spots were identified, where the bearing capacity could not be relied upon.

The building could be supported on ground beams and piles taken down to a firm base, but in this case the solution chosen is to design a wide reinforced strip foundation capable of spanning across a soft area of nominal width. To minimize differential settlements and allow for the soft areas, the allowable bearing pressure will be limited to $n_a = 50 \, \text{kN/m}^2$ throughout. Soft spots encountered during construction will be removed and replaced with lean mix concrete; additionally, the footing will be designed to span 2.5 m across anticipated depressions. This value has been derived from the guidance for local depressions given in Chapter 13 on raft foundations. The ground floor slab is designed to be suspended, although it will be cast using the ground as *permanent formwork*.

Loadings

		(kN/m run)
Dead load from roof and suspended ground floor		= 27
Dead load from wall		= 13
Total dead load,	G	= 40
Imposed load from roof and suspended ground floor,	Q	= 19
Net load = superstructure total load, P		
	$P = G + Q$	= 59 kN/m run

If the foundations and superstructure are being designed to limit state principles, loads should be kept as separate unfactored characteristic dead and imposed values (as above), both for foundation bearing pressure design and for serviceability checks. The loads should then be factored up for the design of individual members at the ultimate limit state as usual.

For foundations under dead and imposed loads only, factoring up loads for reinforcement design is best done by selecting an average partial load factor, γ_P, to cover both dead and imposed superstructure loads from Fig. 11.22 (this is a copy of Fig. 10.20).

$$Q \text{ as a percentage of } P \text{ is } 100Q/P = (100 \times 19)/59$$
$$= 32\%.$$

From Fig. 11.22, the combined partial safety factor for superstructure loads is $\gamma_P = 1.46$.

$$\text{Weight of base and backfill, } f = \text{average density} \times \text{depth}$$
$$= 20 \times 0.9$$
$$= 18.0 \, \text{kN/m}^2$$

This is all dead load, thus the combined partial load factor for foundation loads, $\gamma_F = 1.4$.

Sizing of foundation width
New ground levels are similar to existing ones, thus the (weight of the) new foundation imposes no additional surcharge, and may be ignored.

The minimum foundation width is given by

$$B = \frac{\text{superstructure load}}{\text{net allowable bearing pressure}}$$
$$= \frac{P}{n_a}$$
$$= \frac{59}{50}$$
$$= 1.18 \, \text{m}$$

Adopt a 1.2 m wide × 350 mm deep reinforced strip foundation, using grade 35 concrete (see Fig. 11.23).

Reactive upwards design pressure for lateral reinforcement design

$$\text{Actual superstructure} \atop \text{pressure, } p = \frac{\text{superstructure load}}{\text{width of base}}$$
$$= \frac{P}{B}$$
$$= \frac{59}{1.2}$$
$$= 49.2 \, \text{kN/m}^2$$

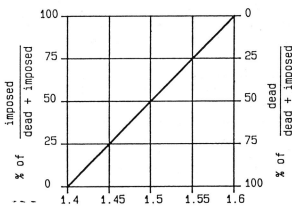

Fig. 11.22 Combined partial safety factor for dead + imposed loads.

Ultimate reactive design pressure $= \dfrac{P_u}{B}$

$$= \dfrac{\gamma_P\, P}{B}$$

$$= \dfrac{1.46 \times 59}{1.2}$$

$$= 71.8\,\text{kN/m}^2$$

Alternatively this can be calculated as

$$p_u = \gamma_P \times (\text{superstructure bearing pressure})$$
$$= \gamma_P p$$
$$= 1.46 \times 49.2$$
$$= 71.8\,\text{kN/m}^2$$

Lateral bending and shear
$b = 1000\,\text{mm}$.

Effective depth, $d = 350 - 50\,(\text{cover}) - 12 - \dfrac{10}{2}$

$$= 283\,\text{mm}$$

Cantilever moment at face of wall

$$M_u = \dfrac{p_u \left(\dfrac{B}{2} - \dfrac{t_w}{2} \right)^2}{2}$$

$$= \dfrac{71.8 \left(\dfrac{1.2}{2} - \dfrac{0.28}{2} \right)^2}{2}$$

$$= 7.6\,\text{kNm/m run}$$

$$\dfrac{M_u}{bd^2} = \dfrac{7.6 \times 10^6}{1000 \times 283^2}$$

$$= 0.09$$

$$A_{s(\text{req})} = 0.02\%\ bd \qquad [\text{BS 8110: Part 3: Chart 2}]$$

This is less than the minimum reinforcement in BS 8110: Part 1: 3.27 given by

$$A_{s(\text{min})} = 0.13\%\ bh$$

$$= \dfrac{0.13}{100} \times 1000 \times 350$$

$$= 455\,\text{mm}^2/\text{m}$$

Provide T10 bars @ $150\,\text{c/c} = 523\,\text{mm}^2/\text{m}$ (see Fig. 11.24)

$$= \dfrac{523 \times 100}{1000 \times 283}$$

$$= 0.18\%\ bd$$

Allowable concrete shear stress

$$v_c = 0.44\,\text{N/mm}^2 \qquad [\text{BS 8110: Part 1: Table 3.9}]$$

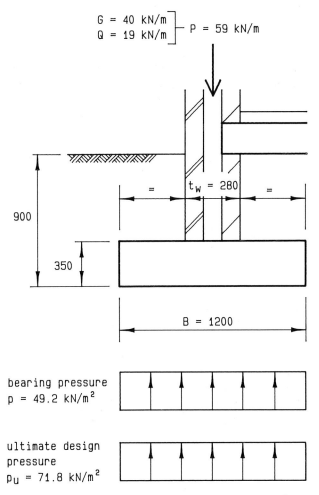

bearing pressure
$p = 49.2\ \text{kN/m}^2$

ultimate design
pressure
$p_u = 71.8\ \text{kN/m}^2$

Fig. 11.23 Reinforced strip foundation design example – loads and bearing pressures.

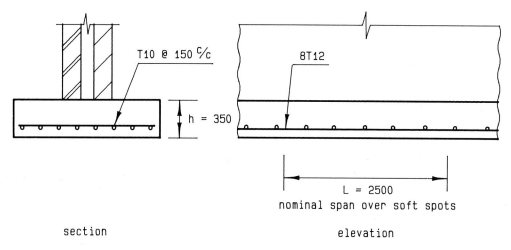

Fig. 11.24 Reinforced strip footing design example — reinforcement.

Shear force at face of wall

$$V_u = p_u \times \left(\frac{B}{2} - \frac{t_w}{2}\right)$$

$$= 71.8 \times \left(\frac{1.2}{2} - \frac{0.28}{2}\right)$$

$$= 33.0 \, \text{kN/m run}$$

Shear stress, $v_u = \dfrac{V_u}{bd}$

$$= \frac{33.0 \times 10^3}{1000 \times 283}$$

$$= 0.12 \, \text{N/mm}^2$$

Thus $v_u < v_c$, therefore no shear reinforcement is required.

Loading for spanning over depressions
Where a local depression occurs, the foundation is acting like a suspended slab. The ultimate load causing bending and shear in the foundation is the total load i.e. superstructure load + foundation load, which is given by

$$T_u = P_u + F_u$$
$$= \gamma_P P + \gamma_F F$$
$$= \gamma_P P + \gamma_F f B$$
$$= (1.46 \times 59) + (1.4 \times 18.0 \times 1.2)$$
$$= 86 + 30$$
$$= 116 \, \text{kN/m}$$

Longitudinal bending and shear due to depressions
Ultimate moment due to foundation spanning — assumed simply supported — over a 2.5 m local depression is

$$M_u = \frac{T_u L^2}{8}$$

$$= \frac{116 \times 2.5^2}{8}$$

$$= 91 \, \text{kNm}$$

Width for reinforcement design is $b = B = 1200 \, \text{mm}$.

Effective depth, $d = 350 - 50 \, (\text{cover}) - \dfrac{12}{2}$

$$= 294 \, \text{mm}$$

$$\frac{M_u}{bd^2} = \frac{-91 \times 10^6}{1200 \times 294^2}$$

$$= 0.87$$

$A_{s(req)} = 0.23\% \, bd$ [BS 8100: Part 3: Chart 2]

$$= \frac{0.23}{100} \times 1200 \times 294$$

$$= 812 \, \text{mm}^2$$

Provide 8 T12 bars $= 905 \, \text{mm}^2$

$$\frac{100 A_s}{bd} = \frac{905 \times 100}{1200 \times 294}$$

$$= 0.26\% \, bd$$

Allowable concrete shear stress, $v_c = 0.49 \, \text{N/mm}^2$
[BS 8110: Part 1: Table 3.9]

Shear force, $V_u = \dfrac{T_u L}{2}$

$$= \frac{116 \times 2.5}{2}$$

$$= 145 \, \text{kN}$$

Shear stress, $v_u = \dfrac{V_u}{bd}$

$$= \frac{145 \times 10^3}{1200 \times 294}$$

$$= 0.41 \, \text{N/mm}^2$$

Thus $v_u < v_c = 0.49 \, \text{N/mm}^2$, therefore shear is not a problem.

Depression at corner of building

The previous calculations have assumed that the depression is located under a continuous strip footing. The depression could also occur at the corner of a building where two footings would meet at right angles. A similar calculation should then be carried out, to provide *top* reinforcement for both footings to cantilever at these corners.

11.2.5 Design Example 4: Reinforced pad base

The axially loaded pad base in Design Example 2 (section 11.1.6) is to be redesigned as a reinforced base, founded in the weathered sandstone. Assuming settlements have been judged to be satisfactory, the base will have an allowable bearing pressure, $n_a = 550 \, \text{kN/m}^2$

Loadings

Dead load,	G	$=$	$2050 \, \text{kN}$
Imposed load,	Q	$=$	$2250 \, \text{kN}$
Net load = superstructure total load,	P	$=$	$4300 \, \text{kN}$

Since dead and imposed loads are approximately equal, a combined partial load factor of $\gamma_P = 1.5$ will be used.

Area of base

$$\text{Required area of foundation} = \frac{\text{superstructure load}}{\text{allowable bearing pressure}}$$

$$= \frac{P}{n_a}$$

$$= \frac{4300}{550}$$

$$= 7.8 \, \text{m}^2$$

Adopt a $3.0 \, \text{m} \times 3.0 \, \text{m}$ square base, i.e. $L = B = 3.0 \, \text{m}$ (see Fig. 11.25).

Reactive design pressure on base for concrete design

$$\begin{aligned}\text{Pressure due to superstructure} \\ \text{loads, } p\end{aligned} = \frac{\text{superstructure load}}{\text{area of base}}$$

$$= \frac{P}{A}$$

$$= \frac{4300}{3.0 \times 3.0}$$

$$= 477 \, \text{kN/m}^2$$

Ultimate design pressure, $p_u = \gamma_P \times$ (superstructure bearing pressure)

$$= \gamma_P p$$

$$= 1.5 \times 477$$

$$= 716 \, \text{kN/m}^2$$

Depth of base

The base and its reinforcement must be capable of resisting bending, beam shear and punching shear. At first glance it is

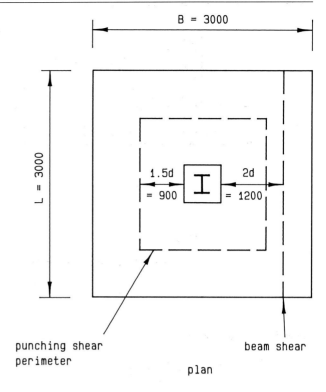

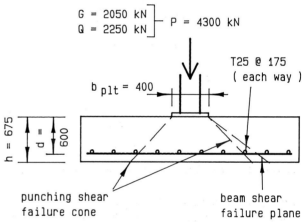

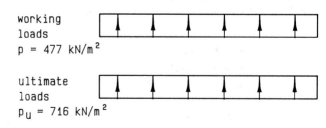

Fig. 11.25 Reinforced pad base design example.

not always possible to judge which is critical. The process of selecting a suitable depth for the base is simplified by use of the charts for estimating effective depths in Appendix H (Figs H.2, H.3, H.4). The effective depth will be checked for each case, assuming a typical reinforcement percentage of between 0.25% and 0.50%. The results are shown in Table 11.1.

Table 11.1 Estimating effective depth for reinforced pad base design example

Design chart for:	Value of y-axis of design chart	$\frac{100A_s}{bd}$	
		0.25	0.50
Bending	$P = 4300\,\text{kN}$	830	570
Beam shear	$\dfrac{P}{B} = \dfrac{4300}{3.0} = 1433\,\text{kN}$	580	560
Punching shear	$\dfrac{PL}{B} = \dfrac{4300 \times 3.0}{3.0} = 4300\,\text{kN}$	approx. 600	

Required effective depth, d (mm)

This indicates that bending is critical, i.e. it requires the greatest effective depth, for low percentages of reinforcement. For this particular example an average effective depth in both directions of $d = 600\,\text{mm}$ will be selected.

Overall depth of base is, $h = 600 + 25$ (bar diameter) + 50 (cover)
$$= 675\,\text{mm}$$

Bending

From Fig. 11.25, the cantilever moment at face of base plate is

$$M_u = \frac{p_u \left(\dfrac{L}{2} - \dfrac{b_{plt}}{2} \right)^2}{2}$$

$$= \frac{716 \left(\dfrac{3.0}{2} - \dfrac{0.4}{2} \right)^2}{2}$$

$$= 605\,\text{kNm/m width}$$

$$\frac{M_u}{bd^2} = \frac{605 \times 10^6}{1000 \times 600^2}$$

$$= 1.68$$

$$A_{s(req)} = 0.45\%\,bd \qquad \text{[BS 8110: Part 3: Chart 2]}$$

$$= \frac{0.45}{100} \times 1000 \times 600$$

$$= 2700\,\text{mm}^2/\text{m}$$

Use T25 bars @ 175 c/c each way = 2805 mm²/m

$$= \frac{2805 \times 100}{1000 \times 600}$$

$$= 0.47\%\,bd$$

Shear

The base should be checked for both beam shear and punching shear, since either may be critical. Grade C40 concrete has been specified.

Local shear at column face

The shear at the face of the column should be checked.

$$v_u = \frac{P_u}{(4 \times b_{plt})\,d}$$

$$= \frac{4300}{(4 \times 400)\,600}$$

$$= 4.5\,\text{N/mm}^2$$

This must not exceed $0.8\sqrt{f_{cu}} \not> 5\,\text{N/mm}^2$

[BS 8110: Part 1: 3.7.7.2]

$$0.8\sqrt{f_{cu}} = 0.8\sqrt{40} = 5\,\text{N/mm}^2$$

$$v_u = 4.5 < 5\,\text{N/mm}^2 \Rightarrow \text{okay.}$$

Beam shear

Allowable concrete shear stress, $v_c = 0.57\,\text{N/mm}^2$

[BS 8110: Part 1: Table 3.9]

From BS 8110: Part 1: 3.4.5.8, the critical location for beam shear is at a distance $2d = 2 \times 600 = 1200\,\text{mm}$ from the face of the load (i.e. from the edge of the base plate in this example). The shear force acting across this failure plane is

V_{beam} = (design pressure) × (area of base beyond critical location)

$$= p_u \left(\frac{L}{2} - \frac{b_{plt}}{2} - 2d \right)$$

$$= 716 \times \left(\frac{3.0}{2} - \frac{0.4}{2} - 2 \times 0.6 \right)$$

$$= 72\,\text{kN/m width}$$

$$\text{Beam shear, } v_{beam} = \frac{V_{beam}}{bd}$$

$$= \frac{72 \times 10^3}{1000 \times 600}$$

$$= 0.12\,\text{N/mm}^2$$

Punching shear

From BS 8110: Part 1: 3.7.7.6, the critical location for punching shear for a square load is a square perimeter a distance $1.5d = 1.5 \times 600 = 900\,\text{mm}$ from the face of the load.

The length of one side of this perimeter is

$$b_{perim} = b_{plt} + 2(1.5d)$$
$$= 400 + (2 \times 1.5 \times 600)$$
$$= 2200\,\text{mm}$$

Area of base outside of perimeter

$$A_{shear} = BL - b_{perim}^2$$
$$= (3.0 \times 3.0) - 2.2^2$$
$$= 4.16\text{m}^2$$

Shear force along perimeter, $V_{punch} = p_u A_{shear}$
$$= 716 \times 4.16$$
$$= 2979\,kN$$

Length of shear perimeter, $u = 4b_{perim}$
$$= 4 \times 2200$$
$$= 8800\,mm$$

Shear stress, $v_{punch} = \dfrac{V_{punch}}{ud}$
$$= \frac{2979 \times 10^3}{8800 \times 600}$$
$$= 0.56\,N/mm^2 < v_c = 0.57\,N/mm^2 \Rightarrow okay.$$

Comparison with $v_{beam} = 0.12\,N/mm^2$ indicates that, in this instance, punching shear is more critical than beam shear. This is normally the case with square pad foundations. If however a foundation size of say $2\,m \times 4\,m$ had been chosen in this example, beam shear may well become critical.

Local bond

Although not covered by BS 8110, local bond can be a problem in foundation design, and should therefore be checked at sections with high shear stress. Local bond is given by

$$f_{bs} = \frac{V_u}{\Sigma u_s d}$$

where Σu_s is the sum of the bar perimeters at the section being considered.

Punching shear, $V_u = 2979\,kN$

The length of the punching shear is $u = 8800\,mm$.
T25 bars @ 175 centres each way are proposed. The total number of bars crossing the shear perimeter is $u/175 = 50$. The local bond stress is

$$f_{bs} = \frac{V_u}{\Sigma u_s l_a}$$

where l_a is the lever arm which CP 110 approximates to the effective depth d.

$$f_{bs} = \frac{2979 \times 10^3}{(50 \times \pi \times 25)600}$$
$$= 1.26\,N/mm^2$$

This is well within the allowable value of $4.1\,N/mm^2$ for grade C40 concrete, given by CP 110: Part 1: Table 21 (BS 8110 does not give allowable local bond stresses).

11.3 PAD FOUNDATIONS WITH AXIAL LOADS AND BENDING MOMENTS

There are various ways of dealing with pad foundations which are subject to both axial loads and bending moments (and sometimes horizontal loads as well). The following design examples will explore the various merits of the differing approaches to the design solutions. The designer should keep in mind at all times the various loading combinations which can apply to any one base. It is not always apparent which is the critical load case, and the base design often develops on an iterative basis.

In each of the following design examples the net allowable bearing pressure of the soil will be taken as $n_a = 300\,kN/m^2$.

These examples concentrate on the analysis of foundation bases to limit bearing pressures arising from combinations of vertical loads, horizontal loads, and bending moments. The design of these bases, to resist bending and shear, should be carried out in a similar manner to Design Examples 1–4 earlier in this chapter.

Calculations for bending moments and shear forces within the base will need to make due allowance for the variation in bearing stresses across the base.

11.3.1 Design Example 5: Pad base – axial load plus bending moment (small eccentricity)

A column pad base is subject to an axial load of 200 kN (dead) plus 300 kN (imposed), and a bending moment of 40 kNm. To suit site constraints, the base is limited to a length of $L = 1.8\,m$.

Load eccentricity

When moments act on a foundation, it is normal to replace them by positioning the vertical load at an equivalent eccentricity. The resultant vertical superstructure load is

$$P = G + Q$$
$$= 200 + 300$$
$$= 500\,kN$$

Q as a percentage of P is $100Q/P = (100 \times 300)/500 = 60\%$. From Fig. 11.22, the combined partial factor for superstructure loads is $\gamma_P = 1.52$.
The resultant eccentricity is given by

$$e_P = \frac{M}{P}$$
$$= \frac{40}{500}$$
$$= 0.08\,m$$

Bearing pressure check – design chart approach

A suitable base size can be checked or calculated using design chart H.1 in Appendix H. For the purpose of this example this is reproduced in Fig. 11.26. Assuming a superstructure bearing pressure of $p = n_a = 300\,kN/m^2$,

$$\frac{P}{p} = \frac{500}{300} = 1.67\,m^2$$

Assuming a base length of $L = 1.8\,m$,

$$\frac{e_P}{L} = \frac{0.08}{1.8} = 0.044$$

From Fig. 11.26, this gives a required base area of

$$A = BL = 2.1m^2$$

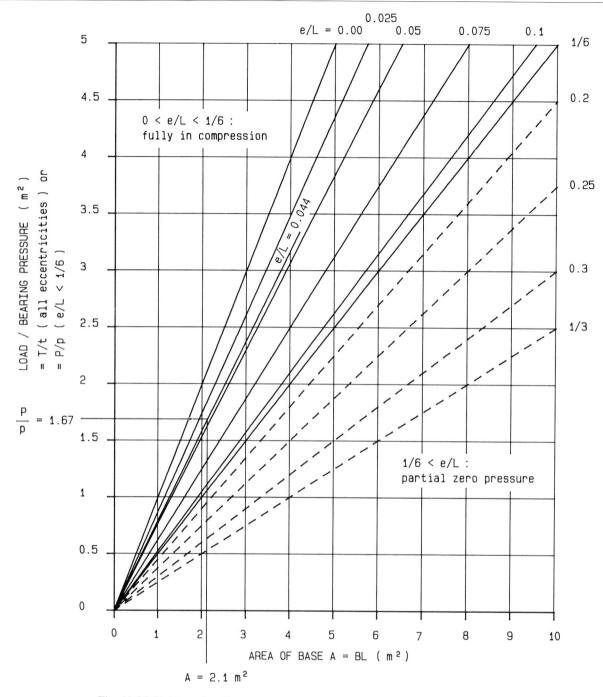

Fig. 11.26 Pad base (small eccentricity) design example – design chart for base size.

Thus

$$B_{min} = \frac{A}{L}$$
$$= \frac{2.1}{1.8}$$
$$= 1.17\,m$$

A width of $B = 1.2\,m$ will be adopted.

Bearing pressure check – calculation approach

The eccentricity $e_P = 0.08\,m$ is less than $L/6 = 1.8/6 = 0.3\,m$, and thus the formation is loaded in compression over the full plan area of the base. Assume a width of $B = 1.2\,m$.

$$p = \frac{P}{A} \pm \frac{M}{Z}$$
$$= \frac{P}{BL} \pm \frac{M}{\left(\dfrac{BL^2}{6}\right)}$$
$$= \frac{500}{1.2 \times 1.8} \pm \frac{40}{\left(\dfrac{1.2 \times 1.8^2}{6}\right)}$$
$$= 231 \pm 62\,kN/m^2$$

Thus $p_{max} = 293\,kN/m^2$ and $p_{min} = 169\,kN/m^2$. These are less than the allowable bearing pressure of $n_a = 300\,kN/m^2$; the width of $B = 1.2\,m$ is therefore satisfactory.

vertical loads
and moments

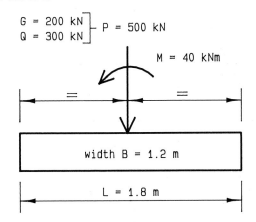

G = 200 kN ⎤
Q = 300 kN ⎦ P = 500 kN

M = 40 kNm

width B = 1.2 m

L = 1.8 m

equivalent vertical
load and eccentricity

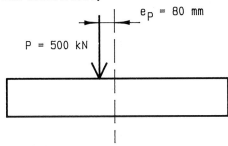

e_P = 80 mm

P = 500 kN

bearing pressure

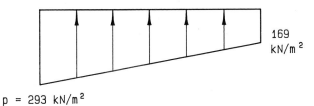

169
kN/m²

p = 293 kN/m²

ultimate design pressure

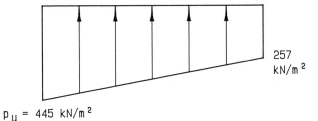

257
kN/m²

p_u = 445 kN/m²

**Fig. 11.27 Pad base (small eccentricity) design example −
loads and bearing pressures.**

Resultant ultimate design pressures

Since the base is fully in compression, ultimate design
pressures, p_u, are obtained by simply factoring up these
pressures using the combined partial safety factor γ_P.

$$p_{u(max)} = \gamma_P p_{max}$$
$$= 1.52 \times 293$$
$$= 445\,kN/m^2$$

$$p_{u(min)} = \gamma_P p_{min}$$
$$= 1.52 \times 169$$
$$= 257\,kN/m^2$$

This is shown in Fig. 11.27.

Effect of offsetting the base

Where the moment always acts in one direction, economies
in the base size can be achieved by positioning the base
eccentric to the vertical load. Thus if the centroid of the
base is offset by $e_P = 0.08$ m, the pressure becomes uniform,
and is simply given by $p = P/A$. This would give

$$A_{req} = \frac{P}{n_a}$$
$$= \frac{500}{300}$$
$$= 1.67\,m^2$$

Compared to $A = 1.8 \times 1.2 = 2.16\,m^2$, this would be a
reduction of 23%. This approach is used in Design Example
8 (section 11.3.4).

11.3.2 Design Example 6: Pad base − axial load plus bending moment (large eccentricity)

A column pad base is subject to an axial load of 100 kN
(dead) plus 100 kN (imposed), and a bending moment of
60 kNm. The bending moment may act in either direction; it
is therefore not possible to reduce the eccentricity by offsetting the base. In addition, site conditions limit the length
of the base to $L = 1.4$ m.

Superstructure loads

The large eccentricity of the applied loading suggests zero
pressure may occur under part of the base. In order to check
this, the bearing pressure calculations should be carried out
in terms of total loads and pressures.

$$\text{Moment, } M = 60\,kNm \text{ (reversible)}$$

of which 75% is due to dead load and 25% is due to imposed
load

$$\text{Superstructure vertical load, } P = G + Q$$
$$= 100 + 100$$
$$= 200\,kN$$

$$\text{Superstructure eccentricity, } e_P = \frac{M}{P}$$
$$= \frac{60}{200}$$
$$= 0.3\,m$$

This is greater than $L/6 = 1.4/6 = 0.23$ m. The resultant load
acts outside of the middle third, indicating zero pressure
over part of the base.

Foundation loads

Foundation distributed load due to the 300 mm deep base
and 200 mm overburden is given by

$$f = \text{average density} \times \text{depth}$$
$$= 20 \times 0.5$$
$$= 10\,\text{kN/m}^2$$

To calculate the foundation load, F, an estimate needs to be made of the base area, A.

An axially loaded foundation has a superstructure bearing pressure of $p = P/A$.

A foundation with the load acting on the edge of the middle third (i.e. $e_P = \frac{1}{6}$) has a superstructure bearing pressure of $p = 2P/A$.

In this case the load is acting outside of the middle third. A reasonable estimate for establishing a trial base size is $p = 2.5P/A$. This gives

$$A = \frac{2.5\,P}{p}$$
$$= \frac{2.5 \times 200}{300}$$
$$= 1.67\,\text{m}^2$$

$$\text{Foundation load, } F = fA$$
$$= 10 \times 1.67$$
$$= 17\,\text{kN}$$

This is all dead load. Thus the partial factor for foundation loads is $\gamma_F = 1.4$.

Total load

Total vertical load, $T = $ superstructure load + foundation load
$$= P + F$$
$$= 200 + 17$$
$$= 217\,\text{kN}$$

$$\text{Total eccentricity, } e_T = \frac{M}{T}$$
$$= \frac{60}{217}$$
$$= 0.277\,\text{m}$$

A check on the eccentricity of the total applied load gives

$$e_T = 0.277\,\text{m} > \frac{L}{6} = \frac{1.4}{6} = 0.23\,\text{m}$$

The total load, T, therefore remains outside of the *middle third*. This confirms that the base is not fully in compression but has zero pressure over part of its length.

Allowable bearing pressure

Net allowable bearing pressure, $n_a = 300\,\text{kN/m}^2$.

The existing overburden pressure, s, is assumed to be approximately equal to $f = 10\,\text{kN/m}^2$.

The allowable bearing pressure will be calculated in terms of total pressures.

From section 10.10,

total allowable bearing = net allowable bearing pressure + pressure, t_a existing surcharge
$$= n_a + s$$

$$= 300 + 10$$
$$= 310\,\text{kN/m}^2$$

Calculation of bearing pressures

The actual stress distribution is triangular, as shown in Fig. 11.28, with the total vertical load, T, located at the one-third point of the pressure diagram for equilibrium. The effective length of the base, L_b (the length over which the bearing stresses occur), is thus given by

$$L_b = 3\left(\frac{L}{2} - e_T\right)$$
$$= 3\left(\frac{1.4}{2} - 0.277\right)$$
$$= 1.27\,\text{m}$$

Since the bearing pressure diagram is triangular, the load T will act at the edge of the middle third of the effective base area $A_b = BL_b$. The maximum total bearing pressure will therefore be

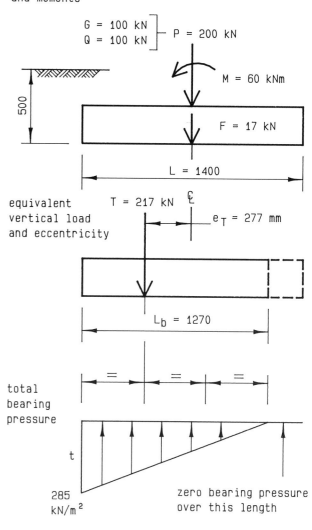

Fig. 11.28 Pad base (large eccentricity) design example – loads and bearing pressures.

$$t_{max} = \frac{2T}{A_b} = \frac{2T}{BL_b}$$

Setting t_{max} equal to the allowable pressure t_a, the equation may be rearranged to give

$$B_{min} = \frac{2T}{L_b t_a}$$

$$= \frac{2 \times 217}{1.27 \times 310}$$

$$= 1.10 \, m$$

A width of $B = 1.2 \, m$ will be adopted. This gives a maximum total bearing pressure of

$$t_{max} = \frac{2T}{BL_b}$$

$$= \frac{2 \times 217}{1.2 \times 1.27}$$

$$= 285 \, kN/m^2$$

Bearing pressure check using design charts

As an alternative, a suitable base size can be checked or calculated using design chart H.1 in Appendix H. For the purpose of this example this is reproduced in Fig. 11.29.

$$\frac{T}{t_a} = \frac{217}{310}$$

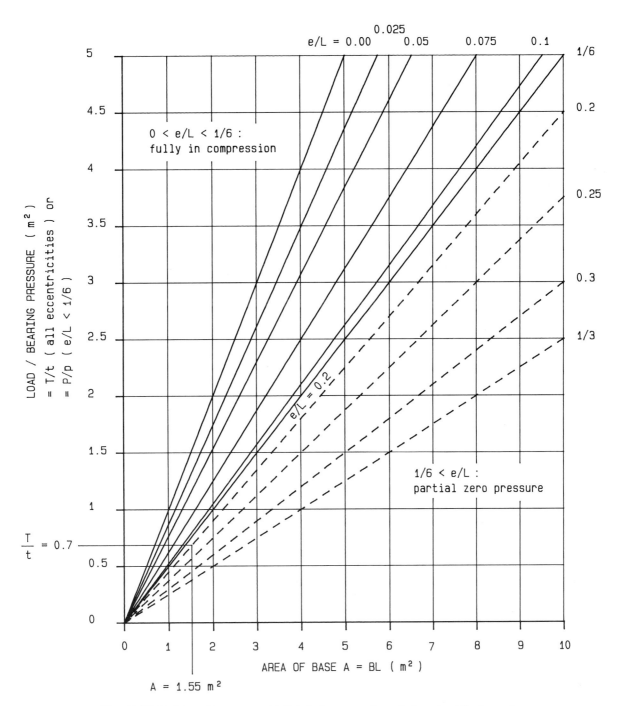

Fig. 11.29 Pad base (large eccentricity) design example — design chart for base size.

$$= 0.70$$

$$\frac{e_T}{L} = \frac{0.277}{1.4}$$

$$= 0.20$$

From the design chart, this gives $A = BL = 1.55\,\text{m}^2$, thus

$$B_{\min} = \frac{A}{L}$$

$$= \frac{1.55}{1.4}$$

$$= 1.11\,\text{m}$$

A width of $B = 1.2\,\text{m}$ will be adopted.

Ultimate loads

Ultimate superstructure load, $P_u = G_u + Q_u$

$$= \gamma_G G + \gamma_Q Q$$

$$= (1.4 \times 100) +$$

$$(1.6 \times 100)$$

$$= 300\,\text{kN}$$

Ultimate foundation load, $F_u = \gamma_F F$

$$= 1.4 \times 17$$

$$= 24\,\text{kN}$$

Total ultimate load, $T_u = P_u + F_u$

$$= 300 + 24$$

$$= 324\,\text{kN}$$

ULTIMATE LIMIT STATE

ultimate applied
loads

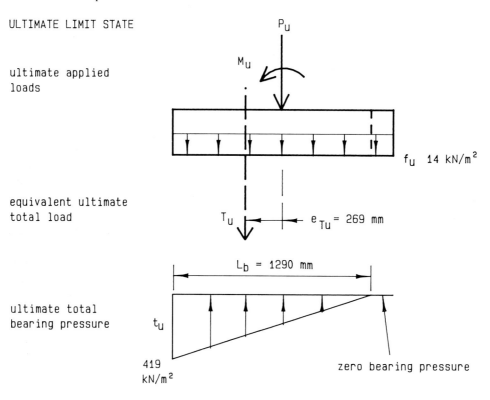

equivalent ultimate
total load

ultimate total
bearing pressure

FOUNDATION MEMBER DESIGN

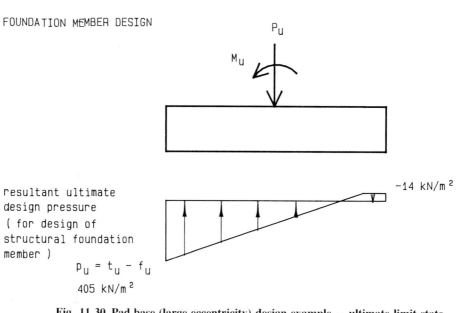

resultant ultimate
design pressure
(for design of
structural foundation
member)

$$p_u = t_u - f_u$$

405 kN/m²

Fig. 11.30 Pad base (large eccentricity) design example – ultimate limit state.

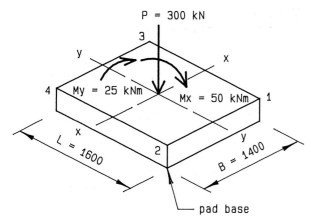

(a) loads and dimensions

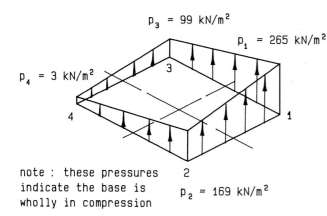

note : these pressures
indicate the base is
wholly in compression

$p_2 = 169 \text{ kN/m}^2$

(b) bearing pressure diagram

Fig. 11.31 Biaxially loaded pad base design example.

As stated previously, 25% of the moment $M = 60\,\text{kNm}$ is due to imposed load. From Fig. 11.22, this gives a combined partial load factor of 1.45. Thus

$$\text{Ultimate moment, } M_u = 1.45 \times 60$$
$$= 87\,\text{kNm}$$

$$\text{Total ultimate eccentricity, } e_{Tu} = \frac{M_u}{T_u}$$
$$= \frac{87}{324}$$
$$= 0.269\,\text{m}$$

(Note this is *not* equal to the working load eccentricity $e_T = 0.23\,\text{m}$ calculated previously.)

Ultimate design pressures
Since there is partial zero pressure below the base, the resultant ultimate design pressure, p_u, for reinforcement design must be obtained, in accordance with section 10.12, by calculating the total ultimate pressure, t_u, and subtracting the foundation ultimate pressure, f_u.

The total ultimate bearing pressure is calculated in a similar manner to the bearing pressure under working loads

$$L_b = 3\left(\frac{L}{2} - e_{Tu}\right)$$
$$= 3\left(\frac{1.4}{2} - 0.279\right)$$
$$= 1.29\,\text{m}$$

For vertical equilibrium, $T_u = t_{u(\text{max})}L_bB/2$, thus

$$t_{u(\text{max})} = \frac{2T_u}{L_bB}$$
$$= \frac{2 \times 324}{1.29 \times 1.2}$$
$$= 417\,\text{kN/m}^2$$

$$p_{u(\text{max})} = t_{u(\text{max})} - f_u$$
$$= t_{u(\text{max})} - \gamma_F f$$
$$= 419 - (1.4 \times 10)$$
$$= 405\,\text{kN/m}^2$$

The actual distribution of p_u is shown in Fig. 11.30.

11.3.3 Design Example 7: Pad base — axial load plus bending moments about both axes

A column pad base is subject to the axial load and biaxial bending moments shown in Fig. 11.31 (a). The bending moments about each axis are reversible, and there is thus no benefit to be gained by offsetting the base relative to the column. The net allowable bearing pressure is again $n_a = 300\,\text{kN/m}^2$.

As a rough guide, in order to select a trial size for a biaxially loaded base, the base area A should be chosen to be between $A = 2.0\,P/p$ and $A = 2.5\,P/p$. In this example this gives

$$A = 2.0\left(\frac{300}{300}\right) \text{ to } 2.5\left(\frac{300}{300}\right)$$
$$= 2.0\,\text{m}^2 \text{ to } 2.5\,\text{m}^2$$

A trial size of $A = 1.6 \times 1.4 = 2.24\,\text{m}^2$ will be checked.

Bearing pressures
Bearing pressures at the corners of the base are calculated in a similar manner to the uniaxial bending case in Design Example 5 (section 11.3.1), taking into account the variation in stress about both axes.

$$p = \frac{P}{A} \pm \frac{M_x}{Z_x} \pm \frac{M_y}{Z_y}$$
$$= \frac{P}{LB} \pm \frac{M_x}{\left(\dfrac{BL^2}{6}\right)} \pm \frac{M_y}{\left(\dfrac{LB^2}{6}\right)}$$
$$= \frac{300}{1.4 \times 1.6} \pm \frac{50}{\left(\dfrac{1.4 \times 1.6^2}{6}\right)} \pm \frac{25}{\left(\dfrac{1.6 \times 1.4^2}{6}\right)}$$
$$= 134 \pm 83 \pm 48\,\text{kN/m}^2$$

$$p_1 = 134 + 83 + 48 = 265\,\text{kN/m}^2 < n_a = 300\,\text{kN/m}^2$$
$$\Rightarrow \text{okay.}$$

$$p_2 = 134 + 83 - 48 = 169 \, \text{kN/m}^2$$
$$p_3 = 134 - 83 + 48 = 99 \, \text{kN/m}^2$$
$$p_4 = 134 - 83 - 48 = 3 \, \text{kN/m}^2$$

The resulting bearing pressure diagram is shown in Fig. 11.31 (b).

11.3.4 Design Example 8: Pad base — axial and horizontal loads

The stanchion of a moment-resisting frame is to be founded on a pad foundation. The stanchion is assumed to be pinned at its base; the vertical loads are $G = 175 \, \text{kN}$ (dead) and $Q = 225 \, \text{kN}$ (imposed). The horizontal thrust due to dead and imposed (working) loads is $H = 50 \, \text{kN}$.

Design approach

In general, for bases where horizontal loads are significant, it is necessary to assume a base size and then check it under combined vertical and horizontal loading, as well as against sliding (and, on occasion, overturning). This may involve a number of *iterations*, to fine-tune the necessary base size.

The designer should first seek ways of cancelling out the horizontal force, e.g. by tying the frame feet together via a tie rod or reinforcement within the slab (see Design Example 1 in section 12.2.4). This example assumes such methods to be impractical in this particular situation.

The loading is from a rigid frame, where in this case the moment from the critical load case (dead + imposed loads) always acts in the same direction. This will be turned to advantage by offsetting the base to cancel out the eccentricity of the applied loads.

Previous iterations have indicated a base size of 1.5 m × 1.5 m is likely to produce an economic answer.

Loadings and eccentricities

$$\text{Superstructure load, } P = G + Q$$
$$= 175 + 225$$
$$= 400 \, \text{kN}$$

The foundation distributed load due to the 1000 mm deep base and 300 mm overburden, assuming an average density of 20 kN/m³, is given by

$$f = 20 \times 1.3$$
$$= 26 \, \text{kN/m}^2$$

$$\text{Foundation total load, } F = fA$$
$$= 26 \times (1.5 \times 1.5)$$
$$= 58 \, \text{kN}$$

The horizontal thrust at the base of the stanchion exerts a moment M_T at the underside of the pad given by

$$M_T = hH$$
$$= 1.0 \times 50$$
$$= 50 \, \text{kNm}$$

$$\text{Resultant total vertical load, } T = P + F$$
$$= 400 + 58$$
$$= 458 \, \text{kN}$$

The corresponding eccentricity, e_T, is given by

$$e_T = \frac{M_T}{T}$$
$$= \frac{50}{458}$$
$$= 0.110 \, \text{m}$$

The centroid of the base will therefore be offset relative to the stanchion by 110 mm, to give a uniform bearing pressure. This is shown in Fig. 11.32.

Allowable bearing pressures

Net allowable bearing pressure, $n_a = 300 \, \text{kN/m}^2$.

The existing overburden pressure, s, is assumed to be approximately equal to $f = 26 \, \text{kN/m}^2$.

As discussed in section 10.11, the total allowable bearing pressures should be used in this situation. From section 10.10,

$$\text{total allowable bearing pressure, } t_a = \text{net allowable bearing}$$
$$\text{pressure} + \text{existing}$$
$$\text{surcharge pressure}$$
$$= n_a + s$$
$$= 300 + 26$$
$$= 326 \, \text{kN/m}^2$$

Vertical bearing pressure

The total vertical bearing pressure is given by

$$t = \frac{T}{BL}$$
$$= \frac{458}{1.5 \times 1.5}$$
$$= 204 \, \text{kN/m}^2$$

This is well below the allowable pressure of $t_a = 326 \, \text{kN/m}^2$, and this base size would therefore be excessively large if it were soley dealing with vertical loads. This margin is however necessary to satisfy the condition for combined vertical and horizontal loading (see below).

It should be noted that the uniform stress condition will only apply if the vertical and horizontal loads from the stanchion continue to act in the same proportion. In practice there will usually be additional load cases (e.g. dead plus wind, dead plus imposed plus wind) which will produce loads of different magnitudes and proportions. If wind loads are significant the bearing pressures for these load cases should also be checked, making due allowance for the 25% increase in allowable bearing pressure under wind loading.

Horizontal resistance to sliding

In this example the base has been cast with the ground acting as permanent shuttering. The ground at the rear of the base is well-compacted granular material forming a road sub-base. Passive resistance may therefore be assumed.

Assume that the soil investigation has indicated a coefficient of friction of $\mu = 0.5$ to be suitable for the base, and a passive lateral pressure coefficient of $K = 2.0$ for the side of the foundation.

The horizontal resistance due to base friction is given by

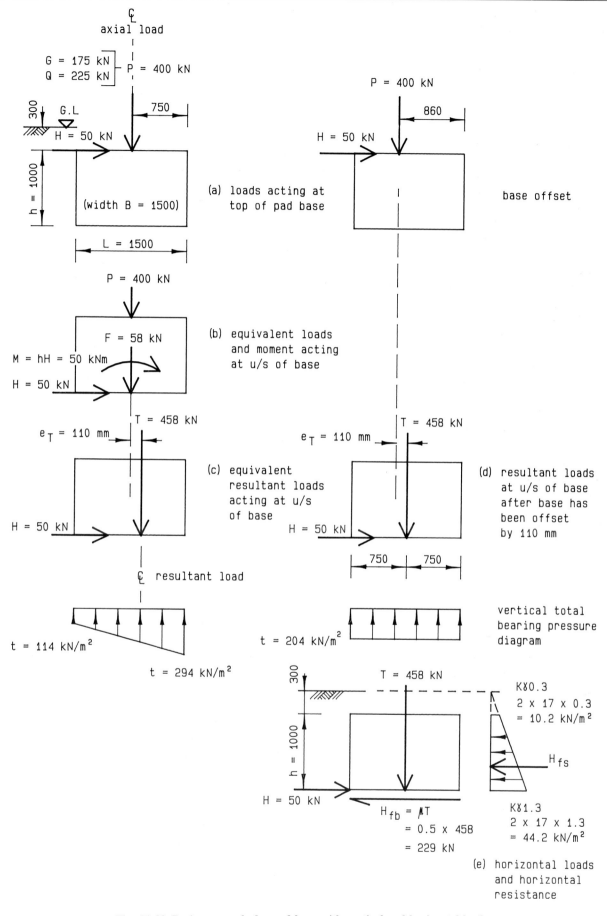

Fig. 11.32 Design example for pad base with vertical and horizontal loads.

$$H_{fb} = \mu T$$
$$= 0.5 \times 458$$
$$= 229 \, \text{kN}$$

From Fig. 11.32 (e), the horizontal resistance due to passive pressure, taking γ as $17 \, \text{kN/m}^2$, is given by

$$H_{fs} = \frac{(10.2 + 44.2) \, hB}{2}$$
$$= 27.2 \times 1.0 \times 1.5$$
$$= 41 \, \text{kN}$$

Total horizontal resistance is

$$H_f = H_{fb} + H_{fs}$$
$$= 229 + 41$$
$$= 270 \, \text{kN}$$

$$\text{Factor of safety against sliding} = \frac{H_f}{H}$$
$$= \frac{270}{50}$$
$$= 5.4$$

A factor of safety of 2.0 is normally adequate, $\Rightarrow$ okay.

With this factor of safety of 2.0, the maximum allowable horizontal load is

$$H_a = \frac{H_f}{\text{factor of safety}}$$
$$= \frac{270}{2.0}$$
$$= 135 \, \text{kN}$$

Check on combined vertical and horizontal loading
From section 1.3.5, the condition which must be satisfied for combined loading is

$$\frac{T}{T_a} + \frac{H}{H_a} < 1$$

which may be rewritten in this case as

$$\frac{t}{t_a} + \frac{H}{H_a} < 1$$
$$\frac{204}{326} + \frac{50}{135} < 1$$
$$0.62 + 0.37 = 0.99 < 1 \quad \Rightarrow \text{okay.}$$

11.3.5 Design Example 9: Shear wall base — vertical loads and horizontal wind loads

The shear wall foundation in Fig. 11.33 is to be designed for the loads shown. The base has an area of $A = BL = 2.0 \, \text{m} \times 8.0 \, \text{m} = 16 \, \text{m}^2$. The net allowable bearing pressure at the level of the foundation is $n_a = 250 \, \text{kN/m}^2$. The foundation is to be cast using the excavated trench sides as a shutter.

Loadings
The vertical superstructure dead and imposed loads are, respectively, $G = 2000 \, \text{kN}$ and $Q = 1600 \, \text{kN}$. The moment

a) loads at ground level

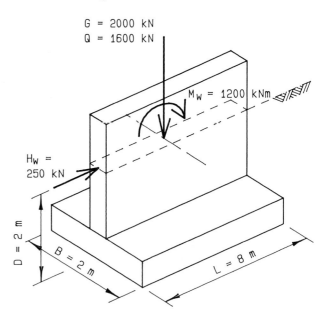

b) loads at underside of foundation level

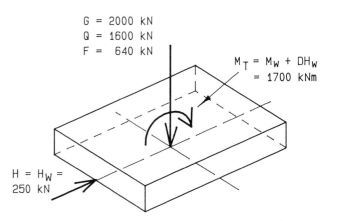

Fig. 11.33 Shear wall base design example — loads.

and horizontal shear at ground level arising from wind loads are $M_W = 1200 \, \text{kNm}$ and $H_W = 250 \, \text{kN}$.

With reference to Fig. 11.33, the moment M at level of underside of foundation — is given by

$$M_T = M_W + DH_W$$
$$= 1200 + (2.0 \times 250)$$
$$= 1700 \, \text{kNm}$$

Individual bearing pressure components

Bearing pressure due to dead loads, $g = \dfrac{G}{A}$

$$= \frac{2000}{16}$$
$$= 125 \, \text{kN/m}^2$$

Bearing pressure due to imposed loads, $q = \dfrac{Q}{A}$

(a) BEARING PRESSURES

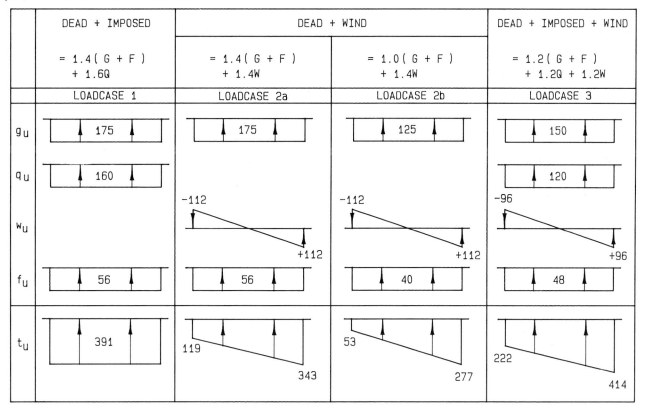

(b) ULTIMATE BEARING PRESSURES

(c) RESULTANT PRESSURE FOR FOUNDATION MEMBER DESIGN

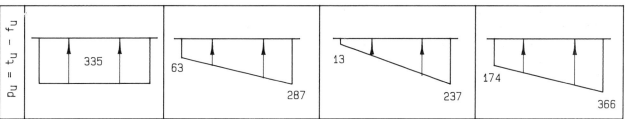

Fig. 11.34 Shear wall base design example − pressures.

$$= \frac{1600}{16}$$

$$= 100 \, \text{kN/m}^2$$

Bearing pressure due to wind loads, $w = \pm \dfrac{M_T}{Z}$

$$= \pm \frac{M_T}{\left(\dfrac{BL^2}{6}\right)}$$

$$= \pm \frac{1700}{\left(\dfrac{2.0 \times 8.0^2}{6}\right)}$$

$$= \pm 80 \, \text{kN/m}^2$$

Assuming an average density of $20 \, \text{kN/m}^3$, the bearing pressure due to weight of foundation and backfill is

$$f = \text{density} \times \text{depth}$$
$$= 20 \times 2.0$$
$$= 40 \, \text{kN/m}^2$$

Allowable bearing pressure
Net allowable bearing pressure, $n_a = 250 \, \text{kN/m}^2$.
Bearing pressure due to existing overburden s is taken as approximately equal to that due to the new foundation and backfill, i.e.

$$s = f = 40 \, \text{kN/m}^2$$

As discussed in section 10.11, the total allowable bearing pressure should be used in this situation. From section 10.9,

total allowable bearing pressure, t_a = net allowable bearing pressure + existing surcharge pressure
$$= n_a + s$$
$$= 250 + 40$$
$$= 290 \, \text{kN/m}^2$$

Total allowable bearing pressure under wind loading = $1.25 t_a = 363 \, \text{kN/m}^2$.

Bearing pressure check
Bearing pressures for the different load cases are given in Fig. 11.34 (a).
Total bearing pressure under dead loads + imposed loads, t_1, is given by

$$t_1 = (g + q) + f$$
$$= (125 + 100) + 40$$
$$= 265 \, \text{kN/m}^2$$

Total bearing pressure under vertical loads + wind load is t_3, where

$$t_3 = (g + q + w) + f$$
$$= (125 + 100 + 80) + 40$$
$$= 345 \, \text{kN/m}^2$$

If t_3 is greater than $1.25 t_1$, then wind load is critical.

$$t_3 = 345 \, \text{kN/m}^2 > 1.25 t_1 = 1.25 \times 265 = 331 \, \text{kN/m}^2$$

Thus wind load is critical in this example.
The allowable bearing pressure under wind loading is

$$1.25 t_a = 363 \, \text{kN/m}^2 > t_3 = 345 \, \text{kN/m}^2 \quad \Rightarrow \text{okay.}$$

Horizontal load resistance
Because the foundation is to be cast using the trench sides as a shutter, it is considered that the horizontal force $H_W = 250 \, \text{kN}$ will be resisted by a combination of passive pressure and friction. In addition, since the proportion of horizontal to vertical loading is small, it is felt that the unity factor need not be separately calculated. This contrasts with Design Example 8 (see section 11.3.4) where a relatively lightly loaded foundation with a large horizontal load was located at a shallow depth, and there was therefore a need to check the foundation for combined vertical and horizontal loading.

Ultimate design pressures
When wind loading is critical, foundations need to be designed for the standard load cases in Table 10.4, i.e. as in Table 11.2.

Table 11.2 Load cases for the ultimate limit state

Load case	Load combination
1	1.4 (dead) + 1.6 (imposed)
2a	1.4 (dead) + 1.4 (wind)
2b	1.0 (dead) + 1.4 (wind)
3	1.2 (dead) + 1.2 (imposed) + 1.2(wind)

The ultimate bearing pressures for these different load cases are shown in Fig. 11.34(b).

For each load case, the individual pressure components are first summated to give the total ultimate bearing pressure $t_u = g_u + q_u + w_u + f_u$. In this example this indicates that the underside of the foundation remains fully in compression under all ultimate load combinations.

The ultimate design pressure $p_u = t_u - f_u$ is then calculated for each load case, for design of foundation members, and is shown on Fig. 11.34 (c). This shows that the maximum ultimate design pressure is $366 \, \text{kN/m}^2$ and that in this instance there is no negative design pressure (the base being in compression under all combinations).

11.4 RECTANGULAR AND TEE-BEAM CONTINUOUS STRIPS

11.4.1 Introduction
Rectangular beam strips are briefly discussed in section 9.3.6 and the inverted T-beam strip in section 9.3.7 where it is mentioned that the main difference in the two beam foundations relates to the relationship between the width of beam required to resist bending moments and shear forces and that required to achieve the allowable bearing pressures.

If the two widths are similar then the rectangular beam tends to be economic. However, on relatively poor-quality sub-strata the beam width required to achieve the allowable bearing pressures often far exceeds that required for bending and shear resistance. In the latter case it tends to prove economic to reduce the beam width and spread the load through a flange slab on the soffit of the beam.

11.4.2 Design decisions

The economic design of continuous beam strips can be greatly affected by the choice of curtailment of the lengths of beams (see section A in Chapter 10 for further information).

They are generally used where longitudinal bending moments are a major problem for the foundation design, i.e. in variable ground, soft sub-strata, or where loading is variable in the length of the beam. They are also used in some areas of mining activity etc., where bending from differential subsidence movement is critical but where tensile and compressive ground strains in the foundation can be controlled.

The decision to use a continuous beam strip usually follows the need to

(1) Reduce differential settlements below framework columns.
(2) Combine foundations which would otherwise tend to overlap.
(3) Ease construction by the use of continuous strips rather than separate pads when they are becoming closely spaced.

The decision to use an inverted T rather than a simple rectangular beam would result from bearing pressure criteria demanding excessive beam widths for bearing when compared to widths required to resist bending and shear.

11.4.3 Sizing of the design

The sizing of the design for the rectangular beam is similar to the sizing of the upstand beam of the inverted T, i.e. based mainly upon bending moments and shear forces. However, the beam width must in this case satisfy that required for allowable bearing pressure criteria for the full contact area of the beam.

For inverted T beams the bearing stresses are reduced to an acceptable amount by the use of the ground strip forming the flange of the inverted T.

The main rib of the T beam is then determined from the design requirements for longitudinal bending and shear forces keeping a reasonably standard profile for shutter reuse to make the beam economic. The flange thickness and reinforcement is determined from the bending moments and shear forces acting on the cantilever slab/flange (see Fig. 11.35).

Casting of the beam is usually carried out in two lifts, the ground slab being cast up to the top of the flange leaving a roughened surface with the cage reinforcement projecting. The main beam is then cast on the cleaned surface of the slab up to its top level (see Fig. 11.36).

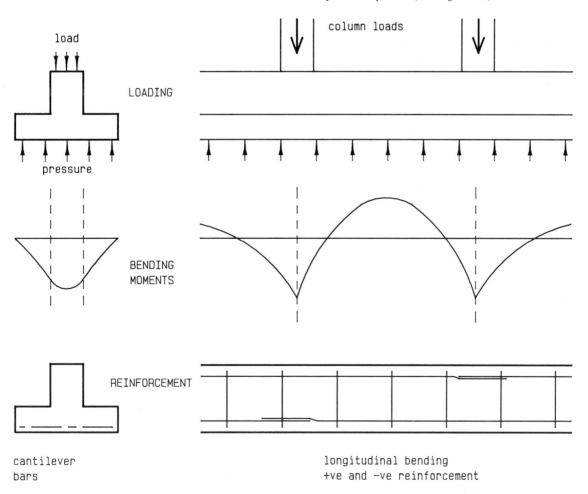

Fig. 11.35 Inverted Tee beam − typical loads, moments and reinforcement.

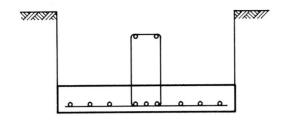

1st stage
ground slab cast

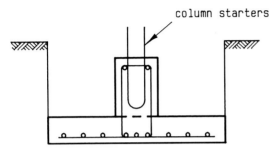

2nd stage
ground beam completed

Fig. 11.36 Typical casting stages for inverted Tee beam.

11.4.4 Design Example 10: Continuous Tee beam footing with uniform bearing pressure

A continuous beam foundation is required to carry the three column loads shown in Fig. 11.37 (a), on poor soil with a net allowable bearing pressure of $n_a = 35\,kN/m^2$. To keep bearing pressures within this limit, a wide flange will be introduced at the bottom, forming an inverted Tee beam.

Size of footing
To minimize differential settlements, the length of the beam has been chosen so that the resultant of the three applied loads falls in the middle of the beam, i.e. there is a uniform bearing pressure under working loads. The superstructure total load is given by

$$\Sigma P = P_B + P_C + P_D$$
$$= (G_B + Q_B) + (G_C + Q_C) + (G_D + Q_D)$$
$$= (200 + 300) + (200 + 300) + (175 + 75)$$
$$= 1250\,kN$$

giving a uniform loading on the foundation of $1250/25 = 50\,kN/m$ run.

$$\text{Minimum width of footing} = \frac{\Sigma P}{n_a L}$$
$$= \frac{1250}{35 \times 25.0}$$
$$= 1.43\,m$$

A footing size of 600 mm wide by 800 mm deep, with a flange 1500 mm wide and 150 mm deep, will be assumed for reinforcement design (see Fig. 11.37 (b)).

Ultimate loads and reactions
Factored ultimate loads become

$$P_{Bu} = 1.4G_B + 1.6Q_B$$
$$= (1.4 \times 200) + (1.6 \times 300)$$
$$= 760\,kN$$

Similarly

$$P_{Cu} = (1.4 \times 200) + (1.6 \times 300) = 760\,kN$$
$$P_{Du} = (1.4 \times 175) + (1.6 \times 75) = 365\,kN$$

$$\Sigma P_u = 760 + 760 + 365 = 1885\,kN$$

Unless dead and imposed loads are in the same proportion for all applied loads – rarely the case – the process of factoring up the loads to ultimate values can cause the resultant of the applied loads to move off the centreline of the footing, and the uniform bearing pressure diagram then changes into a trapezoidal diagram. This will be checked as follows.

Distance of centreline of applied loads from A is given by

$$\bar{x}\,\Sigma P_u = L_{AB}P_{Bu} + L_{AC}P_{Cu} + L_{AD}P_{Du}$$
$$\bar{x} = \frac{[(4.5 \times 760) + (14.5 \times 760) + (24.5 \times 365)]}{1885}$$
$$\bar{x} = 12.4\,m$$
$$\therefore e_{Pu} = 12.5 - 12.4 = 0.1\,m$$

Maximum and minimum ground reactions are given by

$$p_{u(max)}, p_{u(min)} = \frac{\Sigma P_u}{A} \pm \frac{e_{Pu}\,\Sigma P_u}{Z}$$
$$= \frac{1885}{25 \times 1.5} \pm \frac{0.1 \times 1885}{\left(25^2 \times \dfrac{1.5}{6}\right)}$$
$$= 50.2 \pm 1.2$$
$$= 51.4 \quad \text{or} \quad 49.0\,kN/m^2$$

For a genuinely trapezoidal bearing pressure distribution, the shear forces and bending moments would need to be calculated from the ultimate loads and bearing reaction, taking due account of the trapezoidal shape of the pressure diagram. In this case however for all practical purposes these values are near enough equal, and an experienced engineer would carry out the design using an average UDL of $p_u = 50.2\,kN/m^2$ or $50.2 \times 1.5 = 75.3\,kN/m$ run. The ultimate loading diagram, and approximated ultimate bearing reaction, are shown in Fig. 11.38(a) and (b), respectively.

Working load shear forces and bending moments
For a *suspended* continuous beam, only the span loading is initially known. Moments, shear forces and support reactions are derived via moment distribution or some other equivalent elastic analysis, using the relative stiffnesses of the various spans.

It is tempting to invert a continuous foundation beam and analyse it in a similar manner – especially if a continuous beam computer program is at hand – but this would be incorrect. In this situation the *support reactions* are already

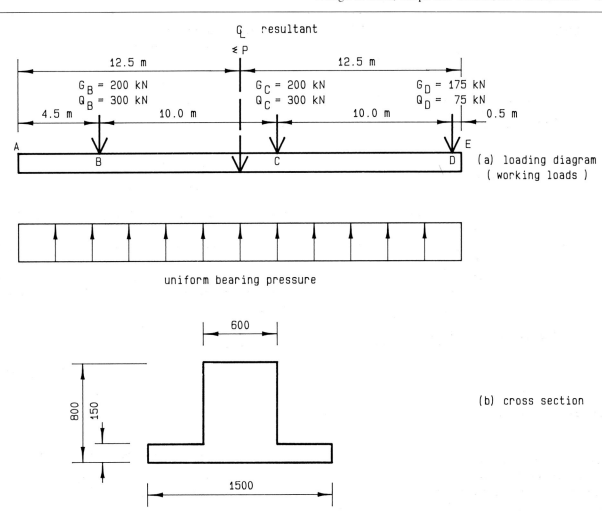

Fig. 11.37 Continuous beam with uniform pressure design example – working loads.

known, and the shear force and bending moment diagrams are derived from simple statics without taking into account the relative stiffnesses of the different spans. Shear forces are obtained by resolving forces vertically at column locations; the bending moments are then equal to the area of the shear force diagram.

If the ratio of dead load to imposed load is the same on all columns then it is acceptable to calculate the design moments directly from the factored loading. In this case however the load ratio is not the same on each column, and to calculate the moments from the approximated factored UDL will produce significant errors. The simplified UDL approach does work satisfactorily if the shear forces and bending moments are first calculated using working loads, and then the load factors applied to the results.

Working load shear forces are calculated from left to right, using simple vertical equilibrium.

$$V_{BA} = pL_{AB} = 50 \times 4.5 = 225\,\text{kN}$$
$$V_{BC} = V_{BA} - P_B = 225 - 500 = -275\,\text{kN}$$
$$V_{CB} = V_{BC} + pL_{BC} = -275 + (50 \times 10.0) = 225\,\text{kN}$$
$$V_{CD} = V_{CB} - P_C = 225 - 500 = -275\,\text{kN}$$
$$V_{DC} = V_{CD} + pL_{CD} = -275 + (50 \times 10.0) = 225\,\text{kN}$$
$$V_{DE} = V_{DC} - P_D = 225 - 250 = -25\,\text{kN}$$

The shear force diagram has been plotted in Fig. 11.38 (b).

Working load bending moments will be calculated from the area of the shear force diagram in Fig. 11.38 (b), again working from left to right.

$$M_B = 4.5 \times \frac{225}{2} = 506\,\text{kNm}$$

$$M_{BC} = M_B + 5.5\frac{(-275)}{2} = 506 - 756 = -250\,\text{kNm}$$

$$M_C = M_{BC} + 4.5\frac{(225)}{2} = -250 + 506 = 256\,\text{kNm}$$

$$M_{CD} = M_C + 5.5\frac{(-275)}{2} = 256 - 756 = -500\,\text{kNm}$$

$$M_D = M_{CD} + 4.5\frac{(225)}{2} = -500 + 506 = 6\,\text{kNm}$$

Bending moments are plotted in Fig. 11.38 (b).

Ultimate shear forces and bending moments
Since both the working loads and ultimate loads are taken as producing a uniform pressure distribution, ultimate shears and moments can be obtained by simply factoring up the working load shears and moments.

The superstructure working load and ultimate load have previously been calculated as $\Sigma P = 1250\,\text{kN}$ and $\Sigma P_u =$

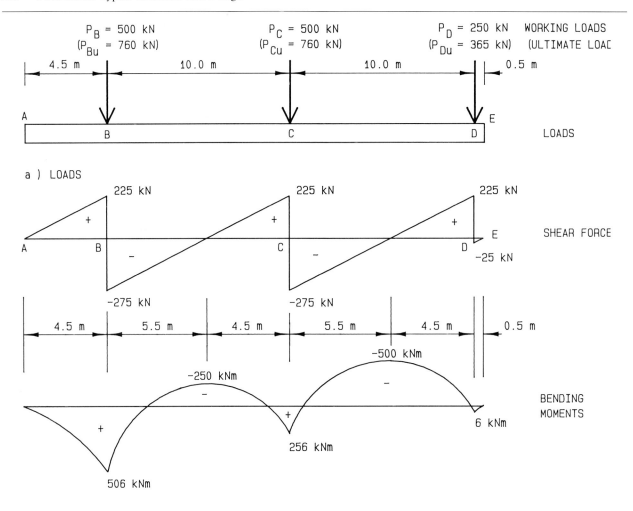

a) LOADS

b) WORKING LOAD SHEARS AND MOMENTS

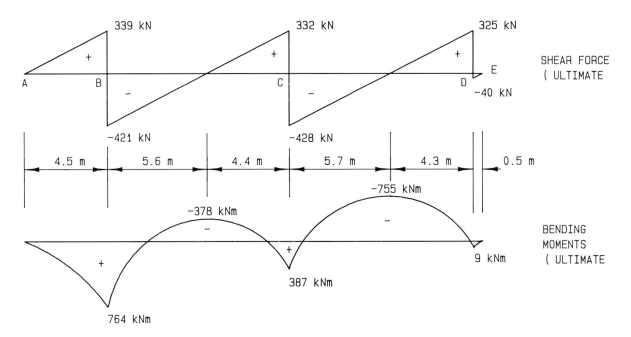

c) ULTIMATE SHEARS AND MOMENTS

Fig. 11.38 Continuous beam with uniform pressure design example — shears and moments.

1885 kN respectively. All working shears and moments should therefore be factored up by

$$\gamma_P = \frac{\Sigma P_u}{\Sigma P} = \frac{1885}{1250} = 1.51$$

This factor has been applied to the shear forces and bending moments in Fig. 11.38 (b), and the resulting ultimate shears and moments are plotted in Fig. 11.38 (c).

Longitudinal bending and shear reinforcement

This example will only look at the reinforcement needed to satisfy the maximum values of bending moment and shear force, in order to confirm that the concrete section size is satisfactory. In a full design, moments and shears along the length of the beam would be considered, and bending reinforcement and shear links would be curtailed to suit.

$$\text{Effective depth, } d = 800 - 40(\text{cover}) - 12 - \frac{25}{2}$$

$$= 735 \, \text{mm}$$

Maximum ultimate moment $= M_{B_u} = 764 \, \text{kNm}$

$$\frac{M_{Bu}}{bd^2} = \frac{764 \times 10^6}{600 \times 735^2}$$

$$= 2.36$$

(*Note*: as the flange is in tension the rectangular section only is used at this location.)

$$A_{s(req)} = 0.65\% \, bd \qquad \text{[BS 8110: Part 3: Chart 2]}$$

$$= \frac{0.65}{100} \times 600 \times 735$$

$$= 2867 \, \text{mm}^2/\text{m}$$

Provide 6 T25 bars $= 2945 \, \text{mm}^2$

$$= \frac{2945 \times 100}{600 \times 735}$$

$$= 0.67\% \, bd$$

$$v_c = 0.60 \, \text{N/mm}^2 \qquad \text{[BS 8110: Part 1: Table 3.9]}$$

$$V_{u(max)} = V_{CDu} = 415 \, \text{kN}$$

$$v_u = \frac{V_{CDu}}{bd} = \frac{415 \times 10^3}{600 \times 735}$$

$$= 0.94 \, \text{N/mm}^2$$

$$v_u - v_c = 0.94 - 0.60 \qquad \text{[BS 8110: Part 1: Table 3.8]}$$

$$= 0.34$$

$$< 0.4 \, \text{N/mm}^2 \quad \Rightarrow \text{use nominal links throughout.}$$

$$A_{sv(req)} = \frac{0.4 \, bs_v}{0.87 \, f_{yv}}$$

using $2 \times \text{T12 legs} = 226 \, \text{mm}^2$,

$$s_v = \frac{226 \times 0.87 \times 460}{0.4 \times 600}$$

$$= 377 \, \text{mm}$$

Use 2 legs of T12 @ 375 centres $= 754 \, \text{mm}^2/\text{m}$ (see Fig. 11.39).

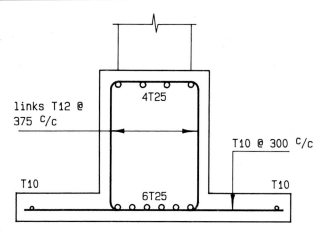

Fig. 11.39 Continuous beam with uniform pressure design example − reinforcement details.

Local bond

Local bond is given by

$$f_{bs} = \frac{V_u}{\Sigma u_s l_a}$$

where Σu_s = sum of the bar perimeters at the section being considered

l_a = lever arm, which CP 110 approximates to the effective depth, d.

Shear force is $V_u = 415 \, \text{kN}$.

The main steel is 6T25 bars.

The local bond stress is

$$f_{bs} = \frac{V_u}{\Sigma u_s \, d}$$

$$= \frac{415 \times 10^3}{(6 \times \pi \times 25)735}$$

$$= 1.20 \, \text{N/mm}^2$$

This is well within the allowable value of 3.75 N/mm² for grade C35 concrete, given by CP 110: Part 1: Table 21 (BS 8110 does not give allowable local bond stresses).

Lateral reinforcement in flange

The bottom flange should in theory be designed to cantilever 450 mm beyond the main 600 mm × 800 mm beam. For this short cantilever span, subjected to the low level of bearing pressure in this example, the resulting reinforcement is expected to be nominal.

11.4.5 Design Example 11: Continuous rectangular beam footing with trapezoidal bearing pressure

A concrete framed building has columns at 5 m centres, with a heavily loaded column located adjacent to the site boundary. The net allowable bearing pressure is $n_a = 400 \, \text{kN/m}^2$. To keep bearing pressures within this limit, a continuous rectangular beam footing will be used, as shown in Fig. 11.40 (a).

It is assumed that site constraints preclude the alternative solution of a trapezoidal balanced foundation − see sections 12.3.3 and 12.3.6.

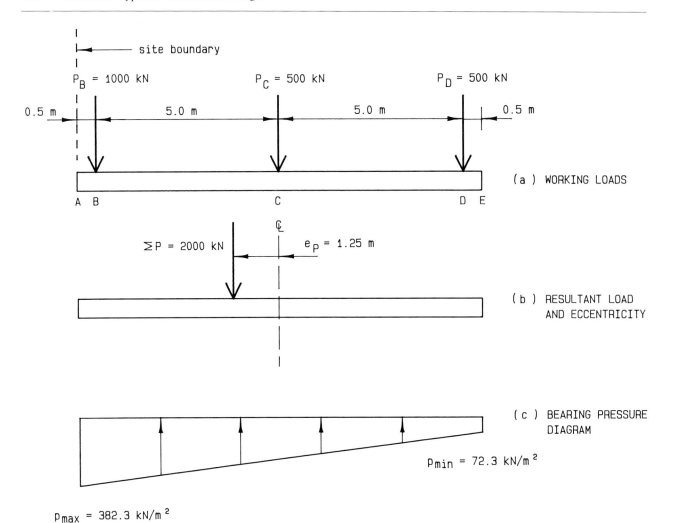

Fig. 11.40 Continuous beam with trapezoidal pressure design example — working loads.

Bearing pressures

The superstructure total load is

$$\Sigma P = P_B + P_C + P_D$$
$$= 1000 + 500 + 500$$
$$= 2000\,\text{kN}$$

Taking moments about the beam centreline, the corresponding eccentricity, e_P, is given by

$$e_P\,\Sigma P = 5.0 P_B - 5.0 P_D$$
$$e_P = \frac{5.0 \times 1000 - 5.0 \times 500}{2000}$$
$$= 1.25\,\text{m}$$

Checking that this is within the middle third:

$$\frac{L}{6} = \frac{11.0}{6} = 1.83\,\text{m} > 1.25\,\text{m}$$

It is, thus the base is fully in compression.
Maximum and minimum bearing pressures are

$$p_{max},\, p_{min} = \frac{\Sigma P}{A} \pm \frac{e_P \Sigma P}{Z}$$

$$= \frac{1}{B}\left(\frac{\Sigma P}{L} \pm \frac{e_P \Sigma P}{(L^2/6)}\right)$$

$$= \frac{1}{B}\left(\frac{2000}{11} \pm \frac{1.25 \times 2000}{(11^2/6)}\right)$$

$$= \frac{1}{B}(181.8 \pm 124.0)$$

$$= \frac{305.8}{B},\, \frac{57.8}{B}$$

For $p_{max} \not> n_a = 400\,\text{kN/m}^2$

$$B = \frac{305.8}{400} = 0.76\,\text{m}$$

A foundation width of $B = 800\,\text{mm}$ will be chosen. This gives

$$p_{max},\, p_{min} = \frac{1}{0.8}(305.8,\, 57.8)$$

$$= 382.3\,\text{kN/m}^2,\, 72.3\,\text{kN/m}^2$$

This pressure distribution is shown in Fig. 11.40 (c).

Ultimate loads and reactions

In this particular example the imposed load is assumed to make up 50% of the superstructure load. Ultimate loads are thus obtained by multiplying working loads by a combined partial load factor of $\gamma_P = 1.5$. The resulting ultimate loads are shown in Fig. 11.42 (a).

Provided this 50% of level of imposed load applies to all columns, ultimate bearing pressures are similarly obtained by factoring the working bearing pressures by 1.5. These have been multiplied by the beam width, $B = 800$ mm, and shown as bearing reactions per unit length in Fig. 11.42 (b).

To reduce the number of subscripts, the u subscript for ultimate loads has been dropped from the ultimate reactions, shears, and moments in the remainder of this example.

Calculation of shears and moments for a trapezoidal bearing pressure distribution

Shear forces are simply calculated by taking vertical equilibrium at any point along the beam. This can be done directly from the loads and reactions (Fig. 11.42 (a) and 11.42 (b)), or by means of the equivalent formulae in Fig. 11.41 (c).

Bending moments at a cross-section are equal to the area of the shear force diagram to one side of the section. These can be determined by calculating these areas in a similar manner to Design Example 10 in section 11.4.4, but taking due account of the curved shape of the shear force diagram arising from the trapezoidal pressures. Alternatively they can be calculated using the equivalent formulae in Fig. 11.41 (d).

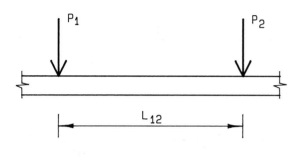

(a) TYPICAL SPAN SHOWING COLUMN LOADS

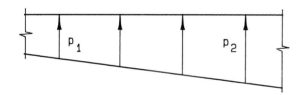

(b) TRAPEZOIDAL BEARING REACTION DIAGRAM

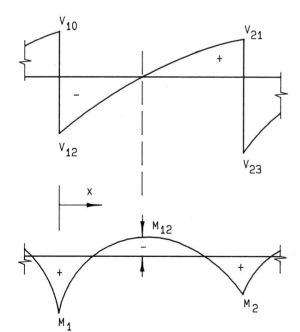

(c) SHEAR FORCE DIAGRAM

$$V_{12} = V_{10} - P_1$$

$$V_{21} = V_{12} + \frac{(p_1 + p_2)L_{12}}{2}$$

$$V_x = V_{12} + p_1 x + \frac{(p_2 - p_1)x^2}{2L_{12}}$$

(d) BENDING MOMENT DIAGRAM

at a distance x from '1'

$$M(x) = M_1 + V_{12}x + \frac{p_1 x^2}{2} + \frac{(p_2 - p_1)x^3}{6L_{12}}$$

$$M_2 = M_1 + V_{12}L_{12} + \frac{(2p_1 + p_2)L_{12}^2}{6}$$

Fig. 11.41 Continuous beam with trapezoidal bearing pressure − formulae for shears and moments.

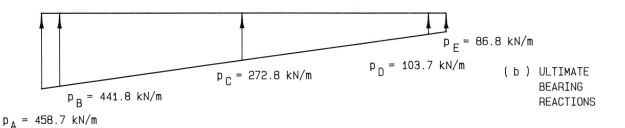

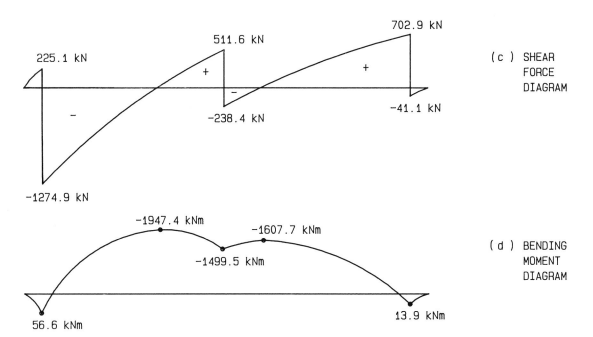

Fig. 11.42 Continuous beam with trapezoidal pressure design example — ultimate loads.

Ultimate shear forces

Shear forces will be calculated from left to right, using the formulae in Fig. 11.41 (c).

$$V_{AB} = 0$$

$$V_{BA} = V_{AB} + \frac{(p_A + p_B)L_{AB}}{2}$$

$$= 0 + \frac{(458.7 + 441.8)\,0.5}{2} \qquad = 225.1\,\text{kN}$$

$$V_{BC} = V_{BA} - P_B$$
$$= 225.1 - 1500 \qquad\qquad = -1274.9\,\text{kN}$$

$$V_{CB} = V_{BC} + \frac{(p_B + p_C)L_{BC}}{2}$$

$$= -1274.9 + \frac{(441.8 + 272.8)\,5.0}{2} = 511.6\,\text{kN}$$

$$V_{CD} = V_{CB} - P_C$$
$$= 511.6 - 750 \qquad\qquad = -238.4\,\text{kN}$$

$$V_{DC} = V_{CD} + \frac{(p_C + p_D)L_{CD}}{2}$$

$$= -238.4 + \frac{(272.8 + 103.7)\,5.0}{2} = 702.9\,\text{kN}$$

$$V_{DE} = V_{DC} - P_D$$
$$= 702.9 - 750 \qquad\qquad = -47.1\,\text{kN}$$

Finally, as a check that $V_{ED} = 0$,

$$V_{ED} = V_{DE} + \frac{(p_D + p_E)L_{DE}}{2}$$

$$= -47.1 + \frac{(103.7 + 86.8)\,0.5}{2} = 0.5\,\text{kN} \Rightarrow \text{near enough.}$$

The resulting shear force diagram is plotted in Fig. 11.42 (c).

Ultimate bending moments

Bending moments at column positions will be calculated, from left to right, using the formula in Fig. 11.41 (d).

$$M_B = M_A + V_{AB}L_{AB} + \frac{(2p_A + p_B)L_{AB}^2}{6}$$

$$= 0 + 0 + \frac{(2 \times 458.7 + 441.8)\,0.5^2}{6}$$

$$= 56.6\,\text{kNm}$$

$$M_C = M_B + V_{BC}L_{BC} + \frac{(2p_B + p_C)L_{BC}^2}{6}$$

$$= 56.6 - 1274.9 \times 5.0 + \frac{(2 \times 441.8 + 272.8)\,5.0^2}{6}$$

$$= -1499.5\,\text{kNm}$$

$$M_D = M_C + V_{CD}L_{CD} + \frac{(2p_C + p_D)L_{CD}^2}{6}$$

$$= -1499.5 - 238.4 \times 5.0 + \frac{(2 \times 272.8 + 103.7)\,5.0^2}{6}$$

$$= 13.9\,\text{kNm}$$

Finally, as a check that $M_E = 0$,

$$M_E = M_D + V_{DE}L_{DE} + \frac{(2p_D + p_E)L_{DE}^2}{6}$$

$$= 13.9 - 47.1 \times 0.5 + \frac{(2 \times 103.7 + 86.8)^2}{6}$$

$$= 2.6\,\text{kNm}$$

This is close to zero when compared with the maximum moment. Reworking the example using an additional decimal place of accuracy would give a value closer to zero.

Maximum 'mid-span' moments are obtained by calculating the point of zero shear from the formula in Fig. 11.41 (c) and calculating the moment at this point using the formula in Fig. 11.41 (d).

For span BC, the point of zero shear is given by

$$V_x = 0 = V_{BC} + P_B x + \frac{(p_{CB} - p_{BC})x^2}{2L_{BC}}$$

$$0 = -1274.9 + 441.8x + \frac{(272.8 - 441.8)x^2}{2 \times 5}$$

$$0 = -16.9x^2 + 441.8x - 1274.9$$

hence $x = 3.3\,\text{m}$ (from the quadratic formula).
For span BC, the maximum moment is found to occur at $x = 3.3\,\text{m}$.

$$M_{BC} = M_B + V_{BC}x + \frac{p_B x^2}{2} + \frac{(p_C - p_B)x^3}{6L_{BC}}$$

$$= 56.6 - 1274.9 \times 3.3 + \frac{441.8 \times 3.3^2}{2} + \frac{(272.8 - 441.8)\,3.3^3}{6 \times 5}$$

$$= -1947.4\,\text{kNm}$$

For span CD, the maximum moment is found to occur for $x = 0.9\,\text{m}$.

$$M_{CD} = M_C + V_{CD}x + \frac{p_C x^2}{2} + \frac{(p_C - p_B)x^3}{6L_{CD}}$$

$$= -1499.5 - 238.4 \times 0.9 + \frac{272.8 \times 0.9^2}{2} + \frac{(103.7 - 272.8)\,0.9^3}{6 \times 5}$$

$$= -1607.7\,\text{kNm}$$

The resulting bending moment diagram is plotted in Fig. 11.42 (d).

Reinforcement

Having calculated the ultimate shear forces and bending moments, a suitable beam depth should be chosen, and bending and shear reinforcement calculated in accordance with BS 8110 (see, for example, Design Example 10 in section 11.4.4).

11.5 GRILLAGE FOUNDATIONS

11.5.1 Introduction

A brief description of the use of grillage foundations is given in Chapter 9 (section 9.6.5) where their use for temporary foundations is discussed together with durability requirements for more permanent use.

11.5.2 Design decisions

As discussed in section 9.6.5 the decision to use a grillage could result from

(1) The need to support very heavy point loads, and/or
(2) To provide a temporary foundation which allows the possibility of simple reuse.

The use of a grillage for temporary bridge works supports is probably one of the most common modern uses for grillage foundations. They are also often encountered as column bases within older existing steel framed buildings.

11.5.3 Sizing of the design

Since the main economic use for grillage bases involves heavy loads the need to provide adequate shear and bending resistance tends to be the major criteria for design. In addition to providing shear resistance it is sometimes necessary to stiffen up the webs of steel beams if concrete surrounds are not being used. The size of the base in terms of plan area will, unless eccentric loads and/or moments are applied, be dependent upon P/n_a as previously shown for the other pad foundations. If however, bending moments or eccentric loads are applied to the foundation an effective eccentricity of the foundation below the stanchion is desirable, the eccentricity of the foundation being made to coincide with that of the applied loadings. By this method an axial/symmetric (assumed uniform) pressure below the base can be achieved (see Fig. 11.43). A check should be made to ensure that other load combinations, for example, the condition of vertical load and maximum bending moments, are adequately catered for within the detail.

The beams within the grillage will generally consist of two layers at right angles positioned below the main steel base plate of the stanchion. An increased number of beam layers would only be adopted if resulting design produced excessively large beam sections to accommodate the resulting stresses. When using steel beams the sizing of the sections can be roughly produced by reference to safe load tables for allowable shear and bending moments. A more accurate analysis for the final design can then be carried out based upon a sketch layout of these preliminary sizes.

11.5.4 Design Example 12: Grillage foundation

A steel grillage foundation has been chosen to provide temporary support during bridge construction. The foundation is required to support a maximum axial load of $P = 1200\,\text{kN}$, of which 25% is imposed load. The soil has an allowable bearing pressure of $n_a = 100\,\text{kN/m}^2$.

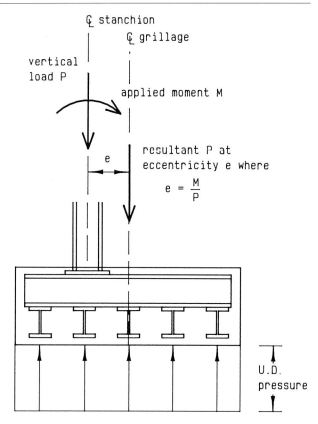

Fig. 11.43 Foundation eccentricity to counteract base moment.

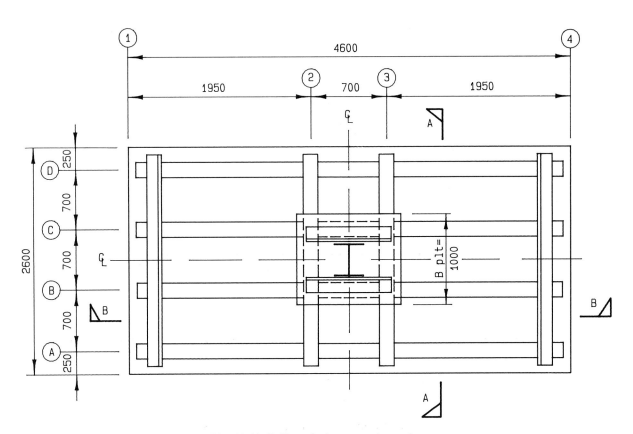

Fig. 11.44 Grillage design example – plan.

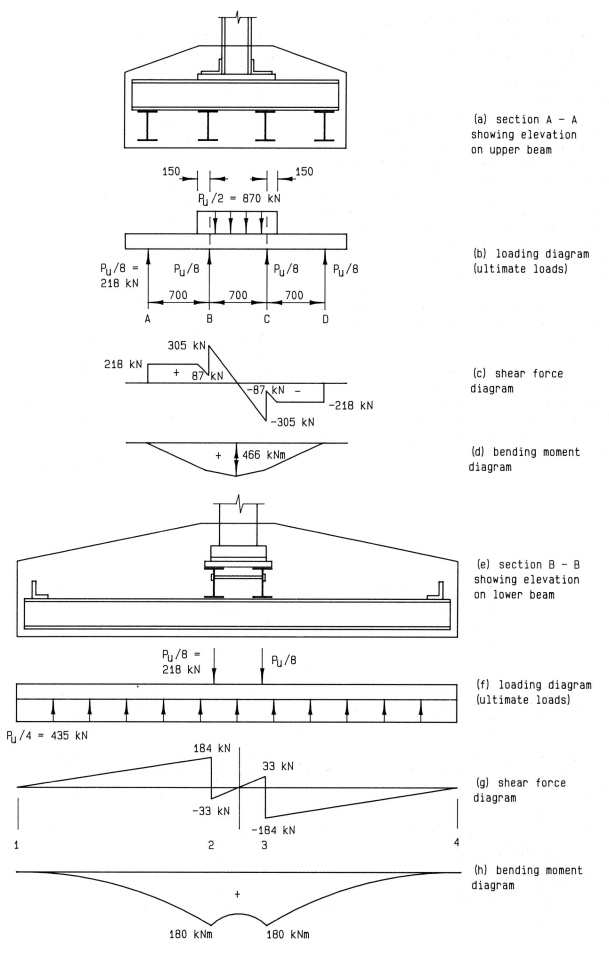

(a) section A – A showing elevation on upper beam

(b) loading diagram (ultimate loads)

(c) shear force diagram

(d) bending moment diagram

(e) section B – B showing elevation on lower beam

(f) loading diagram (ultimate loads)

(g) shear force diagram

(h) bending moment diagram

Fig. 11.45 Grillage design example – sections.

Size of base

$$\text{Required area of base} = \frac{P}{n_a}$$

$$= \frac{1200}{100}$$

$$= 12\,\text{m}^2$$

To suit site conditions, a base size of $4.6\,\text{m} \times 2.6\,\text{m}$ will be chosen, giving a base area of $12.0\,\text{m}$ (see Fig. 11.44).

Ultimate bending moments and shear forces

For 25% imposed load, Fig. 11.22 gives a combined partial load factor of $\gamma_P = 1.45$. Thus

$$P_u = \gamma_P P$$

$$= 1.45 \times 1200$$

$$= 1740\,\text{kN}$$

The upper tier ultimate shear forces are obtained from the loading diagram (Fig. 11.45 (b)).

$$V_{AB} = \frac{P_u}{8} = \frac{1740}{8} = 218\,\text{kN}$$

$$V_{BA} = V_{AB} - 0.15\left(\frac{P_u}{1 \times 2}\right) = 218 - 0.15\left(\frac{1740}{1 \times 2}\right) = 87\,\text{kN}$$

$$V_{BC} = V_{BA} + \frac{P_u}{8} = 87 + \frac{1740}{8} = 305\,\text{kN}$$

The maximum upper tier bending moment is calculated as the area of the shear force diagram (Fig. 11.45 (c)).

$$M_u = (0.7 \times 218) - 0.15\left(\frac{218 - 87}{2}\right) + 0.35\left(\frac{305}{2}\right)$$

$$= 196\,\text{kNm}$$

The lower tier ultimate shear forces are obtained from the loading diagram (Fig. 11.45 (f)).

$$V_{12} = 0$$

$$V_{21} = \frac{1.95}{4.60}\left(\frac{P_u}{4}\right) = 0.42\left(\frac{1740}{4}\right) = 184\,\text{kN}$$

$$V_{23} = V_{21} - \left(\frac{P_u}{8}\right) = 184 - \left(\frac{1740}{8}\right) = -33\,\text{kN}$$

The maximum lower tier bending moment is calculated as the area of the shear force diagram (Fig. 11.45 (g)).

$$M_u = 1.95 \times \left(\frac{184.4}{2}\right)$$

$$= 179\,\text{kNm}$$

Design of steel to BS 5950

Assuming the concrete casing provides lateral restraint to the compression flange, and provided the shear force is less than $0.6P_v$, where P_v is the ultimate shear capacity, then the ultimate moment capacity is given by M_c (BS 5950: Part 1: 4.2.5). P_v and M_c can be obtained from safe load tables published by the Steel Construction Institute (SCI). If the grillage beams were not encased in concrete, then additional checks would be required for lateral torsional buckling.

Design of upper tier beams

To ensure the shear force is within $0.6P_v$, this requires

$$P_v \geqslant \frac{V_u}{0.6}$$

$$\geqslant \frac{305}{0.6}$$

$$\geqslant 508\,\text{kN}$$

From the Steel Construction Institute's *Guide to BS 5950: Volume 1*, a $457 \times 152 \times 52$ UB should be used which has the following properties for grade 43 steel:

$$P_v = 564\,\text{kN} \;(>508\,\text{kN})$$

$$M_c = 300\,\text{kNm} \;(>196\,\text{kNm})$$

Design of lower tier beams

$$P_v \geqslant \frac{V_u}{0.6}$$

$$\geqslant \frac{184}{0.6}$$

$$\geqslant 307\,\text{kN}$$

Again from the SCI publication, a $356 \times 127 \times 39$ UB should be used which has the following properties for grade 43 steel:

$$P_v = 378\,\text{kN} \;(>307\,\text{kN})$$

$$M_c = 180\,\text{kNm} \;(>179\,\text{kNm})$$

Checks should also be carried out for web buckling and bearing with stiffeners being provided as necessary. This is not critical for this design case as the grillage is wholly encased in concrete.

11.6 FLOATING SLABS (GROUND SLABS)

11.6.1 Introduction

A floating slab or ground slab can be thought of as the most common form of raft foundation. It is basically a concrete slab with limited stiffness and reinforcement suitable to disperse the normal floor loads over a greater area of sub-strata and to span over any depressions or soft spots.

11.6.2 Design decisions

The design decisions relate to

(1) The loading anticipated on the slab.
(2) The ground conditions below the slab.
(3) The need to maintain specific levels and finishes for a normal design life within appropriate tolerances.
(4) The required durability.
(5) The control of shrinkage and other movements without excessive cracking.

The floating slab is chosen when the sub-strata or a hardcore layer over the sub-strata is suitable to allow a simple slab to adequately disperse the loads without excessive distortions or cracking. Where such conditions do not exist then a suspended slab may need to be adopted.

A floating slab can be of plain concrete or reinforced concrete depending on the quality of the sub-strata and the loading condition. Generally they are reinforced and while it can be argued that under their loading conditions positive and negative bending moments will be produced, it is common to only reinforce with one layer of reinforcement, usually using a mesh fabric. If one layer of reinforcement is used it can be located in the bottom, top or middle of the slab, depending on the designer's requirements. However, generally a top mesh is usually considered the most suitable.

Cracking of concrete slabs is almost inevitable in some form either as a result of shrinkage or bending tensile stress. Control over such cracking is usually more important on the top surface of the ground floor slab rather than on the underside and by providing the mesh in the top of the slab and accepting some cracking on the soffit the designer can economically control the condition for most ground slabs (see Fig. 11.46).

If, however, there is a need for the slab soffit to be protected then a bottom mesh can also be provided (see Fig. 11.47).

In all cases one of the most important aspects of the design and construction is to maintain adequate cover for both wear and tear of the surface and to provide adequate durability.

The slab is generally sized and reinforced on the basis of experience. However, as with the crust raft, a calculated design can be adopted using nominal rules based upon the ground condition. For example, assumptions for variations in sub-strata and/or hardcore support can be made on the basis of expected diameter of any soft spot which may have to be spanned or cantilevered (see Fig. 11.48).

11.6.3 Sizing of the slab

In general floor slabs are designed by eye from experience and are made up of a sub-base layer of hardcore blinded with either sand or concrete and sealed with a slip membrane upon which the slab is cast (see Table 11.3). However, an alternative approach is to consider the make-up and performance requirements in more detail.

Floor slabs supported directly on the ground are subject to bending and shear forces resulting from differential movements in the ground support during loading. In addition they are subjected to thermal and moisture movements which can produce the critical stresses particularly in slabs on uniform support.

In order to control thermal and moisture movement minimum reinforcement requirements relating to the bay size and joint details should be adopted to embrace such stresses (see Table 11.5).

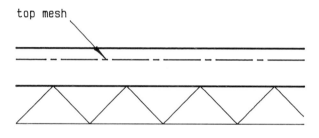

typical slab reinforcement

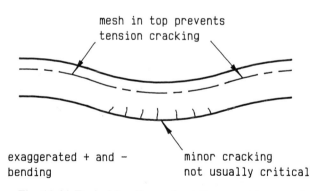

Fig. 11.46 Typical bending and reinforcement in ground bearing slabs.

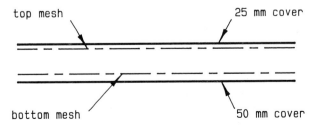

Fig. 11.47 Doubly reinforced ground slab.

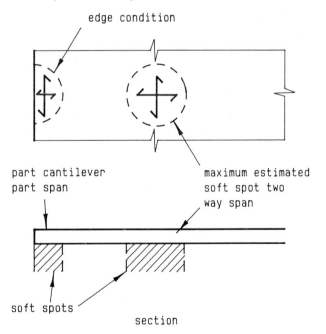

Fig. 11.48 Typical design depression.

Table 11.3 Fabric reinforcement for crack control from moisture and thermal stresses

Slab thickness (mm)	Slab plan size between face joints (m)	BS mesh in top of slab
125 to 150	up to 6 × 6	A98
125 to 200	up to 15 × 10	A142
125 to 200	up to 30 × 10	C283
175 to 200	up to 45 × 10	C385

Table 11.4 Ground floor slabs – typical assumed depressions

Sub-grade classification	Typical soil types	Typical assumed diameter of depression (m)
Consistent firm sub-soil	One only of clay sand sandy clay clayey sand gravel	0.7 to 1.25
Consistent type but variable density i.e. loose-to-firm	One only of clay sand sandy clay clayey sand gravel	1.25 to 1.75
Variable soil type but firm	Two or more of clay silt sandy clay silty clay clayey sand sand firm granular fill gravel	1.75 to 2.25
Variable soil type and variable density	Two or more of clay silt sandy clay clayey sand sand firm granular fill gravel	2.25 to 3.0

It should be noted that the main function of such reinforcement is to distribute the cracking into a larger number of smaller cracks, i.e. to control the cracking rather than to prevent it.

The design process therefore should be to assess the reinforcement requirements for thermal and moisture movement and to compare them with that required for ground support, and to use the worst case requirement. The analysis for ground support can be assessed by the adoption of a design based upon spanning or cantilevering over a depression similar to that adopted for crust rafts (see section 13.1.4). Due to the relatively small loads applied to slabs, the likely settlement depressions tend to be of small diameter when compared with a similar crust raft condition (see Table 11.4).

11.6.4 Design Example 13: Floating slab
A ground floor slab is to be designed for a single-storey supermarket measuring 60 m × 36 m on plan, as shown in Fig. 11.49. The slab is required to carry an imposed load of 25 kN/m². The superstructure is a two bay portal frame on separate foundations, and the soil is a medium dense sand, which the site investigation has indicated to be consistent across the site.

Based on the relatively good ground conditions, a 150 mm concrete slab on 150 mm of hardcore will initially be assumed.

Joints and reinforcement for shrinkage purposes
The slab is intended to be constructed using the *long strip*

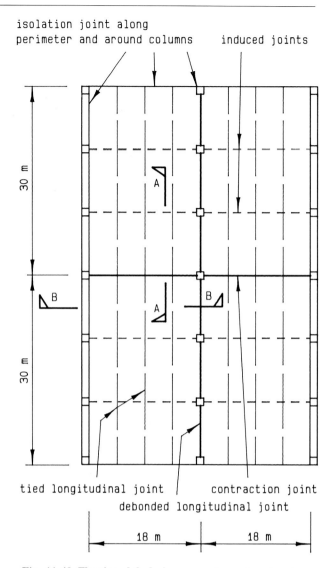

Fig. 11.49 Floating slab design example – plan showing movement joints.

method. The slab will be cast in 60 m × 4.5 m strips, in an alternate bay sequence, as shown in Fig. 11.49. If a C283 mesh is to be used, Table 11.5 indicates that one intermediate contraction joint along the 60 m length of the building will be sufficient. Additional induced crack joints will also be incorporated every 10 m to help control cracking.

In accordance with Table 11.5, one intermediate movement joint will also be required across the width of the building. The various reinforcement and joint details are shown in Fig. 11.50.

Spanning over local depression
The principles are similar to those used for raft design in Chapter 13.

(1) Select a diameter for a local depression from Table 11.4. Modify it if required to take into account the thickness of any compacted granular material/hardcore below the slab, as per Fig. 13.4 (a).

In this example, with the sub-grade comprising a

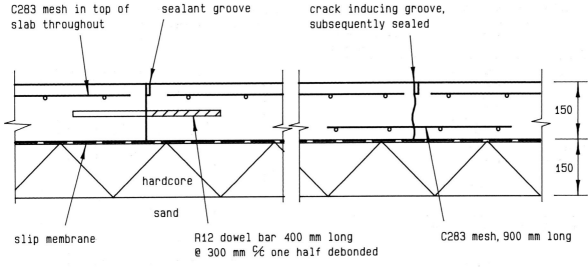

section A - A

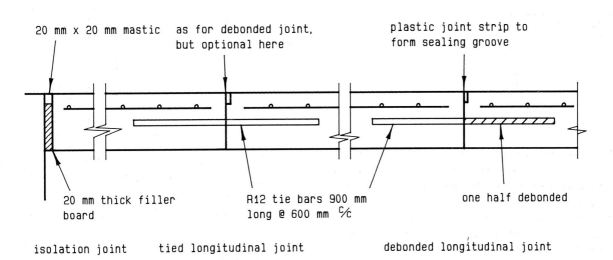

section B - B (sub-base etc omitted for clarity)

Fig. 11.50 Floating slab design example — movement joint details.

consistent medium dense sand, a design span of 0.95 m will be used.

(2) Calculate the loads acting over a depression located at an unsupported slab corner, as per Fig. 11.51. (This is the *worst case* location for a depression.)

In this example, the ultimate foundation loads due to slab self-weight and imposed load of 25 kN/m² is given by

$$f_u = 1.4(24 \times 0.15) + (1.6 \times 25)$$
$$= 45 \text{ kN/m}^2$$

No significant point loads are assumed to act in this particular example.

(3) Calculate the cantilever moment per metre width adjacent to this depression from Fig. 11.51.

A 'C' mesh is proposed, giving a one-way moment of

$$M_u = 0.32 f_u L^2$$
$$= 0.32 \times 45 \times 0.95^2$$
$$= 13.0 \text{ kNm/m}$$

(4) Calculate the corresponding area of mesh reinforcement required.

$$\text{Effective depth, } d = 150 - 20(\text{cover}) -$$
$$\frac{6}{2} \text{ (half bar diameter)}$$
$$= 127 \text{ mm}$$

$$\text{Width, } b = 1000 \text{ mm}$$

$$\frac{M_u}{bd^2} = \frac{13.0 \times 10^6}{1000 \times 127^2}$$

$$= 0.81$$

Table 11.5 Free movement joint spacing and fabric reinforcement for slabs (*Source*: Deacon, R.C. (1986) *Concrete Ground Floors*. Cement and Concrete Association)

Mesh reinforcement according to BS 4483	Longitudinal direction of slab					Transverse direction of slab				
	Maximum spacing of free movement joints (m)					Maximum spacing of free movement joints (m)				
	Slab thickness (mm)					Slab thickness (mm)				
	125	150	175	200	225	125	150	175	200	225
Unreinforced	—	6	6	6	6	—	6	6	6	6
		18^a	18^a	18^a	18^a		18^b	18^b	18^b	18^b
A142	25	21	18	16	14	25	21	18	16	14
A193	34	28	25	21	19	34	28	25	21	19
B196										
$A252^c$	44	37	31	28	25	44	37	31	28	25
B283	49	42	36	31	28	34	28	25	21	19
C283						18^b	18^b	18^b	18^b	18^b
B385	67	56	48	43	37	34	28	25	21	19
C385						18^b	18^b	18^b	18^b	18^b
$A393^c$	69	58	49	44	38	69	58	49	44	38
$B503^c$	—	74	64	55	49	—	37	31	28	25
C503							18^b	18^b	18^b	18^b
C636	—	—	81	74	65	—	—	18^b	18^b	18^b

^{a} Induced joints of 6 m maximum centres
^{b} 6 m maximum strip width
^{c} With these fabrics, high-yield tie-bars of equivalent cross-section to the fabric should be used in longitudinal joints when maximum spacing of movement joints is adopted in the transverse direction.

Example
A slab 175 mm thick is 60 m long between free isolation joints abutting the walls. It is to be laid in a continuous strip with induced joints at 10 m centres. From Table 11.5, long mesh C503 is required. If a debonded dowelled contraction joint is introduced at mid-length, movement joint spacing reduces to 30 m, and the reinforcement could be reduced to square-mesh A252 or long-mesh C283.

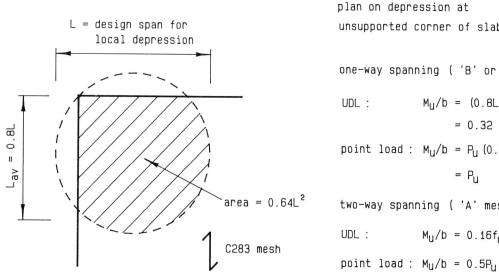

plan on depression at
unsupported corner of slab

one-way spanning ('B' or 'C' mesh)

UDL : $M_u/b = (0.8L/2) f_u (0.64L^2)/(0.8L)$
             $= 0.32 f_u L^2$ kNm/m

point load : $M_u/b = P_u (0.8L)/(0.8L)$
             $= P_u$ kNm/m

two-way spanning ('A' mesh)

UDL : $M_u/b = 0.16 f_u L^2$ kNm/m

point load : $M_u/b = 0.5 P_u$ kNm/m

L = design span for
local depression

$L_{av} = 0.8L$

area $= 0.64L^2$

C283 mesh

Fig. 11.51 Floating slab design example – designing for local depression.

$A_s = 0.22\%\ bd$ [BS 8110: Part 3: Chart 2]

$$= \left(\frac{0.22}{100}\right) \times 1000 \times 127$$

$$= 279\,\text{mm}^2/\text{m}$$

Thus the C283 mesh chosen for shrinkage purposes will be satisfactory.

Normally this reinforcement, calculated for the *worst case* condition at a slab corner, would be provided throughout the slab. In situations where this results in an excessive amount of reinforcement, a separate calculation can be carried out for a depression located in the middle of the slab. This calculation would follow the procedures for raft slabs in section 13.1.5.

CHAPTER 12

TIED AND BALANCED FOUNDATIONS

12.1 GENERAL INTRODUCTION

Tied and balanced foundations are used to combine a number of superstructure loads in order to achieve acceptable bearing pressure. The combined base is used to balance out or tie together difficult eccentricities of loading or horizontal forces. Such foundations usually result from an engineering study of the superstructure loads to be transmitted onto the foundation. The engineer's aim is to make the best use of the magnitude and direction of such forces in balancing out or tying together eccentric reactions and horizontal thrust to economically achieve the required ground bearing pressures.

Particular problems exist where large lateral forces are transferred at the top edge of a foundation from say portal frames or when large column loads occur at or near site boundaries. Fortunately portal frames tend to have an equal and opposite leg with similar opposing horizontal forces which can be reacted against each other. Buildings with large column loads near to the boundary tend also to have other large column loads either internal to the building or on an opposite boundary. The internal or opposite columns can therefore be used to stabilize the moments produced by the eccentricity from the outer perimeter frame. Typical examples are given in the following sections.

12.2 TIED FOUNDATIONS

12.2.1 Introduction

Tied foundations are often adopted as a means of exploiting to advantage opposing forces. This is achieved by linking them together via a tie or tie beam. The effect this has on the design is to reduce the horizontal force requiring to be resisted by the ground (see Fig. 12.1 (a)).

The use of a tie can reduce the amount of movement likely to occur in developing the reaction and reduce the cost of the foundation.

12.2.2 Design decisions

In any situations where horizontal forces, such as thrusts from portal frames, etc., act in opposite directions, consideration should be given to connecting the forces via a tie in order to reduce or totally react a horizontal force. For example, if the forces are equal and opposite then the total force can be reacted. On the other hand, if the forces are

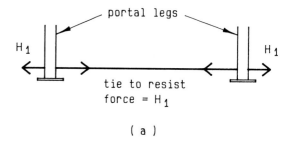

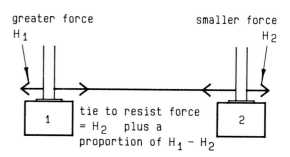

foundations to resist force $H_1 - H_2$ passively + O.T. moments due to eccentricity of the passive reaction

(b)

Fig. 12.1 Tied foundation.

opposite and not equal, the smaller of the two forces can be tied and the remainder left to be reacted by foundation 1 or, if a higher tie force is used, foundation 2 can also be utilized, thereby reducing the force to be taken in passive pressure (see Fig. 12.1 (b)).

12.2.3 Sizing the foundations

The main pad foundations are designed in the same way as those previously discussed in Chapter 11 but taking into account the tie force reaction in accordance with the above considerations. The tie itself must be designed to resist the force H_1 or H_2, as the case may be, and must be detailed to transfer this force without excessive slip or failure between the bases of the stanchions.

This is usually achieved by designing a tie rod for the total force using appropriate permissible tensile stresses for the steel and ensuring that suitable mechanical anchorage or

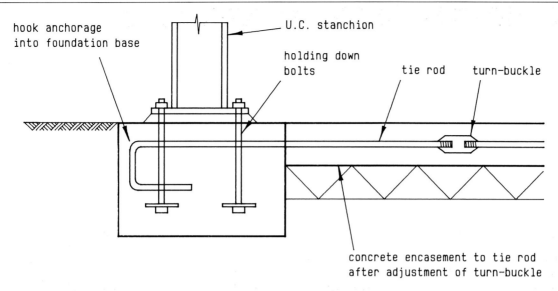

hook anchorage into foundation base

U.C. stanchion

holding down bolts

tie rod

turn-buckle

concrete encasement to tie rod after adjustment of turn-buckle

Fig. 12.2 Tie anchorage.

bond anchorage is achieved in the details between the stanchion and tie (see Fig. 12.2).

In detailing these ties, the detailer should ensure that the tie acts on the centreline of the horizontal thrust force or that any eccentricity produced is designed into the foundation by the designer. The tie rod itself could contain a turn-buckle for tensioning in order to reduce lateral movement due to possible slackness in the rod, alternatively, if adjustment is not required, a reinforced concrete tie beam as shown in Fig. 12.3 could be used. In the case of portal framed factories it is often desirable to construct the floor slab after erection and cladding of the building; in this case the engineer must ensure that all tie members are constructed and covered prior to the erection of the steelwork.

12.2.4 Design Example 1: Tied portal frame base
The pad bases for a single-bay portal frame are to be joined by a horizontal tie to take out the horizontal thrusts from the portal legs. The portal is similar to the one which was designed as an untied portal in section 11.3.4. Loads and dimensions are shown in Fig. 12.4.

Loadings
From section 11.3.4,

$$\text{vertical superstructure load, } P = \text{(dead load)}$$
$$+ \text{(imposed load)}$$
$$= G + Q$$
$$= 175 + 225$$
$$= 400 \, \text{kN}$$

Q as a percentage of P is $100Q/P = (100 \times 225)/400 = 56\%$. From Fig. 10.20, the combined partial factor for dead and imposed loads is $\gamma_P = 1.51$.

$$\text{Horizontal thrust, } H = 50 \, \text{kN}$$

The horizontal thrust H arises from vertical loads G and Q, and will therefore have the same combined partial load factor $\gamma_P = 1.51$.

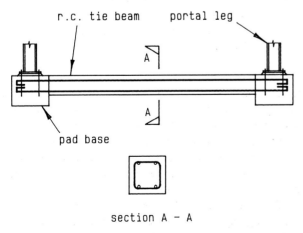

r.c. tie beam   portal leg

A

A

pad base

section A – A

Fig. 12.3 Reinforced concrete tie beam.

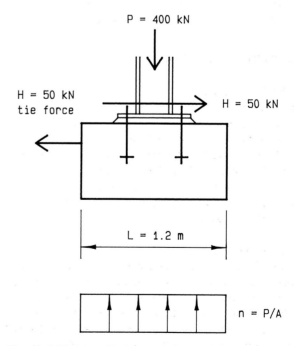

P = 400 kN

H = 50 kN tie force

H = 50 kN

L = 1.2 m

n = P/A

Fig. 12.4 Tied base design example – loads and pressures.

Size of base

From section 11.3.4, the net allowable bearing pressure, $n_a = 300 \, \text{kN/m}^2$.

On the basis that the horizontal thrust will be taken out by the tie joining the portal feet, the minimum area of foundation required is

$$
\begin{aligned}
A_{\text{req}} &= \frac{\text{superstructure load}}{\text{net allowable bearing pressure}} \\
&= \frac{P}{n_a} \\
&= \frac{400}{300} \\
&= 1.33 \, \text{m}^2 \\
&= 1.15 \, \text{m} \times 1.15 \, \text{m}
\end{aligned}
$$

A base $1.2 \, \text{m} \times 1.2 \, \text{m}$ will therefore be chosen. Comparison with the example in section 11.3.4 shows that the introduction of the horizontal tie has reduced the base size.

Design of horizontal tie

The tie will be a mild steel bar, as shown in Fig. 12.5, encased in concrete for durability.

$$
\begin{aligned}
\text{Ultimate tensile force in bar, } H_u &= \gamma_P H \\
&= 1.51 \times 50 \\
&= 76 \, \text{kN}
\end{aligned}
$$

From BS 8110, the characteristic tensile stress $f_y = 250 \, \text{N/mm}^2$ for mild steel. The partial material factor $\gamma_s = 1.15$. The required cross-sectional area of bar is

$$
\begin{aligned}
A_{\text{min}} &= \frac{\text{ultimate tensile force}}{\text{ultimate design stress}} \\
&= \frac{H_u}{\left(\dfrac{f_y}{\gamma_s}\right)} \\
&= \frac{76 \times 10^3}{\left(\dfrac{250}{1.15}\right)} \\
&= 350 \, \text{mm}^2
\end{aligned}
$$

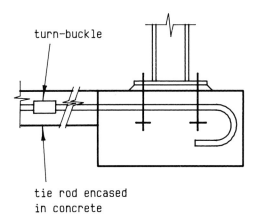

turn-buckle

tie rod encased
in concrete

Fig. 12.5 Tied base design example – tie rod detail.

Provide one number 25 mm diameter mild steel bar (area = $491 \, \text{mm}^2$) to act as the tie. This will need to be adequately anchored into the pad base as shown in Fig. 12.5. To prevent possible foundation spread from lack of fit, the tie will incorporate a turn-buckle or lock nut, to take up any slack prior to steel erection.

12.3 BALANCED FOUNDATIONS (RECTANGULAR, CANTILEVER, TRAPEZOIDAL AND HOLED)

12.3.1 Introduction

A brief discussion regarding balanced foundations is given in Chapter 9 and it is proposed in this chapter to consider these conditions a little further and to give design examples.

12.3.2 Design decisions

The decision to use a combination of column loads to produce a combined balanced foundation would depend upon a number of factors, for example:

(1) The spacing of the point loads.
(2) The combination of loads being considered.
(3) The restrictions of projections due to site boundaries.
(4) The overall eccentricities produced from the resultant of the loads.
(5) The bearing area available.
(6) The need to produce a uniform pressure.
(7) The economics compared to other possible alternatives, if any. For example, in some situations a combination of column loads can be used to balance out eccentric loads which would otherwise extend isolated foundations beyond the boundaries of the site. Balancing out these column loads means that the boundaries can be maintained within a base giving uniform pressure and this may prove more economic than say a piled solution.

In other situations an attempt to balance out the loads may produce cantilevers which would extend beyond the site boundaries therefore making it necessary to look at alternative column combinations or alternative means of support such as piling.

In most cases where these foundations are adopted they relate to: boundaries which are restrictive; foundations which would otherwise overlap; or situations where, by introducing a load from other columns onto the same foundation, bending moments are reduced and pressures become more uniform.

12.3.3 Sizing up the design

(1) RECTANGULAR BALANCED FOUNDATIONS

The foundation base is designed by calculating the position of the resultant applied load and making the centre of gravity of the base coincide with that of the downward load.

This is done by first calculating the area of the base required to resist the resultant load and then finding the most economic rectangular pad to achieve this. The pad is then located so that its centre of gravity is in the same position as the resultant load (see Fig. 12.6).

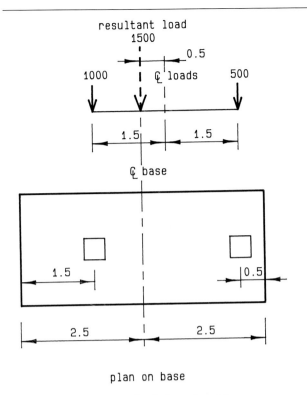

Fig. 12.6 Rectangular balanced pad base.

The base is then designed to resist the bending moments and shear forces produced by the solution, and the depth and reinforcement are determined and detailed accordingly.

(2) CANTILEVER BALANCED FOUNDATIONS

The design of the cantilever balanced foundation is carried out by assuming locations for the pad supports based upon the physical considerations and calculating the reactions from the cantilever beam. The reactions are then accommodated by calculating the required size of rectangular pads for each reaction based upon a uniform bearing pressure. The beam is then designed to support the loads from the superstructure taking account of the induced bending and shear forces, etc. (see Fig. 12.7 for a typical example).

(3) TRAPEZOIDAL BALANCED FOUNDATIONS

The design is carried out by first of all calculating the area of the base required for a uniform pressure to resist the total applied load. The resultant load and its point of application is then calculated. By fixing the dimensions for the length of the base, the dimensions A and B (see Fig. 12.8 (a)) can be calculated to give a centre of gravity which coincides with the location of the resultant load.

The applied bending moments and shear forces are then calculated and the reinforced foundation designed to suit.

(4) HOLED BALANCED FOUNDATIONS

The design is carried out by first calculating the resultant load and its location. The area required for the base is then determined by dividing the resultant load by the allowable bearing pressure. By fixing the length of the base an average width can hence be determined, and by inspection of the eccentricity of the resultant load, an allowance can be made

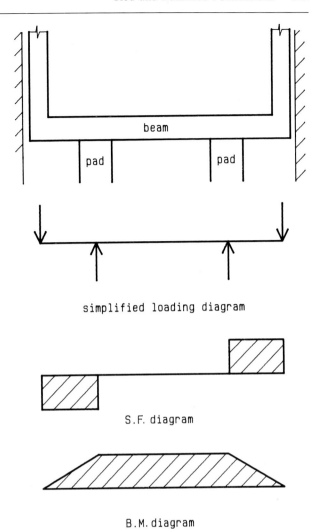

simplified loading diagram

S.F. diagram

B.M. diagram

Fig. 12.7 Bending and shear diagram for typical cantilever base.

for an approximate size of hole and a trial width determined (see Fig. 12.8 (b)).

From this trial width a size and location for the hole can be calculated to give a centre of gravity for the base which will coincide with that of the applied loads and result in a uniform pressure.

Having determined the base dimensions the bending moments and shear forces can be calculated and the foundation design completed.

(5) GENERAL SIZING CONSIDERATIONS

The size of the sections involved is based upon bending moments, shear forces and bond stresses in a similar manner to any other reinforced concrete section. With foundations, however, due to the slightly reduced shuttering cost for concrete below ground compared to elevated sections, it is often more economic to go for slightly larger concrete sections to avoid the use of excessive shear reinforcement or large-diameter bars. Each condition will demand different sizes and therefore the engineer will need to determine the initial size from his feel of engineering, finalizing his design by trial and error.

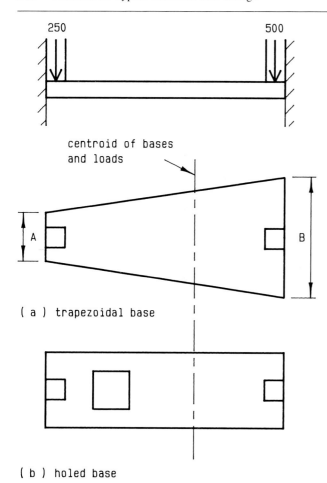

Fig. 12.8 Trapezoidal and holed balanced foundation.

12.3.4 Design Example 2: Rectangular balanced foundation
A five-storey concrete-framed office building has columns located on a regular 6 m × 6 m grid. The soil is a sandy clay with a net allowable bearing pressure, $n_a = 150 \, \text{kN/m}^2$.

Loadings
The column loads are as follows:

Internal column: 2000 kN
Perimeter column: 1000 kN
Corner column: 500 kN

The imposed load may be taken to be 55% of the total load for all columns. Thus, from Fig. 10.20, the combined partial load factor $\gamma_P = 1.51$.

Size of isolated pad bases
Normal internal column foundations have been chosen to be isolated pad foundations, with an area given by

$$A = \frac{\text{superstructure load}}{\text{net allowable bearing pressure}}$$

$$= \frac{P}{n_a}$$

$$= \frac{2000}{150}$$

$$= 13.33 \, \text{m}^2$$

which for a square base gives plan dimensions of 3.65 m × 3.65 m. This size will be used for internal columns, with proportionally smaller sizes for perimeter and corner columns.

The building is however built tight to the site boundary along two sides, as shown in Fig. 12.9. To keep foundations within the site boundary, the four columns adjacent to the corner will share a combined base. The base will be designed as a rectangular balanced foundation in order to minimize bearing pressures and differential settlements.

Size of combined base
Superstructure total load, $\Sigma P = 2000 + 1000 + 1000 + 500$
$$= 4500 \, \text{kN}$$

Taking moments about grid line 2 to calculate the distance of the centroid of the column loads from this grid line,

$$X = \frac{\Sigma Px}{\Sigma P}$$

$$= \frac{(1000 \times 6.0) + (500 \times 6.0)}{4500}$$

$$= 2.0 \, \text{m}$$

Similarly, by symmetry, $Y = 2.0 \, \text{m}$.

To achieve a balanced foundation, it is necessary to provide a base whose centre of gravity coincides with the centroid of the applied loads. The distance, in either direction, from the centroid of loads to the site boundary edge of the base is $6.5 - X = 4.5 \, \text{m}$: therefore if the opposite edge is likewise located 4.5 m from the centroid of loads, the two will coincide. Thus a 9 m × 9 m base will provide a balanced foundation in this situation.

The base will only remain exactly balanced if all four columns have the same level of imposed loading. From a foundation point of view this is unlikely to be critical unless extreme variations in the distribution of imposed loads occur. Where such variations are expected, these should be designed for as a separate load case.

Bearing pressure
The actual bearing pressure will be equal to

$$p = \frac{\text{superstructure total load}}{\text{area of base}}$$

$$= \frac{\Sigma P}{A}$$

$$= \frac{4500}{9.0 \times 9.0}$$

$$= 56 \, \text{kN/m}^2$$

The value of p ($= 56 \, \text{kN/m}^2$) indicates that, although the balanced foundation would limit differential settlement between the four columns sharing the base, it would not, for this particular building example, reduce differential settlements between columns on this base and those on adjacent bases. Adjacent bases would be sized to give bearing pressures close to the allowable value of $n_a = 150 \, \text{kN/m}^2$.

The superstructure would therefore be required to accommodate the differential settlement between the com-

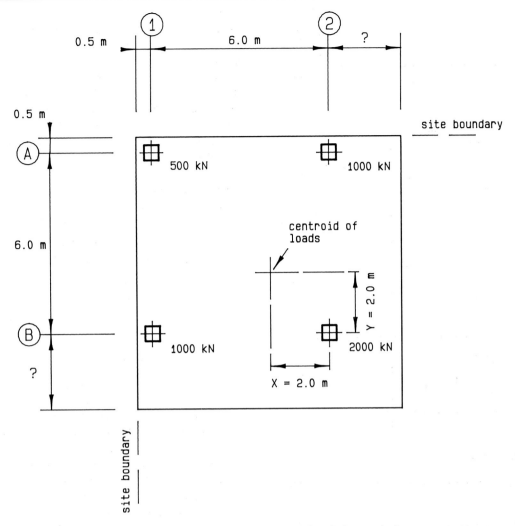

Fig. 12.9 Rectangular balanced foundation design example.

bined corner base and the adjacent isolated bases. If it is unable to accommodate these differential settlements, the bearing pressure on the balanced foundation could be increased, within limits, by turning the foundation into a *holed balanced foundation*. In this particular example this would involve cutting a hole out of the centre of the base, thus reducing the area of the base. Provided the centre of gravity of the base remains in line with the centroid of applied loads, the bearing pressure would remain uniform, but its magnitude would increase. This is illustrated further in design example 5 in section 12.3.7.

Ultimate design pressure
The ultimate design pressure for reinforcement design is given by $p_u = \gamma_P p$, where γ_P is the combined dead and imposed partial load factor.

$$p_u = \gamma_P p$$
$$= 1.51 \times 56$$
$$= 84 \,\text{kN/m}^2$$

12.3.5 Design Example 3: Cantilever balanced foundation
An existing live service run requiring a 1.5 m wide zone is required to pass along one edge of the combined base in the

previous example, as indicated in Fig. 12.10. The design is required to be adjusted accordingly.

Before redesigning the foundation, the designer should explore the possibilities, and relative costs, of either persuading the service engineer to relocate these services, or setting back the two columns on grid line 1, and cantilevering the building out to the site boundary at each floor level. Either solution may well prove more economic than changing the foundation.

If these options fail to bear fruit, the designer will need to design the combined base to cantilever over the service zone without loading it. As in the previous example, the base will be designed as a balanced foundation.

Size of base
The column loads and positions are unchanged, and therefore the centroid of the superstructure loads remains in the same place as in the last example. Again a balanced foundation will be achieved by making the centre of gravity of the effective base (i.e. the centroid of the uniform stress block below the base) coincide with that of the applied loads.

The service zone does not affect the centre of gravity of the base in the Y direction, and the overall dimension in this

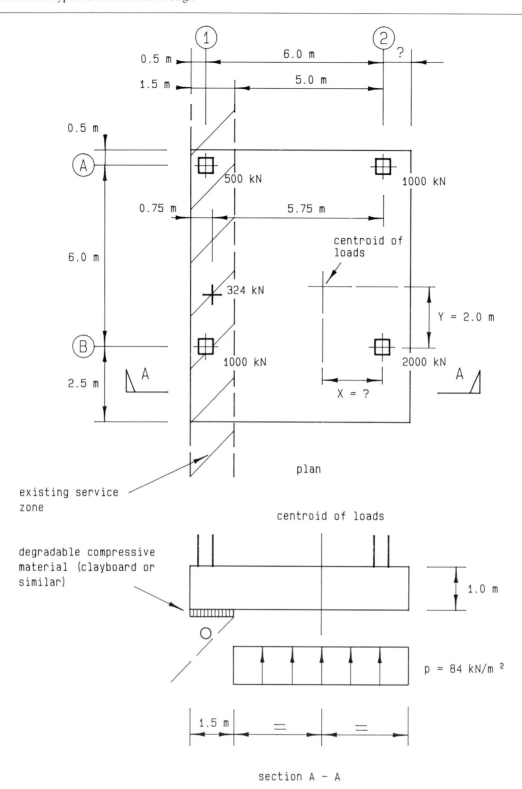

Fig. 12.10 Cantilever balanced foundation design example.

direction for a balanced foundation therefore remains at 9 m. In the x direction, the 1.5 m width of the service zone is discounted in considering the effective base area.

The weight of the cantilever section of the slab acts as a net applied load in this direction and must be taken into account in calculating the centroid of all applied loads. It will therefore be included as part of the superstructure load, P.

The weight of this strip of foundation is

$$24 \, \text{kN/m}^3 \times 1.5 \, \text{m} \times 9.0 \, \text{m} \times 1.0 \, \text{m} = 324 \, \text{kN}$$

Superstructure total load, ΣP = (total column loads)
$$+ \text{(cantilever self-weight)}$$
$$= 4500 + 324$$
$$= 4824 \, \text{kN}$$

Taking moments about grid line 2, the distance to the centroid is given by

$$X = \frac{\Sigma Px}{\Sigma P}$$

$$= \frac{(1000 \times 6.0) + (500 \times 6.0) + (324 \times 5.75)}{4824}$$

$$= 2.25 \, \text{m}$$

With reference to Fig. 12.10, the distance from the centroid to the effective left-hand edge of the base is $5.0 - 2.25 = 2.75 \, \text{m}$.

Thus, in order to align the centre of gravity with the centroid of applied loads, the right-hand edge of the base must also be located at 2.75 m from the centroid of the applied loads. This gives an effective horizontal base width of $2 \times 2.75 \, \text{m} = 5.5 \, \text{m}$, and a total horizontal base width of $5.5 + 1.5 = 7.0 \, \text{m}$. The effective area of the base is given by

$$A_b = 5.5 \times 9.0$$
$$= 49.5 \, \text{m}^2$$

Bearing pressure
The actual bearing pressure will be

$$p = \frac{\Sigma P}{A_b}$$

$$= \frac{4824}{49.5}$$

$$= 97 \, \text{kN/m}^2$$

Ultimate design pressure
From design example 2, the imposed load Q is 55% of the 4500 kN column loads, i.e.

$$Q = 0.55 \times 4500 = 2475 \, \text{kN}$$

Q as a percentage of superstructure load ΣP is given by $100Q/\Sigma P = (100 \times 2475)/4824 = 51\%$.
From Fig. 10.20, the combined partial factor for net loads is $\gamma_P = 1.5$.

$$p_u = \gamma_P p$$
$$= 1.5 \times 97$$
$$= 146 \, \text{kN/m}^2$$

12.3.6 Design Example 4: Trapezoidal balanced foundation
This example deals with the same building considered in the previous two examples, and designs the foundations for the perimeter columns where these occur along a site boundary, as shown in Fig. 12.11. As in the previous examples, the close proximity of the perimeter columns to the site boundary means that isolated pad foundations are not suitable, and that the foundations of the perimeter columns must be combined with those of the adjacent internal columns.

Since the ratio of internal to perimeter loads is 2 : 1, i.e. the same as in Design Example 2 (section 12.3.4), the centroid of loads will again be 2.0 m from grid line B. A 9.0 m long base, as in Design Example 2, would therefore again be required to achieve a balanced rectangular foundation. This relatively long base would however be associated with

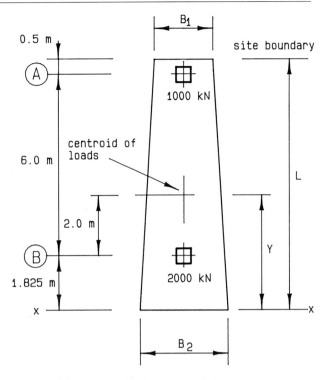

Fig. 12.11 Trapezoidal balanced foundation design example.

comparatively large bending moments and reinforcement areas. A more economic foundation is likely to be achieved using a shorter trapezoidal balanced foundation.

Condition for a balanced trapezoidal foundation
Again the condition for a balanced foundation is for the centre of gravity of the base to coincide with the centroid of the applied loads.

$$\text{Total load, } \Sigma P = 2000 + 1000$$
$$= 3000 \, \text{kN}$$

$$\text{Area of base, } A = \frac{(B_1 + B_2) \, L}{2}$$

Therefore

$$B_1 + B_2 = \frac{2A}{L}$$

With reference to Fig. 12.11, and taking moments about x–x, the location of the centre of gravity of the base of area A is given by

$$YA = \left(\frac{L}{2}\right) LB_1 + \left(\frac{L}{3}\right) \frac{L(B_2 - B_1)}{2}$$

$$= L^2 \left(\frac{B_1}{2} + \frac{B_2 - B_1}{6}\right)^2$$

$$= \frac{L^2}{6} (2B_1 + B_2)$$

This equation may be rewritten to give an expression for B_1 as follows:

$$YA = \frac{L^2}{6} (2B_1 + B_2)$$

$$\frac{6YA}{L^2} = B_1 + (B_1 + B_2)$$

$$B_1 = \frac{6YA}{L^2} - (B_1 + B_2)$$

Substituting for $B_1 + B_2$ gives

$$B_1 = \frac{6YA}{L^2} - \frac{2A}{L}$$

In a similar manner,

$$B_2 = \frac{4A}{L} - \frac{6YA}{L^2}$$

Area of base

The values of B_1 and B_2 would normally be chosen to minimize the size of the base. This would result in a bearing pressure equal to the allowable bearing pressure, n_a, giving a base area

$$A = \frac{\Sigma P}{n_a}$$

$$= \frac{3000}{150}$$

$$= 20 \, m^2$$

Dimensions of base

The end of the base furthest from the site boundary will, in this instance, be chosen to extend beyond grid line B by the same amount as a standard $3.65 \, m \times 3.65 \, m$ internal pad foundation (see section 12.3.4), i.e. extending by $3.65/2 = 1.825 \, m$.

Thus, from Fig. 12.11,

$$L = 6.5 + 1.825 = 8.325 \, m$$
$$Y = 2.0 + 1.825 = 3.825 \, m$$

$$B_1 = \frac{6YA}{L^2} - \frac{2A}{L}$$

$$= \frac{6 \times 3.825 \times 20}{(8.325)^2} - \frac{2 \times 20}{8.325}$$

$$= 1.8 \, m$$

$$B_2 = \frac{4A}{L} - \frac{6YA}{L^2}$$

$$= \frac{4 \times 20}{8.325} - \frac{6 \times 3.825 \times 20}{(8.325)^2}$$

$$= 3.0 \, m$$

These values will give a balanced trapezoidal foundation, with a bearing pressure of $p = 150 \, kN/m^2$.

Ultimate design pressure

The combined dead and imposed partial load factor is $\gamma_P = 1.51$, as in the previous examples. The ultimate design pressure for reinforcement design, p_u, is given by

$$p_u = \gamma_P p$$
$$= 1.51 \times 150$$
$$= 227 \, kN/m^2$$

12.3.7 Design Example 5: Holed balanced foundation

This example again makes use of the same building as in the previous examples, and the same pair of columns in the trapezoidal balanced foundation in section 12.3.6. The trapezoidal shape will be squared off to give a $3.0 \, m \times 8.325 \, m$ rectangular outline, as shown in Fig. 12.12.

To minimize differential settlements – both within the base and between adjacent bases – the combined base will be designed as a balanced foundation, with a bearing pressure equal to the allowable bearing pressure, $n_a = 150 \, kN/m^2$.

A balanced *holed* foundation will be investigated for this example. By inserting a hole off-centre to the centroid of the $3 \, m \times 8.325 \, m$ base, it is possible to cause the centre of gravity of the base to shift until it coincides with that of the applied loads.

The superstructure total load coming onto the base is given by

$$\Sigma P = 2000 + 1000$$
$$= 3000 \, kN$$

Area of hole

The area of the holed base is given by

$$A = (\text{area of } \textit{unholed} \text{ base}) - (\text{area of hole})$$
$$= A_u - A_h$$

From Fig. 12.12, the area of the *unholed* base is given by

$$A_u = 3.0 \times 8.325$$
$$= 24.5 \, m^2$$

The size of the hole would optimally be chosen to give a bearing pressure equal to the allowable bearing pressure, i.e.

$$A = \frac{\Sigma P}{n_a}$$

$$= \frac{3000}{150}$$

$$= 20.0 \, m^2$$

Thus the required area of the hole is

$$A_h = A_u - A$$
$$= 25.0 - 20.0$$
$$= 5.0 \, m^2$$

Condition for a balanced holed foundation

For a balanced foundation, the centre of gravity of the holed base is required to coincide with the centroid of applied loads. Let Y be the distance from the centre of the *unholed* base to the centroid of applied loads, and Y_h the distance to the centre of the hole.

From Figs 12.12 and 12.13, taking moments about x–x, Y is given by

$$YA = 0(A_u) + Y_h(A_h)$$

Thus

$$Y_h = \frac{YA}{A_h}$$

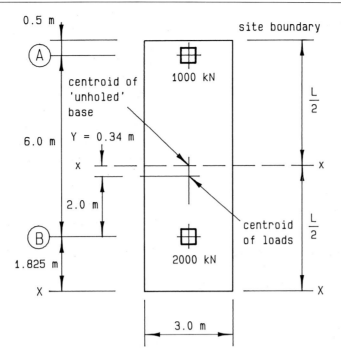

Fig. 12.12 Holed balanced foundation design example – loads.

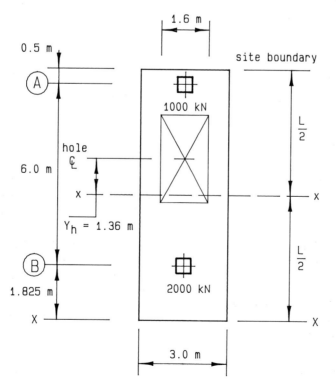

Fig. 12.13 Holed balanced foundation design example – hole size.

This is the condition for a balanced foundation.

From Fig. 12.13, $Y = 0.34$ m. Thus

$$Y_h = \frac{YA}{A_h}$$

$$= \frac{0.34 \times 20.0}{5.0}$$

$$= 1.36 \text{ m}$$

Dimensions of base

To achieve a balanced foundation, the centre of the hole must be at a distance of $Y_h = 1.36$ m from the centre of the *unholed* base. Provided this condition is met, the actual shape of the hole, e.g. square or rectangular, is not critical.

The area of the hole was calculated earlier in this example as $A_h = 5.0 \text{ m}^2$. A rectangular hole will be adopted in this instance, having dimensions of $1.6 \text{ m} \times 3.125 \text{ m} = 5.0 \text{ m}^2$ (see Fig. 12.13).

Ultimate design pressure

The holed base has been sized to give a bearing pressure at working loads of 150 kN/m^2. The ultimate design pressure for reinforcement design, p_u, is calculated as follows, with $\gamma_P = 1.51$ as in the previous example,

$$p_u = \gamma_P p$$

$$= 1.51 \times 150$$

$$= 227 \text{ kN/m}^2$$

CHAPTER 13

RAFT FOUNDATIONS

13.1 DESIGN PROCEDURES FOR SEMI-FLEXIBLE RAFTS

Chapter 9 has briefly discussed the various types of raft foundations and their use, it is intended here to give more detailed design guidance and examples.

13.1.1 Design principles

The design approach is based upon the practical assumption that all soils are variable, and all ground improvement treatments are imperfect. While 'beam on elastic foundation' and similar analyses have their place, they tend to assume a consistent and uniform formation which does not often accord with reality. Local depressions or soft spots will usually occur, and should therefore, particularly in the case of semi-flexible rafts, be designed for.

This approach to raft design consists of five main steps:

(1) Adopt a sensible layout of beam thickenings to avoid stress concentrations and areas of weakness.
(2) Check bearing pressure under concentrated loads on slabs and beam thickenings.
(3) Establish a design span for local depressions, based on the ground conditions and thickness of compacted hardcore filling.
(4) Design the raft slab areas to span over local depressions and to resist shrinkage cracking.
(5) Design the raft beam thickenings to span over local depressions.

It should be emphasized that, as with all other aspects of foundation design, it is not possible to exhaustively cover every imaginable situation, i.e. in this case every possible combination of loading, raft profile and reinforcement, compacted fill, and ground conditions. Therefore, although the guidance which follows should cover most normal situations, any new design situation should be looked at wearing one's 'engineering spectacles' to spot those situations where rules need adapting, or do not apply.

13.1.2 Design of raft layouts

Rafts form a dual *levelling-out* function. Firstly they take concentrated loads (stress concentrations) from the superstructure and spread them more evenly onto the ground below; secondly they mitigate the effects of soft spots (local weaknesses) in the supporting ground through local spanning action.

In order for rafts to fulfil this role it is necessary to avoid areas of undue stress concentration and zones of weakness within the raft structure itself. The arrangement and frequency of raft thickenings – and movement joints where necessary – to achieve this involve as much art as science. Nevertheless, the following guidelines should be considered when creating a suitable raft layout.

(1) CONTINUITY OF THICKENINGS

Thickenings should be continuous wherever possible, with no abrupt terminations or changes in direction. Although a thickening will often be located to coincide with the main load-bearing elements of the superstructure, this is generally not essential. The priority should always be to achieve a consistent and robust arrangement of thickenings. It is recommended that there be sufficient thickenings in both directions to limit the aspect ratio for slab bays (i.e. length : width) to a maximum of 2 : 1.

(2) AVOIDANCE OF AREAS OF WEAKNESS

Areas of weakness generally occur at re-entrant corners and local *necks* within a slab. Re-entrant corners should be dealt with by ensuring that both of the external thickenings meeting at the corner are continued past the corner as internal thickenings on the same lines. Areas of necking within the raft should be suitably strengthened, with additional thickenings if necessary; alternatively, the weakness can be 'acknowledged' by positioning a movement joint at this location (see below).

The other main possible source of weakness in a slab is via poor detailing at thickening junctions and intersections. Suitable tying reinforcement should be provided at all *corner*, *tee*, and *cross* junctions, to ensure thickenings can interact and share load with each other as necessary.

(3) APPROPRIATE USE OF MOVEMENT JOINTS

Movement joints are used to break up a large raft into a number of smaller rafts. This may be done to reduce bending moments and shrinkage stresses in a large raft/superstructure (see section 13.1.5 (1)), or to avoid areas of stress concentration.

Stress concentrations can occur at local *necks* within a slab, or at junctions of *limbs* with the main mass of the raft (see Design Example 2 in section 13.3.3). It is usually easier to deal with the strain energy associated with these stress concentrations by introducing a movement joint and allowing movement to occur, rather than strengthening the area by introducing additional reinforcement and thickenings.

It is good practice to avoid excessively large or excessively elongated rafts, and thus a larger building may well need to incorporate a number of movement joints. It is recommended that rafts be limited to an aspect ratio (length : width) of approximately 4 : 1, and in general a maximum length of 20 m.

It is important that, when a movement joint is introduced into a raft structure, thickenings occur on both sides of the movement joint, to ensure the two halves do both act as independent rafts. The movement joint should be carried up through the superstructure walls and suspended floors; however it is generally not necessary to continue it through a tiled or slated timber roof structure, unless large differential foundation settlements or longitudinal strains are expected.

13.1.3 Bearing pressure design

Rafts are intended to take local stress concentrations, i.e. line and point loads, and spread them over a larger area by the time they reach the formation level, so as not to exceed the allowable bearing pressure. This is done through the combined influence of concrete thickness and profile, reinforcement, and thickness of hardcore/granular fill. Where local bending of the raft is utilized to spread the loads over a wider area, an ultimate limit-state analysis is carried out to size the necessary reinforcement.

(1) SLABS

While the slab thickness does contribute to the spread of concentrated loads, the main factor is the presence or absence of bottom reinforcement. In slabs with top reinforcement only, the load is assumed to spread through the slab at 45°, i.e. as if it were mass concrete. In slabs with bottom reinforcement, the reinforcement can act with the slab to form a local spread footing, distributing the load over a wider distance. These cases are shown in Fig. 13.1.

(2) INTERNAL BEAM THICKENINGS

The width over which the load is assumed to spread is primarily governed by the arrangement of reinforcement. The presence or absence of transverse reinforcement in the thickenings, and bottom reinforcement in the slabs, determines whether the load can be spread merely over the bottom soffit of the thickening, or additionally over any sloping sides to the thickening or the adjacent slab. These cases are shown in Fig. 13.2.

(3) EXTERNAL BEAM THICKENINGS

With external beam thickenings, the effect of eccentric loads is often the most dominant factor. With reference to Fig. 13.3 (a), the bearing pressure is initially checked assuming a uniform pressure distribution of width $2x$ located concentrically below the load. If the bearing pressure must be reduced, this is done either by increasing x (the projection

(a) slab with top reinforcement only

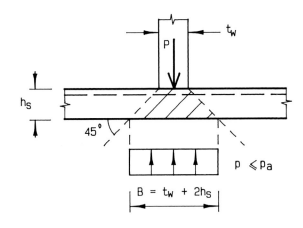

only suitable for $P \not> Bp_a$
$ \not> (t_w + 2h_s)p_a$

(b) slab with top and bottom reinforcement

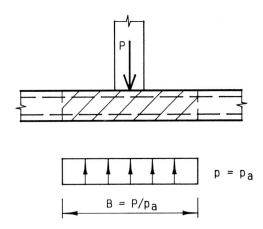

bottom reinforcement must be capable of resisting moment

$$M_u = (\gamma_p p)(B/2)^2/2 = \gamma_p P^2/8p_a$$

and shear force

$$V_u = (\gamma_p p)(B/2) = \gamma_p P/2$$

Fig. 13.1 Bearing pressure design for internal walls on slabs without thickenings.

of the thickening beyond the line of action of the load) or by spreading the load further into the raft, and using the slab reinforcement to transfer a moment to a suitable reaction to balance the vertical loading eccentricity (see Fig. 13.3 (b) and (c)). This latter approach tends to be the more economic provided:

(a) The opposite edge thickening has a similar load intensity, to balance the moment within the slab, and

(b) The slab reinforcement is sufficient for this moment to develop.

(4) EFFECT OF COMPACTED HARDCORE/GRANULAR FILL

The presence of compacted hardcore or granular fill below a

(a) no transverse reinforcement in thickening

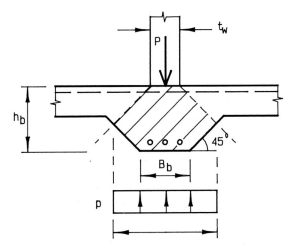

$$B = B_b/2 + t_w/2 + h_b \not> t_w + 2h_b$$

only suitable for $P \not> Bp_a$

$$\not> [(B_b/2 + t_w/2 + h_b) \not> (t_w + 2h_b)]p_a$$

(b) transverse reinforcement in thickening

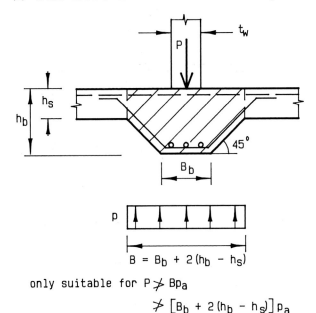

$$B = B_b + 2(h_b - h_s)$$

only suitable for $P \not> Bp_a$

$$\not> [B_b + 2(h_b - h_s)]p_a$$

Fig. 13.2 Bearing pressure design for internal beam thickenings.

(c) transverse reinforcement in thickening and bottom reinforcement in slab

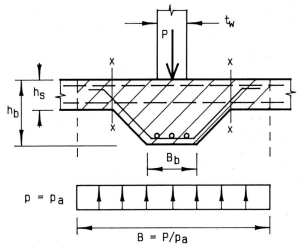

$$B = P/p_a$$

bottom reinforcement at X – X must be designed for

$$M_u = (\gamma_p p) \left[(B - B_b - 2(h_b - h_s))/2 \right]^2 /2$$

$$= \frac{\gamma_p p_a}{8} \left[\frac{P}{p_a} - B_b - 2(h_b - h_s) \right]^2$$

$$V_u = (\gamma_p p) \left[B - B_b - 2(h_b - h_s) \right]/2$$

$$= \frac{\gamma_p p_a}{2} \left[\frac{P}{p_a} - B_b - 2(h_b - h_s) \right]$$

(a) low loading intensity $(P/p_a \leqslant 2x)$

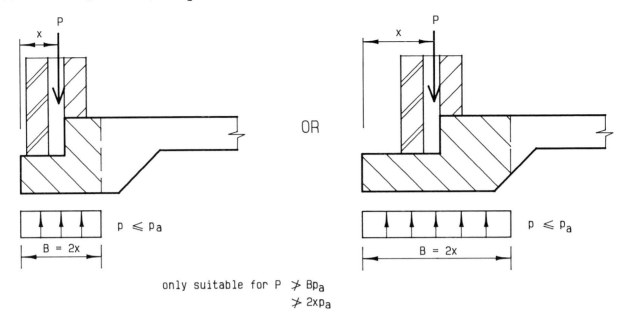

only suitable for $P \not> Bp_a$
$\not> 2xp_a$

(b) high loading intensity $(P/p_a > 2x)$ – thick blinding layer

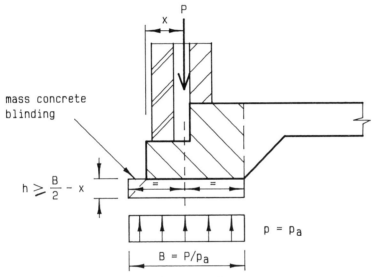

(c) high loading intensity $(P/p_a > 2x)$ – slab restoring moment

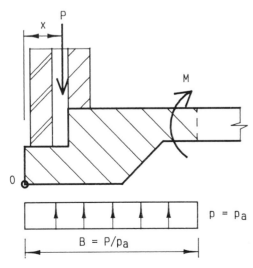

requirements ;
(1) edge thickening on far side of raft must
be similarly loaded to balance moment
(2) slab must be designed for moment
as follows ;

$M(0): \quad xP - (p_aB)B/2 + M = 0$

$\Rightarrow M = P^2/2p_a - xP$

design reinforcement for ;

$M_u = \gamma_p M = \gamma_p P(P/2p_a - x)$

Fig. 13.3 Bearing pressure design for external beam thickenings.

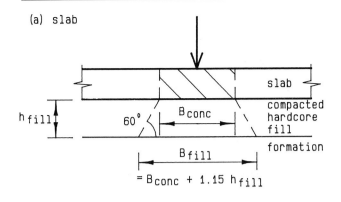

(a) slab

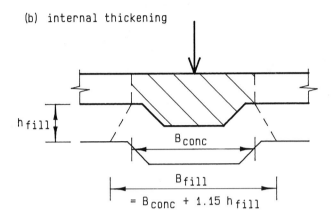

(b) internal thickening

(c) external thickening

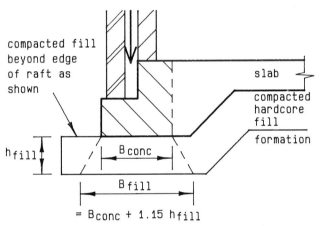

compacted fill
beyond edge
of raft as
shown

notes ;
B$_{fill}$ is to be used for checking bearing
pressure at formation level
B$_{conc}$ is to be used for checking bending within
the concrete

Fig. 13.4 Effect of compacted hardcore fill on bearing pressure design.

raft enables further spreading of concentrated loads, reducing bearing pressures and slab bending moments. Although almost all rafts have some thickness of hardcore/compacted fill, the beneficial effect of this thickness is often ignored in practice except where this thickness is substantial. Where the thickness is taken into account a 60° spread is usually assumed (see Fig. 13.4).

Table 13.1 Design diameter for local depressions

Soil classification	Soil type	Assumed diameter of depression (m)
A Consistent firm sub-soil	One only of: clay, sand, gravel, sandy clay, clayey sand	1.0–1.5
B Consistent soil type but variable density, i.e. loose to firm	One only of: clay, sand, gravel, sandy clay, clayey sand	1.5–2.0
C Variable soil type but firm	Two or more of: clay, sand, gravel, sandy clay, clayey sand	2.0–2.5
D Variable soil type and variable density	Two or more of: clay, sand, gravel, sandy clay, clayey sand	2.5–3.5

13.1.4 Design span for local depressions

(1) The information from the site investigation is used to establish the diameter for local depressions. For typical ground conditions, Table 13.1 may be used for guidance. This table is based on over 30 years' experience by the authors' practice. Where exceptionally large depressions are anticipated, e.g. due to the presence of old mine workings, swallow-holes, pipes and sink holes, their size must be determined from first principles. (See section 4.2.2 and Chapter 6 for further guidance.)

(2) The design span for local depressions is established from Fig. 13.5, based on the local depression diameter and the depth of compacted hardcore or granular fill.

13.1.5 Slab design

Slabs must be designed to span over local depressions. As with normal suspended slabs, raft slabs are designed for the serviceability limit states of deflection and shrinkage (cracking), and the ultimate limit state of bending. In addition, slabs with concentrated loads may need to be designed for the ultimate limit state of shear.

In many instances it is sufficient for raft slabs to contain top reinforcement only. Where there are poor ground conditions (large depressions) and/or concentrated loads, it may be necessary for these slabs to be reinforced both top and bottom. Rafts are normally reinforced using mesh (fabric) reinforcement, with equal quantities of reinforcement in both directions. Thus it is usual for raft slabs to be reinforced using one of the British Standard 'A Series' meshes, i.e. A142, A193, A252 or A393 mesh.

(1) SHRINKAGE

Based on the thickness of the slab and greatest overall length of slab (or maximum distance between movement joints), select an appropriate mesh size from Table 13.2. These are based on the semi-empirical recommendations for ground-bearing slabs in Reference 1; experience has shown these recommendations to be also appropriate for raft slabs.

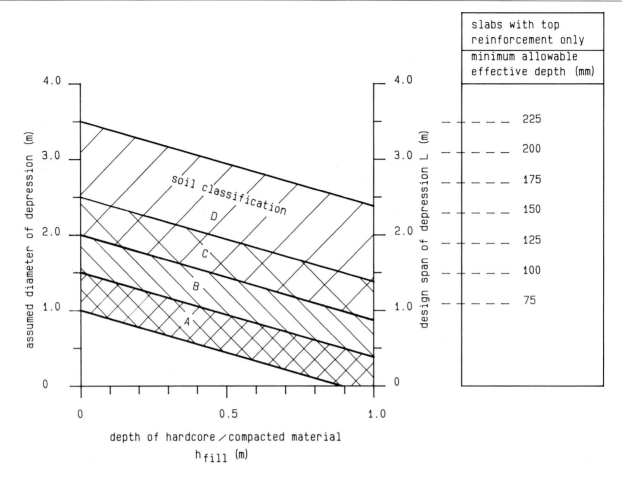

Fig. 13.5 Design span of local depressions.

Table 13.2 Shrinkage reinforcement for raft slabs

Overall slab thickness (mm)	Maximum dimension of raft (m)			
	Fabric reinforcement to BS 4483			
	A142	A193	A252	A393
100	31	42	55	87
125	25	34	44	69
150	21	28	37	58
175	18	25	31	49
200	16	21	28	44
225	14	19	25	38
250	12	17	22	34

(2) DEFLECTION

(a) For slabs reinforced with top reinforcement only, select a minimum average effective depth from Fig. 13.5. Large depressions may require both top and bottom reinforcement to avoid uneconomic slab thicknesses.

(b) For most slabs reinforced with both top and bottom reinforcement, deflection is not a problem for the normal range of depression sizes and slab thicknesses in Tables 13.1 and 13.2, and therefore does not need to be explicitly considered in these situations.

(3) BENDING

For slabs carrying distributed loading in normal domestic or commercial situations, e.g. imposed loads of up to $F_Q = 7.5\,kN/m^2$, bending will not normally be critical, and slabs can simply be reinforced for shrinkage purposes. For slabs carrying concentrated line or point loads, or heavy industrial distributed loads in excess of $7.5\,kN/m^2$, a bending calculation should be carried out as follows:

(a) Using the design span from Fig. 13.5, calculate separately all (ultimate) loads coming on to the plan area of the circular depression shown in Table 13.3. Apply moment factors, K_m, to the various load types, e.g. uniformly distributed, line, concentrated, as per Table 13.3. Calculate $\Sigma(K_m T_u)$, the effective load for bending purposes on the area of the depression.

(b) Using this load and the average effective depth, obtain the required area of reinforcement from Fig. 13.6 (top reinforcement only) or Fig. 13.7 (top and bottom reinforcement) as appropriate.

(c) Any top reinforcement required to deal with out-of-balance loads from external beam thickenings (see section 13.1.3 (3) above) should be calculated separately. This should be provided in addition to the shrinkage reinforcement calculated in (1) above, and the bending reinforcement calculated in (3) (b) above.

Table 13.3 Load types and corresponding moment factors for raft slabs

	top reinforcement only	top and bottom reinforcement	K_m
uniformly distributed load f_S (kN/m^2)	T1 $F_S = f_S \, (\pi L^2/4)$	TB1 $F_S = f_S \, (\pi L^2/4)$	1.0
parallel line load P (kN/m)	T2 $\Sigma P = PL$	TB2 $\Sigma P = PL$	1.5
lateral line load P (kN/m)	T3 $\Sigma P = PL$	TB3 $\Sigma P = PL$	1.5
2 way line load P (kN/m)	T4 $\Sigma P = 2PL$	TB4 $\Sigma P = 2PL$	1.5
point load P (kN)	T5 $\Sigma P = P$	TB5 $\Sigma P = P$	2.0
reinforcement design	see Fig. 13.6	see Fig. 13.7	

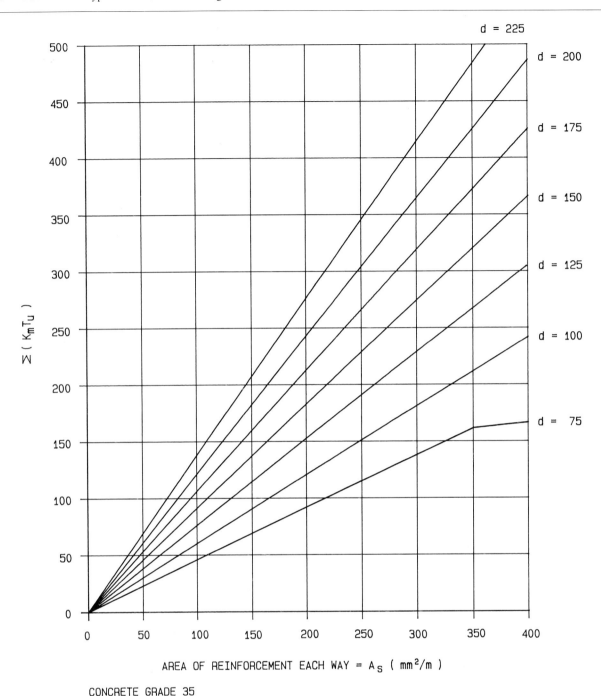

CONCRETE GRADE 35
REINFORCEMENT GRADE 460

Fig. 13.6 Design chart for slabs with top reinforcement only.

(4) SHEAR

(a) Where beam shear due to a heavy line load is considered significant, carry out a normal beam shear check in accordance with BS 8110. Where shear is found to be critical, either a beam thickening should be introduced along the line of the point load, or additional bottom reinforcement should be introduced locally to satisfy the design requirements.

(b) Where it is considered that punching shear due to a point load may be significant, carry out a normal punching shear check in accordance with BS 8110. Where punching shear is found to be critical, either the layout of beam thickenings should be amended so that the point load is positioned on the line of a beam, the slab should be thickened up locally, or the bottom reinforcement increased locally to satisfy the design requirements.

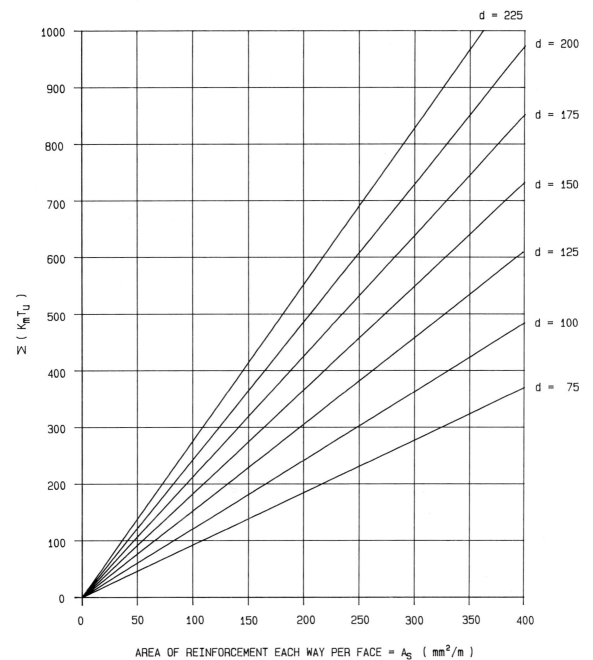

AREA OF REINFORCEMENT EACH WAY PER FACE = A_s (mm^2/m)

CONCRETE GRADE 35
REINFORCEMENT GRADE 460

Fig. 13.7 Design chart for slabs with top and bottom reinforcement.

Table 13.4 Load types and corresponding moment factors for raft beams

	internal beam	edge beam	corner beam	K_m
uniformly distributed load f_S (kN/m²)	I1 $F_S = f_S\,(\pi L^2/4)$	E1 $F_S = f_S\,(\pi L^2/8)$	C1 $\dfrac{L}{\sqrt{2}}$ $F_S = f_S\,(0.64L^2)$	0.5
parallel line load P (kN/m)	I2 $\sum P = PL$	E2 $\sum P = PL$	C2 $\sum P = PL/\sqrt{2}$	1.0
lateral line load P (kN/m)	I3 $\sum P = PL$	E3 $\sum P = PL/2$	C3 $\sum P = PL/\sqrt{2}$	1.0
2 – way line load P (kN/m)	I4 $\sum P = 2PL$	E4 $\sum P = 3PL/2$	C4 $\dfrac{L}{\sqrt{2}}$ $\sum P = 2PL/\sqrt{2}$	1.0
point load P (kN)	I5 $\sum P = P$	E5 $\sum P = P$	C5 $\sum P = P$	2.0
reinforcement design	see Fig. 13.8		see Fig. 13.9	

13.1.6 Beam design

Beam thickenings are designed to span over local depressions, in a similar manner to slabs. Serviceability deflection is not usually a problem for depressions in the range covered by Table 13.1, and serviceability cracking will be adequately covered by the shrinkage reinforcement within the slab. In most circumstances it is therefore only necessary to explicitly carry out calculations for the ultimate limit states of bending and shear. Generally it will be necessary to carry out separate checks for internal beams, edge beams, and corner beams, i.e. edge beams at outside corners.

(1) BENDING

(a) Calculate separately all critical (ultimate) loads coming on to the plan area of the circular depression shown in Table 13.4. The calculation will vary, depending on whether the beam under consideration is an internal beam, an edge beam, or a corner beam.

(b) Apply moment factors K_m to the various load types, e.g. uniformly distributed, line, concentrated, as per Table 13.4.

(c) Calculate $\Sigma(K_m T_u)L/b$, where $\Sigma(K_m T_u)$ is the effective load for bending purposes on the area of the depression, b is the average (typically mid-height) width for internal and edge beams; for corner beams it is taken as the bottom width of the beam.

(d) Calculate the area of reinforcement required from Fig. 13.8 or Fig. 13.9 as appropriate.

(2) SHEAR

(a) Calculate all critical (ultimate) loads coming on to the plan area of the circular depression shown in Table 13.4.

(b) Calculate the shear force as $V_u = T_u/2$.

(c) Calculate the corresponding shear reinforcement, in accordance with BS 8110.

The variations and extra design considerations relating to the various raft types, i.e. over and above these general guidelines, are described in the following sections.

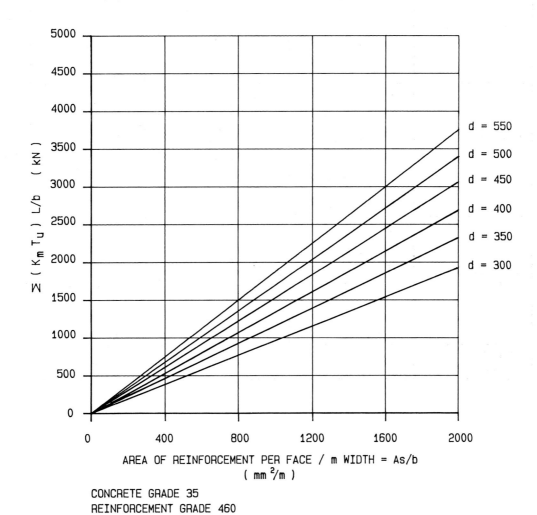

CONCRETE GRADE 35
REINFORCEMENT GRADE 460

Fig. 13.8 Design chart for internal and edge beams.

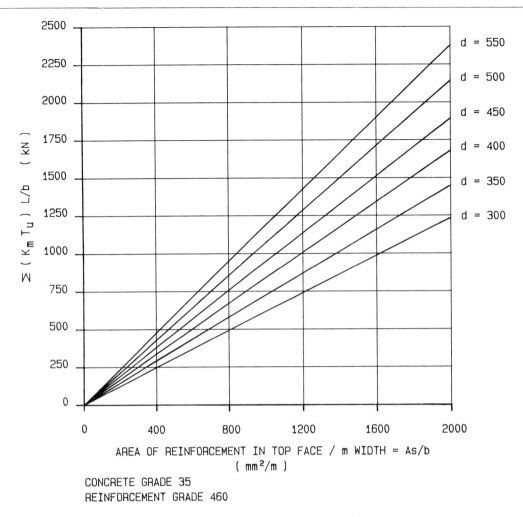

CONCRETE GRADE 35
REINFORCEMENT GRADE 460

Fig. 13.9 Design chart for corner beams.

13.2 NOMINAL CRUST RAFT – SEMI-FLEXIBLE

13.2.1 Design decisions

As discussed in Section 9.4.1 the nominal crust raft is used where loadings are relatively light and ground conditions reasonable. The raft is lightly reinforced and consists of a basic ground slab with nominal thickenings (see Fig. 13.10).

13.2.2 Sizing the design

Such rafts can be designed either from experience simply by adopting a known raft which has performed successfully on similar ground conditions and subjected to similar loadings or by calculation as discussed in sections 13.1–13.4.

The calculated design assumes that the slab and thickening should be capable of spanning and cantilevering over any local depressions which may occur as a result of the loading and/or sub-strata conditions.

Such rafts are used generally for relatively lightly loaded conditions on reasonable ground. In such conditions these lightly reinforced rafts can prove more economic than strip footings particularly where the ground is reasonably level, where the basic ground slab is used as the main body of the raft and where small straight thickenings replace complicated layouts of wall strips.

The layout of the downstands is determined from the overall raft stiffness requirements and while heavy load lines and point loads will have a bearing on the location of such downstands they should not be allowed to dictate the design. For example, the strip wall loadings, shown in Fig. 13.10, *zig-zags* across the building and if the downstand thickenings were made to follow these lines an overall weakness in the thickenings would result at each change in direction and hence the overall behaviour of the raft would be adversely affected.

It is therefore important that a common straight line across the building is used for the downstand which caters for the local heavy loads and overall stiffness (see Figs 13.11 and 13.12).

With regard to overall thickening layouts it may be necessary when considering total raft behaviour to introduce thickenings purely for stiffness and in locations where no vertical load lines exist (see Fig. 13.13).

For raft foundations adequate protection from weathering effects on most granular soils, sandy clays and insensitive clays can be achieved with 450 mm cover similar to road construction. Over-emphasis on clay shrinkage must not be allowed to change the engineering judgement in such soils particularly where past performance has been proven.

The raft is considered as a single element in determining overall behaviour taking account of the stiffness of the raft, and then breaking the foundation down into a number of small elements to simplify the design. These local conditions tend to dictate the cross-section dimensions of the foundations with the overall behaviour being developed and incorporated into the design on the drawing board. For example, if we take the raft shown in Fig. 13.13 and adopt the internal thickening layout discussed it can be seen that the reinforcement details for overall slab behaviour should ensure that beam thickenings can act continuously. In particular the design should avoid local weakenings in the concrete profile or reinforcement in vulnerable locations such as the internal angles of the raft (see Fig. 13.14). The detail therefore must ensure strong intersections at these locations where the overall shape of the raft tends to weaken structural behaviour.

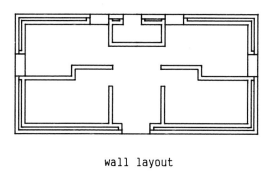

wall layout

Fig. 13.10 Zig-zag wall layout.

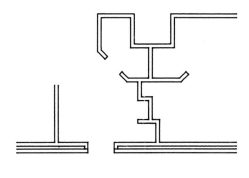

plan on walls

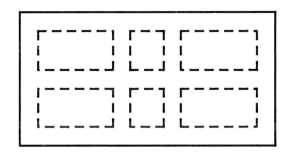

plan on raft

Fig. 13.11 Straight thickenings below raft.

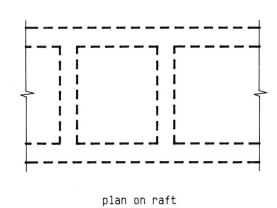

plan on raft

Fig. 13.12 Irregular wall layout but straight thickenings.

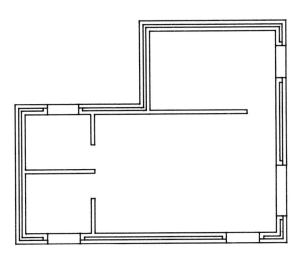

wall layout

raft layout

Fig. 13.13 Thickening layout for raft stiffness.

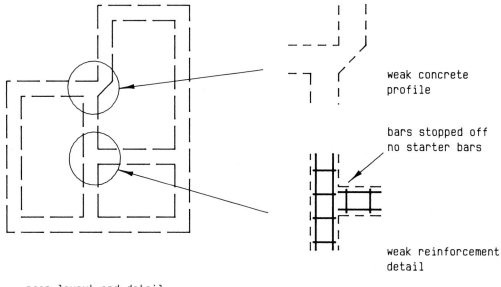

poor layout and detail

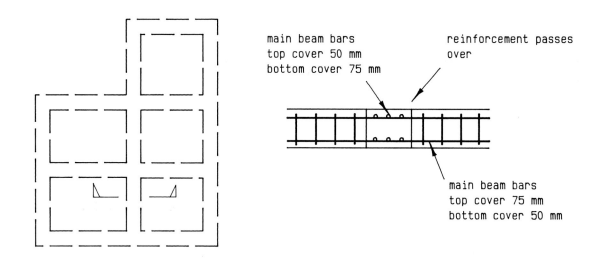

improved layout and detail

Fig. 13.14 Weak and improved thickening layouts.

13.2.3 Design Example 1: Nominal crust raft

A new housing estate, consisting of two-storey semi-detached properties, is to be built on a green field site.

The ground conditions consist of a soft to firm clayey sand. The net allowable bearing pressure for raft design is estimated at $n_a = 100 \, \text{kN/m}^2$.

The foundation for each pair of houses is to be designed as a raft foundation. Taking into account the ground conditions and the relatively light loading, a nominal crust raft is considered adequate. The wall layout, loadings, and corresponding raft layout are shown in Fig. 13.15.

Loadings

The foundation load due to slab self-weight and imposed load is

$$f = f_G + f_Q$$
$$= 3.5 + 2.5$$
$$= 6.0 \, \text{kN/m}^2$$

f_Q as a percentage of f is $100 f_Q / f = 42\%$. From Fig. 10.20, the combined partial safety factor for the foundation load is $\gamma_F = 1.48$.

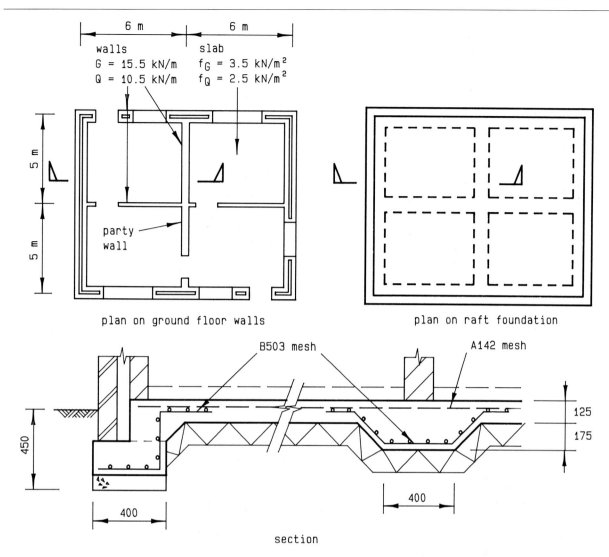

Fig. 13.15 Nominal crust raft design example.

Wall line load, P = (wall dead load) + (wall imposed load)
$$= G + Q$$
$$= 15.5 + 10.5$$
$$= 26.0 \, \text{kN/m}$$

Q as a percentage of P is $100Q/P = 40\%$. From Fig. 10.20, the combined partial safety factor for the superstructure loads is $\gamma_P = 1.48$.

Bearing pressure design

Because of the low level of loading, no explicit check on bearing capacity is considered necessary.

Design span for local depressions

With reference to Table 13.1, the soil conditions are taken to be medium Class B. From Fig. 13.5, assuming 150 mm of hardcore, the design span L is 1.6 m.

Slab design

It is intended for the slab to have top mesh reinforcement only. Figure 13.5 indicates that, for a design span of 1.6 m, a minimum average effective depth of 100 mm is required to comply with the deflection requirements of BS 8110. A slab thickness of 125 mm will therefore be adopted, with 20 mm top cover.

For shrinkage purposes, Table 13.2 indicates, for a 12 m long slab, that A142 mesh is adequate.

Because of the low level of distributed load, and the absence of any significant wall line loads on the slab, there is no need to carry out a local spanning check on the slab under ultimate loads. (If however, the internal thickenings were omitted, and the slab required to carry their load, a check should be carried out in a similar manner to Design Example 2 in section 13.3.3.)

Beam thickening design

Similarly, the low level of loading, and the absence of concentrated point loads on the thickenings, indicate that these can be sized and reinforced on a nominal basis.

For external thickenings, use pre-bent B503 mesh, as shown in Fig. 13.15, with the main T8 bars at 100 mm centres running along the length of the beam. This will result in at least three T8 longitudinal bars in the top and bottom of the beam.

For internal thickenings, again use pre-bent B503 mesh, as shown in Fig. 13.15, with the main bars again running longitudinally to help span over local depressions.

13.3 CRUST RAFT

13.3.1 Introduction

The crust raft is discussed in section 9.4.2 where it is explained that it is a stiffer and stronger version of the nominal crust raft. In this chapter it is intended to take this a stage further through the design procedure and to an actual example.

13.3.2 Design decisions

The crust raft is used where normal ground bearing substrata is relatively poor, where the depth to good load-bearing soils is excessive, but where by dispersing the loads differential settlements can be controlled. It is more attractive where these conditions exist on a relatively level site, i.e. where few steps or changes in level exist.

The considerations for thickening layout and profile are as for the nominal crust (see section 13.2).

The design of the crust raft tends to be on a similar simplified analysis to that of the nominal crust raft but adopting a slightly more analytical approach. The actual calculations for the element cross-sections and raft slab will be carried out on a very similar basis to that of the nominal crust. However, in order to arrive at a suitable span and depression diameter a more detailed analysis of the ground conditions would be carried out. For example, if the raft was to span over possible swallow-holes or shallow mine workings a detailed study of borehole information particularly with regard to the existence of voids below ground combined with historical evidence of previous collapses/depressions would be carried out. Excessively large voids or voids which were creating particular problems in the design of the raft could be considered for grouting in order to reduce the risk of collapse and reduce the diameter of design depressions. If grouting was to be adopted then this would be carried out prior to construction of the raft foundation.

When studying old shallow mine workings reference should be made to Chapter 6. Historical details of pillar and stall workings may be used in the anticipation of the maximum diameter of collapse at cross-over positions, etc. (see Fig. 13.16).

A word of caution should be given at this stage with

regard to the reliance on mining records since, as was emphasized in Chapter 6, while shaft locations are often quite accurate any records of pillar and stall workings tend to be less reliable. This is due to disintegration during oxidation of the pillars, and/or the practice of robbing pillars at the end of the workings' normal life. These actions result in a tendency for larger depressions to occur but usually within a shorter period after completion of the mine workings. The earlier completion of subsidence is an advantage. In some cases, however, if pit props have remained in position un-rotted, early subsidence is prevented. These sort of conditions will be taken into account by the experienced engineer in assessing the borehole records and other records of the possible collapse mechanism and type and size of void or depression to be spanned or grouted.

13.3.3 Design Example 2: Crust raft

A new building is to be built in the grounds of an existing hospital. Ground conditions vary between a firm to soft sandy clay and a clayey sand of variable density. The net allowable bearing pressure for raft design is estimated to be $n_a = 75 \, kN/m^2$.

The wall layout and loadings are shown in Fig. 13.17.

Foundation layout

To avoid the need for deep foundations, a crust raft founded at high level in the sandy clay and clayey sand is to be adopted.

As with all rafts, the layout of thickenings, while taking account of the location of load-bearing walls, is primarily governed by the need to avoid discontinuities or other zones of weakness within the raft. The layout shown in Fig. 13.17 is one way of achieving this goal in this particular situation. Three particular points to note are as follows:

(1) The introduction of a complete movement joint on grid line 2. This splits an awkwardly shaped raft into two approximately rectangular rafts. Note the *doubling-up* of thickenings on both sides of the movement joint, to form two completely separate rafts.
(2) The internal thickening of grid line 3, between grids A and B. This is deliberately positioned to achieve continuity with the external thickening between grids B and D, and does not line with the load-bearing wall to the left of grid line 3.
(3) The internal thickening mid-way between grid lines 2 and 3. This is deliberately positioned to break up the larger areas of raft slab between the thickenings on grids 2 and 3, and to keep the raft slab bays approximately square on plan.

Loadings

Loads are shown in Fig. 13.17, and may be summarized as follows:

Foundation UDL, f = surcharge load, f_S

$$= \text{(slab dead load)} + \text{(slab imposed load)}$$
$$= f_G + f_Q$$
$$= 5.1 + 4.0$$
$$= 9.1 \, kN/m^2$$

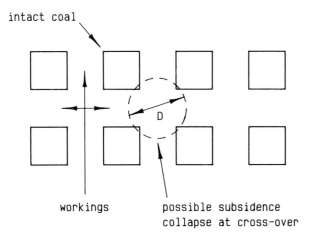

Fig. 13.16 Subsidence prediction for shallow mine workings.

intact coal

workings

possible subsidence
collapse at cross-over

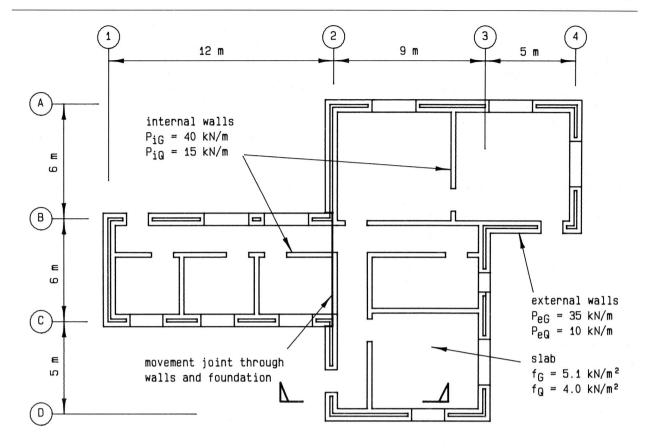

plan on ground floor walls

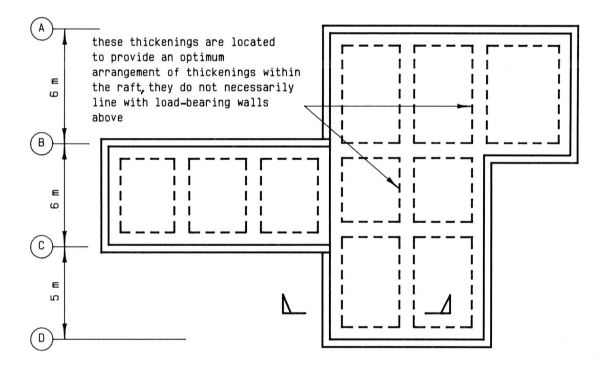

plan on foundation

Fig. 13.17 Crust raft design example – plan layout and loadings.

f_Q as a percentage of f is $100 f_Q / f = 44\%$. From Fig. 10.20, the combined partial safety factor for the foundation loads is $\gamma_F = 1.49$.

$$\begin{aligned}
\text{Internal wall line load, } P_i &= \text{(wall dead load)} + \\
&\quad \text{(wall imposed load)} \\
&= P_{iG} + P_{iQ} \\
&= 40 + 15 \\
&= 55\,\text{kN/m}
\end{aligned}$$

P_{iQ} as a percentage of P_i is $100 P_{iQ} / P_i = 27\%$. From Fig. 10.20, the combined partial safety factor for the superstructure loads is $\gamma_{Pi} = 1.45$.

$$\begin{aligned}
\text{External wall line load, } P_e &= \text{(wall dead load)} + \\
&\quad \text{(wall imposed load)} \\
&= P_{eG} + P_{eQ} \\
&= 35 + 10 \\
&= 45\,\text{kN/m}
\end{aligned}$$

P_{eQ} as a percentage of P_e is $100 P_{eQ} / P_e = 22\%$. From Fig. 10.20, the combined partial safety factor for the superstructure loads is $\gamma_{Pe} = 1.44$.

Allowable bearing pressure

From section 10.10, for zero existing surcharge load, s_S, the net bearing pressure is given by

$$n = p + f_S$$

This may be rearranged to give a superstructure allowable bearing pressure of

$$\begin{aligned}
p_a &= \text{(net allowable pressure)} - \text{(foundation surcharge)} \\
&= n_a - f_s \\
&= 75 - 9.1 \\
&= 66\,\text{kN/m}^2
\end{aligned}$$

Bearing pressure check – slab supporting internal wall

$$\begin{aligned}
\text{Required width of bearing, } B &= \frac{P_i}{p_a} \\
&= \frac{55}{66} \\
&= 0.83\,\text{m}
\end{aligned}$$

Some internal wall loads coincide with raft thickenings, others do not. Consider the latter as the worst case, and design the slab locally in accordance with Fig. 13.1 (b).

$$\begin{aligned}
\text{Ultimate design pressure, } p_u &= \frac{\gamma_{Pi} P_i}{B} \\
&= \frac{1.45 \times 55}{0.83} \\
&= 96\,\text{kN/m}^2
\end{aligned}$$

$$\begin{aligned}
\text{Ultimate bending moment, } M_u &= \frac{p_u (B/2)^2}{2} \\
&= \frac{96(0.83/2)^2}{2} \\
&= 8.3\,\text{kNm/m}
\end{aligned}$$

$$b = 1000\,\text{mm}$$

$$\begin{aligned}
d_{\min} &= 150\,\text{(slab)} - 40\,\text{(bottom cover)} - \\
&\quad 10\,\text{(bar diameter)} - \frac{10\,\text{(bar diameter)}}{2} \\
&= 95\,\text{mm}
\end{aligned}$$

$$\begin{aligned}
\frac{M_u}{bd^2} &= \frac{8.3 \times 10^6}{1000 \times 95^2} \\
&= 0.92
\end{aligned}$$

$$\begin{aligned}
A_{s(req)} &= 0.25\% bd \qquad \text{[BS 8110: Part 2: Chart 2]} \\
&= 238\,\text{mm}^2/\text{m} \qquad \text{(bottom reinforcement)}
\end{aligned}$$

Provide A252 mesh (bottom) $= 252\,\text{mm}^2/\text{m}$.

This minimum level of reinforcement may have to be increased to allow for spanning over local depressions (see below).

Bearing pressure check – external wall thickening

Assume a uniform pressure distribution as in Fig. 13.3 (a)

$$\begin{aligned}
p &= \frac{P_e}{2x} = \frac{45}{2 \times 0.2} \\
&= 113\,\text{kN/m}^2 > p_a = 66\,\text{kN/m}^2
\end{aligned}$$

This pressure is too great.

To keep $p \leq p_a$, one option is to increase the toe width sufficiently to give

$$\begin{aligned}
x &= \frac{P_e}{2p_a} = \frac{45}{2 \times 66} \\
&= 0.34\,\text{m}
\end{aligned}$$

This is an increase in toe width of $340 - 200 = 140\,\text{mm}$. This solution will be adopted in this case (see Fig. 13.18).

A second option would be to provide a thick blinding layer below the edge thickening, as in Fig. 13.3 (b). This would require a width of

$$\begin{aligned}
B &= \frac{P}{p_a} = \frac{45}{66} \\
&= 0.68\,\text{m}
\end{aligned}$$

The required thickness would be

$$\begin{aligned}
h &= \frac{B}{2} - x = \frac{0.68}{2} - 0.2 \\
&= 0.14\,\text{m}
\end{aligned}$$

A third option is for the concrete profile to be left unchanged, and the slab designed to counteract the loading eccentricity, as in Fig. 13.3 (c). Thus the slab is required to provide an ultimate moment of resistance given by

$$\begin{aligned}
M_u &= \gamma_{Pe} P_e \left(\frac{P_e}{2p_a} - x \right) \\
&= 1.44 \times 45 \times \left(\frac{45}{2 \times 66} - 0.2 \right) \\
&= 9.1\,\text{kNm/m}
\end{aligned}$$

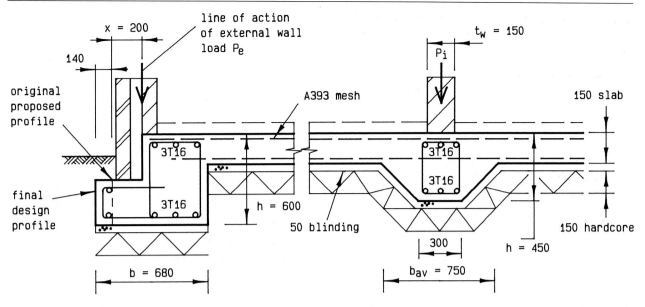

Fig. 13.18 Crust raft design example − section through raft.

$b = 1000\,\text{mm}$

$d_{\text{min}} = 150(\text{slab}) - 20(\text{top}) - 10(\text{bar diameter}) -$

$$\frac{10(\text{bar diameter})}{2}$$

$$= 115\,\text{mm}$$

$$\frac{M_u}{bd^2} = \frac{9.1 \times 10^6}{1000 \times 115^2}$$

$$= 0.69$$

$A_{s(\text{req})} = 0.18\%\,bd$ [BS 8110: Part 2: Chart 2]
$= 207\,\text{mm}^2/\text{m}$ (top reinforcement)

This reinforcement would be in addition to reinforcement calculated below for spanning over local depressions. This solution is relatively uneconomic in this situation, and one of the two previous options − the extended toe or the thick blinding layer − would be preferred.

Design span for local depressions
With reference to Table 13.1, the soil conditions are taken to be medium Class D. From Fig. 13.5, assuming 50 mm blinding and 150 mm hardcore, the design span L is 2.8 m.

Slab spanning over local depression
Consider the worst case situation where an internal load-bearing wall sits directly onto the slab. From Table 13.3, calculate the loading coming onto a 2.8 m diameter depression as follows.

Ultimate load from slab, $F_u = \gamma_F f_s \dfrac{(\pi L^2)}{4}$

$$= 1.49 \times 9.1 \times$$

$$\frac{(\pi \times 2.8^2)}{4}$$

$$= 84\,\text{kN}$$

Ultimate load from wall line load, $P_u = \gamma_{Pi} P_i L$

$$= 1.45 \times 55 \times 2.8$$

$$= 223\,\text{kN}$$

From Table 13.3, cases T1 and T2, the moment factors, K_m, are 1.0 and 1.5, respectively.

$$\Sigma(K_m T_u) = 1.0 F_u + 1.5 P_u$$
$$= (1.0 \times 83) + (1.5 \times 223)$$
$$= 416\,\text{kN}$$

Examining Fig. 13.7, a 150 mm thick slab, with A393 mesh top and bottom is likely to be satisfactory. If the slab has 20 mm top cover and 40 mm bottom cover, it will have average effective depths of 120 mm and 100 mm respectively, giving a combined average effective depth of 110 mm. Figure 13.7 indicates that approximately 320 mm²/m reinforcement is required per face, thus A393 mesh is satisfactory.

This level of reinforcement is relatively heavy for a raft slab, and is due to the large local depression span in this particular example.

Internal beam spanning over local depression
Design the internal beams to carry an internal load-bearing wall, spanning over a local depression. From Table 13.4, the loading coming onto a 2.8 m diameter depression is as follows:

I1: $F_u = \gamma_F f_s \dfrac{(\pi L^2)}{4} = 1.49 \times 9.1 \dfrac{(\pi \times 2.8^2)}{4}$

$$= 83\,\text{kN}$$

I2: $P_{iu} = \gamma_{Pi} P_i L$ $= 1.45 \times 55 \times 2.8$
$$= 223\,\text{kN}$$

From Table 13.4, the total effective load on depression for bending is

$$\Sigma(K_m T_u) = 0.5 F_u + 1.0 P_{iu}$$
$$= (0.5 \times 83) + (1.0 \times 223)$$
$$= 265\,\text{kN}$$

Take average width of internal thickening to be $b = 750\,\text{mm}$.

$$\text{Average effective depth} = 450 - 30(\text{average cover}) -$$
$$10(\text{link}) - \frac{16(\text{bar diameter})}{2}$$
$$= 402\,\text{mm}$$

$$\frac{\Sigma(K_m T_u)L}{b} = \frac{265 \times 2.8}{0.75}$$
$$= 989\,\text{kN}$$

From Fig. 13.8, the area of reinforcement required per face is

$$A_s = 725b = 725 \times 0.75$$
$$= 544\,\text{mm}^2$$

Provide 3T16 top and bottom, giving $A_s = 603\,\text{mm}^2/\text{m}$.

Shear reinforcement is calculated, in accordance with BS 8110, using

$$V_u = \frac{T_u}{2}$$
$$= \frac{F_u + P_{iu}}{2}$$
$$= \frac{83 + 223}{2}$$
$$= 153\,\text{kN}$$

Edge beam spanning over local depression
Consider the worst case, where an internal wall meets the external wall (on grid line A). With reference to Table 13.4, cases E1, E2 and E3, the total ultimate loads coming onto the depression are as follows:

$$\text{E1:} \quad F_u = \gamma_F f_S \frac{(\pi L^2)}{8} = 1.49 \times 9.1 \frac{(\pi \times 2.8^2)}{8}$$
$$= 42\,\text{kN}$$

$$\text{E2:} \quad P_{eu} = \gamma_{Pe} P_e L \quad = 1.44 \times 45 \times 2.8$$
$$= 181\,\text{kN}$$

$$\text{E2:} \quad P_{iu} = \gamma_{Pi} P_i \frac{L}{2} \quad = 1.45 \times 55 \times \frac{2.8}{2}$$
$$= 112\,\text{kN}$$

From Table 13.4, total effective load on depression for bending is

$$\Sigma(K_m T_u) = 0.5 F_u + 1.0 P_{eu} + 1.0 P_{iu}$$
$$= (0.5 \times 42) + (1.0 \times 181) + (1.0 \times 112)$$
$$= 314\,\text{kN}$$

Take average width of edge thickening to be $b = 680\,\text{mm}$.

$$\text{Average effective depth} = 600 - 30(\text{average cover}) -$$
$$10(\text{link}) - \frac{16(\text{bar diameter})}{2}$$
$$= 552\,\text{mm}$$

$$\frac{\Sigma(K_m T_u)L}{b} = \frac{314 \times 2.8}{0.68}$$
$$= 1293\,\text{kN}$$

From Fig. 13.8, the area of reinforcement required per face is

$$A_s = 750b = 750 \times 0.68$$
$$= 510\,\text{mm}$$

Provide 3T16 top and bottom, giving $A_s = 603\,\text{mm}^2/\text{m}$. *Note*: the top reinforcement to adjacent corners may need to be increased (see below).

Shear force for reinforcement design is

$$V_u = \frac{T_u}{2} = \frac{F_u + P_{eu} + P_{iu}}{2}$$

Corner beam spanning over local depression
With reference to Table 13.4, cases C1 and C4, the total ultimate loads coming onto the depression are as follows.

$$\text{C1:} \quad F_u = \gamma_F f_S (0.64 L^2) = 1.49 \times 9.1(0.64 \times 2.8^2)$$
$$= 68\,\text{kN}$$

$$\text{C4:} \quad P_{eu} = \gamma_{Pe} P_e \frac{(2L)}{\sqrt{2}} \quad = 1.44 \times 45 \frac{(2 \times 2.8)}{\sqrt{2}}$$
$$= 257\,\text{kN}$$

From Table 13.4, total effective load on depression for bending is

$$\Sigma(K_m T_u) = 0.5 F_u + 1.0 P_{eu}$$
$$= (0.5 \times 68) + (1.0 \times 257)$$
$$= 291\,\text{kN}$$

$b = 680\,\text{mm}$ and $d = 552\,\text{mm}$, as previously.

$$\frac{\Sigma(K_m T_u)L}{b} = \frac{291 \times 2.8}{0.68}$$
$$= 1198\,\text{kN}$$

From Fig. 13.8, the area of reinforcement required per face is

$$A_s = 950b = 950 \times 0.68$$
$$= 646\,\text{mm}^2$$

Provide 4T16 top, giving $A_s = 804\,\text{mm}^2/\text{m}$. It is recommended that this reinforcement should extend a distance of $L\sqrt{2} = 2.8 \times \sqrt{2} = 4.0\,\text{m}$ from each corner.
Shear force for reinforcement design is

$$V_u = \frac{T_u}{2} = \frac{F_u + P_{eu}}{2}$$

13.4 BLANKET RAFT

13.4.1 Introduction
The design of the reinforced concrete raft which sits on the blanket is similar to that of the crust raft, the variation being the design of the blanket below. The blanket is designed to disperse any heavy point or knife edge loads and to interact with the raft foundation. In addition the action of compacting the stone blanket into the previously reduced level of the sub-strata tends to search out any soft spots and compact the sub-strata during this operation. A typical specification for the blanket would normally require that the compaction of

the sub-strata prior to the stone filling should continue until no further movement is evident and that a similar requirement be applied to each layer of the stone. It should be noted however that this applies mainly to granular soils; should silty materials or waterlogged clay/silts occur, then the compaction would have to begin after laying quite a deep layer of stone, in order to crust up the surface to a suitable degree for operation of the plant. This is due to the need for an immediate equal reaction force to be available from the sub-strata on application of the compaction force. With many sands and free-draining sub-strata no problems exist, however with saturated silts and soft clays porewater pressure cannot dissipate quickly enough. The reaction of such soil if not adequately crusted is to *pudding* or *cow belly* due to the load becoming temporarily supported on the contained water (see Fig. 13.19). The solution is to follow the advice in section 8.2 on surface rolling.

Blanket rafts are usually used for low-rise developments on relatively poor quality ground, for example, where wet conditions exist making excavation difficult or where sub-strata varies (from sands to silts or sands to clays) leading to differential settlement.

In addition they are used in areas where hard spots (such as old foundation brickwork, strips, etc.) exist but can be broken out to a level below the blanket sufficient to reduce the differential stiffness to an acceptable variation.

The design is carried out by sizing from past experience or using calculations from expected differences in ground conditions and loadings.

Construction of the blanket raft is often carried out on a similar basis to that of road construction, by reducing the level in long strips and spreading and compacting the stone in layers in a similar strip construction process. For example, long runs of domestic housing and flats can often best be constructed by this method (see Fig. 13.20).

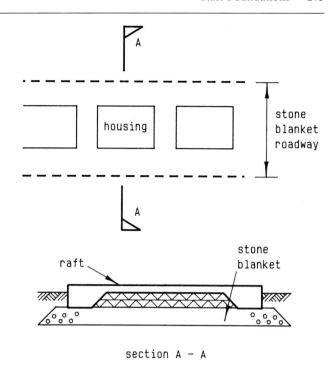

Fig. 13.20 Blanket raft construction.

13.4.2 Design decisions

The decision to use a blanket raft arises from the need to upgrade sub-strata and to disperse concentrated load where the alternative of piling or vibro-compaction is considered to be either unnecessary or excessively expensive in comparison. The blanket is used to reduce the differential settlement and reduce bending moments in the raft foundation. Typical sites would be those on (1) variable sub-strata, (2) granular fill materials, or (3) those for developments with foundations with concentrated loads where the economics of a blanket raft can prove to be very attractive.

13.4.3 Sizing the design

The sizing of the reinforced concrete raft is carried out in a similar manner to that of the crust raft and reference should be made to the design in section 13.3. The sizing of the hardcore blanket however is generally determined by a combination of engineering judgement and the dispersion required for concentrated loads. The blanket is therefore required to reduce the bearing stress at the blanket/sub-strata interface to an allowable pressure. The advantage and added benefits of the interaction which are likely to occur between the blanket and the raft are exploited in the overall behaviour which forms part of the detailing and the experienced engineer's judgement and feel of the total requirements. The authors make no apology for the fact that foundation engineering is partly art and partly analysis, and the two must be blended in order to achieve satisfactory economic results.

It should be noted also that this overall behaviour can be time dependent particularly where rafts span from granular materials to clay materials where the stresses at any given time in the life of the raft can be very different due to the

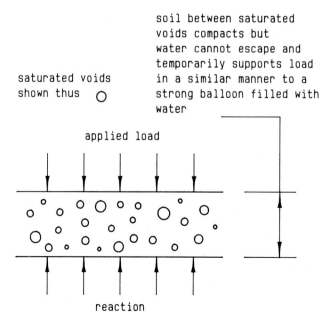

Fig. 13.19 Pudding or cow belly action.

variation in settlement time relationships between the materials. In some cases joints would be incorporated between sections of the foundation to relieve unacceptably high stress concentrations. These pressures tend to be related closely to porewater dispersion during stressing of the substrata. For this reason the typical assumed dispersion for design of local elements is only a design tool to achieve reasonable sizing and the overall behaviour performance of the raft must be considered and assessed by the experienced engineer.

A typical load dispersion assumed in the initial sizing of the blanket depth would relate to the previously mentioned $60°-45°$ dispersion discussed in section 13.1 and consideration is given to the loss of stone during consolidation and the overall behaviour of the raft and blanket interaction over the design life of the building. A typical design is shown in section 13.4.4.

13.4.4 Design Example 3: Blanket raft

The crust raft in Design Example 2 (section 13.3.3) is required to be redesigned for a taller building, with wall line loads increased to twice their previous value as shown in Fig. 13.21. To avoid overstressing the ground, the raft is to be founded on a *blanket* of compacted granular material which will replace existing unsuitable ground. Since the additional depth of blanket will reduce the local depression design span (see Fig. 13.5), shallower external and internal thickenings 450 mm deep and 300 mm deep respectively are proposed, to be excavated into the blanket (see Fig. 13.22).

Loadings

Loads and combined partial load factors are taken from Fig. 13.21 and Design Example 2 as follows:

$$f = 9.1 \, \text{kN/m}^2 \qquad \gamma_F = 1.49$$
$$P_i = 2 \times 55 = 110 \, \text{kN/m} \qquad \gamma_{Pi} = 1.45$$
$$P_e = 2 \times 45 = 90 \, \text{kN/m} \qquad \gamma_{Pe} = 1.44$$

Allowable bearing pressure

From Design Example 2, the superstructure allowable bearing pressure is

$$p_a = 66 \, \text{kN/m}^2$$

Design procedure for the blanket

The purpose of the blanket is to spread the load sufficiently to avoid overstressing the underlying bearing material. The design process therefore involves determining a depth of blanket, h_{fill}, such that the stresses do not exceed the allowable ones.

In this example this will be done for the situation where internal load-bearing walls bear directly on the slab. The width of the external thickenings will then be sized, so that they in turn do not overstress the subgrade material.

Bearing stress in blanket below internal wall bearing on slab

The blanket, composed of compacted granular material, has been judged to have an allowable bearing pressure of $p_{ba} = 200 \, \text{kN/m}^2$. The applied line load from internal walls is

$P_i = 110 \, \text{kN/m}$. To avoid overstressing the blanket, the slab is required to spread this load over a width given by

$$B_{\text{conc}} = \frac{P_i}{P_{ba}}$$
$$= \frac{110}{200}$$
$$= 0.55 \, \text{m}$$

From Fig. 13.1, it can be seen that bottom reinforcement will be needed to spread the load over this width. In this instance the reinforcement provided to span over local depressions will be sufficient to cater for this condition. If a calculation is considered necessary, it will follow the method used in Design Example 2 (see section 13.3.3).

Determination of blanket depth

From Fig. 13.4, assuming a 45° spread of load through the slab, and a 60° spread through the compacted material, the width of bearing at formation level is

$$B_{\text{fill}} = B_{\text{conc}} + 1.15 h_{\text{fill}}$$

To avoid overstressing the subgrade material, this must have a minimum value of

$$B_{\text{fill}} = \frac{\text{wall line load}}{\text{allowable bearing pressure}}$$
$$= \frac{P_i}{p_a}$$

Rearranging, the required thickness of blanket (and blinding), to avoid overstressing the underlying material, is given by

$$h_{\text{fill}} = \frac{B_{\text{fill}} - B_{\text{conc}}}{1.15}$$
$$= \frac{P_i/p_a - B_{\text{conc}}}{1.15}$$
$$= \frac{(110/66) - 0.55}{1.15}$$
$$= 0.97 \, \text{m}$$

A 950 mm thickness of compacted material below slabs plus 50 mm of blinding will be adopted, giving a blanket thickness of $h_{\text{fill}} = 1000 \, \text{mm}$.

Determination of width of external wall thickening

Since the edge thickenings project 300 mm below the underside of the slab, the blanket thickness below the thickenings is given by $h_{\text{fill}} = 1000 - 300 = 700 \, \text{mm}$.

From Fig. 13.4, the width of bearing at formation level is

$$B_{\text{fill}} = B_{\text{conc}} + 1.15 h_{\text{fill}}$$

Again this must have a minimum value of

$$B_{\text{fill}} = \frac{P_e}{p_a}$$

Rearranging gives

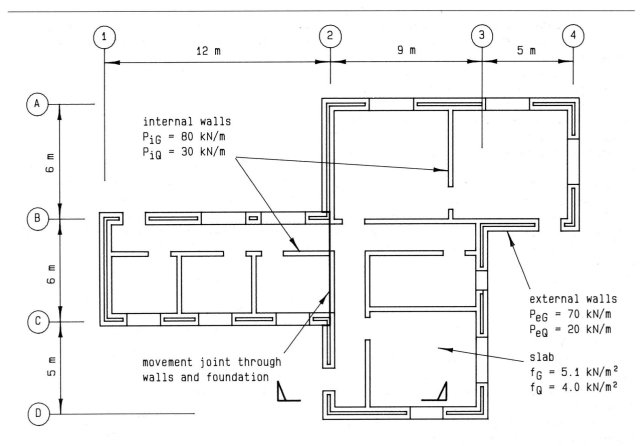

internal walls
P_{iG} = 80 kN/m
P_{iQ} = 30 kN/m

external walls
P_{eG} = 70 kN/m
P_{eQ} = 20 kN/m

slab
f_G = 5.1 kN/m^2
f_Q = 4.0 kN/m^2

movement joint through
walls and foundation

plan on ground floor walls

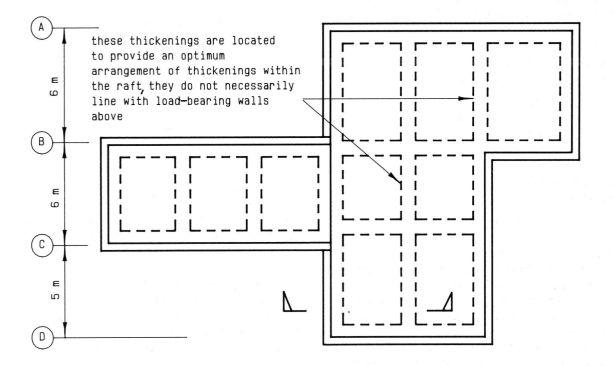

these thickenings are located
to provide an optimum
arrangement of thickenings within
the raft, they do not necessarily
line with load-bearing walls
above

plan on foundation

Fig. 13.21 Blanket raft design example — plan layout and loadings.

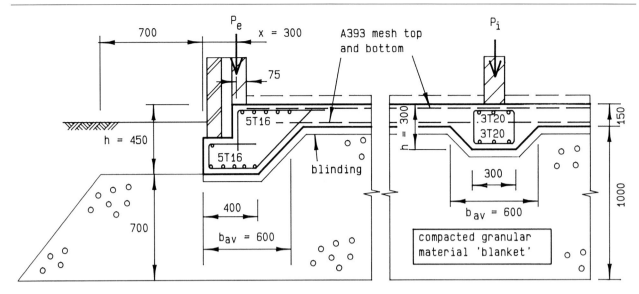

Fig. 13.22 Blanket raft design example — section through raft.

$$B_{conc} = B_{fill} - 1.15h_{fill}$$

$$= \frac{P_e}{p_a} - 1.15h_{fill}$$

$$= \frac{90}{66} - (1.15 \times 0.70)$$

$$= 0.56\,m$$

From Fig. 13.3 (a), the effective bearing width of the edge thickening for bearing pressure design is

$$B_{conc} = 2x$$

Thus the minimum required value for x is

$$x = \frac{0.56}{2}$$

$$= 0.28\,m$$

A projection of $x = 300$ mm will therefore be provided (see Fig. 13.22).

Determination of width of internal wall thickening

A similar calculation could be carried out for internal thickenings. Instead in this instance the bottom slab reinforcement will be fully lapped through the thickenings, using loose bar reinforcement. This will then give a capacity at least as good as those situations where the internal walls bear directly onto the slab without thickenings.

The width of the thickening can thus be chosen to suit the local depression condition. In this instance a value of 300 mm will be assumed, it being a practical minimum value to fit a reinforcement cage.

Design span for local depressions

With reference to Table 13.1, the soil conditions are taken to be medium Class D, as per Design Example 2. From Fig. 13.6, the design span is $L_s = 1.85$ m below slabs ($h_{fill} = 1.0$ m), $L_i = 2.0$ m below internal thickenings ($h_{fill} = 0.85$ m), and $L_e = 2.2$ m below external thickenings ($h_{fill} = 0.70$ m).

Slab spanning over local depression

Consider the worst case situation, where an internal load-bearing wall sits directly onto the slab. Calculate the loading coming onto a 1.85 m diameter depression as follows:

$$\text{Distributed load from slab, } F_u = \gamma_F f_s \frac{(\pi L_s^2)}{4}$$

$$= 1.49 \times 9.1 \frac{(\pi \times 1.85^2)}{4}$$

$$= 36\,kN$$

$$\text{Line load from internal wall, } P_u = \gamma_{Pi} P_i L_s$$

$$= 1.45 \times 110 \times 1.85$$

$$= 295\,kN$$

From Table 13.3, Cases T1 and T2, moment factors K_m are 1.0 and 1.5 respectively.

$$\Sigma(K_m T_u) = 1.0F_u + 1.5P_u$$

$$= (1.0 \times 36) + (1.5 \times 295)$$

$$= 479\,kN$$

If the 150 mm thick slab has 20 mm top cover and 40 mm bottom cover, it will have average effective depths of 120 mm and 100 mm respectively, giving a combined average effective depth of 110 mm. Figure 13.7 indicates approximately 370 mm²/m reinforcement is required per face, thus A393 mesh is adequate.

Internal beam spanning over local depression

Design the internal beams to carry an internal load-bearing wall, spanning over a local depression. From Table 13.4, the loading coming onto a 2.0 m diameter depression is calculated as follows:

$$I1: \quad F_u = \gamma_F f_s \frac{(\pi L_i^2)}{4} = 1.49 \times 9.9 \frac{(\pi \times 2.0^2)}{4}$$

$$= 43\,kN$$

$$I2: \quad P_{iu} = \gamma_{Pi} P_i L_i \quad = 1.45 \times 110 \times 20$$

$$= 319\,kN$$

From Table 13.4, total effective load on depression for bending is

$$\Sigma(K_m T_u) = 0.5F_u + 1.0P_{iu}$$
$$= (0.5 \times 43) + (1.0 \times 319)$$
$$= 340 \, kN$$

Take average width of internal thickening to be $b = 600 \, mm$ (see Fig. 13.22).

$$\text{Average effective depth} = 300 - 30(\text{average cover}) -$$
$$10(\text{link}) - \frac{20(\text{bar diameter})}{2}$$
$$= 250 \, mm$$

$$\frac{\Sigma(K_m T_u)L}{b} = \frac{340 \times 2.0}{0.6}$$
$$= 1134 \, kN$$

From Fig. 13.8, the area of reinforcement required per face is

$$A_S = 1400b = 1400 \times 0.6$$
$$= 840 \, mm^2$$

Provide 3T20 top and bottom, giving $A_s = 943 \, mm^2/m$. Shear force for reinforcement design is

$$V_u = \frac{T_u}{2}$$
$$= \frac{F_u + P_{iu}}{2}$$
$$= \frac{43 + 319}{2}$$
$$= 181 \, kN$$

Edge beam spanning over local depression
Consider the worst case, where an internal wall meets the external wall (on grid line A). With reference to Table 13.4, cases E1, E2 and E3, the total ultimate loads coming onto the depression are as follows:

E1: $F_u = \gamma_F f_S \frac{(\pi L_e^2)}{8} = 1.49 \times 9.1 \frac{(\pi \times 2.2^2)}{8}$
$$= 25 \, kN$$

E2: $P_{eu} = \gamma_{Pe} P_e L_e = 1.44 \times 90 \times 2.2$
$$= 285 \, kN$$

E3: $P_{iu} = \gamma_{Pi} P_i \frac{L_e}{2} = 1.45 \times 110 \times \frac{2.2}{2}$
$$= 175 \, kN$$

From Fig. 13.4, the total effective load on depression for bending is

$$\Sigma(K_m T_u) = 0.5F_u + 1.0P_{eu} + 1.0P_{iu}$$
$$= (0.5 \times 26) + (1.0 \times 285) + (1.0 \times 175)$$

Take average width of edge thickening to be $b = 600 \, mm$ (see Fig. 13.22).

Average effective depth $= 450 - 30(\text{average cover}) -$
$$10(\text{link}) - \frac{16(\text{bar diameter})}{2}$$
$$= 402 \, mm$$

$$\frac{\Sigma(K_m T_u)L}{b} = \frac{473 \times 2.2}{0.6}$$
$$= 1736 \, kN$$

From Fig. 13.8, the area of reinforcement required per face is

$$A_s = 1300b = 1300 \times 0.6$$
$$= 780 \, mm^2$$

Provide 5T16 top and bottom, giving $A_s = 1006 \, mm^2/m$. *Note*: the top reinforcement to adjacent corners may need to be increased (see below).
Shear force for reinforcement design is

$$V_u = \frac{T_u}{2} = \frac{F_u + P_{eu} + P_{iu}}{2}$$

Corner beam spanning over local depression
With reference to Table 13.4, cases C1 and C4, the total ultimate loads coming onto the depression are as follows:

C1: $F_u = \gamma_F f_S(0.64 L_e^2) = 1.49 \times 9.1 \, (0.64 \times 2.2^2)$
$$= 42 \, kN$$

C4: $P_{eu} = \gamma_{Pe} P_e \frac{2L_e}{\sqrt{2}} = 1.44 \times 90 \frac{(2 \times 2.2)}{\sqrt{2}}$
$$= 403 \, kN$$

From Table 13.4, the total effective load on depression for bending is

$$\Sigma(K_m T_u) = 0.5F_u + 1.0P_{eu}$$
$$= (0.5 \times 42) + (1.0 \times 403)$$
$$= 424 \, kN$$

$b = 600 \, mm$ and $d = 402 \, mm$, as previously.

$$\frac{\Sigma(K_m T_u)L}{b} = \frac{424 \times 2.2}{0.6}$$
$$= 1555 \, kN$$

From Fig. 13.8, the area of reinforcement required per face is

$$A_s = 1800b = 1800 \times 0.6$$
$$= 1080 \, mm^2$$

Provide 4T20 top, giving $A_s = 1257 \, mm^2/m$. It is recommended that this reinforcement should extend a distance of $L\sqrt{2} = 2.2 \times \sqrt{2} = 3.1 \, m$ from each corner.
Shear force for reinforcement design is

$$V_u = \frac{T_u}{2} = \frac{F_u + P_{eu}}{2}$$

13.5 SLIP SANDWICH RAFT

13.5.1 Introduction
This raft is mainly used in active mining areas or where clay

is creeping on inclined sand beds where the horizontal ground strains set up during subsidence or creep movements would cause damage to the structure, if allowed to be transferred up to it via the foundation (see Fig. 13.23).

By using a slip-plane of known resistance, the maximum force which can be transferred from the ground to the building before the plane ruptures can be calculated, and the raft designed to resist this force in any direction that it is likely to occur.

The raft can be a flat slab profile thus avoiding the use of

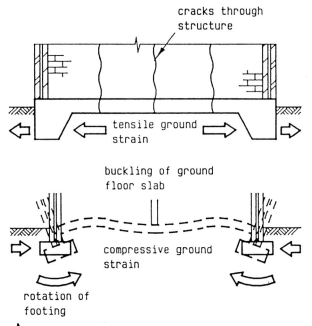

Fig. 13.23 Effects on foundations from horizontal ground strains.

downstand thickenings which may pick up excessive passive load from the ground strain. Alternatively a slab with medium thickenings incorporating a design which provides a slip-plane below the hardcore dumplings (i.e. the raised areas of hardcore protruding up between the beam lines) can be used (see Fig. 13.24).

The ideal ground (i.e. uniform firm layers of, non-frost-susceptible low shrinkability sub-strata) to facilitate a flat slab is rarely on the site to be developed. Therefore to prevent damage from frost, clay heave or differential settlement, thickenings are often necessary. In such situations the ground strains being picked up either have to be designed to be resisted by the raft or a slip-plane layer provided below the level of the downstands to reduce the forces being transferred. The upper raft (above the slip plane layer) of a *slip sandwich raft* can be any of the other rafts already designed and discussed in earlier sections of this chapter. The difference between the slip sandwich raft and the other rafts can be summarized as relating to the slip-plane layer below the slab and the horizontal forces produced from the ground strains transferred through the slip-plane.

The additional stresses are analysed by calculating the forces transferred from the ground strain and these forces are added to the design conditions already discussed for other rafts.

13.5.2 Design decisions

The design decision to use a slip sandwich raft will depend totally on the possible existence of critical horizontal ground strains in the sub-strata during the life of the building and the need to restrict these forces to prevent them being transferred in total to the superstructure. The use of jointing to reduce the overall buildings into small independent robust units is part of the design process. In addition the possible

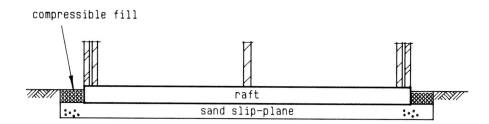

flat slip sandwich raft

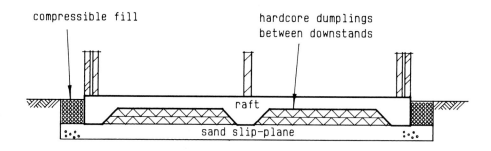

downstand slip sandwich raft

Fig. 13.24 Alternative slip sandwich rafts.

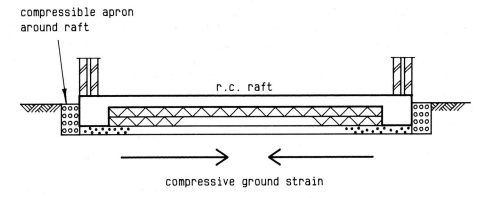

Fig. 13.25 Section through raft and compressive apron.

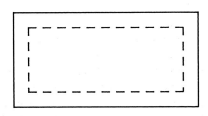

plan on rectangular raft

Fig. 13.26 Simple rectangular raft.

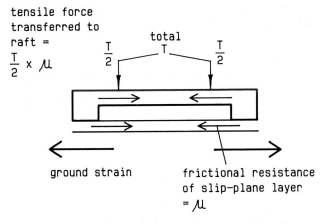

Fig. 13.27 Forces on foundation from ground strains.

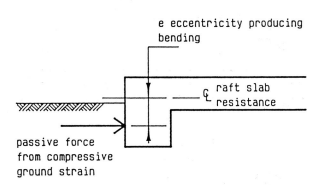

Fig. 13.28 Passive forces on raft downstands.

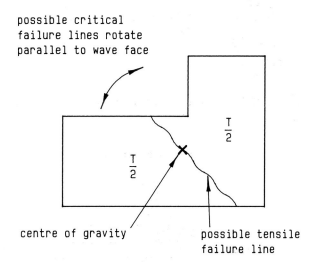

Fig. 13.29 Direction of tensile failure dependent on direction of wave face.

need to incorporate compressible aprons around the raft requires consideration in the design and it is dependent upon the directions and magnitude of the ground strains (see Fig. 13.25).

13.5.3 Sizing the design

The basic sizing of the raft to sit on the slip-plane would follow the principles already discussed in other sections of this chapter.

The additional requirements for the slip sandwich raft however relate to the compressive and tensile forces likely to be transferred through the slip-plane from the ground strains. If a simple rectangular plan shape raft is considered as shown in Fig. 13.26 (which would be the ideal plan shape for such a raft) and a 150 mm thick sand slip-plane, the

following simple analysis can be applied. Assume the total weight of the building and foundations to equal T and the frictional resistance of the sand slip-plane layer to be equal to μ (see Fig. 13.27), the largest horizontal force which can be transferred up from the ground strain through the slip-plane will be equal to $(\mu T)/2$. The reason for the total load acting down being halved is that the maximum force that can be transferred as tension through the building must be reacted by the other half of the load. This formula assumes that no other passive forces are being transferred to the foundation, i.e. that all forces are transferred via the sand as a frictional force.

If however downstands project below the raft then the slip-plane layer should be positioned below such downstands and the downstands kept to a minimum. In addition if compressive ground strains are occurring then an apron must be introduced to prevent or restrict the amount of strain transferred from passive pressure on the raft edges. If such pressures cannot be avoided then they must be added to the force indicated above and allowed for in the design. Any eccentricities of such forces should also be taken into account in the design of the raft since these will produce bending in the raft foundation (see Fig. 13.28) which indicates an eccentric force on a downstand raft thickening.

If the plan shape adopted is not rectangular, for example, the L shape as shown in Fig. 13.29, then the two halves of the building producing the force $(\mu T)/2$ from frictional resistance below the surface will produce tensile or compressive forces across a line which passes through the centre of gravity of the building and which will tend to rotate towards a line parallel to the subsidence wave. Consideration must be given therefore to the additional stresses produced by these forces including any bending moments across this face or on lines parallel to the face (see Fig. 13.29).

13.5.4 Design Example 4: Slip sandwich raft

The nominal crust raft for a pair of semi-detached properties in Design Example 1 (section 13.2.3) is now assumed to be located in a mining area. It will therefore be reworked as a slip sandwich raft, to accommodate the associated ground strains.

The slip sandwich raft is designed on the assumption that the two halves of the raft – on either side of the centreline – are moving away from each other (tension), or towards each other (compression). The maximum horizontal force across the centreline of the raft, arising from the horizontal strains

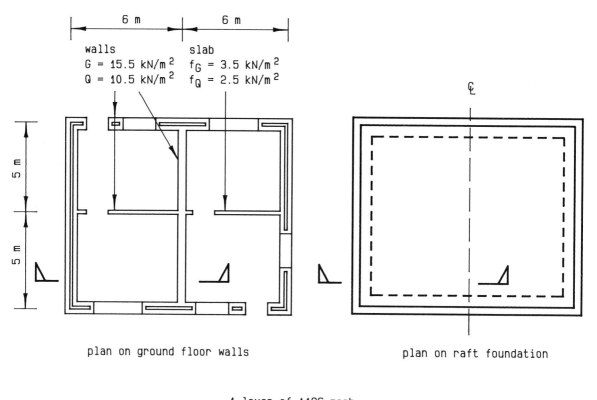

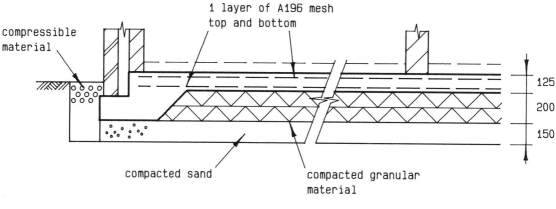

Fig. 13.30 Slip sandwich raft design example.

in the underlying ground, is equal to the maximum frictional force which can be transmitted across the slip-plane into one half of the raft.

Vertical loadings
Loads are as Design Example 1 (see Fig. 13.15):

$$\text{Foundation load, } f = 6.0 \, \text{kN/m}^2; \gamma_F = 1.48$$
$$\text{Wall line load, } P = 26.0 \, \text{kN/m}; \gamma_P = 1.48$$

Horizontal force across raft centreline
The raft is 10.0 m × 12.0 m. With reference to Fig. 13.30, the total ultimate vertical load on one half of the raft is

$$
\begin{aligned}
T_u &= \text{(ultimate foundation load)} + \text{(ultimate wall loads)} \\
&= \gamma_F f(6.0 \times 10.0) + \gamma_P P[(3 \times 6.0) + (1.5 \times 10.0)] \\
&= (1.48 \times 6.0 \, \text{kN/m}^2 \times 60 \, \text{m}^2) + \\
&\quad (1.48 \times 26 \, \text{kN/m} \times 33 \, \text{m}) \\
&= 1802 \, \text{kN}
\end{aligned}
$$

A horizontal layer of 150 mm of compacted sand will be located at the level of the underside of the raft thickenings to act as a slip-plane (see Fig. 13.30). The raft will be assumed to behave as illustrated in Fig. 6.14 and Fig. 13.27.

National Coal Board guidelines[2] recommend the use of a coefficient of friction of $\mu = 0.66$ for a sand slip-plane. The length of the centreline is $B = 10.0$ m. The horizontal force per metre length across the centreline of raft is therefore given by

$$
\begin{aligned}
H_u &= \frac{\mu T_u}{B} \\
&= \frac{0.66 \times 1802}{10.0} \\
&= 119 \, \text{kN/m}
\end{aligned}
$$

Reinforcement design for raft tension
Provide high yield reinforcement to resist this force in tension such that

$$
\begin{pmatrix} \text{ultimate force in} \\ \text{reinforcement} \end{pmatrix} = \begin{pmatrix} \text{ultimate horizontal} \\ \text{tension force} \end{pmatrix}
$$

$$
\begin{aligned}
0.87 f_y A_s &= H_u \\
A_s &= \frac{H_u}{0.87 f_y} \\
&= \frac{119 \times 10^3}{0.87 \times 460} \\
&= 297 \, \text{mm}^2/\text{m}
\end{aligned}
$$

Provide two layers of A196 mesh throughout, as shown in Fig. 13.30. This compares with the single layer of A142 mesh required for the nominal crust raft in Design Example 1.

Design for raft compression
The same tensile force calculated above can also act in compression. By inspection the raft concrete can accommodate this level of compressive stress.

13.6 CELLULAR RAFT

13.6.1 Introduction
Cellular rafts are used where valuable increases in bearing capacity can be achieved by the removal of overburden or where severe bending moments may be induced due to mining activity, seismic loadings, etc. The cellular form in such situations can perform two functions (see Fig. 13.31). The foundation while being economic for such situations is one of the most expensive foundation types used.

13.6.2 Sizing the design
In the case of overburden removal the depth required may relate more to the excavation required to produce adequate reduction in load than to the bending moment resistance of the cellular form (see Fig. 13.32).

On the other hand it is more common for the raft depth to relate to the moments likely to be induced and the reduced overburden load to be a resulting bonus. For example, a raft designed to resist seismic loads or mining subsidence may need to be designed to span two-thirds of its total length and to cantilever one-third of its length. Similarly rafts on

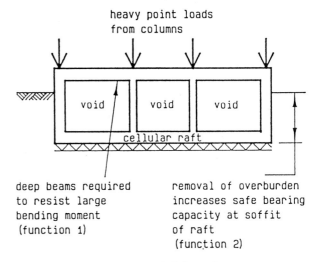

Fig. 13.31 Cellular raft.

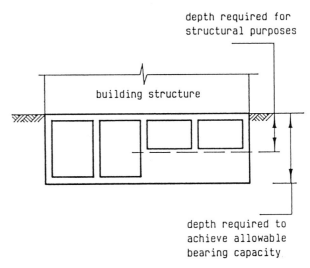

Fig. 13.32 Raft depth dictated by bearing requirements.

variable ground subjected to large differential settlements may require such design parameters. This requirement on a building plan which is restricted to say 20 m on any one side can produce very large shears and bending moments thus requiring deep rectangular, I or box sections.

The size of these beams can often be reduced by jointing buildings into smaller units (see Chapter 6). Cellular rafts or other rafts which are formed from beams crossing at right angles are difficult to assess since loads are resisted in two directions of the framework. The designer must start by calculating his design column loads and relating these to the overall plan and ground pressures.

The calculations for the ground pressures based upon the centre of gravity of the loads and the relative stiffness of the raft foundation is then considered. In the case of the cellular raft a stiff raft would normally be assumed (see Fig. 13.33).

With reference to Fig. 13.33, the centre of gravity of the load would be calculated in the normal way and the resultant load would be the total addition of the loads on the framework for the design conditions being considered.

The ground pressures at the corners of the raft would then be determined. Assume the resultant total load to be T, the total area to be A, the moment in each of the two directions to be M_x and M_y, the eccentricity in each direction to be e_x and e_y, and the section modulus in those directions assuming symmetrical plan to be Z_x and Z_y. The stress at each corner would equal

$$\frac{T}{A} \pm \frac{M_x}{Z_x} \pm \frac{M_y}{Z_y}$$

M_x and M_y, being the resultant moments, equal $T \times e_x$ and $T \times e_y$, respectively (see Fig. 13.34).

As usual in design calculations these theoretical pressures will not necessarily be achieved on site and the difficulty then arises when the engineer tries to assess the actual ground pressures. These pressures will be dependent on the sub-strata, the flexibility of the raft, the actual loads occurring at any time and the time at which these pressures are considered relative to the original application of the load.

None of these conditions can be assessed accurately nor is it necessary to do so. The procedure to be adopted is for the

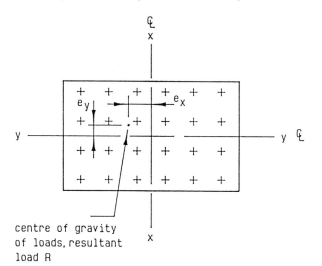

centre of gravity
of loads, resultant
load R

Fig. 13.33 Centre of gravity of load on stiff raft.

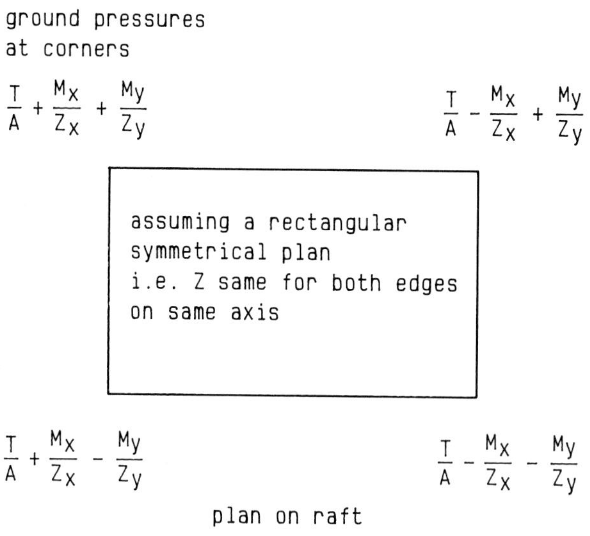

ground pressures
at corners

$$\frac{T}{A} + \frac{M_x}{Z_x} + \frac{M_y}{Z_y} \qquad\qquad \frac{T}{A} - \frac{M_x}{Z_x} + \frac{M_y}{Z_y}$$

assuming a rectangular
symmetrical plan
i.e. Z same for both edges
on same axis

$$\frac{T}{A} + \frac{M_x}{Z_x} - \frac{M_y}{Z_y} \qquad\qquad \frac{T}{A} - \frac{M_x}{Z_x} - \frac{M_y}{Z_y}$$

plan on raft

Fig. 13.34 Corner pressure below stiff raft.

engineer to apply the art of foundation design in producing his raft. This means that he takes consideration of these variables as part of his refinement of the design at the detailing stage. He makes adjustments to sections, reinforcement, location of joints and to the number of beams on plan to produce a more realistic and practical solution. The authors make no apology for suggesting that foundation engineering is an art as well as a science since they have learnt this art from long and bitter experience. The engineer doing this exercise therefore would begin by preparing a rough layout from practical experience indicating rough sizes likely to be required. Assumptions are then made, for example:

(1) The raft will be assumed to be stiff.
(2) The bearing pressure will be assumed to be trapezoidal, as indicated in Fig. 13.34.
(3) The load from the structure will be assumed to be fixed at the design load for the analysis.
(4) The reactions to each beam line will be assumed to be proportional to the bearing area and ground pressure on that line and when both directions are totalled they must be equal to the applied load at that point from the structure.

Many methods of analysing foundations have been proposed, some assuming springs below the foundations, some assuming uniform bearing pressure, some assuming curved bearing pressures, some trying to take into account the stiffness of the raft foundation.

At the end of the day the experienced engineer fully understands that all these calculated methods, while being reasonable and theoretically logical, are not realistic. The foundation which the engineer designs will not sit on the soil which was taken to the laboratory for testing, it will not for its total life be resisting the loads he has calculated, it will not be of the stiffness he has assumed, it will not be subjected to the settlements or movements he has anticipated by calculation. With all this in mind the engineer uses his analysis as one of the tools in his kit bag. He does not ignore the above knowledge produced by design and calculation but tries to adjust his design in a direction which is more realistic.

To achieve this the practical engineer will often make simple assumptions to produce a quicker analysis while satisfying himself of his foundation requirements without the need for complicated and often less accurate methods of analysis. A typical design approach for a cellular raft is shown in the following example.

13.6.3 Design Example 5: Cellular raft

The multi-storey steel-framed building shown in Fig. 13.35 is to be founded in an area where future mining work is anticipated. Ground conditions comprise soft silty clay.

In order to be able to deal with the subsidence wave, and to found within the soft clay layer, it is decided to adopt a cellular raft as shown in Fig. 13.35. Calculations have shown that a maximum net allowable bearing pressure of $n_a = 50 \, kN/m^2$ is necessary to keep differential settlements within acceptable limits.

To calculate the bearing pressure under normal loading, the raft is assumed to be stiff and the pressure distribution is assumed to be uniform or linearly varying. The ground pressure acting on each beam line is taken to be proportional to the width of base slab *carried* by that beam line.

In this example the full superstructure load is assumed to act on the raft. In some situations – particularly narrow rafts – higher bearing pressures can occur when the superstructure imposed load only acts over part of the raft: where appropriate this should be checked as a separate load case. This approach is illustrated in Design Example 8 in section 13.9.3.

Loading

Loads are taken from Fig. 13.35.

$$\begin{aligned} \text{Superstructure dead load, } G &= 3(450 + 810 + 360) + \\ &\quad 2(225 + 405 + 180) \\ &= 6480 \, kN \end{aligned}$$

$$\begin{aligned} \text{Superstructure imposed load, } Q &= 3(675 + 1215 + 540) + \\ &\quad 2(337.5 + 607.5 + 270) \\ &= 9720 \, kN \end{aligned}$$

$$\begin{aligned} \text{Superstructure load, } P &= G + Q \\ &= 6480 + 9720 \\ &= 16\,200 \, kN \end{aligned}$$

Q as a percentage of P is $100Q/P = 60\%$. From Fig. 10.20, the combined partial safety factor for superstructure loads is $\gamma_P = 1.52$

$$\begin{aligned} \text{Superstructure bearing pressure, } p &= \frac{P}{A} \\ &= \frac{16\,200}{24.3 \times 13.8} \\ &= 48.3 \, kN/m^2 \end{aligned}$$

Foundation dead load, f_G

$$\begin{aligned} \text{Slabs} &= 24 \, kN/m^3 \times (0.6 + 0.3 \, m) \\ &= 21.6 \, kN/m^2 \end{aligned}$$

Webs (averaged over an area $3.75 \, m \times 6.0 \, m$)

$$\begin{aligned} &= 24 \, kN/m^3 \times 2.1 \, m \, \frac{[(0.3 \, m \times 3.75 \, m) + (0.6 \times 6.0 \, m)]}{3.75 \, m \times 6.0 \, m} \\ &= 10.6 \, kN/m^2 \end{aligned}$$

$$\begin{aligned} f_G &= 21.6 + 10.6 \\ &= 32.2 \, kN/m^2 \end{aligned}$$

Foundation imposed load, f_Q

Ground floor slab imposed load, $f_Q = 6.0 \, kN/m^2$

$$\begin{aligned} \text{Foundation load} &= f_G + f_Q \\ &= 32.2 + 6.0 \\ &= 38.2 \, kN/m^2 \end{aligned}$$

F_Q as a percentage of f is $100F_Q/f = 16\%$. From Fig. 10.20, the combined partial safety factor for foundation loads is $\gamma_F = 1.43$.

$$\begin{aligned} \text{Total bearing pressure, } t &= \text{(superstructure pressure) +} \\ &\quad \text{(foundation pressure)} \\ &= p + f \\ &= 48.3 + 38.2 \\ &= 86.5 \, kN/m^2 \end{aligned}$$

Bearing pressure check

Net allowable bearing pressure at the formation depth of 3 m is $n_a = 50 \, kN/m^2$. Assuming a unit weight of $18 \, kN/m^3$, the existing overburden pressure is given by

$$s = 18 \times 3.0 = 54 \, kN/m^2$$

From section 10.10, the total allowable bearing pressure at this depth is

$$\begin{aligned} t_a &= \text{(net allowable bearing pressure) + (existing surcharge)} \\ &= n_a + s \\ &= 50 + 54 \\ &= 104 \, kN/m^2 \end{aligned}$$

$t_a > t \, (= 86.5 \, kN/m^2)$, therefore bearing pressure is okay.

Design of bottom slab

Resultant ultimate upwards design pressure on bottom slab is given by

$$\begin{aligned} p_u &= t_u - \gamma_G \text{(self-weight of slab)} \\ &= (\gamma_P p + \gamma_F f) - 1.4(24 \times 0.6) \\ &= (1.52 \times 48.3) + (1.43 \times 38.2) - (1.4 \times 14.4) \\ &= 107.9 \, kN/m^2 \end{aligned}$$

The bottom slab should be designed as two-way spanning in accordance with BS 8110.

Design of beams – introduction

In the normal condition the I-section beams, formed by the walls and slabs of the cellular structure, span 6 m and 7.5 m between the columns under the upward action of the ground pressure. For the 3 m depth of beams involved, these spans are relatively small, and the corresponding reinforcement would be light. It is anticipated that, to ride the predicted subsidence wave, the critical load case comes from the *two-thirds spanning and one-third cantilever* condition,

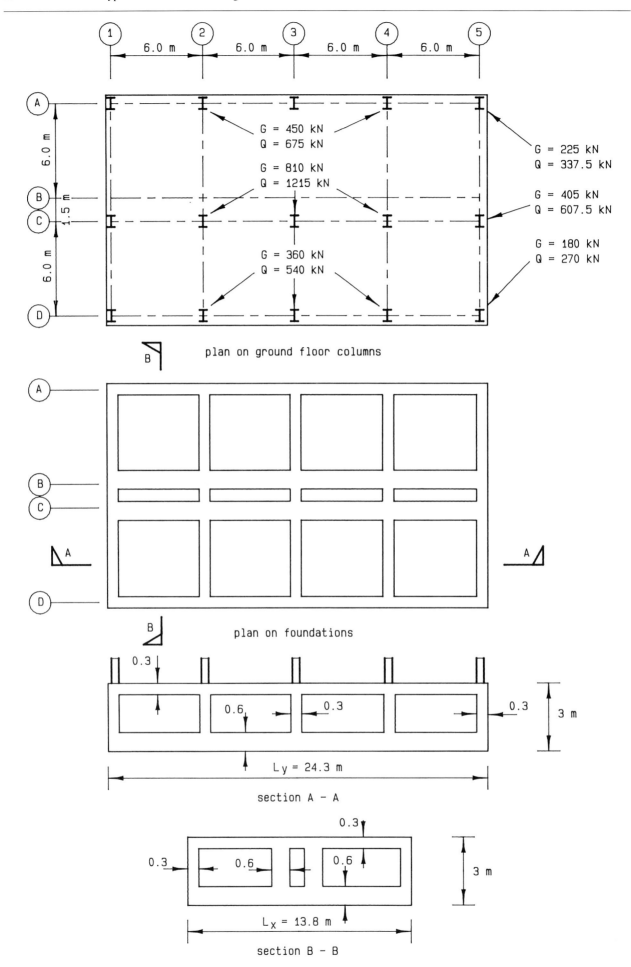

Fig. 13.35 Cellular raft design example – dimensions and loading.

longitudinal beam on grid line C

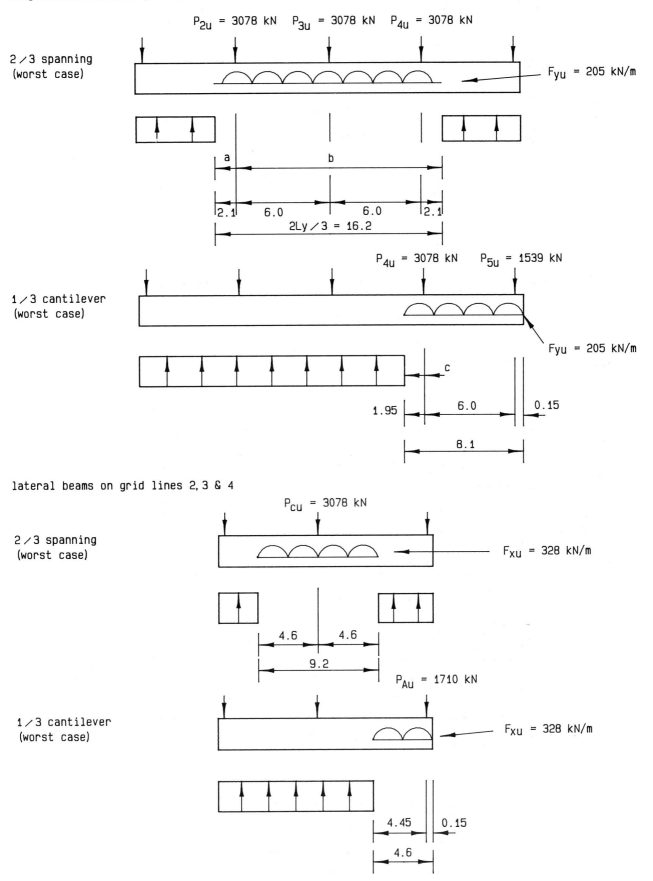

Fig. 13.36 Cellular raft design example — 'two-thirds spanning and one-third cantilever' condition.

described in section 13.6.2, and illustrated in Fig. 13.36.

The longitudinal and lateral beams shown are separately designed for these spanning conditions. In each instance the beam under consideration is designed to carry the super-structure loads over an assumed depression caused by the subsidence wave. In Fig. 13.36 these depressions have each been positioned in a worst case situation relative to the superstructure loads.

The *two-thirds spanning and one-third cantilever* condition would result in greatly increased ground pressures under the parts of the raft which remain in bearing, with correspondingly large settlements. The *two-thirds spanning and one-third cantilever* design condition is intended to produce a stiff raft which can resist excessive differential settlements. Provided the bearing pressures do not exceed the ultimate bearing capacity of the soil, it is not necessary to explicitly check the bearing pressure under this condition.

Design of lateral beams

The ultimate superstructure loads shown in Fig. 13.36 are calculated from the working loads in Fig. 13.35 as follows:

$$P_{Cu} = (1.4 \times 810) + (1.6 \times 1215) = 3078 \, \text{kN}$$
$$P_{Au} = (1.4 \times 450) + (1.6 \times 675) = 1710 \, \text{kN}$$

The ultimate foundation load for a 6 m width of raft foundation is given by

$$F_{xu} = \gamma_F f \times 6.0 \, \text{m}$$
$$= 1.43 \times 38.2 \times 6.0 = 328 \, \text{kN/m}$$

Two-thirds spanning worst case:

$$M_u = \frac{P_{Cu}(2L_x/3)}{4} + \frac{F_{xu}(2L_x/3)^2}{8}$$
$$= \frac{3078 \times 9.2}{4} + \frac{328 \times 9.2^2}{8}$$
$$= 10\,550 \, \text{kNm} \quad \text{(sagging)}$$

$$V_u = \frac{P_{Cu}}{2} + \frac{F_{xu}(2L_x/3)}{2}$$
$$= \frac{3078}{2} + \frac{328 \times 9.2}{2}$$
$$= 3047 \, \text{kN}$$

One-third cantilever worst case:

$$M_u = P_{Au}(L_x/3) + \frac{F_{xu}(L_x/3)^2}{2}$$
$$= (1710 \times 4.6) + \frac{328 \times 4.6^2}{2}$$
$$= 11\,336 \, \text{kNm} \quad \text{(hogging)}$$

$$V_u = P_{Cu} + F_{xu}(L_x/3)$$
$$= 1710 + (328 \times 4.6)$$
$$= 3218 \, \text{kN}$$

Reinforcement for bending and shear needs to be calculated for both cases.

Design of longitudinal beams

Again using the working loads in Fig. 13.35, the ultimate loads in Fig. 13.36 are given by

$$P_{2u} = P_{3u} = P_{4u} = P_{Cu} = 3078 \, \text{kN}$$
$$P_{5u} = (1.4 \times 405) + (1.6 \times 607.5) = 1539 \, \text{kN}$$

The ultimate foundation load for a 3.75 m width of raft foundation is given by

$$F_{yu} = \gamma_F f \times 3.75 \, \text{m}$$
$$= 1.43 \times 38.2 \times 3.75 = 205 \, \text{kN/m}$$

Two-thirds spanning worst case:

$$M_u = \frac{2P_{2u}\,ab}{(2L_y/3)} + \frac{P_{3u}(2L_y/3)}{4} + \frac{F_{yu}(2L_y/3)^2}{8}$$
$$= \frac{2 \times 3078(2.1 \times 14.1)}{16} + \frac{3078 \times 16.2}{4}$$
$$\quad + \frac{205 \times 16.2^2}{8}$$
$$= 30\,442 \, \text{kNm} \quad \text{(sagging)}$$

$$V_u = \frac{(P_{2u} + P_{3u} + P_{Cu})}{2} + \frac{F_{yu}(L_y/3)}{2}$$
$$= \frac{3 \times 3078}{2} + \frac{205 \times 16.2}{2}$$
$$= 6278 \, \text{kN}$$

One-third cantilever worst case:

$$M_u = cP_{4u} + P_{5u}(L_y/3) + \frac{F_{yu}(L_y/3)^2}{2}$$
$$= (1.95 \times 3078) + (1539 \times 8.1) + \frac{205 \times 8.1^2}{2}$$
$$= 25\,192 \, \text{kNm} \quad \text{(hogging)}$$

$$V_u = P_{4u} + P_{Cu} + F_{yu}(L_y/3)$$
$$= 3078 + 1539 + (205 \times 8.1)$$
$$= 6278 \, \text{kN}$$

Reinforcement for bending and shear needs to be calculated for both cases.

13.7 LIDDED CELLULAR RAFT

13.7.1 Introduction

The lidded cellular raft is described in section 9.4.6 and due to its formation tends to be a little less stiff than the true cellular raft. The design calculations however follow similar lines with the exception that the cross-section of the beams tends to be restricted to inverted T and L shapes.

The advantage of this form over the pure cellular raft is that the upper slab can be detailed to allow it to be re-levelled should the floor tilt or distortion become excessive for the building's use. Also in some locations the top of the lidded raft can be constructed in precast units and may prove more economic, avoiding the possible need for permanent formwork.

The raft is usually designed as a number of intersecting

inverted T beams taking advantage of the lower ground slab as the flange of the T but ignoring the upper slab which could be constructed in timber joists and boards or other form to suit the design requirements (see Fig. 13.37).

The detail at the seating of the upper floor depends upon the need for re-levelling and the possible number of times adjustments may need to be made.

As explained in section 9.4.6, the upper floor of the lidded raft is a separate structure to the main inverted T and L beams forming the concrete raft.

13.7.2 Sizing the design

The design procedure is similar to that of the cellular raft except that the upper deck is simply designed to span as a floor between the up-standing ribs. The remainder of the design follows the same procedure as before with the exception already mentioned that the element sections become inverted T or L beams rather that I or box sections.

13.7.3 Design Example 6: Lidded cellular raft

The raft in Design Example 5 (see section 13.6.3) is required to be redesigned as a lidded cellular raft, with the in situ top slab replaced by widespan prestressed concrete planks, for speed of construction.

The analysis remains essentially identical to the closed cellular raft in Design Example 5. The main difference comes in the design of the beams within the raft, for the

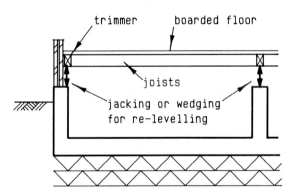

Fig. 13.37 Lidded cellular raft.

two-thirds spanning and one-third cantilever conditions. The I-section beams have been replaced by inverted T-section beams, resulting in the loss of the compression flange in the two-thirds spanning condition, and providing greatly reduced space for positioning tension bars in the one-third cantilever condition. In situations where these beams are heavily loaded, it will be necessary to increase the thickness of the webs, or introduce a narrow top flange as shown in Fig. 13.38.

13.8 BEAM STRIP RAFT

13.8.1 Introduction

The beam strip raft is described in section 9.4.7 and consists of downstand beams in two directions tied together by a ground bearing slab. The beam and the slab are designed as separate elements which are combined together on the drawing board to finalize the rafted design.

These rafts are used where the bearing capacity below the beams is relatively good as is the bearing capacity of the ground below the slab and therefore there is no need to design the total raft foundation when the two are linked together on the drawing board.

The two are generally linked because of the added performance from the two separate elements when they are cast monolithic. The beams may be required due to the point loads from column structures around the edge of the raft or within the body of the raft and beams are designed to span horizontally between these point loads. Similarly the raft slab is designed to float on the ground between the beams but since the bearing capacity where these rafts are adopted is relatively good then a nominal design incorporating a top and/or bottom mesh is all that is required in the slab.

These foundations are generally used in areas where quite shallow sand deposits occur below the topsoil and where there is no need to go to excessive depths around the edges of these rafts for heave or other problems. They can also be used where the strata changes slightly from perhaps clayey sands to sandy/silty clays.

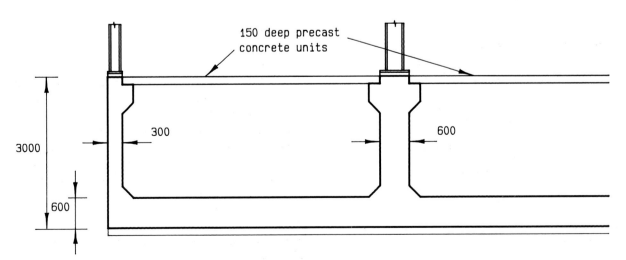

Fig. 13.38 Lidded cellular raft example.

13.8.2 Sizing of the design

The sizing of the sections is carried out by treating the beam strips as independent beams, designed as in Chapter 11. These two parts for the foundation are then tied as shown in Fig. 13.39.

Any necessary adjustments that the engineer may feel are required due to the changes in behaviour resulting from combining the elements are then made in the detailing of the raft. For example, the linking together will generally improve the raft performance by reducing the stresses in the two elements from those applicable if they acted alone. However, there will be some occasions, for example, when a local heavy load occurs on a downstand, where a local detail could become critical due to the change in behaviour, and additional reinforcement or a slight adjustment to a detail may be needed.

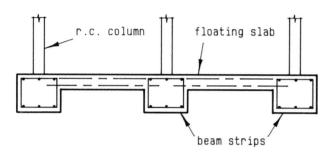

Fig. 13.39 Beam strip raft.

13.8.3 Design Example 7: Beam strip raft

The blanket raft slab in Design Example 3 (see section 13.4.4) is to be redesigned as a beam strip raft, for conditions where local depressions are not anticipated and therefore will not form part of the design (see Fig. 13.40). The net allowable bearing pressure is $n_a = 150 \, \text{kN/m}^2$.

Loadings

Loads and combined partial factors are taken from Fig. 13.21 and Design Example 3, as follows:

$$f = 9.1 \, \text{kN/m}^2; \, \gamma_F = 1.49$$
$$P_i = 110 \, \text{kN/m}; \, \gamma_{Pi} = 1.45$$
$$P_e = 90 \, \text{kN/m}; \, \gamma_{Pe} = 1.44$$

Allowable bearing pressure

In a similar manner to Design Example 2, the superstructure allowable bearing pressure is

$$p_a = n_a - f_S$$
$$= 150 - 9.1$$
$$= 141 \, \text{kN/m}^2$$

Groundbearing slab design

The slab is to be designed simply as a groundbearing slab. From Fig. 13.21 the maximum dimension of the slab is 17 m. From Table 13.2, a 125 mm deep slab, with A142 mesh top reinforcement only, is adequate for shrinkage purposes.

Internal beam strip design

This design is carried out in accordance with Fig. 13.2, in a similar manner to crust rafts. To maintain bearing pressure within that allowed, width of bearing required is

$$B = \frac{\text{wall line load}}{\text{allowable bearing pressure}}$$
$$= \frac{P_i}{p_a}$$
$$= \frac{110}{141}$$
$$= 0.78 \, \text{m}$$

Try a 300 mm deep thickening, with the dimensions shown in Fig. 13.41. Design the bending reinforcement as follows:

Ultimate load on beam strip, $P_{iu} = \gamma_{Pi} P_i$
$$= 1.45 \times 110$$
$$= 160 \, \text{kN/m}$$

Ultimate moment, $M_u = \dfrac{P_{iu}(B/2)}{2}$
$$= \frac{160 \times (0.78/2)}{2}$$
$$= 31 \, \text{kNm/m}$$

$b = 1000 \, \text{mm}$

$d = 300 - 40(\text{cover}) - \dfrac{10(\text{diameter})}{2} = 255 \, \text{mm}$

$$\frac{M_u}{bd^2} = \frac{31 \times 10^6}{1000 \times 255^2}$$
$$= 0.48$$

$$A_{s(\text{req})} = 0.13\% \, bd = 332 \, \text{mm}^2/\text{m}$$
$$[\text{BS 8110: Part 2: Chart 2}]$$

Provide T10 bars at 225 mm c/c = 349 mm²/m.

External beam strip design

This design is carried out in accordance with Fig. 13.3, again in a similar manner to crust rafts. To maintain bearing pressure within that allowed, width of bearing required is

$$B = \frac{\text{wall line load}}{\text{allowable bearing pressure}}$$
$$= \frac{P_e}{p_a}$$
$$= \frac{90}{141}$$
$$= 0.64 \, \text{m}$$

With reference to Fig. 13.3 (a) and (b), this requires the effective projection of the toe beyond the line of the load to be $B/2 = 320 \, \text{mm}$. This is achieved either by increasing x as per Fig. 13.3 (a), or adding a thick blinding layer as per Fig. 13.3 (b). The latter option will be chosen in this case (see Fig. 13.41).

Unlike the internal thickenings, there is no significant lateral bending. The thickening width is therefore sized to suit the cavity wall above, the depth − including that of the

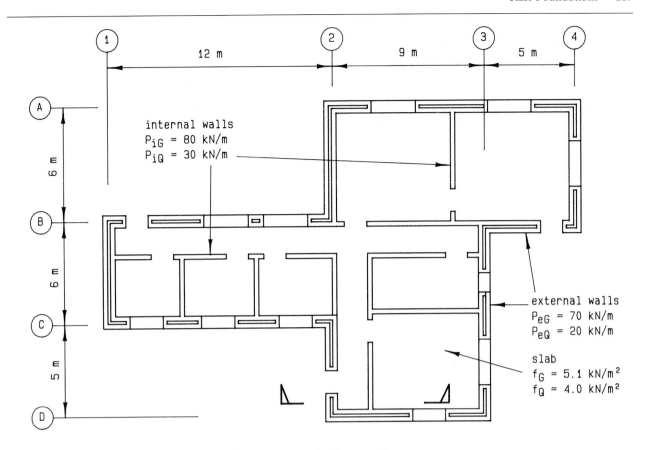

plan on ground floor walls

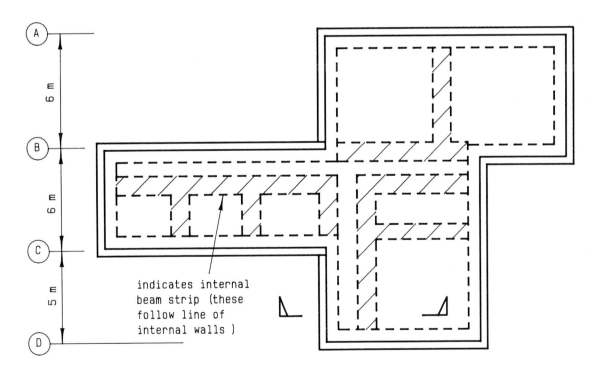

plan on foundation

Fig. 13.40 Beam strip raft design example − plans.

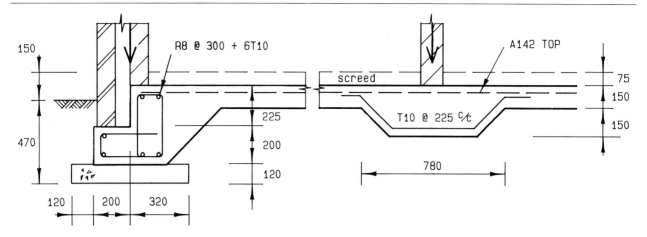

Fig. 13.41 Beam strip raft design example – section.

thick blinding layer – is sized to avoid frost damage, and nominal reinforcement is provided.

13.9 BUOYANCY RAFT

13.9.1 Introduction
The buoyancy raft works on a similar principle to that of a floating structure where the support for the raft is mainly obtained by displacing the weight of earth or overburden by the volume of a large voided foundation. The raft is described in section 9.4.8 and is often economically achieved by making use of the voids as a basement structure (see Fig. 13.42).

It is designed so that sufficient overburden is removed to allow the superstructure load to be applied to the ground with little or no increase in the original stress which existed on the sub-strata prior to excavation and construction. Thus the structure *floats* like a ship – which displaces water equal to its own weight.

The bottom slab can form the basement of the proposed building, and be combined with the ground slab and retaining walls to act as the raft. It can also be of cellular form (see Fig. 13.43).

The raft design takes into account any eccentricity of load and aims to keep differential settlements and tilting within acceptable limits (see Fig. 13.44), which shows how eccentric

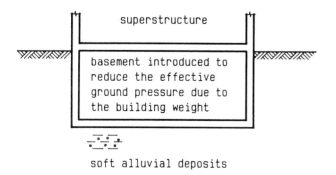

Fig. 13.42 Buoyancy foundation/basement.

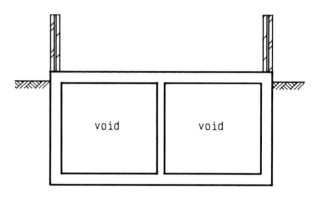

Fig. 13.43 Cellular buoyancy foundation.

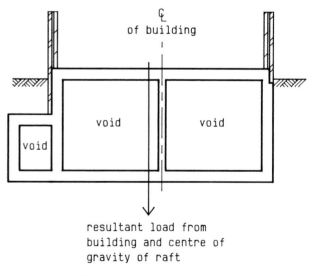

Fig. 13.44 Eccentrically loaded buoyancy raft.

resultant loads can be balanced by the basement projections.

Since buoyancy foundations are expensive compared to more traditional forms they tend only to be used where suitable bearing strata is at too great a depth for other more traditional alternatives. For this reason the foundation tends to be restricted to sites on very deep alluvial deposits such as soft sands and silts and where loads on the foundations can be kept concentric. Examples of such building types would be schemes where deep basements can be economically incorporated into the design or where underground tanks are required. The cases therefore where the engineer would adopt such solutions tend to be limited.

13.9.2 Sizing the design
The overall sizing of the design would generally involve:

(1) Calculating the depth plan size and centre of gravity required for the overburden removal to suit structural buoyancy.
(2) Comparing the results of (1) with the requirements for tanks or basements to suit client's needs.
(3) Calculating the water pressure for (1) to check for flotation.
(4) Combining the requirements of (1), (2) and (3) into a mutually suitable voided foundation.
(5) Designing the external walls, floors, roof and separating wall elements for the pressures, bending moments and shear forces including any projections to prevent flotation.

13.9.3 Design Example 8: Buoyancy raft
A three-storey office building is to be founded in ground conditions consisting of large depths of silty sand. To avoid problems with differential settlements, the building is to be designed as a buoyancy raft with a usable basement, and bearing pressures are to be kept within the level of existing ground stresses. It is to be assumed that the water-table exists at a level below the anticipated basement depth.

Loadings
Loads are shown in Fig. 13.45. Two load cases are considered critical, as shown. The loads are as follows:

$$\text{Foundation UDL, } f = \text{slab load, } f_S = f_G + f_Q$$
$$= 12.6 + 3.0$$
$$= 15.6 \, \text{kN/m}^2$$

f_Q as a percentage of f is $100 f_Q/f = 19\%$. From Fig. 10.20, the combined partial load factor for foundation loads is $\gamma_F = 1.44$.

$$\text{Loadcase 1: } P_1 = G + Q$$
$$= (100 + 100) + (60 + 60)$$
$$= 320 \, \text{kN/m}$$

Q as a percentage of P_1 is $100 Q/P_1 = 38\%$. From Fig. 10.20, the combined partial load factor for superstructure loads is $\gamma_{P1} = 1.48$.

$$\text{Loadcase 2: } P_2 = G + Q$$
$$= (100 + 100) + 60$$
$$= 260 \, \text{kN/m}$$

Q as a percentage of P_2 is $100 Q/P_2 = 23\%$. From Fig. 10.20, the combined partial load factor for superstructure loads is $\gamma_{P2} = 1.45$.

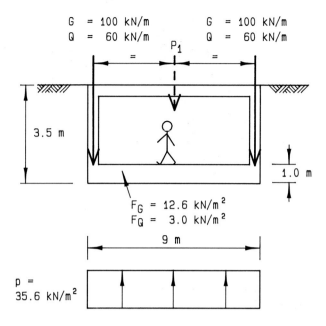

loadcase 1
full dead + imposed load

loadcase 2
full dead load + out of balance imposed load

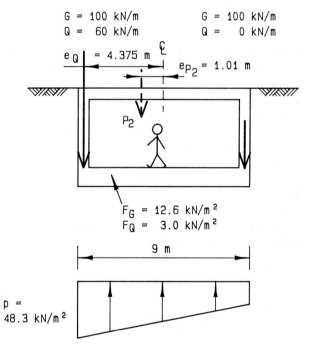

Fig. 13.45 Buoyancy raft design example – section.

The loading eccentricity is given by

$$e_{P2} = \frac{(e_G G + e_Q Q)}{P_2}$$
$$= \frac{(0 \times 200) + (4.375 \times 60)}{260}$$
$$= \frac{262.5}{260}$$
$$= 1.01 \, \text{m}$$

Allowable bearing pressure

Achieving no increase in bearing pressure above existing pressures corresponds to having a net allowable bearing pressure of $n_a = 0$. From section 10.10, net bearing pressure is

$$n = p + f_S - s_S$$

This may be rearranged to give a superstructure allowable bearing pressure of

$$
\begin{aligned}
p_a &= n_a + s_S - f_S \\
&= \text{(net allowable bearing pressure)} + \\
&\quad \text{(existing surcharge)} - \text{(foundation surcharge)} \\
&= 0 + (20\,\text{kN/m}^3 \times 3.5\,\text{m}) - 15.6\,\text{kN/m}^2 \\
&= 54.4\,\text{kN/m}^2
\end{aligned}
$$

Bearing pressure check

$$
\begin{aligned}
\text{Loadcase 1:}\quad p_1 &= \frac{P_1}{L} \\
&= \frac{320}{9.0} \\
&= 35.6\,\text{kN/m}^2
\end{aligned}
$$

$$
\begin{aligned}
\text{Loadcase 2:}\quad p_{2(\text{max})} &= \frac{P_2}{L} + \frac{e_{P2}P_2}{Z} \\
&= \frac{260}{9.0} + \frac{1.01 \times 260}{(9.0^2/6)} \\
&= 28.9 + 19.4 \\
&= 48.3\,\text{kN/m}^2
\end{aligned}
$$

In both cases this is less than the allowable $p_a = 54.4\,\text{kN/m}^2$ > okay.

Ground heave

In this particular example, with ground conditions consisting of silty sands, significant ground heave is not expected. Where the soil is predominantly cohesive (i.e. clays), the reduction in ground stress will result in heave, i.e. settlement in reverse.

As with settlement, heave will have a short-term (*elastic*) component, and a long-term (*consolidation*) component. The amount of heave is calculated on the same basis as settlement (see Chapter 2 on soil mechanics).

From an engineering point of view, short-term heave will normally occur during the excavation period. Where its magnitude is considered significant, the level of the formation can be monitored until the anticipated short-term heave has taken place. The formation is then trimmed down to its required level, and construction of the new substructure proceeds.

Long-term heave is dealt with in the same way as long-term settlement. The anticipated amount of *differential* heave is calculated, and the structure is designed to accommodate this movement.

Foundation slab design

By inspection, the worst case for design of the foundation slab is load case 1.

Ultimate design pressure,
$$
\begin{aligned}
p_u &= \gamma_{P1}p_1 \\
&= 1.48 \times 35.6 \\
&= 52.7\,\text{kN/m}^2
\end{aligned}
$$

Design the slab to span simply supported between the basement walls:

$$
\begin{aligned}
\text{Ultimate moment,}\quad M_u &= \frac{p_u L^2}{8} \\
&= \frac{52.7 \times (9.0 - 0.25)^2}{8} \\
&= 504\,\text{kNm/m}
\end{aligned}
$$

$$b = 1000\,\text{mm}$$

$$d = 450 - 25(\text{top cover}) - \frac{25(\text{diameter})}{2} = 412\,\text{mm}$$

$$
\begin{aligned}
\frac{M_u}{bd^2} &= \frac{504 \times 10^6}{1000 \times 412^2} \\
&= 2.97
\end{aligned}
$$

$$
\begin{aligned}
A_{s(\text{req})} &= 0.83\%\,bd \qquad [\text{BS 8110: Part 3: Chart 2}] \\
&= \frac{0.83}{100} \times 1000 \times 415 \\
&= 3445\,\text{mm}^2/\text{m}
\end{aligned}
$$

Provide T25 at 125 mm c/c (top) = 3927 mm²/m.

The remaining reinforcement in the basement slab will be sized to comply with shrinkage stresses and detailing requirements in accordance with BS 8110 and other relevant codes of practice.

13.10 JACKING RAFT

13.10.1 Introduction

Raft foundations suitable for jacking are specifically designed versions of crust rafts, cellular rafts, lidded cellular rafts, beam strip rafts or other foundations whose stiffness and behaviour is designed to resist the jacking forces and moments involved in the process of re-levelling.

The raft is designed to cater for the bending and shear forces likely to be produced during subsidence and re-levelling activity. Jacking points are built into the foundation to allow for re-levelling and the type of raft and number of jacking points depends to a large extent on the proposed use, the size of the structure and the predicted subsidence likely to occur.

The need for a jacking raft tends to be determined by the unpredictability of subsidence and the practicalities for the building user of re-levelling within the life of the building. As mentioned in section 9.4.9, domestic sites in areas of brine mining are typical of sites where such foundations have been adopted and basically two design conditions need to be considered.

(1) To design for the normal subsidence condition for the site including bending moments and forces as previously discussed for the raft type.

(2) Additional analysis and design to incorporate the structural requirements to resist stresses and distortions during the jacking operations.

13.10.2 Sizing the design

From section 13.10.1 it can be seen that the initial sizing of the foundation for trial design would be to adopt generous sizes for a standard raft of the type being proposed in anticipation of embracing the jacking stresses.

The stresses produced during jacking are dependent on the restrictions and methods of jacking and therefore tend to be one-off designs for set conditions.

13.11 REFERENCES

1. Deacon, R.C. (1986) *Concrete ground floors*. Cement and Concrete Association, Wexham Springs, Slough, UK.
2. National Coal Board (1975) *Subsidence Engineers' Handbook*. NCB, Mining Department.

CHAPTER 14

PILES

14.1 INTRODUCTION

Piling is one of the oldest foundation techniques known to mankind. The authors' practice has uncovered timber piles used by the Romans, and, in its structural survey of the Albert Docks, Liverpool, found extensive use of piling made by Victorian engineers. Piles are used to transfer the applied loads from the structure through the upper level strata to the soils at depth. The purpose of this transfer varies from site to site as is shown in the following applications.

14.2 APPLICATIONS

Typical applications of piling are:

(1) Where soil of low bearing capacity of significant depth is underlain by strong strata.

   Piling which transfers the foundation load to the strong strata is frequently a more economic solution than alternative foundations (see Fig. 14.1).

(2) Where the surface strata is susceptible to unacceptable settlement, and is underlain by stiff material (see Fig. 14.2).

   On a low-cost site the authors' practice has installed 18 m long in situ concrete piles to support two-storey domestic housing. The increased cost of the piles was more than compensated for by the low cost of the site.

(3) Where surface foundation would impose unacceptable increase in bearing pressure, or surcharge, on existing foundations (see Fig. 14.3).

(4) Where the foundation is subject to lateral loads which

Fig. 14.1 Stiff strata at depth.

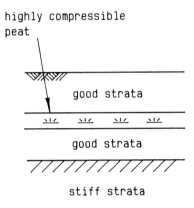

Fig. 14.2 Compressible strata.

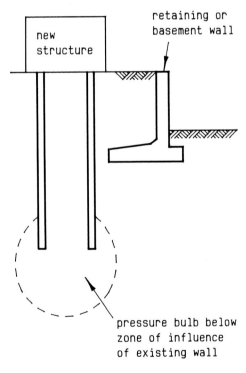

Fig. 14.3 Load transfer below existing foundation.

can be more economically resisted by raking piles (see Fig. 14.4).

(5) Where variations in the compressibility of the soil would lead to excessive differential settlement of sur-

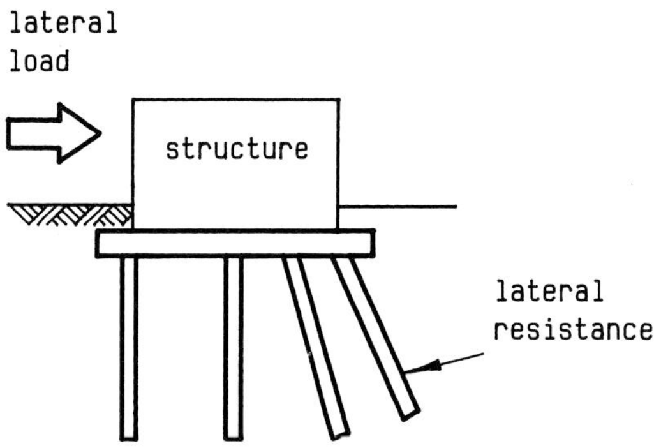

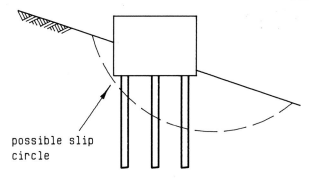

Fig. 14.4 Lateral resistance.

Fig. 14.6 Load transfer below critical slip circle.

face foundations (see Fig. 14.5). The leaning tower of Pisa is a classic example of differential settlement.

(6) Where excavation to firm strata would prove expensive and difficult, e.g. soft waterlogged alluvial deposits. On one important contract, near the coast in North Wales, it was found that the proposed structure was sited over a glaciated channel filled to a depth of 15 m with a soft, highly saturated silt which was impossible to dewater. Excavation and foundation construction would have been difficult and expensive.

(7) Where, on sloping sites, it is necessary to transfer the additional load to a level below the possible slip circle (see Fig. 14.6).

(8) Where anchoring of a flotation foundation by tieing down or tension pile is necessary (see Fig. 14.7).

(9) Where heave and swelling of clay could exert excessive forces and movements on surface spread foundations.

(10) Whenever piling is a more economic solution. In the past few decades there have been advances in piling manufacture and construction which have considerably reduced the cost of piling. Piling is no longer a last resort but can be considered as an economic alternative in foundation design.

Piles should be used with caution, if at all, where the ground is subject to significant lateral movement, e.g. in areas affected by mining, as such movements can shear off the piles leaving the structure unsupported (see Chapter 6).

14.3 TYPES OF PILES

There is a wide variety of pile types, materials, methods of placing, etc., which are outlined below. The criteria for choice of pile is discussed in section 14.5.

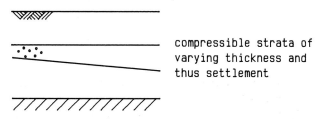

Fig. 14.5 Variable compressibility.

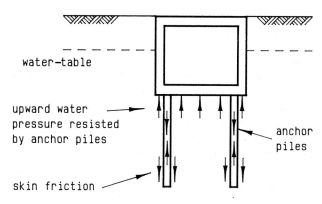

Fig. 14.7 Anchorage against flotation.

14.3.1 Load-bearing characteristics

There are two broad classifications of piles, as follows:

(1) *End bearing*. The pile is driven through weak soil to rock, dense gravel or similar material and the pile's load-bearing capacity is derived from the assistance of the stratum at the toe of the pile (see Fig. 14.8 (a)).

(2) *Skin friction*. Skin friction develops between the surface area of the pile and the surrounding soil (similar to driving a nail into timber). The frictional resistance developed must provide an adequate factor of safety for the pile load (see Fig. 14.8 (b)).

It is not uncommon for piles to rely on both types of load-bearing capacity. For example, if the 'stiff strata' shown in Fig. 14.2 is compact gravel and the 'good strata' above is a firm sand, then a pile driven into the gravel could rely both on end bearing from the gravel and skin friction from the sand.

14.3.2 Materials

Both classes of piles can be made of various materials.

(1) TIMBER PILES

The oldest material used for piles is timber and it is still in use, particularly in developing countries, today. It has proved surprisingly durable and provided care is given to the detail and treatment of the toe and head of the pile and the durability conditions it should still be used where it is economical. The toe can be subject to splintering during

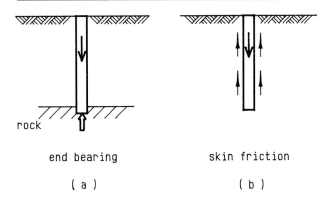

end bearing   skin friction

(a)   (b)

Fig. 14.8 Pile resistance.

driving and should be tapered to a blunted point and, if necessary, encased in steel. The head of the pile during driving may also need protection from splintering and this is usually provided by placing a driving cap or helmet over the head of the pile.

Where the top of the pile is below the lowest water level and is in permanently wet conditions, experience shows that there are few durability problems. However, when the level of the top of the pile or any part of it is in the area of a fluctuating water-table and is therefore subjected to alternate wetting and drying, this section of the pile should be treated with preservatives and water repellents and even then may still have a limited life.

If the length of the pile is found on site to be too long there is little problem in cutting off the excess length but there is a problem if the pile is found to be too short. Extending the pile involves splicing on an added length and any metal connections must be corrosion-resistant.

The choice of timber is not restricted to greenheart or oak except in conditions where the pile is subject to alternate wetting and drying such as in piles supporting jetties in tidal conditions. In numerous structural surveys of Victorian buildings the authors' practice has discovered, in good condition below the water-table level, timbers such as birch, larch, pine, etc.

(2) CONCRETE PILES

Concrete piles are the most widely used in the developed countries and may be cast in situ, precast, reinforced and prestressed. The choice of type is discussed more fully in section 14.5.

(a) Precast

This type is commonly used where:

(i)   The length required can be realistically predicted.
(ii) Lateral pressure from a strata within the soil profile is sufficient to squeeze (*neck*) an in situ pile.
(iii) Where there are large voids in sections of the soil which would possibly have to be filled with an excessive amount of in situ concrete or could cause loss of support for wet concrete prior to setting.
(iv) For structures such as piers and jetties above water level on coastal, estuary and river sites.

Though precast piles can be manufactured on site it is more common to have them designed, manufactured and installed by specialist subcontractors.

There are disadvantages in the use of precast concrete piles as follows:

(i)   It is not easy to extend their length.
(ii)   They are liable to fracture when driven into such obstacles as large boulders in boulder clay and the damage can remain out of sight.
(iii) Obstructions can cause the pile to deflect from the true vertical line.
(iv)   There is an economic limit, restricted by buckling, of the unrestrained length of the pile.
(v)   Noise and vibration caused by driving can cause nuisance and damage.
(vi)   There can be large wastage due to the need to cut off the projecting length after driving.
(vii) The accuracy of the estimated length is only proved on site when short piles can be difficult to extend and long piles can prove to be expensive and wasteful.
(viii) The relatively large rig required for driving often needs extensive hard-standings to provide a suitable surface for pile driving.

The advantages of precast concrete piles are:

(i)   It is easier to supervise the initial quality of construction in precast than in situ.
(ii) The pile is not driven until the concrete is matured.
(iii) Stresses due to driving are usually higher than those due to foundation loading so that manufacturing faults are more easily discovered and, in effect, the pile is preload tested (provided the defects can be detected).

The reinforcement, while adding to the load-bearing capacity, is mainly designed to cope with handling, transporting and driving stresses.

(b) Cast in situ

There is an ever increasing variety of cast in situ piles offered by specialist piling subcontractors. The piles are usually circular in cross-section and are regarded as small-diameter piles when their diameters are from 250–600 mm and larger-diameter piles when their diameters exceed 600 mm; large-diameter piles are now possible with diameters up to and exceeding 1.5 m.

The advantages of cast in situ piles are:

(i)   They can be constructed immediately, thus cutting out the time required for casting, maturing and delivering of precast piles.
(ii) There is no need to cut off or extend excessive lengths of the piles as they can be cast in situ to the required level.
(iii) They can be cast to longer lengths than is practical with precast piles.
(iv) Most obstructions can be hammered and broken through by the pile-driving techniques.

(v)   The placing can cause less noise vibration and other disturbance compared to driving precast piles.

(vi) Soil taken from boring can be inspected and compared with the anticipated conditions.

The disadvantages of cast in situ piles are:

(i)   It can be difficult to place and ensure positioning of any necessary reinforcement.

(ii) Concrete quality control is more difficult.

(iii) There is a danger of *necking* from lateral earth pressure.

(iv) Young concrete is susceptible to attack from some soil chemicals before it has set and hardened.

(c) Prestressed

Prestressed concrete in superstructure design is made of higher strength concrete, requires smaller cross-sectional area and can be made impact-resistant. The same results apply to prestressed piles relative to comparison with precast reinforced piles. Their advantages compared to precast reinforced are:

(i)   Handling stresses can be resisted by a smaller cross-section which can result in a more economical pile.

(ii) It is easier with the smaller section to achieve longer penetration into load-bearing gravels.

(iii) Tensile stresses that are generated up from the toe of the pile after the hammer blow can be compensated for by prestress.

(iv) The reduction of tensile cracking of the concrete can lead to greater durability.

The disadvantages of prestressed piles are:

(i)   The smaller section provides less end bearing and total peripheral skin friction.

(ii) Deeper penetration into end-bearing strata (gravel, compact sand, etc.) may be necessary.

(iii) It is more difficult to extend the length of a precast driven pile.

(iv) As in prestressed concrete superstructure elements, stricter quality control in manufacture is necessary.

(3) STEEL PILES

Though most piling is carried out in some form of concrete there are situations where steel piles should be considered. There is considerable experience of the use of sheet piling in civil engineering which can be applied to piles for structures. The piles are generally a standard H section (see Table 14.1) or, for longer or more heavily loaded sections, tubular box section piles are used (see Table 14.2). In some cases they can form a structurally efficient and cost-effective solution. The advantages of steel piles are:

(i)   They have a lighter weight for a required load-bearing capacity.

(ii) They can be used in longer lengths and extending them by butt welding is relatively simple. Similarly excess length is easily cut off.

(iii) They are more resistant to handling and driving stresses.

(iv) They can have good resistance to lateral forces, bending stresses and buckling.

(v)   They are particularly useful for marine structures (piers, jetties, etc.) above water where the piles may be subjected to impact forces, ships docking, etc.

The disadvantages of steel piles are:

(i)   There is a need for corrosion-protection.

(ii) The pile cost per metre run can be relatively high.

(iii) There are, as yet, fewer piling sub-contractors competent to carry out the work.

(iv) There is a tendency for H sections to bend about the weak axis when being driven − the resulting curved pile has a lower bearing capacity.

(4) STONE PILES/VIBRO-STABILIZATION

The use of stones or large gravel in piling is a relatively recent development although the authors have been involved in extensive use of this form of piling for over 25 years. The method is used mainly as a geotechnical process to consolidate and compact soils and/or to improve their drainage. The technique is discussed in detail in Chapter 8 and reference should be made to that chapter for further information.

14.4 METHODS OF PILING

There is a wide variety of methods used for piling and every piling contractor has a number of variations for their system − improvements in method and equipment continues. The main classes only are discussed below.

14.4.1 Driven piles

This method is used for piles of timber, precast concrete, prestressed concrete and the various types of steel piles. The pile is hammered into the ground by pile-driving plant shown in outline in Fig. 14.9 (a). Methods of protecting the head of the pile from shattering are shown in Fig. 14.9 (b).

Driven piles are classified as *displacement piles* and, where the soil can enter during driving, as *small displacement piles* (e.g., open ended tubular or other hollow sections often in steel).

14.4.2 Driven cast-in-place piles

A closed ended hollow steel or concrete casing is driven into the ground and then filled with fresh concrete. The casing may be left in position to form part of the whole pile or withdrawn for reuse as the cast concrete is placed. The cast concrete is rammed into position by a hammer as the casing is withdrawn ensuring firm contact with the soil and compaction of the concrete. Care must be taken to see that the cast concrete is not over-rammed or the casing withdrawn too quickly. There is a danger that as the liner tube is withdrawn it can lift up the upper portion of in situ concrete leaving a void or necking in the upper portions of the pile. This can be avoided by good quality control of the concrete and slow withdrawal of the casing.

Table 14.1 Universal bearing piles – dimensions and properties (*Steel Piling Products*, British Steel Corporation, 1985)

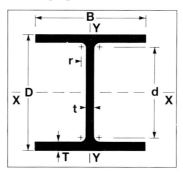

Description		Depth of section	Width of section	Thickness		Root radius	Depth between fillets	Area of section	Moment of inertia		Radius of gyration		Elastic modulus	
Serial Size	Mass per unit length			Web	Flange				X-X axis	Y-Y axis	X-X axis	Y-Y axis	X-X axis	Y-Y axis
		D	B	t	T	r	d							
mm	kg/m	mm	mm	mm	mm	mm	mm	cm²	cm⁴	cm⁴	cm	cm	cm³	cm³
356 x 406	340	406.4	403.0	26.5	42.9	15.2	290.2	432.7	122,474	46,816	16.8	10.4	6,027	2,32.4
	287	393.7	399.0	22.6	36.5	15.2	290.2	366.0	99,994	38,714	16.5	10.3	5,080	1,94.0
	235	381.0	395.0	18.5	30.2	15.2	290.2	299.8	79,110	31,008	16.2	10.2	4,153	1,57.0
356 x 368	174	361.5	378.1	20.4	20.4	15.2	290.1	222.2	51,134	18,444	15.17	9.11	2,829	975.7
	152	356.4	375.5	17.9	17.9	15.2	290.1	193.6	43,916	15,799	15.06	9.03	2,464	841.5
	133	351.9	373.3	15.6	15.6	15.2	290.1	169.0	37,840	13,576	14.96	8.96	2,150	727.4
	109	346.4	370.5	12.9	12.9	15.2	290.1	138.4	30,515	10,900	14.85	8.87	1,762	588.4
305 x 305	223	338.0	325.4	30.5	30.5	15.2	246.5	284.8	52,821	17,571	13.62	7.85	3,125	1,080
	186	328.3	320.5	25.6	25.6	15.2	246.5	237.3	42,628	14,109	13.40	7.71	2,597	880.4
	149	318.5	315.6	20.7	20.7	15.2	246.5	190.0	33,042	10,869	13.19	7.56	2,075	688.8
	126	312.4	312.5	17.7	17.7	15.2	246.5	161.3	27,483	8,999	13.06	7.47	1,760	575.8
	110	307.9	310.3	15.4	15.4	15.2	246.5	140.4	23,580	7,689	12.96	7.40	1,532	495.6
	95	303.8	308.3	13.4	13.4	15.3	246.5	121.4	20,113	6,530	12.87	7.33	1,324	423.7
	88	301.7	307.2	12.3	12.3	15.2	246.5	111.8	18,404	5,959	12.83	7.30	1,220	388.0
	79	299.2	306.0	11.1	11.1	15.2	246.5	100.4	16,400	5,292	12.78	7.26	1,096	345.9
254 x 254	85	254.3	259.7	14.3	14.3	12.7	200.2	108.1	12,264	4,188	10.65	6.22	964.5	322.6
	71	249.9	257.5	12.1	12.1	12.7	200.2	91.1	10,153	3,451	10.56	6.15	812.7	268.1
	63	246.9	256.0	10.6	10.6	12.7	200.2	79.7	8,775	2,971	10.49	6.11	710.9	232.1
203 x 203	54	203.9	207.2	11.3	11.3	10.2	160.8	68.4	4,987	1,683	8.54	4.96	489.2	162.4
	45	200.2	205.4	9.5	9.5	10.2	160.8	57.0	4,079	1,369	8.46	4.90	407.6	133.4

Notes: Rolling margin ± 2.5% on theoretical mass. Length tolerance ± 50mm.

Driven cast in situ piles can prove to be economic for sands, loose gravels, soft silts and clays particularly when large numbers of piles are required. For small numbers of piles the on-site costs can however prove to be expensive.

14.4.3 Bored cast-in-place piles
The hole for the pile shaft is formed by drilling or augering and the toe of the hole can be enlarged by under-reaming in stiff clays to provide greater end-bearing capacity for the pile. The method tends to be restricted to clayey soils and, as with the driven cast-in-place pile, care must be exercised to prevent necking of the cast concrete. If they are used in loose sand or silt the inflow of soil into the bore must be prevented. They can be installed in very long lengths and be of large diameter.

The relatively small on-site cost of bored piles means that smaller sites can be more economically piled than they can using a driven piling system. The bored pile is not usually economic in granular soils where loosening and disturbance of surrounding ground can cause excessive removal of soil and induce settlement in the surrounding area. During piling operations the hole can be lined with a casing which can be driven ahead of the bore to overcome difficulties caused by groundwater and soft sub-soil but sometimes difficulties of withdrawing the casing after casting can prove expensive.

14.4.4 Screw piles
Screw piles of steel or concrete cylinders with helical blades attached are screwed into the ground by rotating the blades.

Table 14.2 Larssen box piles — dimensions and properties (*Steel Piling Products*, British Steel)

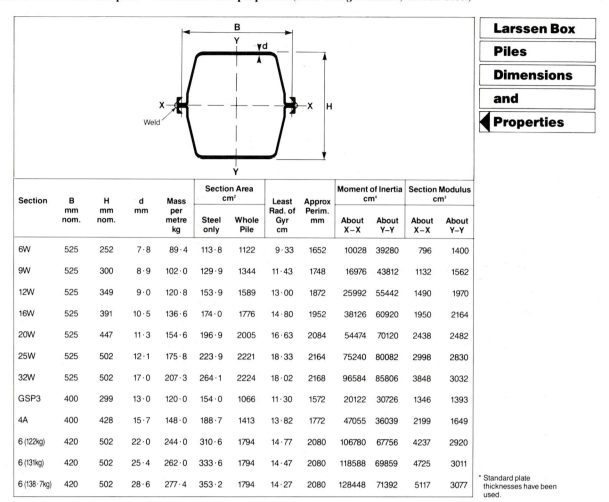

Section	B mm nom.	H mm nom.	d mm	Mass per metre kg	Section Area cm²		Least Rad. of Gyr cm	Approx Perim. mm	Moment of Inertia cm⁴		Section Modulus cm³	
					Steel only	Whole Pile			About X–X	About Y–Y	About X–X	About Y–Y
6W	525	252	7·8	89·4	113·8	1122	9·33	1652	10028	39280	796	1400
9W	525	300	8·9	102·0	129·9	1344	11·43	1748	16976	43812	1132	1562
12W	525	349	9·0	120·8	153·9	1589	13·00	1872	25992	55442	1490	1970
16W	525	391	10·5	136·6	174·0	1776	14·80	1952	38126	60920	1950	2164
20W	525	447	11·3	154·6	196·9	2005	16·63	2084	54474	70120	2438	2482
25W	525	502	12·1	175·8	223·9	2221	18·33	2164	75240	80082	2998	2830
32W	525	502	17·0	207·3	264·1	2224	18·02	2168	96584	85806	3848	3032
GSP3	400	299	13·0	120·0	154·0	1066	11·30	1572	20122	30726	1346	1393
4A	400	428	15·7	148·0	188·7	1413	13·82	1772	47055	36039	2199	1649
6 (122kg)	420	502	22·0	244·0	310·6	1794	14·77	2080	106780	67756	4237	2920
6 (131kg)	420	502	25·4	262·0	333·6	1794	14·47	2080	118588	69859	4725	3011
6 (138·7kg)	420	502	28·6	277·4	353·2	1794	14·27	2080	128448	71392	5117	3077

* Standard plate thicknesses have been used.

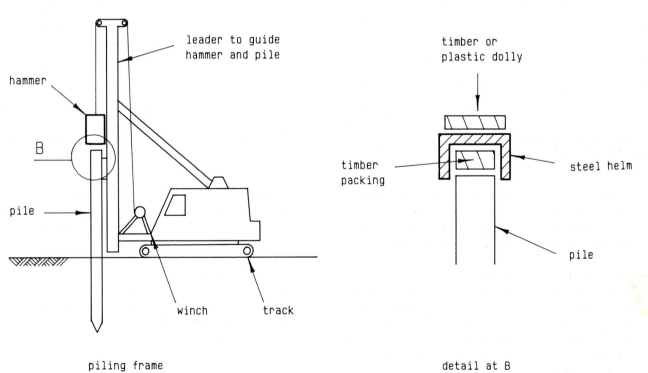

Fig. 14.9 Driven piling.

Their best application is in deep stratum of soft alluvial soils underlain by firm strata. Due to the large diameter of the blades the piles have increased resistance to uplift forces.

14.4.5 Jacked piles

Jacked piles are used where headroom for the pile and pile driver are limited as in underpinning within an existing building. The pile is jacked in short sections using the existing superstructure as a reaction frame.

14.4.6 Continuous flight auger piles

The flight auger pile system uses a hollow stem auger mounted on a mobile rig. The auger is drilled into the ground with very little vibration and spoil removal. When the required depth has been reached (see Fig. 14.10), concrete or grout is injected through the auger shaft. Usually the concrete or grout mixing plant and the pumping equipment are located nearby but can, if such areas are sensitive, be located well away from such positions. Pile lengths of up to 25 m can generally be achieved with pile diameters from 300 mm to 600 mm. Piles can be raked up to an angle of 1 in 6 from vertical. The system is suitable for use in most virgin soils and fine granular fills and rigs can operate in areas with restricted headroom.

14.4.7 Mini or pin piles

There are a number of mini or pin piles on the market. The systems range from water- or air-flushed rotary percussion augers to small-diameter driven steel cased piles which are driven to a set.

The pile diameters generally vary between 90 mm and 220 mm and can be used in most soils and with restricted access/limited headroom. Where necessary, noise and vibration can be kept to a minimum and piles can be driven within a few hundred millimetres of adjacent properties. In underpinning they can be used to penetrate existing concrete or masonry foundations, and are bonded into the existing elements.

Slenderness of such small-diameter piles must however be taken into account and the need for good quality control particularly with regard to filling such small bores with concrete.

The piles are not generally suitable in mining areas where surface movements and strains may be expected to distort or shear the piles.

14.5 CHOICE OF PILE

Having found a satisfactory pile and a reliable and co-operative piling contractor for a particular site and conditions, there is a temptation for a busy designer, with inadequate time to investigate the wide choice of piles and systems, to use the same piling contractor for all future projects. This understandable reaction does not make for cost effectiveness nor structural efficiency. A guide to the choices available is given below:

(1) The piling system must provide a safe foundation with an adequate factor of safety (see section 14.6.1) against failure of the foundation on supporting soil.
(2) The total settlement and differential settlement must be limited to that which the structure can tolerate.
(3) The pile should be the right type of pile for the ground conditions and structure (see section 14.5.1).
(4) The driving of the piles and the load they impose on the soil must not damage neighbouring structures.
(5) The piles must be economic (see section 14.5.3) and durable (see section 14.5.2), and where speed of construction is important, quick to place.

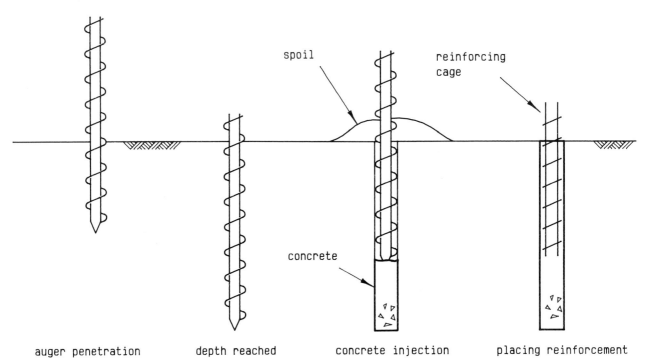

spoil

reinforcing cage

concrete

auger penetration   depth reached   concrete injection   placing reinforcement

Fig. 14.10 Flight auger pile.

14.5.1 Ground conditions and structure

(1) When invited to tender for the contract the piling contractor should be provided with a soil report, the position and magnitude of structural loads and the location of the structure together with information on adjoining properties. He should also be asked to visit the site to inspect the access for piling plant movements.

(2) Driven and cast-in-place piles, where the shell is left in, are used on sites over water (jetties, piers, etc.), on sites known to contain large voids, and sites subject to high water pressure. Driven piles should not be chosen where the ground is likely to contain large boulders but they are one of the best piles for loose-to-compact wet sands and gravels.

These types of piles are frequently the cheapest to use on building sites with light-to-moderate pile loadings and where the charges for moving onto site are spread over a large number of piles.

(3) Bored piles are frequently the lowest cost piles when piling into firm clays or sandstone and when vibration and ground heave would cause problems to existing adjacent buildings.

(4) Jacked piles need something to jack against and tend to be expensive. Their main use is therefore in underpinning when they can prove to be cost-effective.

(5) Steel H piles are often chosen when long length piles with deep penetration into sands and gravels are required.

14.5.2 Durability

The ground conditions can affect the choice and method of protection of piling material. Sulphate and acids will attack poor-quality concrete, some acids will cause problems with steel piles and alternative wetting and drying can cause timbers to rot.

14.5.3 Cost

Piles are, or should be, chosen as the economic and safe alternative to strip and raft foundations but there is more to cost analysis than comparing the cost per metre run of piles; there are on-costs. In comparing piling contractors' estimates it can be unwise to accept the lowest cost per metre run. Examination of extra over-costs for such items as extending lengths of piles, conducting check loading tests, etc. is prudent. The designer should examine the piling contractor's resources available to complete the project on time, the length of notice required to start work and the contractor's experience in piling on similar sites. The contractor's reputation should be investigated and proof obtained of adequate insurance to indemnify building owners for any claims or damage to adjoining buildings or failure of piles due to design and construction faults.

To the cost of the piles must be added the cost of excavation for constructing pile caps and any necessary tie beams. This increases the cost of construction supervision and design.

Decisions must be taken early so that design, detailing, construction and planning can be completed well in advance of starting the contract. Too often the time is restricted by delays in site investigations, change of design brief, recent changes in contractors' prices, etc.

14.6 DESIGN OF PILED FOUNDATIONS

The design of piles has become increasingly specialized and relies upon a detailed knowledge of ground conditions, properties of the types of pile, effects produced by loading, possible imperfections in the pile and the effect on the structure. The design is commonly the responsibility of the piling contractor but it is advisable that the structural designer appreciates the basic principles and checks the piling design.

As discussed in section 14.3.1 piles develop their load-bearing capacity from skin friction and end bearing. Values of skin friction and end bearing can be estimated from soil mechanics tests, past experience of similar conditions and on-site driving resistance. Since none of these methods can be totally relied upon it is often advisable to carry out load tests on a sample of piles and apply a suitable factor of safety in the design.

The ultimate bearing capacity equals the sum of the ultimate end-bearing capacity and ultimate skin-friction capacity (see section 14.6.2).

14.6.1 Factor of safety

BS 8004 recommends a factor of safety of between 2 and 3 for a single pile. The factor of safety is not a fixed constant and depends on the allowable settlement of the pile which is dependent on the pile's surface and cross-sectional area, the compressibility of the soil, and the reliability of the ground conditions. The factor should be increased when:

(1) The soil is variable, little is known of its behaviour or its resistance is likely to deteriorate with time.
(2) Small amounts of differential settlement are critical.
(3) The piles are installed in groups.

The factor may be decreased when:

(1) As a result of extensive loading tests, the resistance can be confidently predicted.
(2) As a result of extensive local experience, the soil properties are fully known.

A common factor of safety taken in design is 2.5. A properly designed single 500 mm diameter pile driven into non-cohesive soil is unlikely to settle more than about 15 mm.

In a load test (see section 14.6.3) the settlement is noted for increasing increments of load and a settlement/load graph is plotted. The graph resembles that of the stress/strain graph for many structural materials (see Fig. 14.11). Up to working load there tends to be practically full recovery of settlement on removal of load but beyond that loading there is likely to be a permanent set (as in steel loaded beyond the elastic limit) and at ultimate load there is likely to be no recovery at all.

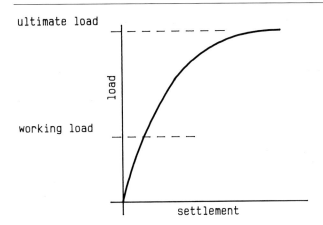

Fig. 14.11 Load/settlement graph.

14.6.2 Determination of ultimate bearing capacity

SOIL TESTS

The pile load and its own weight are supported by the surface skin friction between the soil and the pile plus the end-bearing resistance, i.e.:

$$T_f = t_{sf}A_s + t_{bf}A_b$$

where T_f = ultimate bearing capacity
t_{sf} = average surface skin friction per unit area of the surface
A_s = surface area of the pile shaft
t_{bf} = ultimate value of resistance of the base per unit area
A_b = plan area of base

The 'f' subscript for T_f, t_{sf} and t_{bf} indicates that these values relate to *ultimate bearing capacity* not *ultimate limit state*.

It will be noted that, for simplicity, the own weight of the pile has been omitted in the above formula since the base area times the overburden pressure at the base are approximately equal to the own weight.

For tubular box sections or other open ended piles, A_b would normally be ignored. The values of t_{sf} and t_{bf} are usually assessed from laboratory tests or in situ tests.

(a) Laboratory tests on soil samples

Laboratory tests on non-cohesive soils (sands and gravels) can be unreliable due to the difficulty of taking truly undisturbed samples. Determination of skin friction of non-cohesive soils is better assessed by in situ tests (see below).

In cohesive soils, t_{sf} is dependent on c (the cohesion), the effect of remoulding of the soil during piling, the type of pile surface and the period of time after driving the pile. In driving the pile there is a remoulding effect on the soil — its moisture content around the pile surface changes and takes time to recover. The larger the surface area of the pile then the higher the total skin friction.

The term t_{sf} is commonly expressed as αc_s, where c_s is the average cohesion over the pile length and α, the adhesion factor, varies from 0.3–0.6, and is dependent on the factors mentioned above. Advice on the value of α is better obtained from piling contractors with knowledge of their piles and the

soil; for example, α is generally assumed at about 0.45–0.5 for bored piles in London clay.

(b) In situ tests

In situ tests are generally used in non-cohesive soils. Loose sands are compacted by driving (but not by boring) and their skin friction is improved but little improvement is noted in dense compact sands. Piles in non-cohesive soils rely for bearing capacity more on end bearing than skin friction. A common rule of thumb assessment of skin friction is given in Table 14.3.

The end-bearing resistance can be determined from the standard penetration test or static penetration test (Dutch penetration) where the force required to push the cone down is measured. The application of the test results to the calculation of end bearing is both empirical and dependent on experience and any uncertainty should be decided by pile loading tests.

(c) Dynamic pile driving formulae

Basically the harder it is to drive a pile then the greater is its bearing capacity. The number of hammer blows to drive a pile, say 100 mm, into dense compact sand would be far more than driving it into soft clay. So the resistance to the impact of the hammer when driving the pile is related to the resistance of the pile to penetration under working load. The depth of penetration of the pile per hammer blow is known as the *set* and generally the deeper the pile and the stronger the ground then the less the *set*. This magnitude of the set, S, is affected by:

- the resistance overcome in driving (tons), R
- the weight of the ram (tons), W
- the free fall of the hammer (inches), h
- the final set of penetration of pile per blow (inches), S
- the sum of the temporary elastic compressions of the pile, dolly, packings and ground (inches), C (see Table 14.6)
- the efficiency of the blow, η (see Tables 14.4 and 14.5).

Probably the best known or most widely used formula for determining the driving resistance, R, is the Hiley formula:

$$R = \frac{Wh \cdot \eta}{S + \dfrac{C}{2}}$$

(For this equation the symbols are those defined above and not those indicated in the Notation list at the front of the book.)

The Hiley formula is based on two assumptions which are not fully applicable to soils and pile driving.

Table 14.3 Rule of thumb assessment of skin friction

Relative density	Average unit skin friction (kN/m²)
up to 0.35 (loose)	10
0.35–0.65 (medium)	10–25
0.65–0.85 (dense)	25–70
0.85+ (very dense)	70–100

Table 14.4 Pile driving data – coefficient of restitution, e (*Pile Driving* by W.L. Dawson. ICE Works Construction Guides, Telford)

The value of the coefficient of restitution e has been determined experimentally for different materials and conditions and is approximately as follows:

Piles driven with double-acting hammer

Steel piles without driving cap	0.5
Reinforced concrete piles without helmet but with packing on top of pile	0.5
Reinforced concrete piles with short dolly in helmet and packing	0.4
Timber piles	0.4

Piles driven with single-acting and drop hammer

Reinforced concrete piles without helmet but with packing on top of piles	0.4
Steel piles or steel tube of cast-in-place piles fitted with driving cap and short dolly covered by steel plate	0.32
Reinforced concrete piles with helmet and packing, dolly in good condition	0.25
Timber piles in good condition	0.25
Timber piles in poor condition	0.0

The efficiency of the blow can be obtained from Table 14.5 for various combinations of e with the ratio P/W, provided that W is greater than Pe and the piles are driven into penetrable ground.

Table 14.5 Pile driving data – efficiency of blow, η (*Civil Engineering Code of Practice No. 4, Institution of Civil Engineers, 1954*)

P/W	e = 0.5	e = 0.4	e = 0.32	e = 0.25	e = 0.0
½	0.75	0.72	0.70	0.69	0.67
1	0.63	0.58	0.55	0.53	0.50
1½	0.55	0.50	0.46	0.44	0.40
2	0.50	0.44	0.40	0.37	0.33
2½	0.45	0.40	0.36	0.33	0.28
3	0.42	0.36	0.33	0.30	0.25
4	0.36	0.31	0.28	0.25	0.20
5	0.31	0.27	0.25	0.21	0.16
6	0.27	0.24	0.23	0.19	0.14

Table 14.6 Pile driving data – temporary compression, C (mm) (*Pile Driving* by W.L. Dawson, ICE Works Construction Guides, Telford.)

Form of compression	Material	Easy driving	Medium driving	Hard driving	Very hard driving
Pile head and cap: C_c	Head of timber pile	1	2.5	4	5
	Short dolly in helmet or driving cap[a]	1	2.5	4	5
	75 mm packing under helmet or driving cap[a]	2	4	5.5	7.5
	25 mm pad only on head of reinforced concrete pile	1	1	2	2.5
Pile length: C_p	Timber pile. E: 10.5×10^3 N/mm^2	0.33H	0.67H	1.00H	1.33H
	Precast concrete pile. E: 14.0×10^3 N/mm^2	0.25H	0.5H	0.75H	1.00H
	Steel pile, steel tube or steel mandrel for cast-in-place pile. E: 210×10^3 N/mm^2	0.25H	0.5H	0.75H	1.00H
Quake: C_q	Ground surrounding pile and under pile point	1	2.5 to 5	4 to 6	1 to 4

[a] If these devices are used in combination, the compressions should be added together.
Pile length H measured in metres.

(1) The resistance to penetration of the pile can be determined by the kinetic energy of the hammer blow, and
(2) The resistance to driving equates to the ultimate bearing capacity, but kinetic energy is lost in elastic strains, vibration of the pile, its dolly and helmet and the soil. Further, the soil properties can change during and after piling.

These variants have produced a mash of empirical constants and variation in formulae – and as in all structural design the more formulae then the less is known with accuracy, so that the calculated results of the ultimate bearing capacity lie in a range ±40%. Some experienced engineers are opposed to the use of driving formula but a simple formula backed by statistical data and checked by test loading is acceptable to many engineers.

Piling has become a very specialized brand of engineering and the designer is advised to consult a reputable piling contractor for advice when difficult sites are encountered and, before awarding a piling contract to a contractor, should obtain from them indemnity and guarantees.

14.6.3 Pile loading tests
Loading tests are carried out to:

(1) Determine the settlement at working load.
(2) Determine the ultimate bearing capacity of the pile.
(3) Check the soundness of the pile.

Load tests are expensive particularly if kentledge has to be brought to the site, erected, left for the test (involving standing time for piling plant), dismantled and removed from the site. Over-enthusiasm for pile testing on a small site could cost more and take longer than the actual piling. In such cases it is better to employ a conservative factor of safety.

Like boreholes, pile tests relate to the isolated individual pile and the data must be applied with care and judgement to the total piled foundation. The test pile is loaded in 25% increments or other agreed increases of the working load and the rate and magnitude of settlement noted. When the rate of settlement has decreased to a negligible amount the next increment of load is added. After working load is reached, the pile is sometimes loaded to ultimate failure in 10% increments of working load. At shear failure of the soil, the pile penetrates relatively rapidly. A rough guide for ultimate bearing capacity is the place on the load/settlement curve where there is a clear indication of rapidly increased settlement (see Fig.14.11) or where the load causes the pile to penetrate 10% of the pile diameter.

The method of loading described above is known as the *maintained load method*. Another common method, known as the *constant rate penetration* (CRP), is used where the pile is continuously loaded by jacks to make it penetrate the soil at constant speed. Failure is defined when either:

(1) The pile penetrates without increase in load, or
(2) The penetration equals 10% of the pile base diameter.

The CRP method is a rapid method useful for research or preliminary testing in a series of other tests when the factor of safety against ultimate failure is needed. But the method does not give the elastic settlement under the working load and requires heavy kentledge and very strong anchor piles.

14.6.4 Pile groups
It is sometimes necessary to drive a group of piles to support heavy loadings and it is important to notice two effects:

(1) The pressure bulb of the group affects deeper layers of soils than a single pile (see Fig. 14.12) in a similar manner to a wide foundation.
(2) The load-bearing capacity of a group is not necessarily the product of the capacity of the single pile times the number of piles. There can be a pressure 'overlap' (see Fig. 14.13) and the capacity of the group could decrease as the difference between a pad and strip foundation.

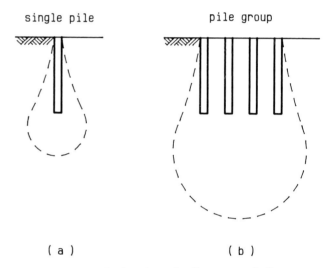

Fig. 14.12 Section through pile pressure bulbs.

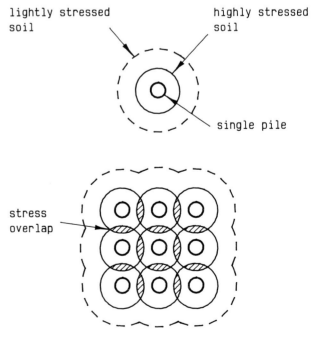

Fig. 14.13 Plan on pile pressure bulbs.

A single pile, in driving, displaces soil which can result in heave at ground level and a group can cause greater heave and displacement; this fact should be checked and considered. Driving a single pile, too, in loose sand and fills will compact the soil around the pile to a diameter of approximately 5.5 times the pile diameter and make it denser. (This is the principle of vibro-stabilization, see Chapter 8.) If a group of piles is driven it could create such a compact block of soil as to prevent driving of all the piles in the group. The central piles should be driven first and then, working out to the perimeter of the group, the remaining piles should be driven.

14.6.5 Spacing of piles within a group
Approximate values for centre-to-centre spacing are as follows:

(1) Friction piles − not less than the perimeter of the pile.
(2) End-bearing piles − not less than twice the diameter of the pile.
(3) Screw piles − not less than 1.5 times the diameter of the blades.
(4) Piles with enlarged bases − at least pile diameter between enlarged bases.

These values are affected by the soil conditions, the group behaviour of the piles, the possible heave and compaction, and the need to provide sufficient space to install the piles to the designed penetration without damage to the pile or group.

14.6.6 Ultimate bearing capacity of group
The group can be considered to act as a buried raft and will spread the load in varying conditions, as shown in Fig. 14.14.

As in raft foundations, so with pile groups, it is limiting the magnitude of settlement rather than applying a factor of safety to the ultimate bearing capacity which is the major design criterion. In Fig. 14.14 (a) and (b) the ultimate bearing capacity is likely to equal the number of piles, n,

times the bearing capacity of an individual pile, T_f, i.e. $n T_f$. In Fig. 14.14 (c) the capacity can be around 25% less than $n T_f$.

14.6.7 Negative friction
Compressible fills and sensitive soft clays consolidating under pressure will exert a downward drag on the surface of the pile − negative skin friction. For example, if piling into firm strata through decaying fill material, the settlement of the decaying fill will apply downward friction forces onto the outer surface of the pile and these forces need to be taken into account as an additional load on the pile.

14.7 PILE CAPS

14.7.1 Introduction
The design of pile caps had at one time become a mathematician's delight − and a designer's nightmare. Highly complex formulae with numerous empirical variants could result in expensive design and construction to save a couple of reinforcing bars. As in all design and construction the aim must be 'to keep it simple'.

14.7.2 The need for pile caps − capping beams
It is frequently not possible to sit a superstructure column direct on to a pile because:

(1) It is practically impossible to drive piles in the exact position and truly vertical. Piles wander in driving and deviate from their true position. A normal specification tolerance for position is ±75 mm and for verticality not more than 1 in 75 for a vertical pile or 1 in 25 for a raking pile. A column sitting directly on a pile with an eccentricity of 75 mm will exert bending as well as direct stresses in the pile.
(2) A single, heavily loaded column supported by a pile group will need a load spread (pile cap) to transmit the load to all the piles.

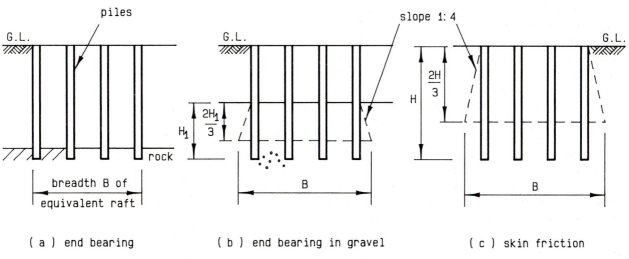

(a) end bearing   (b) end bearing in gravel   (c) skin friction

Fig. 14.14 Pile groups.

(3) A line of piles supporting a load-bearing wall will need a capping beam to allow both for tolerance of pile positioning and load spreading of the piles' concentrated load to the wall.

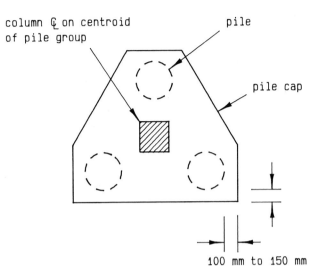

Fig. 14.15 Plan on triple pile cap.

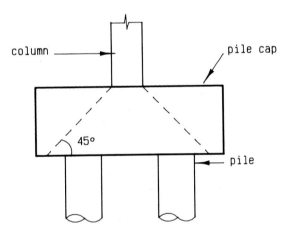

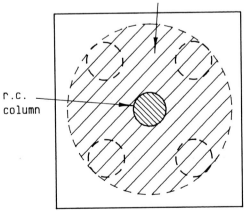

Fig. 14.16 Load transfer from column to piles.

14.7.3 Size and depth

Pile caps are usually of concrete but can be large slabs of rock or mats of treated timber. This discussion is limited to the more common use of concrete.

To allow for the pile deviation the pile cap should extend 100–150 mm beyond the outer face of the piles. The pile group centroid should ideally coincide with the column's position (see Fig. 14.15).

The depth must be adequate to resist the high shear force and punching shear and to transmit the vertical load (see Fig. 14.16). The shaded area of the pile cap plan in Fig. 14.16 is the area where the column load is directly transferred to the piles. For such a condition the shear stresses are generally small but bending moments need to be catered for.

Alternatively, peripheral steel as a ring tension around a cone shaped compression block may be considered to be a suitable equilibrium of forces (see Fig. 14.17), however, full tension laps must be provided for the peripheral steel.

Single column loads supported on larger pile groups can create significant shear and bending in the cap which will need top and bottom reinforcement as well as shear links (see Fig. 14.18).

The heads of r.c. piles should be stripped and the exposed reinforcement bonded into the pile cap for the necessary bond length. Pile caps to steel piles can be reduced in depth if punching shear is reduced by capping and/or reinforcing the head of the pile, as shown in Fig. 14.19.

Piles for continuous capping beams supporting load-bearing walls can be alternately staggered to compensate for

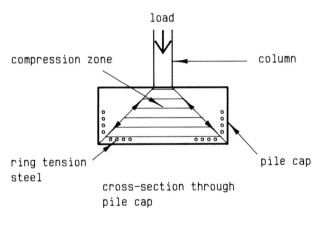

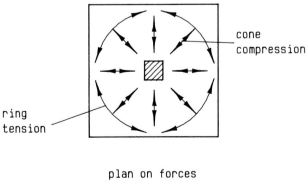

Fig. 14.17 Ring tension pile cap.

University Centre Library
The Hub at Blackburn College

Customer ID: *******37

Title: Structural foundations designers manual
ID: BC80588

Due: Mon, 07 Nov 2016

Total items: 1
31/10/2016 08:50

Please retain this receipt for your records
Contact Tel. 01254 292165

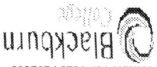

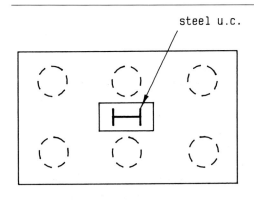

plan

50 mm cover top
100 mm cover bottom

100 mm

section

Fig. 14.18 Pile cap, typical reinforcement.

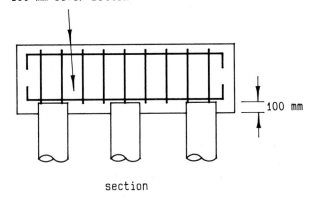

reinforcement

cap plate   plan

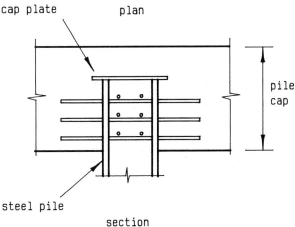

pile
cap

steel pile

section

Fig. 14.19 Reinforced pile head.

the eccentricity of loading due to the 75 mm out-of-line tolerance (see Fig. 14.20).

14.8 DESIGN OF FOUNDATIONS AT PILE HEAD

A general description of ground beams and pile caps is discussed in Chapter 9 (see section 9.5.8) and restraints and cap/beam details are briefly mentioned.

In addition to providing restraint, the ground beam is also used to transfer loads from the superstructure to the pile and can be used with or without pile caps. For example, two alternative layouts are shown in Fig. 14.21 indicating a wide ground beam solution and a narrow beam using pile caps.

Where the increased width of the beam needed to accommodate the pile diameter, plus the total of all necessary tolerance, is only slight and where a reduction in beam depth helps to compensate for the additional concrete, a wider beam omitting the pile caps can be more economic.

Often the ground beam can be designed compositely with the walls above and by using *composite* beams a standard nominal size ground beam, dictated mainly by the practicalities for construction, can be used. This has the advantage of standardizing shuttering, reinforcement and excavation, making site construction simple, economic and quicker than the traditional solution. Many different beams designed ignoring the benefit of the contribution from the structure above can severely complicate the foundations (see Fig. 14.22).

When considering the use of composite action, consideration must be given to services which may pass through below ground level in these zones. It is often the case that in adopting composite beams the resulting shallow beams can be more easily made to pass over the services.

A further help in standardizing a smaller and more economic section is that composite action often makes it possible to precast the beams alongside the excavation and roll them into position, speeding up construction.

For building structures the basic alternative foundations for support on piles generally adopted consists of one or a combination of the following:

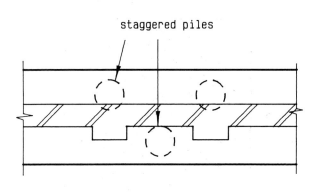

plan on capping beam

Fig. 14.20 Continuous capping beam.

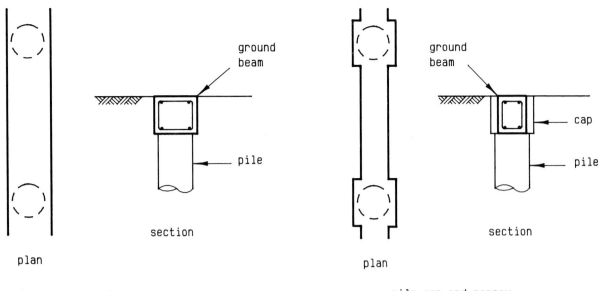

Fig. 14.21 Alternative beam/cap layouts.

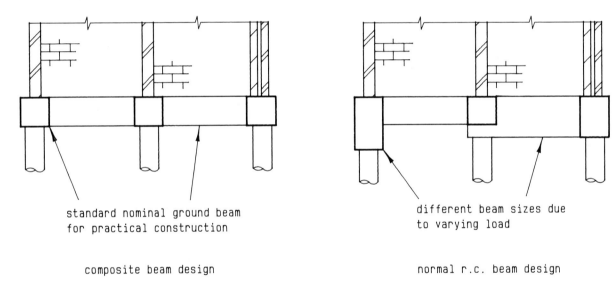

Fig. 14.22 Composite action versus normal design.

Type 1 Concrete ground beams with or without caps supporting the main superstructure load but with a floating ground floor slab between the main wall (see Fig. 14.23).

Type 2 Concrete ground beams and suspended in situ or precast concrete floor slabs (see Fig. 14.24).

Type 3 Flat slab construction (see Fig. 14.25).

Type 4 Suspended slab and beam foundations with voids or void formers (see Fig. 14.26).

The economic viability of the pile solutions for the above foundations will differ depending on many variables but, by applying the following basic principles, realistic cost comparisons can be made and piling options exploited:

(1) Minimizing pile numbers relative to pile length/cost and beam length/cost ratio.

(2) Maintaining axial loads on piles and ground beams wherever practical.

(3) Providing pile restraints from other necessary structures wherever practical.

(4) Standardizing on the minimum beam size which can accommodate pile driving tolerances, restraint stresses and pile eccentricity while exploiting any possible composite action.

(5) Minimizing the depth of excavations.

(6) Minimizing the required bending of reinforcement.

(7) Minimizing the shuttering costs by simple standard beam profiles.

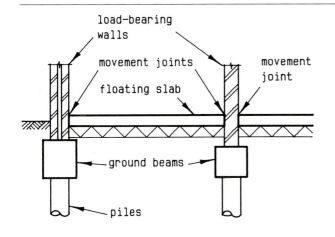

typical section through piled
foundation with floating ground slab

Fig. 14.23 Piles and *floating* ground slab.

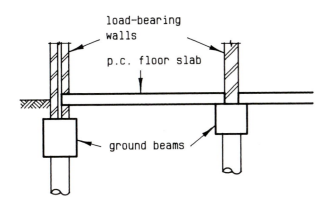

piled foundation and suspended
p.c. floor slab

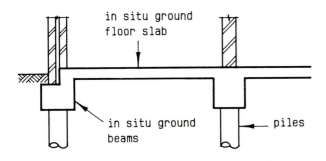

piled foundation and in situ
suspended floor slab

Fig. 14.24 Piles and suspended ground slab.

(8) Use of simply supported design and simple beam cages wherever possible unless some small cantilever action can greatly reduce the number of piles per unit.

(9) Minimizing the need for pile caps wherever practical by the use of slightly wider beams.

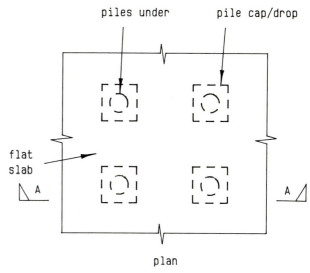

plan

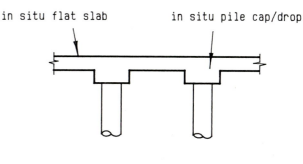

section A – A

Fig. 14.25 Piles and flat slab construction.

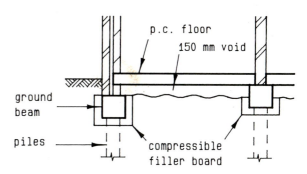

Fig. 14.26 Piled suspended slab and beam construction.

14.9 DESIGN EXAMPLES

14.9.1 Design Example 1: Calculation of pile safe working loads

A site investigation indicates 5 m of variable fill overlying 4 m of medium dense gravel, overlying stiff silty clay (see Fig. 14.27). The fill is unsuitable for treatment by the ground improvement methods discussed in Chapter 8.

Capacity of individual soil layers

The following parameters are assumed for the design of a bored piled foundation, based upon the soils test results:

(1) *Fill from 0 m to −5 m*: The fill is considered to have

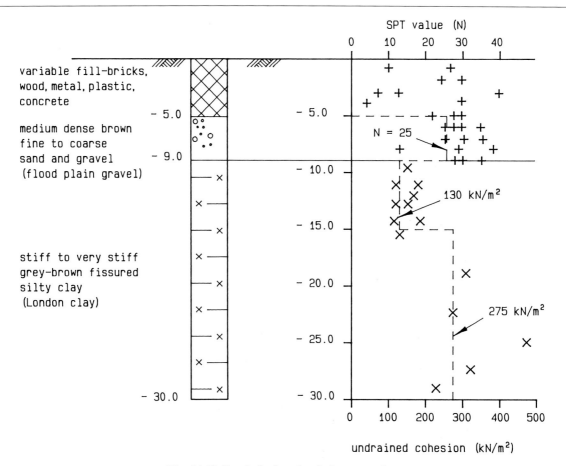

Fig. 14.27 Borehole data for design examples.

negligible contribution to the skin friction capacity of the pile.

(2) *Sand and gravel from −5m to −9m*: The SPT value is to be taken as $N = 25$ (see Fig. 14.27). This corresponds to an angle of shearing resistance of $\phi = 35°$. The skin friction capacity at failure is normally taken to be

$$T_{sf} = (Ks' \tan \delta) A_s$$

where K = an earth pressure coefficient = 1.0, based on the soils investigation

s' = effective overburden pressure = γz = $20 \text{ kN/m}^3 \times 7 \text{ m}$ average depth in this example

δ = angle of wall friction = 0.75 ϕ

A_s = perimeter area of pile = πD per metre depth, for a circular pile of diameter D.

Thus, $T_{sf} = (Ks' \tan \delta) A_s$
$= (1.0) \times (20 \times 7) \times [\tan (0.75 \times 35)] \times (\pi D)$
$= 217D$ kN/m depth

(3) *Clay from −9m to −15m*: An average undrained shear strength of $c_s = 130 \text{ kN/m}^2$ is assumed for design purposes. The skin friction capacity at failure is normally taken as

$$T_{sf} = (\alpha c_s) A_s$$

where α = adhesion factor = 0.45, for London clay

Thus, $T_{sf} = (\alpha c_s) A_s$
$= (0.45) \times (130) \times (\pi D)$
$= 184D$ kN/m depth

If the pile is founded within this clay stratum, the end-bearing capacity at failure is given by

$$T_{bf} = N_c c_b A_b$$

where N_c = bearing capacity factor = 9, for circular piles

c_b = undisturbed shear strength at base = 130 kN/m^2, from the soils investigation (see Fig. 14.27)

A_b = base area of pile = $\pi D^2/4$, for a circular pile of diameter D.

Thus, $T_{bf} = N_c c_b A_b$
$= (9) \times (130) \times (\pi D^2/4)$
$= 919D^2$ kN

(4) *Clay from −15m to −30m*: An average undrained shear strength of $c_s = 275 \text{ kN/m}^2$ is assumed for design purposes. As above, the skin friction capacity is given by

$$T_{sf} = (\alpha c_s) A_s$$
$$= (0.45) \times (275) \times (\pi D)$$
$$= 389D \text{ kN/m depth}$$

As above, if the pile is founded within this clay stratum, the end-bearing capacity at failure is normally given by

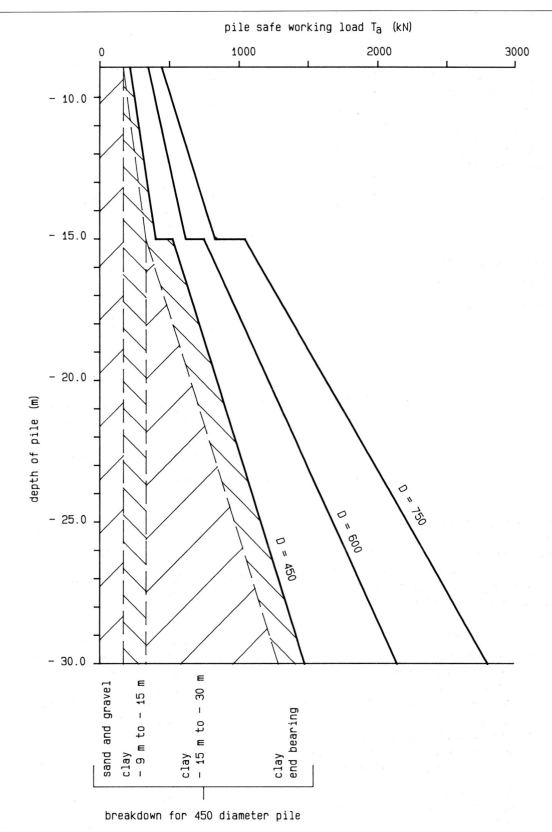

Fig. 14.28 Pile safe working loads for design examples.

$$T_{bf} = N_c c_b A_b$$
$$= (9) \times (275) \times (\pi D^2/4)$$
$$= 1944 D^2 \, kN$$

Total capacity of piles between 9 m and 15 m in length

For a pile between 9 m and 15 m in length, of total length H, the total capacity is derived by summing the capacities from the various soil strata as follows:

Capacity due to sand/gravel $\quad T_{sf} = 217D \times (9\,m - 5\,m)$
$$= 868D \, kN$$
Capacity due to clay $\quad T_{sf} = 184D \times (H - 9\,m)$
$$T_{bf} = 919 D^2 \, kN$$

Thus, adding these together gives

$$T_f = 868D + 184D(H - 9) + 919D^2$$
$$= [868 + 184(H - 9)]D + 919D^2$$

Applying a typical factor of safety of 2.5 gives the allowable load (safe working load) as

$$T_a = \frac{T_f}{2.5}$$
$$= [347 + 74(H - 9)]D + 368D^2$$

T_a is plotted in Fig. 14.28 for a range of pile diameters and depths.

Total capacity of piles in excess of 15 m in length

For a pile in excess of 15 m in length, of total length H, the total capacity is derived as follows:

Capacity due to sand/gravel $\quad T_{sf} = 217D \times (9\,m - 5\,m)$
$$= 868D \, kN$$
Capacity, clay down to 15 m $\quad T_{sf} = 184D \times (15\,m - 9\,m)$
$$= 1104D \, kN$$
Capacity, clay below 15 m $\quad T_{sf} = 389D \times (H - 15\,m)$
$$T_{bf} = 1944 D^2 \, kN$$

Adding these components together gives

$$T_f = 868D + 1104D + 389D(H - 15) + 1944D^2$$
$$= [1972 + 389(H - 15)]D + 1944D^2$$

Again applying a typical factor of safety of 2.5 gives the allowable load (safe working load) as:

$$T_a = \frac{T_f}{2.5}$$
$$= [789 + 156(H - 15)]D + 778D^2$$

T_a is plotted in Fig. 14.28 for a range of pile diameters and depths.

14.9.2 Design Example 2: Pile cap design

A pile cap is required to transfer the load from a 400 mm × 400 mm column to four 600 mm diameter piles, as shown in Fig. 14.29.

Pile caps can be designed either by the truss analogy or by bending theory (see BS 8110: Part 1: 3.11.4.1). In this example bending theory will be used.

For a pile cap with closely spaced piles, in addition to bending and bond stress checks, a check should be made on

the local shear stress at the face of the column, and a *beam shear* check for shear across the width of the pile cap. For more widely spaced piles, a punching shear check should also be carried out.

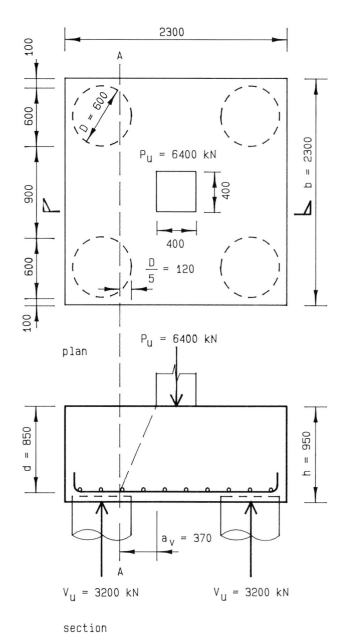

Fig. 14.29 Pile cap design example.

Local shear check

The ultimate column load is $P_u = 6400 \, kN$.

Length of column perimeter is $u = 2(400 + 400) = 1600 \, mm$.

The shear stress at the face of the column is

$$v_u = \frac{P_u}{ud}$$

BS 8110: Part 1: 3.11.4.5 requires this to be limited to $0.8\sqrt{f_{cu}} < 5 \, N/mm^2$.

For grade C35 concrete, $0.8\sqrt{f_{cu}} = 4.73 \, N/mm^2$. This gives a requirement that

$$v_u \leqslant 4.73$$

$$\frac{P_u}{ud} \leqslant 4.73$$

$$d \geqslant \frac{P_u}{4.73u}$$

$$\geqslant \frac{6400 \times 10^3}{4.73 \times 1600}$$

$$d \geqslant 846\,mm$$

Bending shear check

In accordance with BS 8110: Part 1: 3.11.4.3, shear is checked across a section 20% of the diameter of the pile (i.e. $D/5$) inside the face of the pile. This is section $A-A$ in Fig. 14.29.

The shear force across this section – ignoring the self-weight of the pile cap, which is small in comparison – is given by

$$V_u = \frac{P_u}{2}$$

$$= \frac{6400}{2}$$

$$= 3200\,kN$$

The corresponding shear stress is given by $v_u = V_u/bd$. In accordance with BS 8110: Part 1: 3.11.4.4, this must not exceed $(2d/a_v)v_c$ where a_v is defined in Fig. 14.29 and v_c is the design concrete shear stress from BS 8110: Part 1: Table 3.9. Thus

$$\frac{V_u}{bd} \leqslant \left(\frac{2d}{a_v}\right)v_c$$

$$d^2 \geqslant \frac{a_v V_u}{2bv_c}$$

$$d \geqslant \sqrt{\left(\frac{a_v V_u}{2bv_c}\right)}$$

For grade C35 concrete, from BS 8110: Part 1: Table 3.9 the minimum value of v_c is $0.38\,N/mm^2$, giving

$$d \geqslant \sqrt{\left(\frac{370 \times 3200 \times 10^3}{2 \times 2300 \times 0.38}\right)}$$

$$d \geqslant 823\,mm$$

Thus, provided the average effective depth exceeds $d = 846\,mm$ (the local shear check), minimum reinforcement to satisfy bond and bending tension requirements will be adequate in this instance.

The necessary depth for the pile cap is

$$h = d + 25(\text{diameter bar}) + 75(\text{cover})$$

$$= 846 + 100$$

$$= 946\,mm \Rightarrow \text{use } h = 950\,mm$$

14.9.3 Design Example 3: Piled ground beams with floating slab

A two-storey terrace of four office units is to be founded in the ground conditions described in Design Example 1 (see section 14.9.1). Wall and ground beam line loads are shown in Fig. 14.30.

Examination of the soil profiles indicates that, while ground floor slab loads can be carried on the existing ground as a floating slab, the main superstructure loads need to be supported on piled foundations. The proposed pile and ground beam layout is shown in Fig. 14.30.

Pile design

Pile loads have been calculated from the wall (and ground beam) loads, and are shown in Fig. 14.30. For the purpose of deriving pile loads, the ground beams have been assumed to be simply supported.

Based on the safe working loads calculated in Design Example 1 (see Fig. 14.28), the required pile lengths are given in Table 14.7. The choice between 450 mm and 600 mm diameter piles would be based on economic considerations, bearing in mind that the larger pile diameter will require wider ground beams.

For the purpose of this example, 450 mm diameter piles will be adopted.

Table 14.7 Pile lengths and diameter for Design Example 3

Pile load (kN)	Pile length (m)	
	450 mm diameter pile	600 mm diameter pile
415	14.5	10.7
473	15.0	12.0
583	16.0	14.5
765	18.6	15.1

Check on strength of pile cross-section

Piles usually only carry vertical loading. Concrete piles are typically only nominally reinforced, unless carrying lateral loads or tension. A check does however need to be carried out, to ensure the concrete is not overstressed.

From Fig. 14.30, the maximum pile working load is $T = 765\,kN$ (on grid lines 2, 3 and 4).

From Fig. 14.30, the imposed load Q as a percentage of T is $100Q/T = 100 \times 80/(90 + 80) = 47\%$.

From Fig. 10.20, the combined partial safety factor for loads is $\gamma_T = 1.49$.

The corresponding ultimate load is

$$T_u = \gamma_T T$$

$$= 1.49 \times 765$$

$$= 1140\,kN$$

For a 450 mm diameter pile, assuming all the load is carried by the concrete, this gives an ultimate concrete stress of

$$f_c = \frac{T_u}{A}$$

$$= (1140 \times 10^3)\left(\frac{\pi\,450^2}{4}\right)$$

$$= 7.2\,N/mm^2$$

From BS 8110: Part 1: 3.8.4.4, the allowable concrete stress is $0.35f_{cu}$.

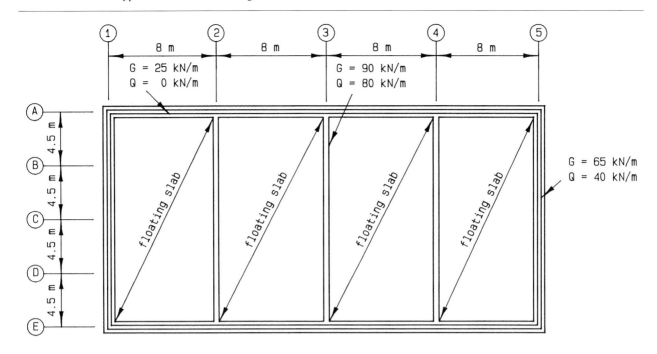

plan on ground floor walls and slab

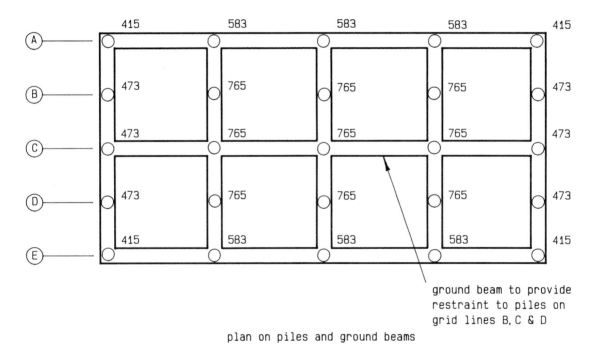

plan on piles and ground beams

Fig. 14.30 Piled ground beam and floating slab design example.

$$\Rightarrow f_{cu(req)} = \frac{f_c}{0.35}$$

$$= \frac{7.2}{0.35}$$

$$= 20.5\,\text{N/mm}^2$$

Therefore grade 25 concrete with nominal reinforcement is satisfactory, and for this example meets the durability

requirements, i.e. no sulphates or other contamination in the groundwater − revealed by the soils investigation.

Ground beam bending moments and shear forces for reinforcement design

(1) *Ground beam below internal load-bearing wall (Grid lines 2, 3, 4)*
From Fig. 14.30, the characteristic dead and imposed loads

are $G = 90\,\text{kN/m}$ and $Q = 80\,\text{kN/m}$ respectively. The ultimate design load is given by

$$T_u = 1.4G + 1.6Q$$
$$= (1.4 \times 90) + (1.6 \times 80)$$
$$= 254\,\text{kN/m}$$

From BS 8110: Part 1: Table 3.6,

$$M_{u(\text{max})} = 0.11T_uL^2 = 0.11 \times 254 \times 4.5^2 = 566\,\text{kNm}$$
$$V_{u(\text{max})} = 0.6T_uL = 0.6 \times 254 \times 4.5 = 686\,\text{kN}$$

(2) *Ground beam below front or rear wall (Grid lines A and E)*

From Fig. 14.30, the characteristic dead and imposed loads are $G = 25\,\text{kN/m}$ and $Q = 0\,\text{kN/m}$ respectively. The ultimate design load is given by

$$T_u = 1.4G + 1.6Q$$
$$= (1.4 \times 25) + (16 \times 0)$$
$$= 34\,\text{kN/m}$$

From BS 8110: Part 1: Table 3.6,

$$M_{u(\text{max})} = 0.11T_uL^2 = 0.11 \times 35 \times 8.0^2 = 246\,\text{kNm}$$
$$V_{u(\text{max})} = 0.6T_uL = 0.6 \times 35 \times 8.0 = 168\,\text{kN}$$

Thus the ground beams under the internal load-bearing walls are the critical design case.

Sizing of ground beam

If, as in this case, there are no pile caps, then ground beams should be a minimum of 150 mm wider than the piles, to allow for standard $\pm 75\,\text{mm}$ tolerances. In this example therefore, a ground beam width of $b = 450 + 150 = 600\,\text{mm}$ will be adopted.

Having chosen a width, the designer needs a procedure for calculating a suitable depth. In practice, the critical limit state for most ground beams is the ultimate limit state of bending. Typical reinforcement percentages for ground beams are in the range $0.5\% - 1.5\%$. Assuming bending is the critical condition, the BS 8110 design equations for grade C35 concrete give the following relationship between effective depth, reinforcement percentage, and allowable bending moment (Table 14.8).

The maximum ultimate bending moment is $M_u = 566\,\text{kNm}$. This gives

Table 14.8 Estimation of effective depth for ground beams using grade C35 concrete

| Effective depth, d (mm) | Allowable bending moment per unit width, M_u/b (kNm/m) | | |
| | Reinforcement percentage, $100A_s/bd$ | | |
	0.5%	0.75%	1.0%
300	162	243	315
400	288	432	560
500	450	675	875
600	648	972	1260
700	882	1323	1715
800	1152	1728	2240

$$\frac{M_u}{b} = \frac{566}{0.6}$$
$$= 943\,\text{kNm/m}$$

From Table 14.8, this indicates an effective depth in the range 525 mm to 725 mm, depending on the reinforcement percentage. For this particular example, the ground beam will be chosen to be 600 mm wide by 625 mm deep. This gives an effective depth of

$$d = 625(\text{depth}) - 40(\text{cover}) - 12(\text{link}) - 25/2(\text{main bar})$$
$$= 560\,\text{mm}$$

Bending reinforcement

The percentage of reinforcement could be estimated by interpolation from Table 14.8. In this case, however, it will be calculated more accurately, for the section with the highest moment.

$$\frac{M_u}{bd^2} = \frac{566 \times 10^6}{600 \times 560^2}$$
$$= 3.01$$

$$A_{s(\text{req})} = 0.85\%\,bd \qquad [\text{BS 8110: Part 3: Chart 2}]$$
$$= \frac{0.85}{100} \times 600 \times 560$$
$$= 2856\,\text{mm}^2/\text{m}$$

Provide 6T25 $= 2945\,\text{mm}^2/\text{m}$, giving

$$\frac{100A_s}{bd} = \frac{100 \times 2945}{600 \times 560}$$
$$= 0.88\%$$

This amount of reinforcement will be reduced for less heavily loaded sections.

Shear reinforcement

The maximum value of shear is $V_u = 686\,\text{kN}$. This occurs at sections where $A_s = 0.88\%\,bd$, giving an allowable concrete shear stress of $v_c = 0.68\,\text{N/mm}^2$ (BS 8110: Part 1: Table 3.9).

$$\text{Shear stress, } v_u = \frac{V_u}{bd}$$
$$= \frac{686 \times 10^3}{600 \times 560}$$
$$= 2.04\,\text{N/mm}^2$$

$$A_{sv(\text{req})} = \frac{b(v_u - v_c)10^3}{0.87f_{yv}}$$
$$= \frac{600 \times (2.04 - 0.68) \times 10^3}{0.87 \times 460}$$
$$= 2039\,\text{mm}^2/\text{m} \quad \text{(high yield bars)}$$

Provide 4 legs of T12 links @ 200 c/c $= 2264\,\text{mm}^2/\text{m}$.

14.9.4 Design Example 4: Piled ground beams with suspended slab

Design Example 3 is to be reworked on the assumption that the building is now to be relocated in an area where the 5 m

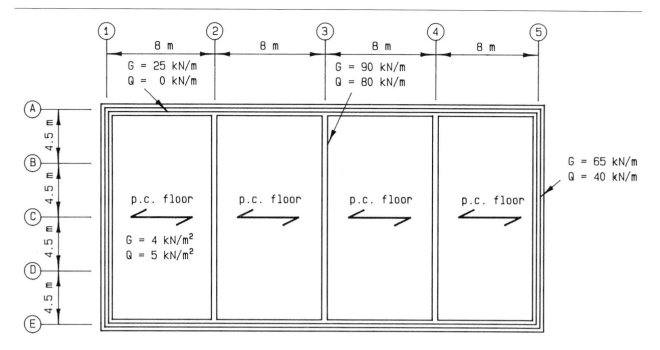

plan on ground floor walls and slab

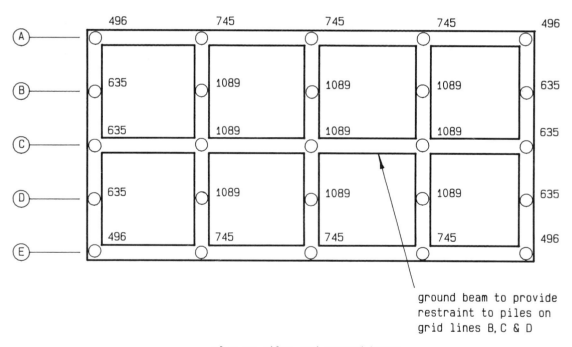

ground beam to provide
restraint to piles on
grid lines B, C & D

plan on piles and ground beams

Fig. 14.31 Piled ground beam and suspended slab design example.

depth of fill is of a much poorer quality, and is considered unsuitable for supporting a floating ground floor slab. The ground floor slab is therefore to be replaced by wide plank precast concrete floors, spanning 8 m parallel to grid lines A–E.

The additional loads due to this suspended floor are shown in Fig. 14.31, and the increased pile loads are indicated. The increased loads could be catered for by increasing the number of piles along each load-bearing internal wall (parallel to grid lines 1–5). In this case however, it has been decided to maintain the same pile and ground beam layout as in Design Example 3.

Pile capacities

As previously, the pile capacities given in Table 14.9 are derived from Design Example 1 (Fig. 14.28).

Piles of 450 mm diameter will again be used. Comparison with Design Example 3 indicates increases in length of between 0.5 m and 4.6 m.

Check on strength of pile cross-section

A check on the stresses in the pile cross-section, carried out in a similar manner to Design Example 3, indicates that grade 30 concrete is required (grade 25 was sufficient in Design Example 3).

Table 14.9 Pile lengths and diameters for Design Example 4

Pile load (kN)	Pile length (m)	
	450 mm diameter pile	600 mm diameter pile
496	15.0	12.5
635	16.7	15.0
745	18.3	15.0
1089	23.2	18.6

Ground beam size

The ground beams are designed in a similar manner to Design Example 3, taking due account of the additional loading from the suspended ground floor.

The calculations will be found to indicate that the 600 mm × 625 mm deep ground beams in Design Example 3 will need to be deepened by approximately 200 mm to accommodate this additional loading.

14.9.5 Design Example 5: Piled foundation with suspended flat slab

A five-storey office building is to be founded on the same ground conditions as the previous design examples in this chapter. The building is steel framed, with columns on a 6 m × 6 m grid. To simplify the pile caps, it was decided to use one pile per column (see Fig. 14.32).

The ground floor slab is to be designed as a two-way spanning suspended flat slab. Separate pile caps are provided to the top of each pile, both to avoid the need for punching shear reinforcement in the flat slab, and to enable steel superstructure erection to commence prior to construction of the flat slab.

Working loads

From Fig. 14.32, the superstructure dead and imposed column loads per pile are

$$P = G + Q$$
$$= 1000 + 1100$$
$$= 2100 \, \text{kN}$$

From Fig. 14.32, the foundation dead and imposed slab loads are given by

$$f = f_G + f_Q$$
$$= 7.5 + 6.0$$
$$= 13.5 \, \text{kN/m}^2$$

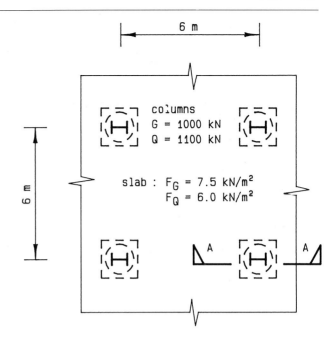

plan on foundations

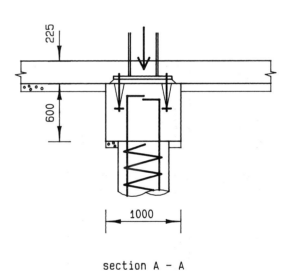

section A – A

Fig. 14.32 Piled flat slab design example.

The foundation imposed load as a percentage of f is $100 f_Q / f = 100 \times 6.0/13.5 = 44\%$.

From Fig. 10.20, the combined partial load factor for foundation loads is $\gamma_f = 1.49$.

The foundation load per pile is

$$F = fA$$
$$= 13.5 \times 6.0 \times 6.0$$
$$= 486 \, \text{kN}$$

The total load per pile is

$$T = (\text{superstructure load}) + (\text{foundation load})$$
$$= P + F$$
$$= 2100 + 486$$
$$= 2586 \, \text{kN}$$

Total imposed load as a percentage of T is $100(Q + F_Q)/T = 100 \times (1100 + 216)/2586 = 51\%$.

From Fig. 10.20, the combined partial load factor for total loads is $\gamma_T = 1.50$.

Ultimate loads

The foundation ultimate distributed load is

$$
\begin{aligned}
f_u &= \gamma_F f \\
&= 1.49 \times 13.5 \\
&= 20.1 \, \text{kN/m}^2
\end{aligned}
$$

The foundation ultimate load is

$$
\begin{aligned}
F_u &= f_u A \\
&= 20.1 \times 6.0 \times 6.0 \\
&= 724 \, \text{kN}
\end{aligned}
$$

The total ultimate pile load is

$$
\begin{aligned}
T_u &= \gamma_T T \\
&= 1.50 \times 2586 \\
&= 3879 \, \text{kN}
\end{aligned}
$$

Pile design

Since the ground conditions are the same as in previous examples, the pile may be designed from Fig. 14.28 in Design Example 1. This indicates that, for a pile working load of $T = 2586 \, \text{kN}$, a 750 mm diameter pile, 28.5 m long, is required.

Check on strength of pile cross-section

The pile cross-section is checked, to ensure the concrete is not overstressed, in a similar manner to Design Example 3. The maximum pile ultimate load is $T_u = 3879 \, \text{kN}$.

For a 750 mm diameter pile, this gives an ultimate concrete stress of

$$
\begin{aligned}
f_c &= \frac{T_u}{A} \\
&= \frac{3879 \times 10^3}{(\pi \, 750^2/4)} \\
&= 8.8 \, \text{N/mm}^2
\end{aligned}
$$

From BS 8110: Part 1: 3.8.4.4, based on the concrete section alone, the allowable concrete stress is $0.35 f_{cu}$.

$$
\begin{aligned}
\Rightarrow f_{cu(req)} &= \frac{f_c}{0.35} \\
&= \frac{8.8}{0.35} \\
&= 25.1 \, \text{N/mm}^2
\end{aligned}
$$

Therefore grade 30 concrete (with nominal reinforcement) will be used.

Flat slab design

A ground floor slab thickness of 225 mm, together with pile caps 1000 mm × 1000 mm × 600 mm deep, are to be initially assumed for design purposes.

Bending design and reinforcement

The flat slab will be designed in accordance with BS 8110: Part 1: Table 3.19, assuming grade C35 concrete. Since the size of the drop heads (i.e. the pile caps) are less than one-third of the bay size, they enhance the punching shear but not the bending capacity (see BS 8110: Part 1: 3.7.1.5).

The foundation ultimate load on a 6 m × 6 m bay was calculated earlier as $F_u = 724 \, \text{kN}$.

From BS 8110, the flat slab panel bending moments are:
Centre of interior span:

$$
M_{\text{sagging}} = +0.071 F_u L = 0.071 \times 724 \times 6.0 = 308 \, \text{kNm}
$$

Interior support:

$$
M_{\text{hogging}} = -0.055 F_u L = -0.055 \times 724 \times 6.0 = -239 \, \text{kNm}
$$

The moments in the individual strips are given by:
Mid-span – column strip:

$$
M_u = \frac{0.55 M_{\text{sagging}}}{3} = \frac{0.55 \times 308}{3} = 56 \, \text{kNm/m}
$$

Mid-span – middle strip:

$$
M_u = \frac{0.45 M_{\text{sagging}}}{3} = \frac{0.45 \times 308}{3} = 46 \, \text{kNm/m}
$$

Support – column strip:

$$
M_u = \frac{0.75 M_{\text{hogging}}}{3} = \frac{0.75 \times (-239)}{3} = -60 \, \text{kNm/m}
$$

Support – middle strip:

$$
M_u = \frac{0.25 M_{\text{hogging}}}{3} = \frac{0.25 \times (-239)}{3} = -20 \, \text{kNm/m}
$$

The average effective depth in the two directions is

$$
\begin{aligned}
d &= 225(\text{slab}) - 40 \, (\text{cover}) - 16(\text{bar diameter}) \\
&= 169 \, \text{mm}
\end{aligned}
$$

The reinforcement will be calculated for a unit width of $b = 1000 \, \text{mm}$. For the purpose of this example, this will only be done for the worst case bending moment of $M_u = 60 \, \text{kNm}$.

$$
\begin{aligned}
\frac{M_u}{bd^2} &= \frac{60 \times 10^6}{1000 \times 169^2} \\
&= 2.10
\end{aligned}
$$

The required area of tension reinforcement is

$$
A_s = 0.57\% \, bd \qquad \text{[BS 8110: Part 3: Chart 2]}
$$

$$
= \frac{0.57}{100} \times 1000 \times 169
$$

$$
= 963 \, \text{mm}^2/\text{m}
$$

Provide T16 @ 200 mm c/c = $1010 \, \text{mm}^2/\text{m}$.

$$
\begin{aligned}
\frac{100 A_s}{bd} &= \frac{100 \times 1010}{1000 \times 169} \\
&= 0.60\%
\end{aligned}
$$

Punching shear design

The intention is for the area of the pile cap to be sufficiently large to avoid the need for shear reinforcement around the pile supports.

From BS 8110: Part 1: 3.7.7.6, the critical location for punching shear for a square load is a square perimeter a distance $1.5d = 1.5 \times 169 = 254\,mm$ from the face of the load (i.e. from the face of the 1000 mm × 1000 mm pile cap in this instance).

The length of one side of this perimeter is

$$b_{perim} = 1000 + 2(1.5d)$$
$$= 1000 + 2(254)$$
$$= 1508\,mm$$

The area within the shear perimeter is

$$A_p = b_{perim}{}^2$$
$$= 1508^2$$
$$= 2.27\,m^2$$

The ultimate support reaction due to slab loads is $F_u = 724\,kN$.

Shear force along perimeter, $V_u = F_u -$ (load within shear perimeter)
$$= F_u - f_u A_p$$
$$= 724 - (20.1 \times 2.27)$$
$$= 678\,kN$$

Length of shear perimeter, $u = 4b_{perim}$
$$= 4 \times 1508$$
$$= 6032\,mm$$

Shear stress, $v_u = \dfrac{V_u}{ud}$
$$= \dfrac{679 \times 10^3}{6032 \times 169}$$
$$= 0.67\,N/mm^2$$

For a reinforcement percentage of 0.60%, in grade C35 concrete, the allowable concrete shear stress is $v_c = 0.74\,N/mm^2$ (BS 8110: Part 1: Table 3.9). This is greater than the actual shear stress, $v_u = 0.67\,N/mm^2$, therefore the adopted pile cap size is sufficient to avoid the need for shear links.

CHAPTER 15

RETAINING WALLS, BASEMENT WALLS, SLIP CIRCLES AND UNDERPINNING

15.1 INTRODUCTION

The subject matter of this chapter is peripheral to the main work of the structural foundation designer, but nonetheless demands consideration on certain sites.

The topics covered were touched upon in Chapter 4, where topography was related to proposed site development. They will be expanded further here, but not at great length, since they are covered in detail by many excellent text books where they are of primary rather than secondary concern.

15.2 RETAINING WALLS AND BASEMENTS

Retaining walls and peripheral walls to basements are subject to lateral (i.e. horizontal) pressure from retained earth, liquids or a combination of soil and water. They are normally made, in structural work, of concrete or brick (plain, reinforced or prestressed).

Basements are relatively expensive to construct (the cost per square metre is higher than for normal floor construction) so the client should be advised to carry out a cost evaluation of, say, adding a further storey to the structure and eliminating the basement. However, basements can be made cost-effective when they are used as cellular buoyancy rafts or where increased height is restricted by planning.

The walls are basically vertical cantilevers, either free or propped (at the top by a floor slab). Where the ground floor slab can be made continuous with the top of the wall (and not merely be propped) the basement can be designed as a continuous box. The walls can be constructed with either a base slab extending under the retained earth (see Fig. 15.1 (a)), which is generally the more economical form for cuttings, or projecting forward (see Fig. 15.1 (b)), the more economical form for basements.

While propped cantilevers (e.g. basement wall propped by ground floor slab) have a maximum bending moment (for a udl) of $pH^2/8$, compared to that of a free cantilever of $pH^2/2$, they are not frequently used in building structures. This is because the wall must either be temporarily propped, or not backfilled, until the ground floor can act as the prop.

However, in the authors' experience it can be worth considering the use of the more economical propped cantilever, especially for design-and-build contracts, where a close relationship is developed with the contractor from an early stage, and construction methods can be programmed into the design.

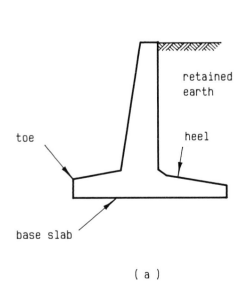

(a)

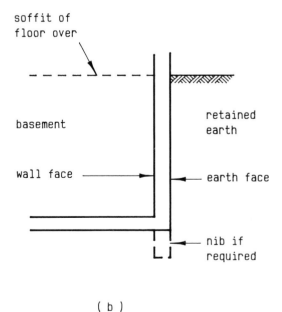

(b)

Fig. 15.1 Typical retaining walls.

The bending moment diagrams for triangular pressure (i.e. no surcharge) for the three cases: free cantilever, propped cantilever and cellular (fixed), are shown in Fig. 15.2.

As derived in section 15.6.4, it can be seen that partially filling a basement with water can equalize the external earth pressure on the basement wall. The authors' practice has used this method of temporary *propping*[1], raising the water level as backfill is placed. Where the basement is constructed in waterlogged ground, filling the basement in this way can also be utilized to avoid flotation before the weight of the rest of the building is added.

Walls to swimming pools are a special case since they can be subject to reversal of stress. With the pool empty, the wall is subject to earth/water pressure on its back face and with the pool full and earth pressure absent (either due to shrinkage of backfill or water testing for leaks, before backfilling), the wall is subject to water pressure alone on its water face (see Fig. 15.3).

Walls to culverts can similarly be subject to reversal of stress under the two conditions of earth pressure acting alone or when the water pressure is acting alone. Service ducts, boiler houses, inspection chambers and similar excavated substructures can unwittingly be subject to internal water pressure acting alone (and not designed for). This has happened when heavy rainfall during construction has flooded and filled the substructures with water before the backfill has been placed.

15.3 STABILITY

Retaining walls are subject to forces other than earth or water pressure, as shown in Fig. 15.4.

Even when the wall is designed to withstand the pressures without being overstressed, a check must be taken to determine its resistance to rotation with a possible slip circle (see Fig. 15.5 and section 15.7).

The pressure of the retained earth can be considerably increased if water pressure is allowed to build up. Where suitable drainage can be provided to prevent this increase of pressure, it is advisable to drain the earth face as shown in

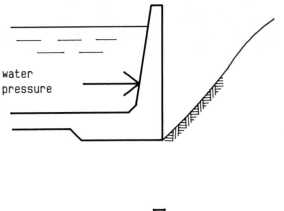

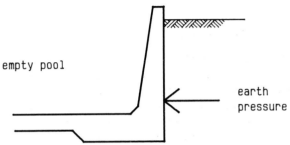

Fig. 15.3 Pressures acting on swimming pool walls.

Fig. 15.6, although in many cases the water-table cannot be easily drained. This problem is causing increasing concern in London and other cities as the water-table rises, this being mainly due to the cessation of artesian well extraction.

15.4 FLOTATION

Swimming pools, large tanks, pits and similar structures can be subject to *upward* vertical pressure in waterlogged ground (see Fig. 15.7). To prevent flotation it is necessary to *anchor* down the structure with ground anchors, or by extending the base (Fig. 15.8).

15.5 BUOYANCY

A ship floating in the sea displaces a weight of water equal

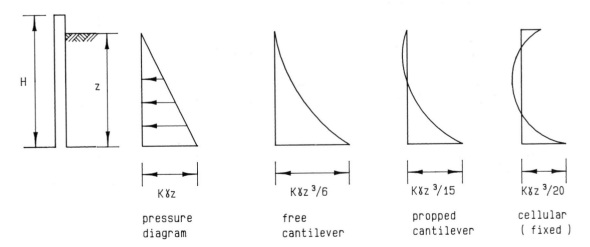

Fig. 15.2 Bending moment diagrams for retaining walls.

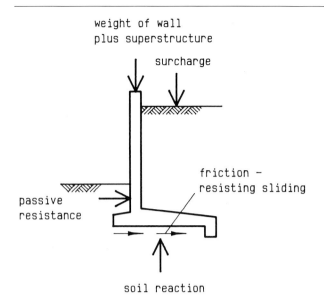

Fig. 15.4 Additional forces on retaining walls.

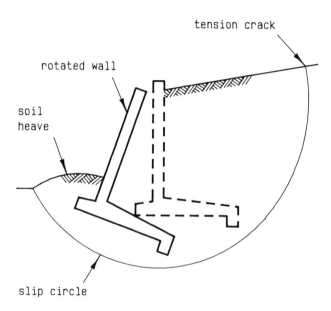

Fig. 15.5 Slip circle failure of retaining wall.

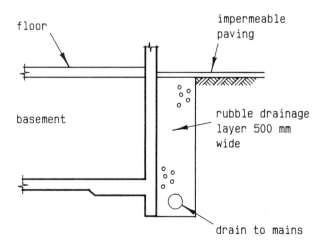

Fig. 15.6 Drainage behind basement retaining wall.

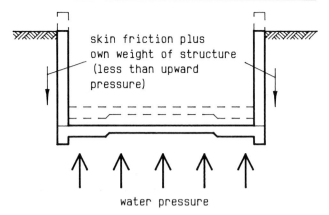

Fig. 15.7 Uplift water pressure (flotation) on tank structure.

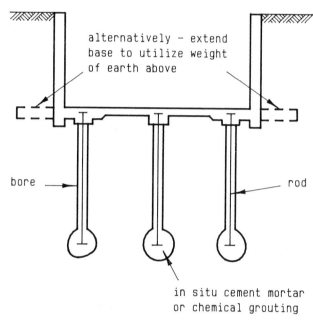

Fig. 15.8 Measures to counteract flotation.

to its own weight and contents. If the ship's contents are increased in weight then the ship displaces more water and sinks a little. (If the contents are too heavy the ship will be sunk since it cannot displace more than its volume.)

$$\text{Factor of safety} = \frac{\text{weight of ship}}{\text{weight of water displaced}} = 1.0$$

Exactly the same basic principle (with, of course, a higher factor of safety than 1.0) applies to cellular basements. A floor load (dead + imposed) is commonly about $10\,kN/m^2$ and a roof load is generally about $6\,kN/m^2$, so a five-storey building over a basement exerts a pressure of $56\,kN/m^2$. Soil of density $16\,kN/m^3$ exerts a vertical pressure of $56\,kN/m^2$ at a depth of 3.5 m. The five-storey building with a 3.5 m deep basement would *float* without exerting further pressure on the soil.

15.6 PRESSURES

The earth pressure can, to a limited extent, be determined

by soil mechanics. There is need for caution for despite the valuable advances in the last half century, soil mechanics (like much structural engineering) is not an exact science. It has been reported that world experts in soil mechanics, given the same detailed site investigation reports, predicted the limiting heights of an earth embankment on a soft clay ranging from 2.8 m to 9.2 m. The embankment actually failed at a height of 5.2 m.

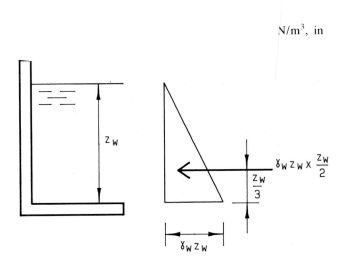

Fig. 15.9 Lateral pressure from retained liquid.

terms of the unit weight of water, the expression becomes

$$\frac{1.8\,\gamma_w z^3}{18} = \frac{\gamma_w z^3}{10}$$

15.6.3 Surcharge

Assuming a surcharge loading for an area adjacent to the main entrance of a building (with access for vehicles not exceeding 2.5 tonnes total weight) of 10 kN/m²,

$$\text{total pressure} = \frac{10}{3} \times z = \frac{10z}{3}$$

$$\text{moment of pressure} = \frac{10z}{3} \times \frac{z}{2} = \frac{10z^2}{6}$$

15.6.4 Simplified expressions (for preliminary estimates)

By substituting $\gamma_w = 10\,\text{kN/m}^3$ in the equations in sections 15.5.1 and 15.5.2, expressions in terms of z_w and z can be derived:

water pressure: $\gamma_w z_w^3/6$ becomes $1.67 z_w^3$ kNm/m run of wall.

earth pressure: $\gamma_w z^3/10$ becomes $1.0 z^3$ kNm/m run.

surcharge: $10z^2/6$ becomes $1.67 z^2$ kNm/m run.

The expressions above can be useful in producing initial calculations prior to a detailed analysis.

15.7 SLIP CIRCLE EXAMPLE

Slip circles have been mentioned in Chapter 4, and a simple example is included here for completeness.

A detached house, 9 m × 9 m on plan, is to be constructed on a sloping site; a section through the proposal is as shown in Fig. 15.10.

For the purposes of the structural foundation designer, initially the slip circle including the whole building should be considered. If this proves unacceptable, or close to the acceptable factor of safety, then other trial arcs should be investigated.

Assuming an average value of $c_u = 50\,\text{kN/m}^2$, consider a one metre wide strip for the case where $\phi_u = 0$, i.e. the undrained condition immediately following construction.

$$\text{Factor of safety} = \frac{c_u l_a r}{F_1 d_1 + P d_2 - F_3 d_3}$$

where c_u = undrained cohesion
l_a = arc length
r = arc radius
F_1 = weight of ground causing slip
d_1, d_2, d_3 = lever arms
P = weight of house
F_3 = weight of ground resisting slip.

Next set up a circular arc using compasses, to pass through the edge of the excavation for the basement of slab/footing and close to the bottom of the new embankment. Measure the radius, and compute the arc length, $r = 12.5$ m. The angle subtended by the arc = 90°. Therefore

$$l_a = 2\pi 12.5 \times \frac{90°}{360°} = 19.6\,\text{m}$$

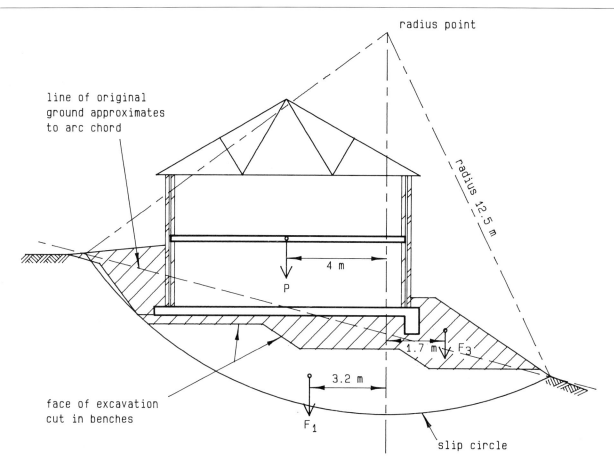

Fig. 15.10 Slip circle design example.

Deduct a length disturbed by excavation and subsequent filling, i.e. $19.6 - 3.6 = 16$ m.

Weight of ground is F_1, assuming the small area of fill above the chord line equals the area omitted within the house.

$$\text{Area of arc} = \left(\pi\, 12.5^2 \times \frac{90°}{360°} \right) - \frac{17.7}{2} \times 9$$

$$= 122.7 - 79.7$$

$$= 43\,\text{m}^2$$

Using $16\,\text{kN/m}^3$ for existing ground and compacted fill

$$\text{weight, } F_1 = 43 \times 16 = 688\,\text{kN/m}$$

The weight of a detached house of two storeys, including external and internal load-bearing walls, when averaged per metre run, equates to $170\,\text{kN/m}$. Therefore

$$P = 170\,\text{kN/m}$$

$$F_3 = 16 \times \frac{9.5}{2} \times 1.5$$

$$= 114\,\text{kN/m}$$

By simple geometry, the centroids of the areas are located, and scaling their lever arms

$$F_1 d_1 = 688 \times 3.2 = 2202\,\text{kNm}$$
$$P d_2 = 170 \times 4 \;\;= \;\;680\,\text{kNm}$$
$$F_3 d_3 = 114 \times 1.7 = \;\;194\,\text{kNm}$$

Combining these gives

$$2202 + 680 - 194 = 2688\,\text{kNm}$$

Therefore

$$\text{factor of safety} = \frac{50 \times 16 \times 12.5}{2688} = 3.7$$

Since 3.7 is greater than 1.5 or 2, then there is an adequate factor of safety against slip circle failure.

15.8 CONTINUOUS UNDERPINNING

In this chapter the authors have considered only underpinning of existing buildings adjacent to new developments and not underpinning required due to structural settlement or subsidence, which is a separate subject beyond the scope of this book.

All foundation types may require underpinning when development takes place alongside or under an existing structure. The possible combinations of ground conditions, foundation details and levels is endless and complex. The basic methods and principles are quite simple. Where a new foundation or structure is to be constructed with its foundation soffit below that of an adjoining foundation, underpinning is usually necessary. The exception to this is where the adjoining building is built upon a substantial ground strata such as hard rock. Where underpinning is needed it is generally carried out in sequenced construction and in short lengths (commonly 1.0 to 1.2 mm). The sequence is arranged to allow limited undermining to the structure at any one time. The limit of this undermining is dependent upon the structure's capability of spanning over the undermined section and the stability of the short section of unrestrained earth. In some cases beam underpinning may be provided to help the structure to span over greater distances. Typical underpinning is shown in Fig. 15.11.

The simplest and most common form of underpinning is to remove a series of short lengths of sub-soils from below the adjoining building in a sequenced operation. As each section is excavated it is replaced immediately with mass concrete, which is allowed adequate time for curing prior to the construction of the adjoining section. The top of the concrete is either cast with a pressure head so that it rises to the underside of the foundation, or is cast low to allow wedging with dry pack or slate. Figure 15.12 gives a typical example of mass concrete underpinning.

In the authors' opinion the preferred method of construction is to cast whenever possible with a pressure head. Concrete shrinks, and so theoretically this method encourages some slight settlement as the building above follows this shrinkage downwards. However, in the authors' experience such settlement is usually negligible and is offset by the following advantages of the pressure head method:

(1) The underpinning is completed in one operation, rather than waiting up to seven days before dry packing. Also since concrete continues to shrink for weeks, even months, the logic of dry packing is inconsistent.
(2) The workmanship of the dry packing process is often of poor quality due to the difficulty of the technique. This requires increased supervision, and slows the whole sequence down even further.

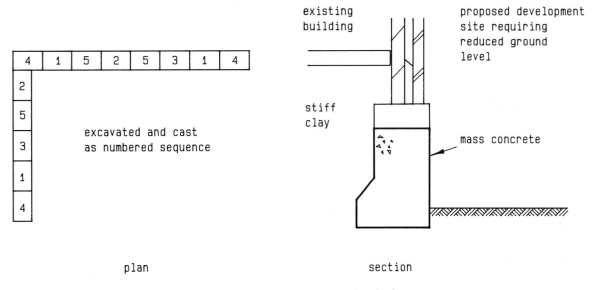

Fig. 15.11 Typical continuous underpinning.

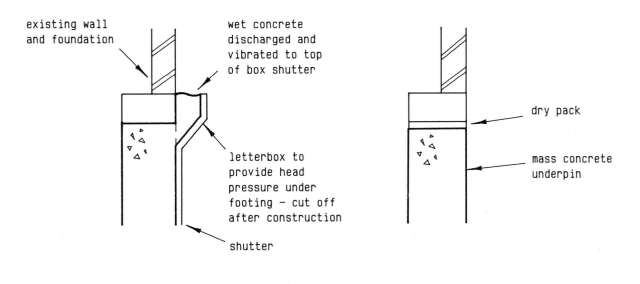

Fig. 15.12 Construction methods for mass concrete underpinning.

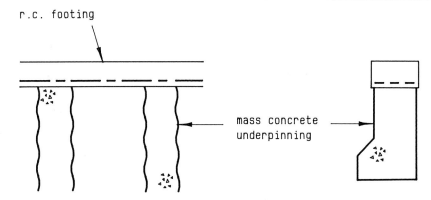

Fig. 15.13 Typical discontinuous underpinning.

It is rarely necessary to mechanically key mass concrete underpinning across the joints and the majority of mass concrete underpinning will perform successfully without a key. The need for keying depends upon the requirement for vertical shear and/or tensile strength across the face, neither of which is usually necessary.

15.9 DISCONTINUOUS UNDERPINNING

Where the existing foundation has reasonable spanning capability it is sometimes possible to excavate and install piers in mass concrete or concrete and brick at a spacing to suit the spanning capability of the original foundation. The area of the base of this underpinning needs to be capable of distributing the ground pressure from vertical and horizontal loading into the sub-strata without allowable limits being exceeded (see Fig. 15.13 for typical details).

In other situations where good ground exists but the foundation is not capable of spanning, a pier and underpinned beam can be used, the beam being inserted in sections in a similar manner to that of the mass concrete underpinning. This operation tends to be more tedious and more time consuming, but where excavations are deep it can prove economic (see Fig. 15.14).

It is particularly useful for foundation jacking where subsidence or settlement requires re-levelling, the jacks being inserted between the soffit of the beam and the top of the piers. In some cases, particularly where the building to be underpinned forms part of the new construction, piles can be inserted on either side of the structure to support needle beams inserted through the existing structure to bear onto the piles. This is particularly useful where a basement extension is to be added to an existing building; the piles form the basement columns and the beams the framework for the ground floor structure. Typical pile beam underpinning is shown in Fig. 15.15.

Temporary lowering of the water-table by sump-pumping for underpinning operations requires careful consideration relative to the effect on new and existing foundations. As previously discussed, there is a danger that soils such as fine sands may suffer from loss of fines and may cause settlement of adjoining structures. There is also the possibility that in

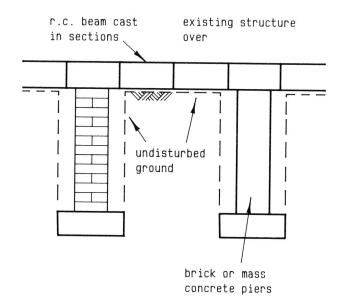

Fig. 15.14 Typical pier and beam underpinning.

certain soils when the dewatering process stops, running sand or clay softening may occur. It is therefore important under these circumstances that the effects of the temporary works and methods of construction are considered at design stage.

There are numerous ingenious piling systems available which minimize disruption of the existing structure, while maximizing economy and practicality of construction and a reputable specialist contractor should be approached at an early stage where appropriate.

15.10 SPREAD UNDERPINNING

Occasionally, due to site constraints, underpinning is achieved by spreading the foundation load over a greater area of ground, rather than transferring to a bearing strata at a lower level.

An example undertaken by the authors' practice was in the restricted cellars of a series of large Victorian properties

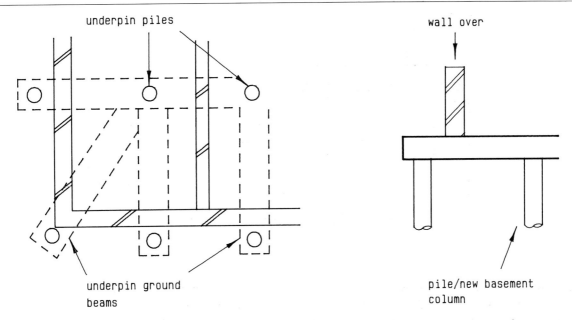

Fig. 15.15 Typical pile and beam underpinning.

being redeveloped as office accommodation. The load-bearing walls sat on stepped brick footings, just beneath a cellar floor of compacted earth. In this case it was possible to cut pockets out of these footings, run reinforcement through the holes, and cast the whole of the cellar floor area as a reinforced concrete raft slab. This proved a very cost-effective and practical way of enhancing the load-bearing capacity of the premises and providing a basement slab at the same time.

15.11 REFERENCE

1. Adams, S. (Curtins Consulting Engineers plc) (1989) *Buildability*. CIRIA, Butterworths.

APPENDICES

INTRODUCTION TO APPENDICES

Appendices are an important and useful part of any practical design manual provided that they are related to the main text and that their use is clearly defined and understood.

These appendices have largely been gathered from the many tables, graphs and charts developed in the main text in its presentation of practical design examples. They have then been added to and tabulated into a quick reference form for easy access by the knowledgeable reader.

The authors therefore hope that all readers will progress via the main text to familiarity with the appendices and use them in their pursuit of the practical design of economical foundations.

PROPERTIES AND PRESUMED BEARING PRESSURES OF SOME WELL KNOWN ENGINEERING SOILS AND ROCKS

There is a tendency on the part of more experienced engineers, particularly those who have become well acquainted over the years with a variety of soil mechanics problems on a wide variety of sites, to assume a working knowledge of certain commonly encountered engineering soils, while less experienced engineers, or those who have worked for most of their lives in only one area of the country, are less well versed in this area than their colleagues assume.

With this in mind, the following is a brief list of the properties of well known engineering soils/rocks common in certain parts of the country. (Presumed bearing pressures quoted for clays assume a 1 metre to 2 metre wide footing.) Further information is given in Fig. A.1 and Table A.1.

Kimmeridge clay

Kimmeridge clay, like London and Oxford clays, can contain naturally occurring sulphates. These clays are stiff fissured heavy clays which swell with moisture increase and shrink with moisture decrease.

Presumed bearing pressure (stiff) = 150 to 300 kN/m^2

London clay

London clay is an overconsolidated clay with either a red, brown or greenish-blue colour due to the presence of iron oxide. The estimated consolidation load in the central London area was about 3500 kN/m^2 in previous geological ages.

The depths of the clay beds vary, but, including the underlying sands, gravels and boulder clay, are typically 50 metres thick over the underlying chalk.

London clay can have a high plasticity index and is often highly shrinkable.

Presumed bearing pressure (stiff *blue* clay) = 200 to 400 kN/m^2

Presumed bearing pressure (firm *brown* clay) = 100 to 200 kN/m^2

Oxford clay

As its name suggests, Oxford clay is found in thick beds between Oxford and Cirencester, and is extensively worked for brick production.

Presumed bearing pressure (stiff) = 150 to 300 kN/m^2

Wealden clay

As its name suggests, Wealden clay is found in The Weald, between the North and South Downs, in a crescent running west from Eastbourne to Horsham and Haslemere, and then east as far as Hythe on the south coast.

Presumed bearing pressure (stiff) = 150 to 300 kN/m^2

As discussed in Chapter 2, many structures are founded on engineering soils classed as *drift* on the geological maps, varying in depth from 2 to 200 metres and sometimes more. But larger structures are often founded on rock, and certain rock types, termed *solid* on the geological maps, are mentioned below.

Keuper Marl (a red-brown Mercia mudstone)

Keuper Marl is an argillaceous rock, often interbedded with sandstone, which can be highly fissured — this encourages water percolation which leads to softening. It outcrops on either side of the Pennines and extends as a single band down through the Midlands to the Bristol area. It reaches its greatest thicknesses (1200 m to 1500 m) in the Cheshire/Shropshire basin.

Although a stable mudstone at depth, when disturbed in shallow foundations, Keuper Marl behaves as a clay susceptible to swelling and softening by the action of groundwater.

CIRIA Report 47 distinguishes zones of Keuper Marl, varying in presumed bearing pressure from 100 to 1200 kN/m^2.

Greywacke

Greywacke consists of badly sorted muddy sedimentary rocks with much coarse material compressed in deep troughs, sometimes thousands of metres in depth.

Presumed bearing pressure = 550 to 1200 kN/m^2

Bunter sandstone

Bunter sandstone is a softish rock consisting of cemented particles of sand with 10% to 20% passing the 75 micron sieve.

Presumed bearing pressure = 450 to 900 kN/m^2

Chalk

Chalk is a sound, soft white limestone, but is susceptible to softening when subject to percolating water. As covered in the text, in extreme cases this can result in the formation of swallow-holes.

Presumed bearing pressure = 125 to 1200 kN/m^2

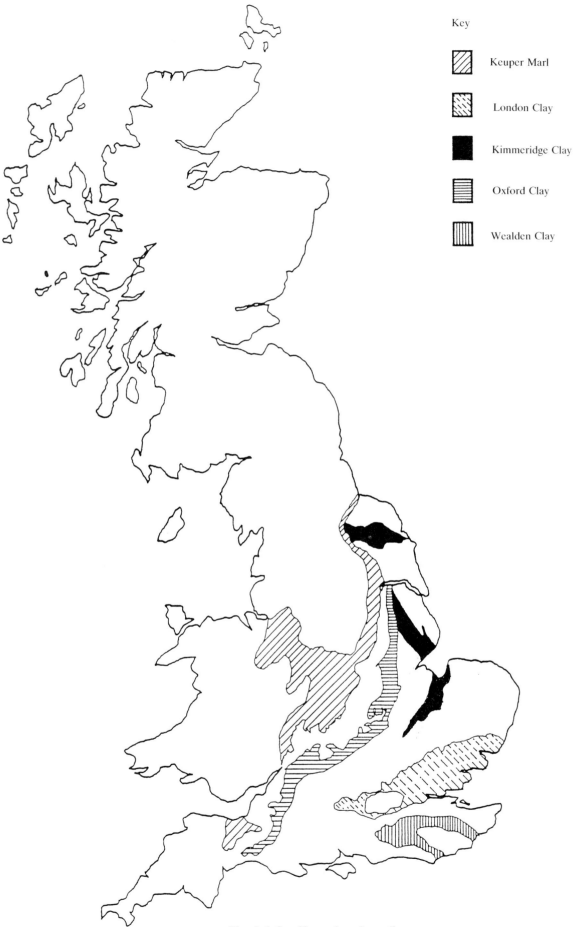

Fig. A.1 Specific engineering soils.

Table A.1 Presumed allowable bearing values (BS 8004, Table 1)

NOTE. These values are for preliminary design purposes only, and may need alteration upwards or downwards. No addition has been made for the depth of embedment of the foundation (see 2.1.2.3.2 and 2.1.2.3.3).

Category	Types of rocks and soils	Presumed allowable bearing value		Remarks
		kN/m^2	kgf/cm^2* tonf/ft^2	
Rocks	Strong igneous and gneissic rocks in sound condition Strong limestones and strong sandstones Schists and slates Strong shales, strong mudstones and strong siltstones	10 000 4 000 3 000 2 000	100 40 30 20	These values are based on the assumption that the foundations are taken down to unweathered rock. For weak, weathered and broken rock, see 2.2.2.3.1.12
Non-cohesive soils	Dense gravel, or dense sand and gravel Medium dense gravel, or medium dense sand and gravel Loose gravel, or loose sand and gravel Compact sand Medium dense sand Loose sand	>600 <200 to 600 <200 >300 100 to 300 <100 Value depending on degree of looseness	>6 <2 to 6 <2 >3 1 to 30 <1	Width of foundation not less than 1 m. Groundwater level assumed to be a depth not less than below the base of the foundation. For effect of relative density and groundwater level, see 2.2.2.3.2
Cohesive soils	Very stiff boulder clays and hard clays Stiff clays Firm clays Soft clays and silts Very soft clays and silts	300 to 600 150 to 300 75 to 150 <75 Not applicable	3 to 6 1.5 to 3 0.75 to 1.5 <0.75	Group 3 is susceptible to long-term consolidation settlement (see 2.1.2.3.3). For consistencies of clays, see table 5
Peat and organic soils		Not applicable		See 2.2.2.3.4
Made ground or fill		Not applicable		See 2.2.2.3.5

* 107.25 kN/m^2 = 1.094 kgf/cm^2 = 1 tonf/ft^2.
All references within this table refer to the original document.

MAP SHOWING AREAS OF SHRINKABLE CLAYS

Fig. B.1 Areas of firm shrinkable clays.

MAP SHOWING AREAS OF COAL AND SOME OTHER MINERAL EXTRACTIONS

Key

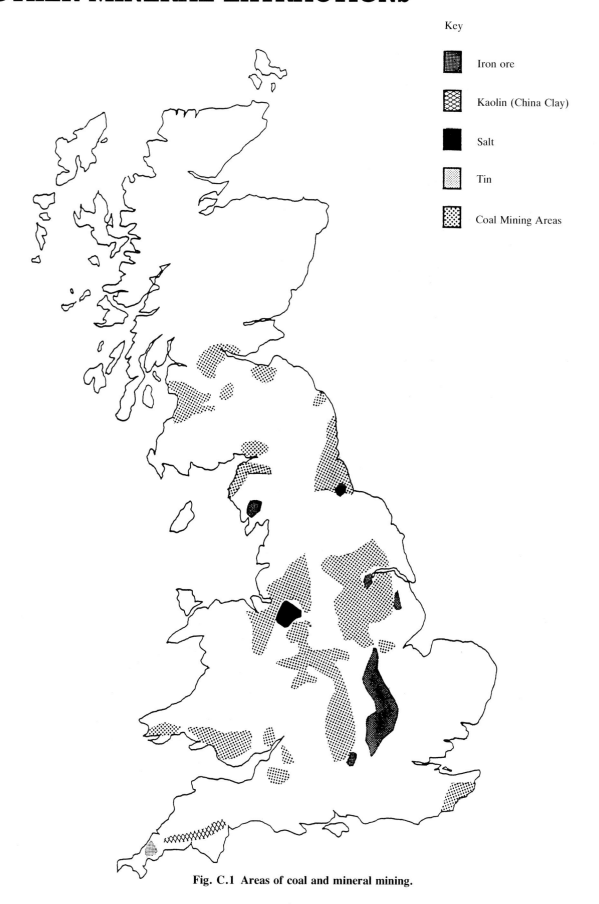

Fig. C.1 Areas of coal and mineral mining.

FOUNDATION SELECTION TABLES

Tables 10.3, 10.1 and 10.2 from the main text are reproduced here as Tables D.1, D.2 and D.3 respectively for ease of reference in the foundation selection process.

Each of the tables gives details of suitable foundations to suit varying site and sub-soil conditions with guidance notes for factors to be considered during the selection process.

Table D.1 Foundation selection – bearing strata strength and depth

Sub-soil conditions	Suitable foundation
Condition 1 Suitable bearing strata within 1.5 m of ground surface	**Strips** **Pads** **Rafts** When loading on pads is relatively large and pad sizes tend to join up or the foundation needs to be **balanced** or connected then **continuous beam** foundations are appropriate. Strip foundations are usually considered the norm for these conditions but rafts can prove more economical in some cases.
Condition 2 Suitable bearing strata at 1.25 m and greater below ground surface	**Strips** **Pads** } on improved ground using vibro or dynamic **Rafts** } consolidation techniques
Condition 3 Suitable bearing strata at 1.5 m and greater below ground surface	As Condition 2 plus the following **Piles** and **ground beams** **Pier** and **ground beams** **Piles** and **raft**
Condition 4 Low bearing pressure for considerable depth	As Condition 2 plus the following **Buoyant rafts**
Condition 5 Low bearing pressure near surface	As Condition 2 plus the following **Rafts** Ground improvement using preloading to support reinforced strips on rafts

Table D.2 Foundation selection − sub-soil type

Sub-soil type	Suitable foundation	Factors to be considered
Group 1 Rock; hard sound chalk; sand and gravel, sand and gravel with little clay content, dense silty sand	Strips/Pads/Rafts	(1) Minimum depth to formation for protection against frost heave 450 mm for frost susceptible soils. (2) Weathered rock must be assessed on inspection. (3) Beware of swallow-holes in chalk. (4) Keep base of strip or trench above groundwater level where possible. (5) Sand slopes may be eroded by surface water − protect foundation by perimeter drainage. (6) Beware of running sand conditions.
Group 2 Uniform firm and stiff clays (a) where existing nearby vegetation is insignificant	Strips/Pads/Rafts	(1) Trench fill likely to be economic in this category. (2) Minimum depth to underside of foundation 900 mm. (3) When strip foundations are cast in desiccated clay in dry weather, they must be loaded with the structure before heavy rains return.
(b) where trees, hedges or shrubs exist close to the foundation position or are to be planted near the building at a later date	Concrete piles supporting reinforced ground beams and precast concrete floor units OR Concrete piles supporting a suspended reinforced in situ concrete slab OR Specially designed trench fill (possibly reinforced) in certain clay soils depending on position of foundation relative to trees OR Rafts	(1) Clay type and shrinkage potential, distance of trees from foundation and spread of roots dictate necessity or otherwise of piling. (2) Type and dimensions of pile depend on economic factors. (3) Where a suspended in situ concrete ground slab is used a void must be formed under it if laid in very dry weather over clay which is desiccated. (4) Where existing mature trees grow very close (e.g. within quarter of mature tree height) to the position in which piles will be installed. It might be prudent to design for sub-soil group 2(c). (5) Where trees have been or will be planted at a distance of at least one to two times the tree height from the foundation, a strip foundation may be suitable. (6) In marginal cases, i.e. with clay of low to medium shrinkage potential and in the perimeter zone of the tree root system, reinforced trench fill can be used.

Table D.2 *Continued*

Sub-soil type	Suitable foundation	Factors to be considered
(c) Where trees and hedges are cut down from area of foundations shortly before construction	Reinforced concrete piles (in previous tree root zone) OR Strip foundations as in groups 2(a) and 2(b) (outside previous root zone) OR Rafts	(1) Piles must be tied adequately into ground beams or the suspended reinforced concrete slab. An adequate length of pile must be provided to resist clay heave force, and the top section of the pile possibly sleeved to reduce friction and uplift. (2) Special pile design may be required for clay slopes greater than 1 in 10 where soil creep may occur and it is necessary to design for lateral thrust and cantilever effects. (3) In marginal cases, i.e. with clay of low to medium shrinkage potential and in the perimeter zone of the tree root system, reinforced trench fill can be used.
Group 3 Soft clay, soft silty clay, soft sandy clay, soft silty sand	Wide strip footings if bearing capacity is sufficient and predicted settlement allowable OR Rafts OR Piles to firmer strata below – for small projects consider pier and beam foundations to firm strata	(1) Strip footings should be reinforced depending on thickness and projection beyond wall face. (2) Service entries to building should be flexible.
Group 4 Peat	Concrete piles taken to firm strata below. For small projects, consider pad and beam foundations taken to firm strata below. Where no firm strata exist at a reasonable depth below ground level but there is a thick (3–4 m) hard surface crust of suitable bearing capacity, consider raft.	(1) Pile types used are bored cast in place with temporary casing; driven cast in place; and driven precast concrete. (2) Allow for peat consolidation drag on piles. (3) Where peat layer is at surface and shallow over firm strata, dig out and replace with compacted fill. Then use raft or reinforced wide-strip foundations depending on expected settlement. (4) Where raft is used, service entries should be flexible. Special high-grade concrete and protection may be necessary in some aggressive peat soils.

Ground improvements of sub-soil Groups 3 and 4 by vibro treatments can often be achieved and can be an effective and economical solution when used in conjunction with raft or strip foundations

Table D.3 Foundation selection — varying site conditions

Site condition	Suitable foundation	Factors to be considered

Filled site	Concrete piles taken to firm strata below. For small projects consider beam and pier foundations taken to firm strata below. If specially selected and well compacted fill has been used consider
	(1) Raft or
	(2) Reinforced wide-strip footings
	(3) Strip/pad/raft on ground improved using vibro or dynamic consolidation depending on fill type
		(1) Allow for fill consolidation drag on piles, piers or deep trench fill taken down to firm strata below.
		(2) Proprietary deep vibro and dynamic compaction techniques can with advantage improve poor fill before construction of surface or shallow foundations.
		(3) If depth of poorly compacted and aggressive fill is small remove and replace with inert compacted fill, then use reinforced strip or raft foundations.
		(4) Deep trench fill taken down to a firm stratum may be economic if ground will stand with minimum support until concrete is placed.
		(5) Allow flexible service entries to building.
		(6) Avoid building a unit partly on fill and partly on natural ground.
		(7) Take precautionary measures against
		   (a) combustion on exposure to atmosphere,
		   (b) possible toxic wastes,
		   (c) production of methane gas.

Mining and other subsidence areas	Slip-plane raft	(1) Where a subsidence wave is expected, building should be carried on individual small rafts. Avoid long terrace blocks and L-shaped buildings. (2) In older mining areas, locate buildings to avoid old mining shafts and bell-pits. (3) In coal mining areas, consult British Coal in all cases. (4) Avoid piled foundations.
Sloping site	Foundations to suit normal factors and soil conditions, but designed for special effect of slope	(1) Strip foundations act as retaining walls at steps. With clay creep downhill, design and reinforce for horizontal forces on foundations. Provide good drainage behind retaining wall steps. (2) Foundations are deeper than normal, so keep load-bearing walls to a minimum. Keep long direction of building parallel to contours.

Table D.3 *Continued*

Site condition	Suitable foundation	Factors to be considered
		(3) In addition to local effects of slope on foundations, consider total ground movement of slopes including stability of cohesionless soils, slip and sliding of cohesive soils.
		(4) Make full examination of all sloping sites inclined more than 1 in 10.
		(5) The presence of water can increase instability of slope.
		(6) Special pile design may be required for clay soil slopes greater than 1 in 10 where soil creep may occur and it is necessary to design for lateral thrust and cantilever effects.
Site containing old building foundations	Normal range of foundations. It is possible to use strips, piling, and pads but beware of varying depths of fill in old basements, causing differential settlement, and old walls projecting into fill over which slabs may break their back.	(1) Notes relating to 'filled site' apply.
		(2) Where possible, dig out badly placed or aggressive fill and replace with inert compacted material.
		(3) Remove old walls in filled basements, or use piers or piles carrying ground beams to span such projections.
		(4) Deep trench fill down to firm strata at original basement level may be economic.
		(5) Trench fill depths may vary greatly as old basement depth varies. Some formwork may be required in loose fill areas.
		(6) Remove old timber in demolition material − a source of dry rot infection.
Site with ground-water problems	Normal range of foundation types can be used. Consider piling through very loose saturated sand to denser stratum to provide support for raft or strip foundation at high level above groundwater. Consider use of proprietary vibro-replacement ground techniques to provide support for raft or strip foundation at high level above groundwater.	(1) In sand and gravel soil, keep foundation above groundwater level where possible.
		(2) Avoid forming steep cuttings in wet sand or silty soil.
		(3) Consider use of sub-surface *shelter* drains connected to surface water drains, and allow for resulting consolidation or loss of ground support.
		(4) Take precautions against lowering of groundwater level which may affect stability of existing structures.

GUIDE TO USE OF GROUND IMPROVEMENT

Surface rolling of imported granular materials and vibro-stabilization are probably the most commonly used forms of ground improvement. Table 8.1 and Figs 8.7 and 8.10 from the main text are repeated here as Table E.1 and Figs E.1 and E.2 respectively for ease of reference when considering these options.

Table E.1 (Table 8.1) gives details of grading and compaction of hardcore when considering surface rolling.

The soils most suited to improvement by vibro compaction range from medium-to-fine gravel to fine uniform sand as shown in Fig. E.1 (Fig. 8.7). Cohesive soils require further considerations to achieve improvements from vibro methods. Figure E.2 (Fig. 8.10) shows some typical examples of vibro treatments in a range of soils to support various types of developments.

Table E.1 Hardcore grading and compaction

Hardcore material should be composed of granular material and shall be free from clay, silt, soil, timber, vegetable matter and any other deleterious material and shall not deteriorate in the presence of water. The material shall be well graded and lie within the grading envelope below:		Hardcore material should be placed and spread evenly. Spreading should be concurrent with placing and compaction carried out using a vibrating roller as noted below:	
BS sieve size	**Percentage by weight passing**	**Category of roller (mass per metre width of roll)**	**Number of passes for layers not exceeding 150 mm thick**
75.0 mm	90−100	Below 1300 kg	not suitable
37.5 mm	80−90	Over 1300 kg up to 1800 kg	16
10.0 mm	40−70	Over 1800 kg up to 2300 kg	6
5.0 mm	25−45	Over 2300 kg up to 2900 kg	5
600 μm	10−20	Over 2900 kg up to 3600 kg	5
75 μm	0−10	Over 3600 kg up to 4300 kg	4
		Over 4300 kg up to 5000 kg	4
Note: This grading falls between the recommendations of the Department of Transport for sub-base Type 1 and Type 2 gradings.		Over 5000 kg	3
		Compaction should be completed as soon as possible after material has been spread	

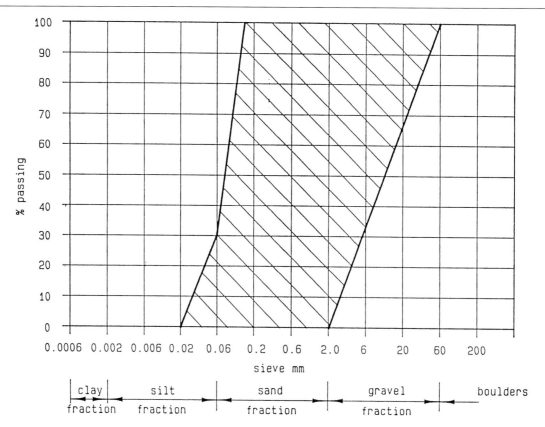

Fig. E.1 Soil grading for vibro treatment.

GROUND CONDITIONS	DEVELOPMENT	FOUNDATION SOLUTION	VIBRO TREATMENT
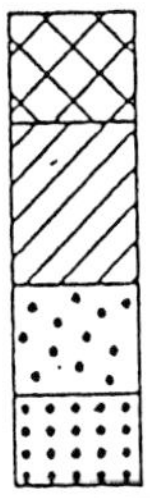 **1.75 m** Demolition fill **1.75–2.35 m** Compact fill (Mainly sub-soil) **2.35–3.2 m** Compact red sand **3.2–3.6 m** Hard red sandstone (trial pit dry)	Two- and three-storey housing of traditional construction	Traditional strip footings on vibro-improved ground Vibro treatment on load-bearing wall lines	Dry process adopted Probes at **1.5 m** centres on centreline of load-bearing walls Probes carried through fill to sand layer Depth of treatment **2.5 m** Allowable bearing pressure **150 kN/m²**
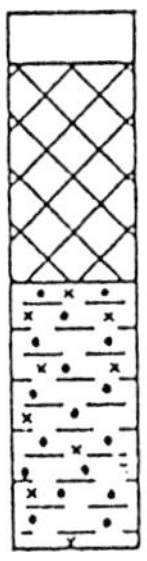 **0–0.1 m** Topsoil **0.1–2.4 m** Fill Soft to firm brown and grey sandy silty clay with ash and bricks **2.4–6.0 m** Firm to stiff dark brown slightly sandy to sandy silty clay Becoming stiffer with depth (borehole dry)	Five-storey residential building Load-bearing masonry construction with suspended concrete floor slab (including ground floor)	Traditional strip footings on vibro-improved ground Vibro treatment on load-bearing wall lines Footings 0.7 to 1.20 m wide reinforced with two layers of B785 mesh	Dry process adopted Two lines of probes at **0.95** to **1.5 m** staggered centres on centreline of load-bearing walls Probes carried through fill to clay Depth of treatment **3 m** Allowable bearing pressure **150 kN/m²**
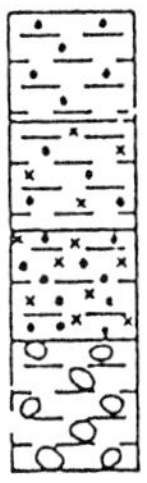 **0–1.0 m** Sandy clay probable fill **1.0–2.2 m** Firm, sandy, silty clay **2.2–3.8 m** Soft very sandy silty clay **3.8–6.0 m** Stiff boulder clay	Tall single-storey factory/warehouse Steel portal frame with steel sheeting and dado masonry	Pad bases beneath columns, with masonry walls on strip footings between bases Vibro-improved ground beneath foundations and ground slab	Dry (bottom-feed) process adopted Probes on **1.5 m** grid under pad bases (**2.8 m** square pad on nine probes) Probes at **1.6 m** centres on centreline of footings Probes at **2.0 m** grid beneath slab area Depth of treatment **4 m** Allowable bearing pressure: **100 kN/m²** to pads/strips; **25 kN/m²** to slabs
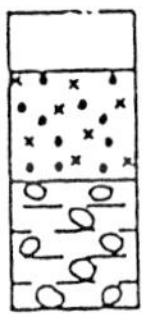 **0–0.15 m** Topsoil **0.15–2.4 m** Loose saturated silty sand **2.4–6.0 m** Firm to stiff boulder clay	Two-storey institutional building, part load-bearing masonry part r.c. frame	Pad bases to columns, strip footings to load-bearing walls Vibro-improved ground beneath foundations and ground slab	Wet process adopted Probes on **1.5 m** grid under pad bases (**2.0 m** square base on four probes) Probes at **1.5 m** centres on centreline of footings Depth of treatment **2.5 m** Allowable bearing pressure: **150 kN/m²** to pads/strips; **25 kN/m²** to slabs
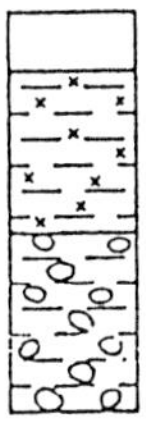 **0–0.3 m** Topsoil and sub-soil **0.3–2.7 m** Soft to very soft bands of clay and silts saturated **2.7–6.0 m** Firm to stiff boulder clay	Tall single-storey load-bearing masonry sports hall	Wide strip footings on vibro-improved ground 1.5 m wide footing reinforced with C785 mesh	Dry (bottom-feed) process adopted two lines probes at **1.25 m** staggered centres on centreline of load-bearing walls Probes at **1.8 m** staggered centres under slab Depth of treatment **2.8 m** Allowable bearing pressure: **150 kN/m²** to footings; **25 kN/m²** to slab *Note* Following testing programme the treatment centres reduced to **0.75 m** in localized area of very soft ground to achieve settlement test criteria
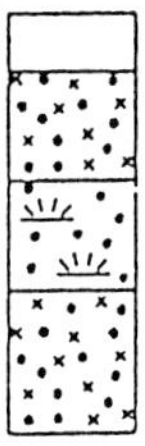 **0–0.2 m** Topsoil **0.2–1.8 m** Loose brown fine silty sand **1.8–2.2** Loose moist dark brown peaty sand **2.2–9.5** Greyish brown fine silty sand	Two-storey teaching block, load-bearing masonry construction	Crust raft on vibro-improved ground Raft slab incorporated internal thickening under load-bearing wall lines	Wet process adopted[a] Probes at **1.7 m** centres on centreline of raft edge and internal thickenings Probes at **2.5 m** grid under floor areas Depth of treatment **4.8 m** Allowable bearing pressure **110 kN/m²** [a]This project was undertaken in late 1970s before bottom-feed dry vibro-treatment was available (it is considered that the dry bottom-feed method would have proved effective in this case)

Fig. E.2 Examples of vibro treatment/foundation solutions.

APPENDIX F

TABLES RELATING TO CONTAMINATED SITES/SOILS

Tables 5.1 to 5.9 inclusive from the main text are repeated here as Tables F.1 to F.9 respectively for ease of reference when considering the implications of the development of contaminated land.

Detailed discussion on the contents of each table is given in Chapter 5. The tables give guidance on the types of contamination, where it is likely to occur, and potential hazards. The tables also give suggestions for the chemical tests of samples and classification of contaminated soils.

Table F.1 Contaminant type and hazard (reproduced with permission from ICRCL Guidance Note 59/83, 2nd edn, 1987, Table 1)

Type of contaminant	Likely to occur on	Principal hazards
Toxic metals e.g. cadmium, lead, arsenic, mercury Other metals e.g. copper, nickel, zinc	Metal mines, iron and steel works, foundries, smelters; electroplating, anodizing and galvanizing works; engineering works, e.g. shipbuilding; scrap yards and shipbreaking sites	Harmful to health of humans or animals if ingested directly or indirectly. May restrict or prevent the growth of plants.
Combustible substances e.g. coal and coke dust	Gas works, power stations, railway land	Underground fires
Flammable gases e.g. methane	Landfill sites, filled dock basins	Explosions within or beneath buildings
Aggressive substances e.g. sulphates, chlorides, acids	*Made ground* including slags from blast furnaces	Chemical attack on building materials e.g. concrete foundations
Oily and tarry substances, phenols	Chemical works, refineries, by-products plants, tar distilleries	Contamination of water supplies by deterioration of service mains
Asbestos	Industrial buildings; waste disposal sites	Dangerous if inhaled

Table F.2 Characteristics and effects of hazardous gases (Leach & Goodger: *Building on Derelich Land*, CIRIA SP78 (1991))

Gas	Characteristics	Effect	Special features
Methane	colourless odourless lighter than air	non-toxic asphyxiant	flammable limits 5−15% in air can explode in confined spaces toxic to vegetation due to deoxygenation of root zone
Carbon dioxide	colourless odourless denser than air	toxic asphyxiant	can build up in pits and excavations corrosive in solution to metals and concrete comparatively readily soluble
Hydrogen sulphide	colourless 'rotten egg' smell denser than air	highly toxic	flammable explosive limits 4.3−4.5% in air causes olfactory fatigue (loss of smell) at 20 p.p.m. toxic limits reached without odour warning soluble in water and solvents toxic to plants
Hydrogen	colourless odourless lighter than air	non-toxic asphyxiant	highly flammable explosive limits 4−7.5% in air
Carbon monoxide	colourless odourless	highly toxic	flammable limits 12−75% in air product of incomplete combustion
Sulphur dioxide	colourless pungent smell	respiratory irritation toxic	corrosive in solution
Hydrogen cyanide	colourless faint 'almond' smell	highly toxic	highly flammable highly explosive
Fuel gases	colourless 'petrol' smell	non-toxic but narcotic	flammable/explosive may cause anoxaemia at concentrations above 30% in air
Organic vapours (e.g. benzene)	colourless 'paint' smell	carcinogenic toxic narcotic	flammable/explosive can cause dizziness after short exposure has high vapour pressure

Table F.3 Potential contaminated sites related to past usage (BSI, DD175, 1988)

IMPORTANT: This table should not be taken to mean that other types of site need not be investigated nor to mean that other contaminants are absent (see table footnote below).

Industry	Examples of sites	Likely contaminants
Chemicals	Acid/alkali works Dyeworks Fertilizers and pesticides Pharmaceuticals Paint works Wood treatment plants	Acids; alkalis; metals; solvents (e.g. toluene, benzene); phenols, specialized organic compounds
Petrochemicals	Oil refineries Tank farms Fuel storage depots Tar distilleries	Hydrocarbons; phenols; acids; alkalis and asbestos
Metals	Iron and steel works Foundries, smelters Electroplating, anodizing and galvanizing works Engineering works Shipbuilding/shipbreaking Scrap reduction plants	Metals, especially Fe, Cu, Ni, Cr, Zn, Cd and Pb; asbestos
Energy	Gas works Power stations	Combustible substances (e.g. coal and coke dust); phenols; cyanides; sulphur compounds; asbestos
Transport	Garages, vehicle builders and maintenance workshops Railway depots	Combustible substances; hydrocarbons; asbestos
Mineral extraction Land restoration (including waste disposal sites)	Mines and spoil heaps Pits and quarries Filled sites	Metals (e.g. Cu, Zn, Pb); gases (e.g. methane); leachates
Water supply and sewage treatment	Water works Sewage treatment plants	Metals (in sludges) Micro-organisms
Miscellaneous	Docks, wharfs and quays Tanneries Rubber works Military land	Metals; organic compounds; methane; toxic, flammable or explosive substances; micro-organisms

Ubiquitous contaminants include hydrocarbons, polychlorinated byphenyls (PCBs), asbestos, sulphates and many metals used in paint pigments or coatings. These may be present on almost any site.

Table F.4 Principal hazards and contaminants (Based upon Table 2 in ICRCL 59/83, 2nd edn, 1987, Table 2)

WARNING. Consideration should be given to the inclusion of other contaminants if the site history has identified former uses likely to have introduced them.

Hazard (see Note 1)	Typical end uses where hazard may exist	Contaminants
Direct ingestion of contaminated soil by children	Domestic gardens, recreational and amenity areas	total arsenic total cadmium total lead free cyanide polycyclic aromatic hydrocarbons phenols sulphate
Uptake of contaminants by crop plants (see Note 2)	Domestic gardens, allotments and agricultural land	total cadmium (see Note 3) total lead (see Note 3)
Phytotoxicity (see Notes 2 and 3)	Any uses where plants are to be grown	total copper total nickel total zinc
Attack on building materials and services (see Note 2)	Housing developments Commercial and industrial buildings	sulphate sulphide chloride oily and tarry substances phenols mineral oils ammonium ion
Fire and explosion	Any uses involving the construction of buildings	methane sulphur potentially combustible materials (e.g. coal dust, oil, tar, pitch)
Contact with contaminants during demolition, clearance, and construction work	Hazard mainly short-term (to site workers and investigators)	polycyclic aromatic hydrocarbons phenols oily and tarry substances asbestos radioactive materials
Contamination of ground and surface water (see Note 2)	Any uses where possible pollution of water may occur	phenols cyanide sulphate soluble metals

Note 1. The hazards listed are not mutually exclusive. Combinations of several hazards may need consideration.

Note 2. The soil pH should be measured as it affects the importance of these hazards.

Note 3. Uptake of harmful or phytotoxic metals by plants depends on which chemical forms of these elements are present in the soil. It may therefore be necessary to determine the particular forms, if the total concentrations present indicate that there exists a possible risk.

Table F.5 Contaminant testing

Previous site usage	Analysis
Metal mines Iron and steel works Lead works Foundries and smelters Electroplating and galvanizing works Scrap yards Shipbreaking/shipbuilding yards	Toxic metals e.g. cadmium, lead, arsenic, mercury Other metals e.g. copper, nickel, zinc
Chemical works Refineries By-product plants	— chromium — lead, zinc and cadmium — arsenic — other extractable material — mercury — selenium — nickel — boron — oils — hydrocarbons
Tar and turpentine distillers Gas works	— phenols and tar residues — free and complex cyanides — thiocyanates — sulphur — sulphides
Refuse and landfill areas Filled dock basins	— methane gas — carbon dioxide
Tanneries and hide merchants	— chromium — sulphide — arsenic — biological contamination i.e. anthrax — other extractable material
All solid samples	pH; volatile matter; sulphate content; asbestos
All water samples	pH; sulphate; lead; zinc; cadmium; chromium; BOD (Biochemical Oxygen Demand); COD (Chemical Oxygen Demand); phenols; chlorides; iron; total dissolved solids; volatile total solids

Table F.6 Tentative *trigger concentrations* for inorganic contaminants (ICRCL 59/83, 2nd edn, 1987, Table 3)

Conditions

1. This table is invalid if reproduced without the conditions and footnotes.

2. All values are for concentrations determined on 'spot' samples based on an adequate site investigation carried out prior to development. They do not apply to analysis of averaged, bulked or composited samples, nor to sites which have already been developed. All proposed values are tentative.

3. The lower values in Group A are similar to the limits for metal content of sewage sludge applied to agricultural land. The values in Group B are those above which phytoxicity is possible.

4. If all sample values are below the threshold concentrations then the site may be regarded as uncontaminated as far as the hazards from these contaminants are concerned and development may proceed. Above these concentrations, remedial action may be needed, especially if the contamination is still continuing. Above the action concentration, remedial action will be required or the form of development changed.

Contaminants	Planned uses	Trigger concentrations (mg/kg air-dried soil)	
		threshold	action
Group A: *Contaminants which may pose hazards to health*			
Arsenic	Domestic gardens, allotments	10	*
	Parks, playing fields, open space.	40	*
Cadmium	Domestic gardens, allotments	3	*
	Parks, playing fields, open space.	15	*
Chromium (hexavalent)[1]	Domestic gardens, allotments		
	Parks, playing fields, open space.	25	*
Chromium (total)	Domestic gardens, allotments	600	*
	Parks, playing fields, open space.	1000	*
Lead	Domestic gardens, allotments	500	*
	Parks, playing fields, open space.	2000	*
Mercury	Domestic gardens, allotments	1	*
	Parks, playing fields, open space.	20	*
Selenium	Domestic gardens, allotments	3	*
	Parks, playing fields, open spaces	6	*
Group B: *Contaminants which are phytotoxic but not normally hazards to health*			
Boron (water-soluble)[3]	Any uses where plants are to be grown[2,6].	3	*
Copper[4,5]	Any uses where plants are to be grown[2,6].	130	*
Nickel[4,5]	Any uses where plants are to be grown[2,6].	70	*
Zinc[4,5]	Any uses where plants are to be grown[2,6].	300	*

Notes:

 * Action concentrations will be specified in the next edition of ICRCL 59/83.

1. Soluble hexavalent chromium extracted by 0.1M CH1 at 37°C; solution adjusted to pH 1.0 if alkaline substances present.

2. The soil pH value is assumed to be about 6.5 and should be maintained at this value. If the pH falls, the toxic effects and the uptake of these elements will be increased.

3. Determined by standard ADAS method (soluble in hot water).

4. Total concentration (extractable by $HNO_3/CH1O_4$).

5. The phytotoxic effects of copper, nickel and zinc may be additive. The trigger values given here are those applicable to the 'worst-case': phytotoxic effects may occur at these concentrations in acid, sandy soils. In neutral or alkaline soils phytotoxic effects are unlikely at these concentrations.

6. Grass is more resistant to phytotoxic effects than are most other plants and its growth may not be adversely affected at these concentrations.

Table F.7 Tentative *trigger concentrations* for contamination associated with former coal carbonation sites (ICRCL 59/83, 2nd edn, 1987, Table 4)

Conditions

1. This table is invalid if reproduced without the conditions and footnotes.
2. All values are for concentrations determined on 'spot' samples based on an adequate site investigation carried out prior to development. They do not apply to analysis of averaged, bulked or composited samples, nor to sites which have already been developed.
3. Many of these values are preliminary and will require regular updating. They should not be applied without reference to the current edition of the report 'Problems Arising from the Development of Gas Works and Similar Sites'[1].
4. If all sample values are below the threshold concentrations then the site may be regarded as uncontaminated as far as the hazards from these contaminants are concerned, and development may proceed. Above these concentrations, remedial action may be needed, especially if the contamination is still continuing. Above the action concentrations, remedial action will be required or the form of development changed.

Contaminants	Proposed uses	Trigger concentrations (mg/kg air-dried soil)	
		threshold	action
Polyaromatic hydrocarbons[1,2]	Domestic gardens, allotments, play areas.	50	500
	Landscaped areas, buildings, hard cover.	1000	10 000
Phenols	Domestic gardens, allotments.	5	200
	Landscaped areas, buildings, hard cover.	5	1 000
Free cyanide	Domestic gardens, allotments, landscaped areas.	25	500
	Buildings, hard cover.	100	500
Complex cyanides	Domestic gardens, allotments.	250	1 000
	Lanscaped areas.	250	5 000
	Buildings, hard cover.	250	NL
Thiocyanate[2]	All proposed uses.	50	NL
Sulphate	Domestic gardens, allotments, landscaped areas.	2000	10 000
	Buildings[3].	2000[3]	50 000[3]
	Hard cover.	2000	NL
Sulphide	All proposed uses.	250	1 000
Sulphur	All proposed uses.	5000	20 000
Acidity (pH less than)	Domestic gardens, allotments, landscaped areas.	pH 5	pH 3
	Buildings, hard cover.	NL	NL

Notes:

NL: No limit set as the contaminant does not pose a particular hazard for this use.

1. Used here as a marker for coal tar, for analytical reasons. See 'Problems Arising from the Redevelopment of Gasworks and Similar Sites' Annex A1[1].
2. See 'Problems Arising from the Redevelopment of Gasworks and Similar Sites' for details of analytical methods[1].
3. See also BRE Digest 250: Concrete in sulphate-bearing soils and groundwater[4].

Table F.8 Greater London Council guidelines for classification of contaminated soils

Parameter	Typical values for uncontaminated soils	Slight Contamination	Contaminated	Heavy contamination	Unusually heavy Contamination
pH (acid)	6−7	5−6	4−5	2−4	(less than 2)
pH (alkaline)	7−8	8−9	9−10	10−12	12
Antimony	0−30	30−50	50−100	100−500	500
Arsenic	0−30	30−50	50−100	100−500	500
Cadmium	0−1	1−3	3−10	10−50	50
Chromium	0−100	100−200	200−500	500−2500	2500
Copper (available)	0−100	100−200	200−500	500−2500	2500
Lead	0−500	500−1000	1000−2000	2000−1.0%	1.0%
Lead (available)	0−200	200−500	500−1000	1000−5000	5000
Mercury	0−1	1−3	3−10	10−50	50
Nickel (available)	0−20	20−50	50−200	200−1000	1000
Zinc (available)	0−250	250−500	500−1000	1000−5000	5000
Zinc (equivalent)	0−250	250−500	500−2000	2000−1.0%	1.0%
Boron (available)	0−2	2−5	5−50	50−250	250
Selenium	0−1	1−3	3−10	10−50	50
Barium	0−500	500−1000	1000−2000	2000−1.0%	1.0%
Beryllium	0−5	5−10	10−20	20−50	50
Manganese	0−500	500−1000	1000−2000	2000−1.0%	1.0%
Vanadium	0−100	100−200	200−500	500−2500	2500
Magnesium	0−500	500−1000	1000−2000	2000−1.0%	1.0%
Sulphate	0−2000	2000−5000	5000−1.0%	1.0%−5.0%	5.0%
Sulphur (free)	0−100	100−500	500−1000	1000−5000	5000
Sulphide	0−10	10−20	20−100	100−500	500
Cyanide (free)	0−1	1−5	5−50	50−100	100
Cyanide (total)	0−5	5−25	25−250	250−500	500
Ferricyanide	0−100	100−500	500−1000	1000−5000	5000
Thiocyanate	0−10	10−50	50−100	100−500	2500
Coal Tar	0−500	500−1000	1000−2000	2000−1.0%	1.0%
Phenol	0−2	2−5	5−50	50−250	250
Toluene extract	0−5000	5000−1.0%	1.0%−5.0%	5.0%−25.0%	25.0%
Cyclohexane extract	0−2000	2000−5000	5000−2.0%	2.0%−10.0%	10.0%

Note: All numerical values expressed in milligrams per kilogram (mg/kg), which equates to parts per million (ppm), unless noted otherwise.

Table F.9 Netherlands Standards for soil contaminants

Element/compound	Concentration in soil (mg/kg dry weight)		
	A[a]	B[a]	C[a]
Arsenic	20	30	50
Barium	200	400	2000
Cadmium	1	5	20
Chromium	100	250	800
Cobalt	20	50	300
Copper	50	100	500
Lead	50	150	600
Mercury	0.5	2	10
Molybdenum	10	40	200
Nickel	50	100	500
Tin	20	50	300
Zinc	200	500	3000
Bromine (total)	20	50	300
Fluorine (total)	200	400	2000
Sulphur (total)	2	20	200
Cyanide (free)	1	10	100
Cyanide (complex)	5	50	500
Benzene	0.01	0.5	5
Ethyl benzene	0.05	5	50
Toluene	0.05	3	30
Xylene	0.5	5	50
Phenols	0.02	1	10
Naphthalene	0.1	5	50
Anthracene	0.1	10	100
Phenanthrene	0.1	10	100
Fluoranthrene	0.1	10	100
Pyrene	0.1	10	100
Benz(a)pyrene	0.05	1	10
Chlorinated aliphatics	0.1	5	50
Chlorobenzenes	0.05	1	10
Chlorophenols	0.01	0.05	5
PCB (total)	0.05	1	10
Organic chlorinated pesticides	0.1	0.5	5
Pesticides (total)	0.1	2	20
Tetrahydrofuran	0.1	4	40
Pyridine	0.1	2	20
Tetrahydrothiophene	0.1	5	50
Cyclohexanone	0.1	6	60
Styrene	0.1	5	50
Fuel	20	100	800
Mineral oil	100	1000	5000

[a] Below level A the soil is regarded as unpolluted.
   Above level A a preliminary site investigation is required.
   Above level B further investigation is required.
   Above level C remove soil or clean-up to level A values.

APPENDIX G

FACTORS OF SAFETY

In Chapter 1, section 1.5.1, it is stated that 'the designer should exercise his or her judgement in choice of safety factor'. Since judgement is built up by experience over many years, Table G.1 is included here as a guide to younger engineers.

Figure 10.20 is also repeated here as Fig. G.1. It shows the relationship between the ratio of dead or imposed load to total load and the combined partial load factors γ_P (for superstructure loads only), γ_F (for foundation loads only) and γ_T (for total loads). Figure G.1 is only suitable for use with the dead plus superimposed loading condition.

Table G.1 Typical safety factors

	Material	$\times$ Load	$=$ Overall
(1) Engineering soils			
(a) Safe bearing pressure	—	—	2.5 to 3
(b) Piles	—	—	3
(c) Retaining wall (i) sliding	—	—	2 to 3
(ii) overturning	—	—	1.5 to 2
(d) Stability of slopes	—	—	1.5 to 2
(2) Structural materials			
Designed for dead and superimposed loading			
(a) Structural steel (BS 5950)	1.0	1.4 to 1.6	1.4 to 1.6
(b) Concrete (BS 8110)	1.5	1.4 to 1.6	2.1 to 2.4
(c) Reinforcement (BS 8110)	1.15	1.4 to 1.6	1.61 to 1.84
(d) Masonry (BS 5628)	2.5 to 3.5	1.4 to 1.6	3.5 to 5.6
(e) Timber (BS 5268)	2.5 to 2.9	—	2.5 to 2.9
Designed for wind loading			
(a) Structural steel (BS 5950)	1.0	1.4	1.4
(b) Concrete (BS 8110)	1.5	1.4	2.1
(c) Reinforcement (BS 8110)	1.15	1.4	1.61
(d) Masonry (BS 5628)	2.5 to 3.5	1.4	3.5 to 4.9
Designed for disproportionate collapse			
(a) Structural steel (BS 5950)	1.0	1.05	0.9 to 1.05
(b) Concrete (BS 8110)	1.3	1.05	1.37
(c) Reinforcement (BS 8110)	1.0	1.05	1.05
(d) Masonry (BS 5628)	1.25 to 1.75	1.0	1.25 to 1.75
(3) Wind uplift on lightweight roofs	—	—	1.4

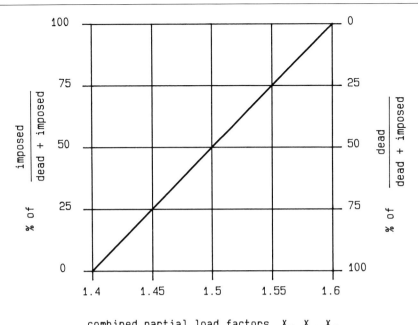

combined partial load factors γ_P, γ_F, γ_T

Fig. G.1 Combined partial load factors.

DESIGN CHARTS FOR PAD AND STRIP FOUNDATIONS

Figure H.1 − sizing of pad and strip bases

This design chart gives a quick analysis of the relationship between axial loading and bending moments to find the required size of base. The chart works equally for total loads with total allowable bearing pressures and for superstructure loads with net allowable bearing pressures (providing the weight of foundation and backfill is approximately equal to the weight of soil removed and that the resultant eccentricity is less than $L/6$ if superstructure loads only are considered).

The chart is used by first calculating the value of T/t_a (total load/total allowable bearing pressure) or P/p_a (superstructure load/net allowable bearing pressure) and the corresponding eccentricity e_T or e_P. The length of base has to be assumed and the value e/L calculated. The value of T/t_a or P/p_a is read on the y-axis and a line is taken horizontally to meet the appropriate e/L line from where a vertical line is dropped and the value of area required is read from the x-axis.

Note that the dotted line area indicates where there is partial zero pressure under the base and that no values are given for bases where $e/L > \frac{1}{3}$, as overturning of the base is likely to become critical in this area.

Examples of the use of the chart in Fig. H.1 are given in Design Examples 5 and 6 in Chapter 11.

Figures H.2, H.3 and H.4 − preliminary estimation of effective depth required for reinforced pad bases in bending

These design charts give a preliminary estimate of the effective depth required of a pad foundation in bending, beam shear and punching shear respectively, given the superstructure load and the desired $100(A_s/bd)$. Values of $100(A_s/bd)$ range from the minimum reinforcement requirement of 0.13 to a moderate value of 1.0 (values greater than 1.0 can be used in the design of reinforced concrete elements

but are outside the scope of these charts). The charts are based on a uniform bearing pressure and the requirements of BS 8110 and incorporate an average partial safety factor $\gamma_p = 1.5$ (the superstructure loads used in the charts do not therefore need to be factored).

Note that these charts are intended to enable the effective depth to be estimated and are *not* for direct calculations of $100(A_s/bd)$; this should be derived by calculation of the design moments and shears and the use of BS 8110.

Figure H.2 − bending

The value of PL/B (superstructure load × length/breadth) is calculated and read on the y-axis. A line is then taken horizontally to meet the curve for the required value of $100(A_s/bd)$ and a vertical line taken from this point to read the required effective depth on the x-axis.

Figure H.3 − beam shear

The value of P is read on the y-axis and a line taken horizontally to meet the required $100(A_s/bd)$ curve applicable to the base length L, then a vertical line is dropped from this point to read the required value of effective depth from the x-axis.

Figure H.4 − punching shear

The chart in this figure is used in the same manner as the chart in Fig. H.3.

Choice of effective depth

The highest value of effective depth from Figs H.2 to H.4 should then be used in the design of the pad foundation.

An example of the use of the charts in Figs H.2 to H.4 is given in Design Example 4 in Chapter 11.

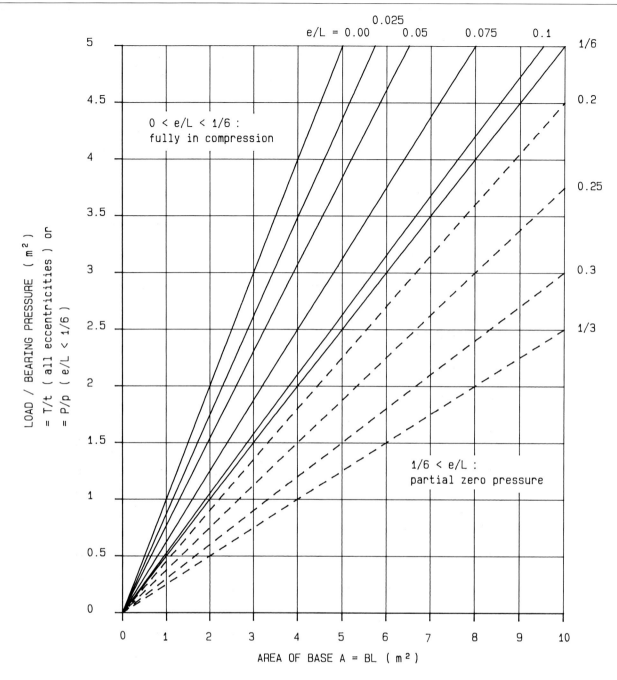

Fig. H.1 Sizing of pad and strip bases.

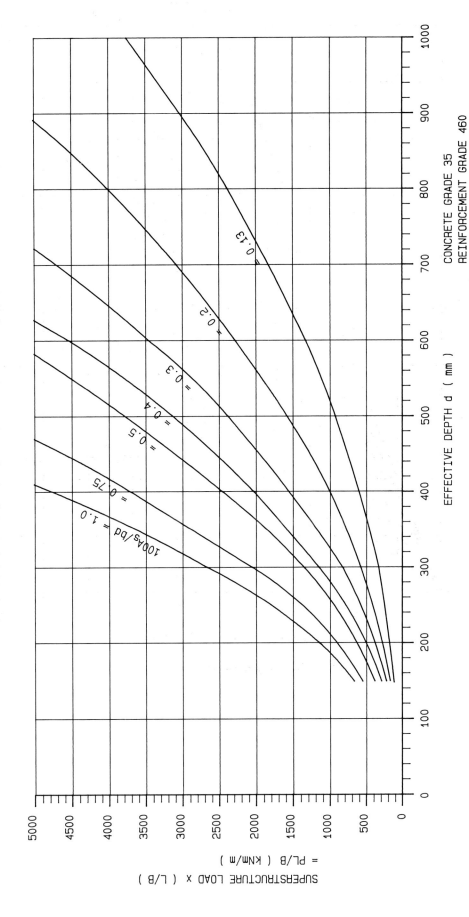

Fig. H.2 Estimation of effective depth – reinforced pad base in bending.

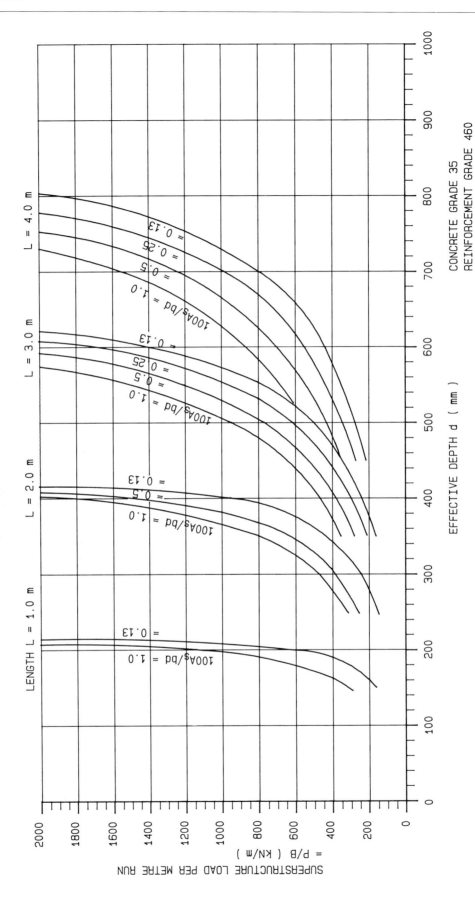

Fig. H.3 Estimation of effective depth – reinforced pad base in beam shear.

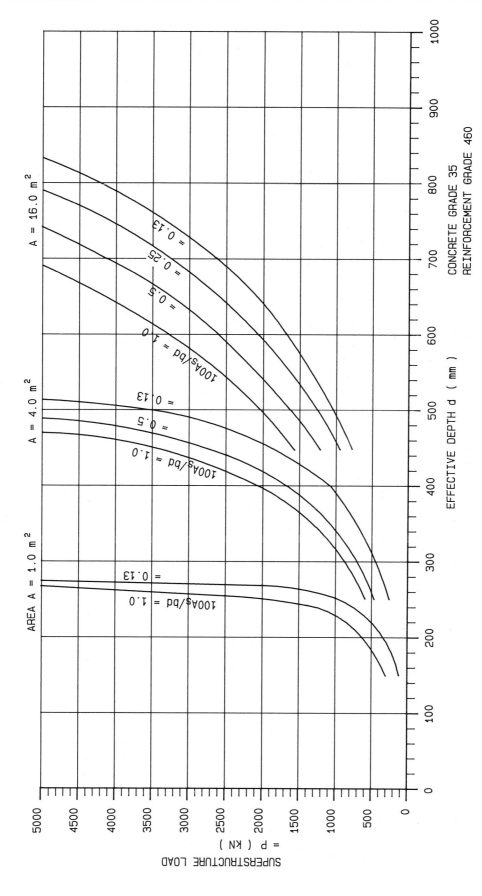

Fig. H.4 Estimation of effective depth – reinforced pad base in punching shear.

TABLE OF GROUND BEAM TRIAL SIZES

Table J.1 quotes bending moments (for factored loads) which produce a balanced design (i.e. no compression reinforcement) based on Clause 3.4.4.4 of BS 8110, using the formula:

$$M_u = 0.156 \times bd^2 \times f_{cu}$$

where $f_{cu} = 30\,kN/mm^2$.

Having selected a trial section, the tension reinforcement can be calculated using $A_s = M_u/(0.87 f_y z)$.

Table J.1 Ground beam bending moment capacities (in kNm) − balanced design

Effective depth, d (mm)	Width of proposed beam (mm)		
	300	600	900
200	56	112	168
300	126	253	379
400	225	449	674
500	351	702	1053
600	505	1011	1516
700	688	1376	2064
800	898	1797	2696
900	1137	2274	3411
1000	1404	2808	4212
1100	1699	3398	5097
1200	2022	4044	6066

Characteristic (unfactored) loads can be factored for use in calculating bending moments for this table by using the combined partial safety factor derived from the graph shown in Fig. G.1 in Appendix G.

DESIGN GRAPHS AND CHARTS FOR RAFT FOUNDATIONS SPANNING LOCAL DEPRESSIONS

The reader is advised to read the text in Chapter 13 before using these charts. The charts and figures are repeated here for quick reference and the following is an *aide-mémoire* for the experienced user of the procedure.

(1) Select the required design span of the depression based on the engineer's experience and using Table K.1 and/or Fig. K.1 as a guide.

(2) If there is no bottom reinforcement in the slab, use the chart on the right of Fig. K.1 to select the minimum effective depth required.

(3) For each loading type on the area under consideration, calculate the total factored load on the design span from Table K.2 and multiply each by the appropriate moment factor K_m which is also obtained from Table K.2.

(4) Sum the results from (3) above to give $\Sigma(T_u K_m)$ and use Fig. K.2 for slabs with top reinforcement only and Fig. K.3 for slabs with top and bottom reinforcement to determine the area of reinforcement required for the selected effective depth. Note that the area of reinforcement is required in both directions (i.e. a square mesh is needed).

(5) If heavy point or line loads are present, a shear capacity check should also be undertaken.

(6) A similar design process is adopted for designing raft beams, using Table K.3 and Figs K.4 and K.5.

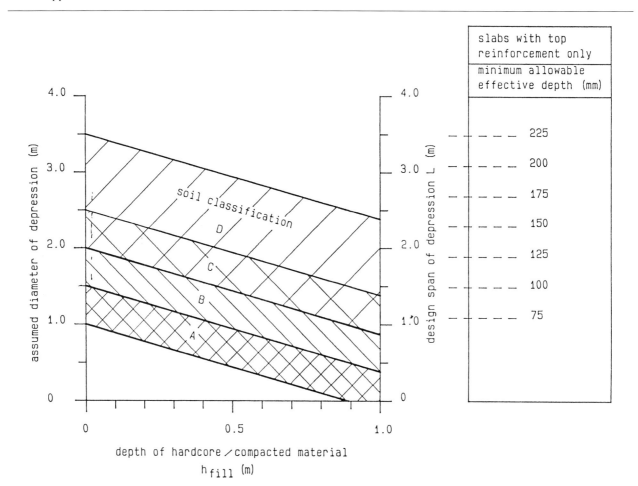

Fig. K.1 Design span for local depression.

Table K.1 Design diameter for local depression

Soil classification	Soil type	Assumed diameter of depression (m)
A Consistent firm sub-soil	One only of: clay, sand, gravel, sandy clay, clayey sand	1.0−1.5
B Consistent soil type but variable density, i.e. loose to firm	One only of: clay, sand, gravel, sandy clay, clayey sand	1.5−2.0
C Variable soil type but firm	Two or more of: clay, sand, gravel, sandy clay, clayey sand	2.0−2.5
D Variable soil type and variable density	Two or more of: clay, sand, gravel, sandy clay, clayey sand	2.5−3.5

Table K.2 Load types and moment factors for raft slabs spanning over a depression of diameter _L_

	top reinforcement only	top and bottom reinforcement	K_m
uniformly distributed load f_S (kN/m^2)	T1 $F_S = f_S (\pi L^2/4)$	TB1 $F_S = f_S (\pi L^2/4)$	1.0
parallel line load P (kN/m)	T2 $\Sigma P = PL$	TB2 $\Sigma P = PL$	1.5
lateral line load P (kN/m)	T3 $\Sigma P = PL$	TB3 $\Sigma P = PL$	1.5
2 way line load P (kN/m)	T4 $\Sigma P = 2PL$	TB4 $\Sigma P = 2PL$	1.5
point load P (kN)	T5 $\Sigma P = P$	TB5 $\Sigma P = P$	2.0

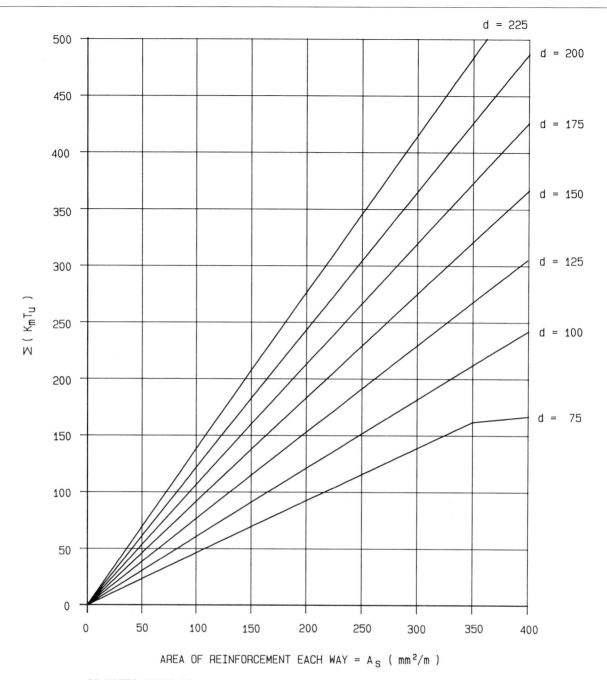

CONCRETE GRADE 35
REINFORCEMENT GRADE 460

Fig. K.2 Design chart for slabs with top reinforcement only.

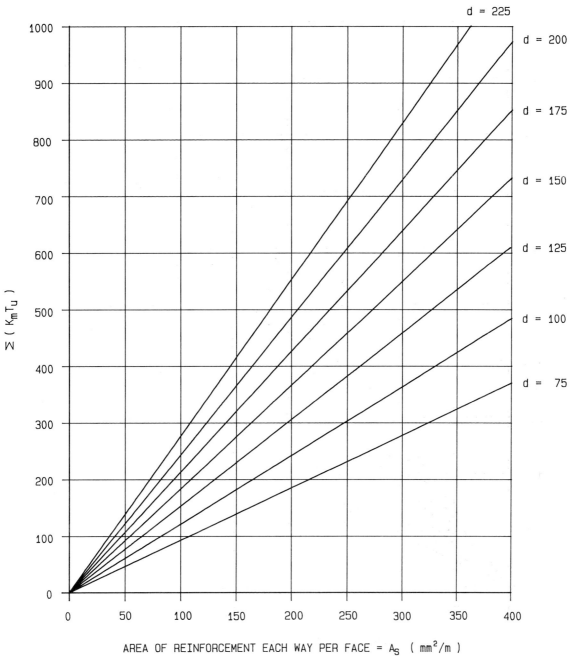

Fig. K.3 Design chart for slabs with top and bottom reinforcement.

Table K.3 Load types and moment factors for raft beams

	internal beam	edge beam	corner beam	K_m
uniformly distributed load f_S (kN/m²)	I1 $F_S = f_S (\pi L^2/4)$	E1 $F_S = f_S (\pi L^2/8)$	C1 $F_S = f_S (0.64 L^2)$	0.5
parallel line load P (kN/m)	I2 $\Sigma P = PL$	E2 $\Sigma P = PL$	C2 $\Sigma P = PL/\sqrt{2}$	1.0
lateral line load P (kN/m)	I3 $\Sigma P = PL$	E3 $\Sigma P = PL/2$	C3 $\Sigma P = PL/\sqrt{2}$	1.0
2 – way line load P (kN/m)	I4 $\Sigma P = 2PL$	E4 $\Sigma P = 3PL/2$	C4 $\Sigma P = 2PL/\sqrt{2}$	1.0
point load P (kN)	I5 $\Sigma P = P$	E5 $\Sigma P = P$	C5 $\Sigma P = P$	2.0

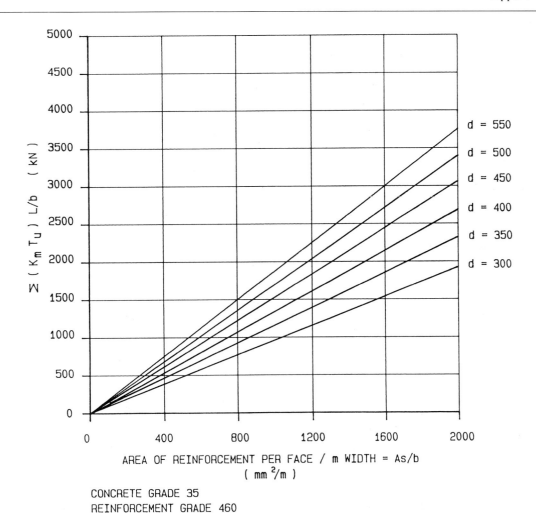

CONCRETE GRADE 35
REINFORCEMENT GRADE 460

Fig. K.4 Design chart for internal and edge beams.

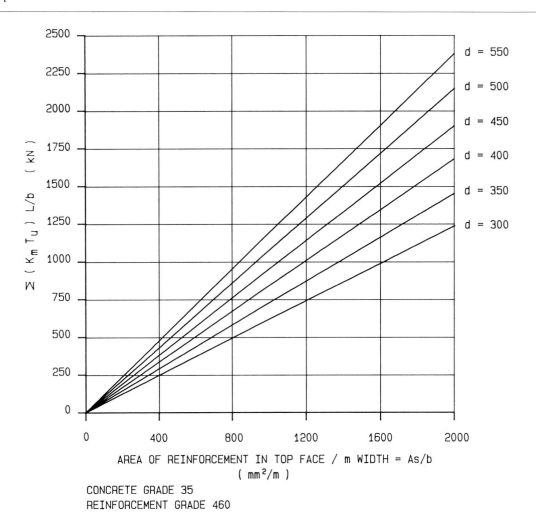

CONCRETE GRADE 35
REINFORCEMENT GRADE 460

Fig. K.5 Design chart for corner beams.

TABLE OF MATERIAL FRICTIONAL RESISTANCES

Frictional resistances are used in calculations within the
text. These are repeated here, in Table L.1, together with
resistances for other structural materials, for comparison
and completeness.

Table L.1 Typical frictional resistances

Materials	Frictional resistance, μ
In situ concrete base on sand	0.6
In situ concrete base on clay	[a]
Precast concrete on steel	0.3
Timber on timber, fibres parallel to the motion	0.4
Timber on timber, fibres at 90° to the motion	0.5
Metal on timber	0.2
Metal on metal	0.15 to 0.2
Timber on stone	0.4
Metal on masonry	0.3 to 0.5
Masonry on masonry (hard)	0.2 to 0.3
Masonry on masonry (soft)	0.4 to 0.6
Well lubricated hard smooth surfaces (bearings)	0.05

[a] Frictional resistance unreliable therefore use cohesion only

COST INDICES FOR FOUNDATION TYPES

Throughout the text, reference is made in differing contexts to the relative costs of various foundation types. Every construction site is different, and is often affected by local factors which are not relevant generally. Every contractor is different, and what suits one method of working will not necessarily be as economic with another method. Prices vary throughout the country and often depend upon site accessibility.

Despite these reservations, the many qualitative statements made in discussing relative foundation costs demand a quantitative treatment. Thus the following table of cost indices (Table M.1) has been produced to give a feel, in general terms, for cost differentials.

A number of the types are not comparable directly, and it is not the authors' intention to give any more than general guidance within the context of the discussion chapters of this book (for example, item A6 should not be compared directly with item A9, and the conclusion reached that piles are only marginally more expensive than pad bases; item A6 is derived from three-storey and higher, framed buildings for comparison with item A8, whereas A9 is derived from a domestic scale of construction for comparison with item A4).

For specific project cost considerations the engineer should always instigate cost comparisons made on the criteria of the particular site, utilizing the services of contractors and/or quantity surveyors as necessary, and should not place any reliance upon the very general information incorporated in Table M.1.

Table M.1 Foundation cost indices

Foundation type	Index[a]
A Index expressed per metre run of typical footing	
(1) Unreinforced strip footing (underside of footing at 600 mm deep)	1
(2) Reinforced strip footing	1.1
(3) Unreinforced strip footing (underside of footing at 900 mm deep)	1.3
(4) Trench fill footing	
1000 mm deep	1.2
1500 mm deep	1.4
2000 mm deep	1.7
2500 mm deep	2.0
(5) Semi-raft downturn on edge of light raft	1.2 to 1.4
(6) Reinforced pad base, and ground beam	1.9 to 2.1
(7) Mass concrete pad base, and ground beam	1.7 to 1.9
(8) Reinforced concrete piles and ground beams used in domestic type applications (typically 150/300 mm diameter piles, 3 m to 8 m deep)	1.7 to 3.0
(9) Vibro treatment and r.c. footings at 600 mm depth used in domestic type applications (typically 500 mm diameter stone columns at 2 m centres, and 3 m deep)	1.4 to 1.7
(10) Reinforced concrete piles and ground beams used in non-domestic applications (typically 600 mm diameter piles, 10 m to 15 m deep)	3 to 4
B **Underpinning expressed per metre run of footing**	
(1) Conventional mass concrete 1 m deep	3
(2) Conventional mass concrete 2 m deep	5
(3) Small-diameter piles at regular close centres (3 m to 6 m deep) utilizing existing footings	4
(4) Small-diameter piles (3 m to 6 m deep) utilizing new r.c. ground beams	7
C **Index expressed per individual foundation (i.e. at one column position)**	
(1) Mass concrete pad base from 2 m × 2 m to 2.5 m × 2.5 m on plan (excluding side shutters)	4.0
(2) Reinforced concrete base from 2 m × 2 m to 2.5 m × 2.5 m on plan with a base thickness of 0.75 m at a formation level of 1.8 m (including side shutters and working space)	6.8
(3) In situ concrete pile cap and twin 600 mm diameter piles, 10 m to 15 m long	10.8

[a] Values taken for the index will depend upon the engineer's experience and also upon the area of the country and accessibility of the site

APPENDIX N:

ALLOWABLE BEARING PRESSURE FOR FOUNDATIONS ON NON-COHESIVE SOIL

The general route for establishing the *allowable bearing pressure* is as follows.

(1) Divide the *ultimate bearing capacity* by a factor of safety (typically 3.0), to obtain the *safe bearing capacity*.
(2) By looking at predicted values for settlement, determine the bearing pressure which corresponds to an acceptable level of settlement.
(3) The *allowable bearing pressure* is the *lower* of the two values obtained from (1) and (2).

In section 2.3.5 *Safe bearing capacity – cohesionless soils*, it states that:

> 'foundation design on non-cohesive soil is usually governed by acceptable settlement, and this restriction on bearing pressure is usually much lower than the ultimate bearing capacity divided by the factor of safety of 3. Generally only in the case of narrow strip foundations on loose submerged sands it is vital to determine the ultimate bearing capacity, since this may be more critical than settlement'.

This indicates that a *settlement* rather than a *bearing capacity* calculation would be the normal route for establishing the allowable bearing pressure for sands and gravels. Because of the uncertainties and assumptions involved in detailed settlement calculations, this approach is normally short-circuited by use of the Terzaghi and Peck allowable bearing pressure chart in Fig. N.1.

The allowable bearing pressures in the chart assume a maximum settlement of 25 mm, which experience has shown is a satisfactory value for maintaining total and differential settlements within acceptable limits. They also assume the water-table is at least a depth of B below foundation level; if the water table is at or close to the foundation level then the allowable bearing pressures indicated should be halved.

To use the chart, an SPT value is obtained from the soils investigation report. The proposed width of base, together with the SPT value, are used to read off an allowable bearing pressure. This is then checked to ensure it exceeds the applied bearing pressure; if not the base length and/or width is increased, and the process repeated until a satisfactory base size is obtained.

This process is illustrated in the following worked examples.

Worked Example 1: Square pad base

A pad foundation is required to support a superstructure load of $P = 1500$ kN. The soils investigation indicates 0.9 m of topsoil and silty clay overlying a considerable depth of medium dense sand. Average SPT values for the top metre of sand are in the range $N = 22-41$; a conservative average value of $N = 25$ will be assumed for determining the allowable bearing pressure.

A square base of $2 \text{ m} \times 2 \text{ m}$ is initially assumed. From Fig. N.1, the allowable bearing pressure is $n_a = 280$ kN. This gives a capacity of

$$P_a = n_a \, BL$$
$$= 280 \times 2.0 \times 2.0$$
$$= 1120 \text{ kN} \qquad <P = 1500 \text{ kN} \qquad \Rightarrow \text{Not enough}$$

At this allowable bearing pressure, the required area of a square base would be

$$A = P/n_a$$
$$= 1500/280$$
$$= 5.36 \text{ m}^2$$
$$= 2.31 \text{ m} \times 2.31 \text{ m}$$

However examination of Fig. N.1 indicates that a larger width of base will result in a lower allowable bearing pressure. A $2.4 \text{ m} \times 2.4 \text{ m}$ base will therefore be assumed; from Fig. N.1 this gives an allowable bearing pressure of $n_a = 270$ kN. The actual bearing pressure is

$$n = P/A$$
$$= 1500/(2.4 \times 2.4)$$
$$= 260 \text{ kN/m}^2 \qquad <n_a = 270 \text{ kN} \qquad \Rightarrow \text{OK}$$

and the area of the base is

$$A = BL$$
$$= 2.4 \times 2.4$$
$$= 5.76 \text{ m}^2$$

Worked Example 2: Rectangular pad base

The previous example will be reworked for a rectangular base, whose width is limited by site constraints to $B = 2.0$ m. A base size of $2.0 \text{ m} \times 2.5 \text{ m}$ is initially assumed. From Fig. N.1, the allowable bearing pressure is $n_a = 270$ kN. This gives a capacity of

$$P_a = n_a \, BL$$
$$= 270 \times 2.0 \times 2.5$$
$$= 1350 \text{ kN} \qquad < P = 1500 \text{ kN} \qquad \Rightarrow \text{Not enough}$$

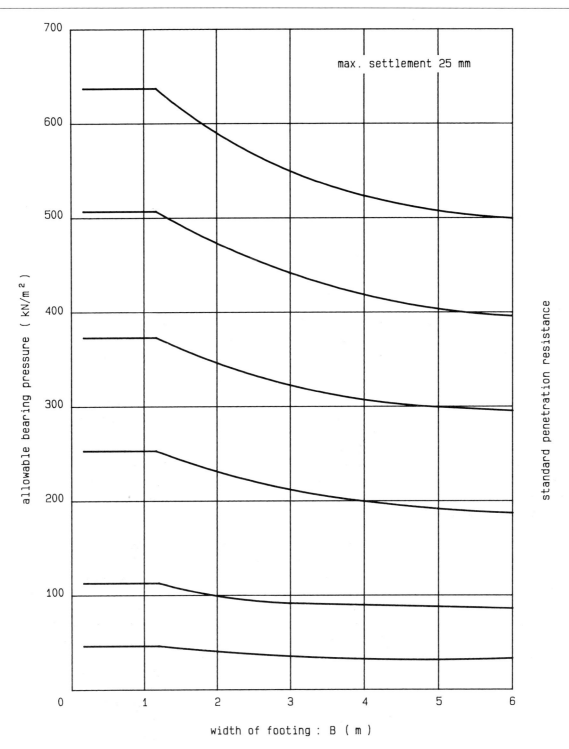

Fig. N.1 Allowable bearing pressure on sands (Reproduced from K. Terzaghi & R. B. Peck (1967) *Soil Mechanics in Engineering Practice*, **by permission of John Wiley & Sons, Inc., © authors)**

Increase length of base by the ratio P/P_a

$$L = 2.5 \times (1500/1350)$$
$$= 2.8\,m$$

This results in a bearing pressure of $n = n_a = 270\,kN/m^2$, and a base area of

$$A = BL$$
$$= 2.0 \times 2.8$$
$$= 5.6\,m^2$$

It will be noted that the rectangular base results in a slightly lower area than the square base. (The opposite is true for cohesive soils where bearing capacity rather than settlement is critical − see sections 2.3.2 and 2.3.4.) Nevertheless a square column base should be preferred to a rectangular one; mass concrete square bases will require a lesser thickness for load dispersion, while reinforced square bases will have lower bending moments and thus require less reinforcement.

INDEX